TOPOI
THE CATEGORIAL ANALYSIS OF LOGIC

TOPOI
THE CATEGORIAL ANALYSIS OF LOGIC

REVISED EDITION

ROBERT GOLDBLATT
Victoria University
Wellington, New Zealand

DOVER PUBLICATIONS, INC.
Mineola, New York

Bibliographical Note

This Dover edition, first published in 2006, is a slightly corrected unabridged republication of the revised (second) edition of the work originally published in 1984 by Elsevier Science Publishers B.V., Amsterdam, The Netherlands, as Volume 98 in the North-Holland Series, "Studies in Logic and the Foundations of Mathematics." A new Preface and updated Bibliography have been specially prepared for this reprint.

International Standard Book Number

ISBN-13: 978-0-486-45026-1
ISBN-10: 0-486-45026-0

Manufactured in the United States by LSC Communications
45026007 2018
www.doverpublications.com

To

My Parents

"Though we may never see precisely how the protean dancing stuff of everything endlessly becomes itself, we have no choice, being human and full of desire, but to go on perpetually seeking clarity of vision. The ultimate form within forms, the final shape of change, may elude us. The pursuit of the idea of form—even the form of force, of endlessly interacting process—is man's inevitable, crucial need"

John Unterecker

PREFACE

No doubt there are as many reasons for writing books as there are people who write them. One function served by this particular work has been the edification of its author. Translations can sometimes create a sense of *explanation*, and this seemed to me particularly true of the alternative account of mathematical constructions being produced by category theory. Writing the book gave me a framework within which to confirm that impression and to work through its ramifications in some detail. At the end I knew a great deal more than when I began, so that the result is as much a recording as a reconstruction of the progress of my own understanding. And at the end it seemed to me that much that I had dwelt on had finally fallen into place.

As to the more public functions of the book – I hope that it provides others with the prospect of a similar experience. Less presumptiously, I have tried to write an exposition that will be accessible to the widest possible audience of logicians – the philosophically motivated as well as the mathematical. This, in part, accounts for the style that I have adopted. There is a tendency in much contemporary literature to present material in a highly systematised fashion, in which an abstract definition will typically come before the list of examples that reveals the original motivation for that definition. Paedogogically, a disadvantage of this approach is that the student is not actually *shown* the genesis of concepts – how and why they evolved – and is thereby taught nothing about the mechanisms of creative thinking. Apart from lending the topic an often illusory impression of completedness, the method also has the drawback of inflating prerequisites to understanding.

All of this seems to me particularly dangerous in the case of category theory, a discipline that has more than once been referred to as "abstract nonsense". In my experience, that reaction is the result of features that are not intrinsic to the subject itself, but are due merely to the style of some of its expositors. The approach I have taken here is to try to move always from the particular to the general, following through the steps of the abstraction process until the abstract concept emerges naturally. The starting points are elementary (in the "first principles" sense), and at the finish it would be quite appropriate for the reader to feel that (s)he had just arrived at the subject, rather than reached the end of the story.

As to the specific treatment of category theory, I have attempted to play down the functorial perspective initially and take an elementary (in the sense of "first-order") approach, using the same kind of combinatorial manipulation of algebraic structure that is employed in developing the basic theory of any of the more familiar objects of pure-mathematical study. In these terms categories as structures are no more rarified than groups, lattices, vector-spaces etc.

I should explain that whereas the bulk of the manuscript was completed around May of 1977, the sections 11.9, 14.7 and 14.8 were written a year later while I was on leave in Oxford (during which time I held a Travelling Fellowship from the Nuffield Foundation, whose assistance I am pleased to acknowledge). The additional material was simply appended to Chapters 11 and 14, since, although the arrangement is less than ideal, it was impractical at that stage to begin a major reorganisation. I imagine however that there will be readers interested in the construction of number-systems in 14.8 who do not wish to wade through the earlier material in Chapter 14 on Grothendieck topologies, elementary sites etc. In fact in order to follow the definition of Dedekind-reals in the topos of Ω-sets, and their representation as classical continuous real-valued functions, it would suffice to have absorbed the description of that topos given in 11.9. The full sheaf-theoretic version of this construction depends on the theory of Ω-sheaves developed in 14.7, but a sufficient further preparation for the latter would be to read the first few pages of 14.1, at least as far as the introduction of the axiom COM on page 362.

A point of terminology: – I have consistently used the word "categorial" where the literature uniformly employs "categorical". The reason is that while both can serve as adjectival forms of the noun "category", the second of them already has a different and long established usage in the domain of logic, one that derives from its ordinary-language meaning of "absolute". Logicians have known since the work of Gödel that set theory has no categorical axiomatisation. One function of this book will be to explain to them why it does have a categorial one.

There are a number of people who I would like to thank for their help in the production of the book. I am indebted to Shelley Carlyle for her skilful typing of the manuscript; to the Internal Research Committee and the Mathematics Department of the Victoria University of Wellington for substantially subsidising its cost; to Pat Suppes for responding favourably to it, and supporting it; and to Einar Fredriksson and Thomas van den

Heuvel for the expertise and cooperation with which they organised its editing and publishing.

My involvement with categorial logic gained impetus through working with Mike Brockway on his M.Sc. studies, and I have benefited from many conversations with him and access to his notes on several topics. In obtaining other unpublished material I was particularly helped by Gonzalo Reyes. Dana Scott, by his hospitality at Oxford, performed a similar service and provided a much appreciated opportunity to aquaint myself with his approach to sheaves and their logic. In preparing the material about the structure of the continuum I was greatly assisted by discussions with Scott, and also with Charles Burden.

Finally, it is a pleasure to record here my indebtedness to my teachers and colleagues in the logic group at VUW, particularly to my doctoral advisors Max Cresswell and George Hughes, and to Wilf Malcolm, for their involvement in my concerns and encouragement of my progress throughout the time that I have been a student of mathematical logic.

Where did topos theory come from? In the introduction to his recent book on the subject, Peter Johnstone describes two lines of development in the fields of algebraic geometry and category theory. It seems to me that a full historical perspective requires the teasing out of a third strand of events in the area of specific conern to this book, i.e. logic, especially model theory. We may begin this account with Cohen's work in 1963 on the independence of the continuum hypothesis et. al. His forcing technique proved to be the key to the universe of classical set theory, and led to a wave of exploration of that territory. But as soon as the method had been reformulated in the Scott–Solovay theory of Boolean-valued models (1965), the possibility presented itself of replacing "Boolean" by "Heyting" and thereby generalising the enterprise. Indeed Scott made this point in his 1967 lecture-notes and then took it up in his papers (1968, 1970) on the topological interpretation of intuitionistic analysis.

Meanwhile the notion of an *elementary topos* had independently emerged through Lawvere's attempts to axiomatise the category of sets. The two developments became linked together by the concept of a *sheaf:* the study of cartesian-closed categories with subobject classifiers (topoi) got under way in earnest once it was realised that they included all the Grothendieck sheaf-categories, while the topological interpretation was seen to have provided the first examples for a general axiomatic theory of sheaf-models over Heyting algebras that was subsequently devised by Scott and developed in association with Michael Fourman (cf. 14.7 and

14.8). In this latter context (many of whose ideas have precursors in the initial Boolean work), the earlier problem (Scott 1968, p. 208) of dealing with partially defined entities is elegantly resolved by the introduction of an *existence predicate*, whose semantical interpretation is a measure of the extent to which an individual is defined (exists). To complete the picture, and round out this whole progression of ideas, some unpublished work of Denis Higgs (1973) demonstrated that the category of sheaves over **B** (a complete Boolean algebra) is equivalent to the category of **B**-valued sets and functions in the original Scott–Solovay sense.

And what of the future? What, for instance, is the likely impact of the latest independence results to the effect that there exist topoi in which the Heine–Borel Theorem fails, the Dedekind-reals are not real-closed, complex numbers lack square-roots etc.? Predictions at this stage would I think be premature – after all today's pathology may well be dubbed "classical" by some future generation. The intellectual tradition to which topos theory is a small contribution goes back to a time when mathematics was closely tied to the physical and visual world, when "geometry" for the Greeks really had something to do with land-measurement. It was only relatively recently, with the advent of non-Euclidean geometries, that it became possible to see that discipline as having a quite independent existence and significance. Analogously, that part of the study of structure that is concerned with those structures called "logics" has evolved to a point that lies beyond its original grounding (the analysis of principles of reasoning). But the separation from this external authority has no more consequences as to the true nature of reasoning than does the existence of non-Euclidean geometries decide anything either way about the true geometrical properties of visual space.

The laws of Heyting algebra embody a rich and profound mathematical structure that is manifest in a variety of contexts. It arises from the epistemological deliberations of Brouwer, the topologisation (localisation) of set-theoretic notions, and the categorial formulation of set theory, all of which, although interrelated, are independently motivated. This ubiquity lends weight, not to the suggestion that the correct logic is in fact intuitionistic instead of classical, but rather to the recognition that thinking in such terms is simply inappropriate – in the same way that it is inappropriate to speak without qualification about *the* correct geometry.

At the same time, these developments have shown us more clearly than ever just how the properties of the structures we study depend on the principles of logic we employ in studying them. Particularly striking is the fine-tuning that has been given to the modern logical/set-theoretical articulation of the structure of the intuitively conceived continuum (which to Euclid was not a set of points at all, let alone an object in a topos). Indeed it seems that the deeper the probing goes the less will be the currency given to the definite article in references to "the continuum".

Other areas of mathematics (abstract algebra, axiomatic geometry) have long since become autonomous activities of mental creation, just as painting and even music have long since progressed beyond the representational to aquire substantial (in some cases all-consuming) subjective and intellectual components. A similar situation could be said to be arising in mathematical logic. In the absence of that external authority (the representation of things "out there") we may not so readily determine what is worthwhile and significant, just as it is no longer so easy to understand and make judgements about many contemporary aesthetic developments. Were we to identify the valuable with that whose value is lasting, a considerable period of winnowing might well be required before we could decide what is wheat and what is chaff. Looking back over the progress of the last two decades or so we see several strands that weave together to present the current interest in Heyting-valued structures as the natural product of the evolution of a substantial area of mathematical thought. Wherever it may be heading, we may already locate its permanent importance in the way it has brought a number of disciplines (logic, set theory, algebraic geometry, category theory) together under one roof, and in the contribution it has thereby made to our understanding of the house that we mentally build for ourselves to live in.

No doubt these remarks will be thought contentious by some. I hope that they will be found provocative as well. Should it inspire, or incite, anybody to respond to them, this book will have fulfilled one of its intended functions.

Wellington R. I. Goldblatt
Autumnal Equinox, 1979

PREFACE TO THE SECOND EDITION

This edition contains a new chapter, entitled Logical Geometry, which is intended to introduce the reader to the theory of *geometric morphisms* between Grothendieck topoi, and the model-theoretic rendering of this theory due to Makkai and Reyes. The main aim of the chapter is to explain why a theorem, due to Deligne, about the existence of geometric morphisms from **Set** to certain "coherent" topoi is equivalent to the classical logical Completeness Theorem for a certain class of "geometric" first-order formulae.

I have also taken the opportunity to correct a number of typographical errors, and false assertions, most of which have been kindly supplied by readers. In particular there are changes to Exercises 9.3.3, 11.5.3, 11.5.4, 14.3.4, 14.3.6, 14.3.7. Also, the statement as to the nature of the Cauchy reals in Ω-**Set** on page 414 requires qualification – it holds only for certain **CHA**'s Ω. For spatial **CHA**'s (topologies), Fourman has given a necessary and sufficient condition for the statement to be true, which, in spaces with a countable basis, is equivalent to local connectedness (cf. M. P. Fourman, Comparison des réelles d'un topos; structures lisses sur un topos élémentaire, Cahiers top. et géom. diff., XVI (1976), 233–239).

No doubt more errors remain: for these I can only crave the indulgence of the reader.

Wellington, 1983 R. I. Goldblatt

PREFACE TO THE DOVER EDITION

I am gratified by the continuing interest in this book three decades after I began its writing, and grateful to the many people who have contacted me over the years in appreciation and to encourage its ongoing availability. I am particularly pleased that it will be readily accessible as a Dover paperback, and thank John Grafton for proposing this. I also wish to acknowledge the congenial cooperation I received from the Cornell University Library, and David Ruddy in particular, in making the manuscript available through the internet.

This is a book about logic, rather than category theory per se. It aims to explain, in an introductory way, how certain logical ideas are illuminated by a category-theoretic perspective. There are now many more monographs that include material related to categorical logic and topoi, at various levels. I have added a list of some of these below, followed by a few more minor corrections.

January 2006 Rob Goldblatt

Further Reading:

Andrea Asperti and Giuseppe Longo, *Categories, Types, and Structures.* MIT Press, 1991. Downloadable from `www.di.ens.fr/~longo/download.html`

Michael Barr and Charles Wells, *Toposes, Triples and Theories.* Springer-Verlag, 1985. Republished in *Reprints in Theory and Applications of Categories*, no. 12, 2005, `www.tac.mta.ca/tac/reprints`

Michael Barr and Charles Wells, *Category Theory for Computing Science.* Prentice Hall, 1990. 3rd edition, Centre de Recherches Mathématiques, Montreal, 1999, `www.crm.umontreal.ca/pub/Ventes/CatalogueEng.html`

John L. Bell, *Toposes and Local Set Theories : An Introduction.* Oxford University Press, 1988.

John L. Bell, *A Primer of Infinitesimal Analysis.* Cambridge University Press, 1998.

Andreas Blass and Andre Scedrov, *Freyd's Models for the Independence of the Axiom of Choice.* Memoirs of the American Mathematical Society, no. 404, 1989.

Francis Borceux, *Handbook of Categorical Algebra 3: Categories of Sheaves.* Cambridge University Press, 1994.

Jonathan Chapman and Frederick Rowbottom, *Relative Category Theory and Geometric Morphisms: A Logical Approach.* Oxford University Press, 1992.

Roy L. Crole, *Categories for Types.* Cambridge University Press, 1993.

Peter J. Freyd and Andre Scedrov, *Categories, Allegories.* North-Holland, 1990.

Silvio Ghilardi and Marek Zawadowski, *Sheaves, Games, and Model Completions : A Categorical Approach to Nonclassical Propositional Logics.* Kluwer Academic Publishers, 2002.

John W. Gray, editor, *Mathematical Applications of Category Theory.* Contemporary Mathematics, v. 30, American Mathematical Society, 1984.

John W. Gray and Andre Scedrov, editors, *Categories in Computer Science and Logic.* Contemporary Mathematics, v. 92, American Mathematical Society, 1989.

Bart Jacobs, *Categorical Logic and Type Theory.* Elsevier, 1999.

Peter T. Johnstone, *Stone Spaces.* Cambridge University Press, 1982.

Peter T. Johnstone, *Sketches of an Elephant: A Topos Theory Compendium. Volumes 1 and 2.* Oxford University Press, 2002.

Andre Joyal, Ieke Moerdijk, *Algebraic Set Theory*, Cambridge University Press, 1995.

Andre Joyal and Myles Tierney, *An Extension of the Galois Theory of Grothendieck.* Memoirs of the American Mathematical Society, no. 309, 1984.

Anders Kock, *Synthetic Differential Geometry.* Cambridge University Press, 1981.

J. Lambek and P. J. Scott, *Introduction to Higher Order Categorical Logic.* Cambridge University Press, 1986.

René Lavendhomme, *Basic Concepts of Synthetic Differential Geometry.* Kluwer Academic Publishers, 1996.

F. William Lawvere and Robert Rosebrugh, *Sets for Mathematics.* Cambridge University Press, 2003.

F. William Lawvere and Stephen H. Schanuel, *Conceptual Mathematics : A First Introduction to Categories.* Cambridge University Press, 1997.

Michael Makkai, *Duality and Definability in First Order Logic.* Memoirs of the American Mathematical Society, no. 503, 1993.

Michael Makkai and Robert Paré, *Accessible Categories : The Foundations of Categorical Model Theory.* Contemporary Mathematics, v. 104, American Mathematical Society, 1989.

Saunders Mac Lane and Ieke Moerdijk, *Sheaves in Geometry and Logic : A First Introduction to Topos Theory.* Springer-Verlag, 1992.

Colin McLarty, *Elementary Categories, Elementary Toposes.* Oxford University Press, 1992.

Ieke Moerdijk, *Classifying Spaces and Classifying Topoi.* Lecture Notes in Mathematics 1616, Springer-Verlag, 1992.

Ieke Moerdijk and Gonzalo E. Reyes, *Models for Smooth Infinitesimal Analysis.* Springer-Verlag, 1991.

Ieke Moerdijk and J. J. C. Vermeulen, *Proper Maps of Toposes.* Memoirs of the American Mathematical Society, no. 705, 1992.

Maria Cristina Pedicchio and Walter Tholen (editors), *Categorical Foundations : Special Topics in Order, Topology, Algebra and Sheaf Theory.* Cambridge University Press, 2004.

Andrew M. Pitts, *Categorical Logic.* In *Handbook of Logic in Computer Science, Volume 5: Algebraic and Logical Structures*, edited by S. Abramsky, Dov M. Gabbay and T. S. E. Maibaum, Oxford University Press, 2000.

David E. Rydeheard and Rod M. Burstall, *Computational Category Theory.* Prentice-Hall, 1988.

Andre Scedrov, *Forcing and Classifying Topoi.* Memoirs of the American Mathematical Society, no. 295, 1984.

Steven Vickers, *Topology Via Logic*, Cambridge University Press, 1989.

Oswald Wyler, *Lecture Notes on Topoi and Quasitopoi.* World Scientific, 1991.

Corrections:

Page 10, line 10: latter
Page 12, line 9 from below: led
Page 38, line after Fig. 3.1: establish
Page 50, first line of DEFINITION: $a \times c \to b \times$
Page 57, line 11 from below: $i \circ 1_e$
Page 57, line 10 from below: 1_e
Page 57, lines 6 and 5 from below: Im f
Page 202, EXERCISE 3: Add "The converse is true if $\mathcal{D}$ has pullbacks."
Page 359, line beginning (i): id_{FU}
Page 415, line 5 from below: quantified
Page 441, 2nd line of EXERCISE 2: $a \downarrow G$
Page 465, last line: $G \circ F \dot{\to} G' \circ F$
Page 472, 2nd line from below: codomain
Page 475, line 3: site
Page 488, 2nd line of text: If v is a variable of sort a, then
Page 505, 3rd line from below: $\varphi(\mathbf{v}') \wedge \psi(\mathbf{w}')$

CONTENTS

TOPOI
THE CATEGORIAL ANALYSIS OF LOGIC

PROSPECTUS

> *"... all sciences including the most evolved are characterised by a state of perpetual becoming."*
>
> Jean Piaget

The purpose of this book is to introduce the reader to the notion of a *topos*, and to explain what its implications are for logic and the foundations of mathematics.

The study of topoi arises within *category theory*, itself a relatively new branch of mathematical enquiry. One of the primary perspectives offered by category theory is that the concept of *arrow*, abstracted from that of *function* or *mapping*, may be used instead of the set membership relation as the basic building block for developing mathematical constructions, and expressing properties of mathematical entities. Instead of defining properties of a collection by reference to its members, i.e. *internal* structure, one can proceed by reference to its *external* relationships with other collections. The links between collections are provided by functions, and the axioms for a category derive from the properties of functions under composition.

A category may be thought of in the first instance as a universe for a particular kind of mathematical discourse. Such a universe is determined by specifying a certain kind of "object", and a certain kind of "arrow" that links different objects. Thus the study of topology takes place in a universe of discourse (category) with topological spaces as the objects and continuous functions as the arrows. Linear algebra is set in the category whose arrows are linear transformations between vector spaces (the objects); group theory in the category whose arrows are group homomorphisms; differential topology where the arrows are smooth maps of manifolds, and so on.

We may thus regard the broad mathematical spectrum as being blocked out into a number of 'subject matters' or categories (a useful way of lending coherence and unity to an ever proliferating and diversifying discipline). Category theory provides the language for dealing with these domains and for developing methods of passing from one to the other. The subject was initiated in the early 1940's by Samuel Eilenberg and Saunders Maclane. Its origins lie in algebraic topology, where constructions are developed that connect the domain of topology with that of algebra, specifically group theory. The study of categories has rapidly become however an abstract discipline in its own right and now constitutes a substantial branch of pure mathematics. But further than this it has had a considerable impact on the conceptual basis of mathematics and the language of mathematical practice. It provides an elegant and powerful means of expressing relationships across wide areas of mathematics, and a range of tools that seem to be becoming more and more a part of the mathematician's stock in trade. New light is shed on existing theories by recasting them in arrow-theoretic terms (witness the recent unification of computation and control theories described in Manes [75]). Moreover category theory has succeeded in identifying and explicating a number of extremely fundamental and powerful mathematical ideas (universal property, adjointness). And now after a mere thirty years it offers a new theoretical framework for mathematics itself!

The most general universe of current mathematical discourse is the category known at **Set**, whose objects are the sets and whose arrows are the set functions. Here the fundamental mathematical concepts (number, function, relation) are given formal descriptions, and the specification of axioms legislating about the properties of sets leads to a so called *foundation* of mathematics. The basic set-theoretic operations and attributes (empty set, intersection, product set, surjective function e.g.) can be described by reference to the arrows in **Set**, and these descriptions interpreted in any category. However the category axioms are "weak", in the sense that they hold in contexts that differ wildly from the initial examples cited above. In such contexts the interpretations of set-theoretic notions can behave quite differently to their counterparts in **Set**. So the question arises as to when this situation is avoided, i.e. when does a category look and behave like **Set**? A vague answer is – when it is (at least) a topos. This then gives our first indication of what a topos is. It is a category whose structure is sufficiently like **Set** that in it the interpretations of basic set-theoretical constructions behave much as they do in **Set** itself.

The word *topos* ("place", or "site" in Greek) was originally used by Alexander Grothendieck in the context of algebraic geometry. Here there is a notion called a "sheaf" over a topological space. The collection of sheaves over a topological space form a category. Grothendieck and his colleagues extended this construction by replacing the topological space by a more general categorial structure. The resulting generalised notion of category of sheaves was given the name "topos" (cf. Artin et al. [SGA4]).

Independently of this, F. William Lawvere tackled the question as to what conditions a category must satisfy in order for it to be "essentially the same" as **Set**. His first answer was published in 1964. A shortcoming of this work was that one of the conditions was set-theoretic in nature. Since the aim was to categorially axiomatise set theory, i.e. to produce set-theory out of category theory, the result was not satisfactory, in that it made use of set-theory from the outset.

In 1969 Lawvere, in conjunction with Myles Tierney, began the study of categories having a special kind of arrow, called a "subobject classifier" (briefly, this is an embodiment of the correspondence between subsets and characteristic functions in **Set**). This notion proved to be, in Lawvere's words, the "principle struggle" – the key to the earlier problem. He discovered that the Grothendieck topoi all had subobject classifiers, and so took over the name. The outcome is the abstract axiomatic concept of an *elementary topos*, formulated entirely in the basic language of categories and independently of set theory. Subsequently William Mitchell [72] and Julian Cole [73] produced a full and detailed answer to the above question by identifying the elementary topoi that are equivalent to **Set**.

As mentioned earlier set theory provides a general conceptual framework for mathematics. Now, since category theory, through the notion of topos, has succeeded in axiomatising set-theory, the outcome is an entirely new *categorial foundation of mathematics*! The category-theorists attitude that "function" rather than "set membership" can be seen as the fundamental mathematical concept has been entirely vindicated. The pre-eminent role of set theory in contemporary mathematics is suddenly challenged. A revolution has occurred in the history of mathematical ideas (albeit a peaceful one) that will undoubtedly influence the direction of the path to the future.

The notion of topos has great unifying power. It encompasses **Set** as well as the Grothendieck categories of sheaves, and so brings together the domains of set theory and algebraic geometry. But it also has ramifications for another area of rational inquiry, namely *logic*, the study of the canons of deductive reasoning. The principles of classical logic are represented in **Set** by operations on a certain set – the two element Boolean algebra. Each topos has an analogue of this algebra and so one can say that each topos carries its own logical calculus. It turns out that this calculus may differ from classical logic, and in general the logical princples that hold in a topos are those of *intuitionistic* logic. Now Intuitionism is a constructivist philosophy about the nature of mathematical entities and the meaning and validity of mathematical statements. It has nothing to do, per se, with logic in a topos, since the latter arises from a reformulation in categorial language of the set-theoretical account of classical logic. And yet we have this remarkable discovery that the two enterprises lead to the same logical structure. An inkling of how this can be comes on reflection that there is a well-known link between intuitionistic logic and topology, and that sheaves are initially topological entities. Furthermore the set-theoretical modelling of intuitionistic logic due to Saul Kripke [65] can be used to construct topoi in which the logic, as generalised from **Set**, turns out to be a reformulation of Kripke's semantic theory. Moreover these topoi of Kripke models can be construed as categories of sheaves.

These developments have yielded significant insights and new perspectives concerning the nature of sets and the connection between intuitionistic and classical logic. For example, one property enjoyed by the arrows in **Set** is *extensionality*; a function is uniquely determined by the values it gives to its arguments. Now the individuals of a topos may be thought of as 'generalised' sets and functions that may well be non-extensional. Interestingly, the imposition of extensionality proves to be one way of ensuring that the topos logic is classical. Another way, equally revealing, is to invoke (in arrow language) the axiom of choice.

Our aim then is to present the details of the story just sketched. The currently available literature on topoi takes the form of graduate level lecture notes, research papers and theses, wherein the mathematical sophisticate will find his needs adequately served. The present work on the other hand is an attempt at a fully introductory exposition, aimed at a wide audience. The author shares the view that the emergence of topos theory is an event of supreme importance, that has major implications for the advancement of conceptual understanding as well as technical knowledge in mathematics. It should therefore be made available to the philosopher–logician as well as the mathematician. Hence there are very few prerequisites for this book. Everything – set theory, logic, and category theory – begins at square one. Although some material may be very familiar, it should be remembered that one of our main themes is the development of new perspectives for familiar concepts. Hence it would seem quite appropriate that these concepts be re-appraised and that explicit discussion be provided of things that to many will have become second nature.

There are a number of proofs of theorems whose length and detail may be discouraging. A similar comment applies to the *verification* of the structural properties of some of the more complex categories (sheaves, Kripke models). The reader is recommended to skip over all of this detail initially and concentrate on the flow of ideas. It can often happen that although the verifications are long and tedious, the facts and ideas are themselves clear and readily comprehensible. Hopefully by steering a judiciously chosen course through elementary expositions that will bore the cognoscente, abstruse constructions that will tax the novice, and detailed justifications that will exhaust anyone, the reader will emerge with some insight into the "what" and "why" of this fascinating new area of logical-mathematical–philosophical study.

CHAPTER 1

MATHEMATICS = SET THEORY?

"No one shall drive us out of the paradise that Cantor has created"

David Hilbert

1.1. Set theory

The basic concept upon which the discipline known as *set theory* rests is the notion of *set membership.* A set may be initially thought of simply as a collection of objects, these objects being called *elements* of that collection. *Membership* is the relation that an object bears to a set by dint of its being an element of that set. This relation is symbolised by the Greek letter $\in$ (epsilon). "$x \in A$" means that A is a collection of objects, one of which is x, i.e. x is a *member* (element) of A. When x is not an element of A, this is written $x \notin A$. If $x \in A$, we may also say that x *belongs* to A.

From these fundamental ideas we may build up a catalogue of definitions and constructions that allow us to specify particular sets, and construct new sets from given ones. There are two techniques used here.

(a) *Tabular form:* this consists in specifying a set by explicitly stating all of its elements. A list of these elements is given, enclosed in brackets. Thus

$$\{0, 1, 2, 3\}$$

denotes the collection whose members are all the whole numbers up to 3.

(b) *Set Builder form:* this is a very much more powerful device that specifies a set by stating a property that is possessed by all the elements of the set, and by no other objects. Thus the property of "being a whole number smaller than four" determines the set that was given above in tabular form. The use of properties to define sets is enshrined in the

PRINCIPLE OF COMPREHENSION. *If $\varphi(x)$ is a property or condition pertaining to objects x, then there exists a set whose elements are precisely the objects that have the property (or satisfy the condition) $\varphi(x)$.*

The set corresponding to the property $\varphi(x)$ is denoted

$$\{x: \varphi(x)\}$$

This expression is to be read "the set of all those objects x such that φ is true of x".

EXAMPLE 1. If $\varphi(x)$ is the condition "$x \in A$ and $x \in B$" we obtain the set

$$\{x: x \in A \text{ and } x \in B\}$$

of all objects that belong to both A and B, i.e. the set of objects that A and B have in common. This is known as the *intersection* of the sets A and B, and is denoted briefly by $A \cap B$.

EXAMPLE 2. The condition "$x \in A$ or $x \in B$" yields, by the Comprehension Principle the set

$$\{x: x \in A \text{ or } x \in B\}$$

consisting of all of the elements of A together with all of those of B, and none others. It is called the *union* of A and B, written $A \cup B$.

EXAMPLE 3. The condition "$x \notin A$" determines $-A$, the *complement* of A. Thus

$$-A = \{x: x \notin A\}$$

is the set whose members are precisely those objects that do not belong to A.

These examples all yield new sets from given ones. We may also directly define sets by using conditions that do not refer to any particular sets. Thus from "$x \neq x$" we obtain the set

$$\emptyset = \{x: x \neq x\}$$

of all those objects x such that x is not equal to x. Since no object is distinct from itself, there is nothing that can satisfy the property $x \neq x$, i.e. $\emptyset$ has no members. For this reason $\emptyset$ is known as the *empty* set. Notice that we have already "widened our ontology" from the original conception of a set as something with members to admit as a set something that has no members at all. The notion of an empty collection is often difficult to accept at first. One tends to think initially of sets as objects built up in a rather concrete way out of their constituents (elements). The introduction of $\emptyset$ forces us to contemplate sets as abstract "things-in-themselves". One could think of references to $\emptyset$ as an alternative form of words, e.g. that "$A \cap B = \emptyset$" is a short-hand way of saying "A and B have no elements in common". Familiarity and experience eventually show that the admission of $\emptyset$ as an actual object enhances and simplifies the theory. The justification for calling $\emptyset$ *the* empty set is that there can be only one set with no members. This follows from the definition of equality of sets as embodied in the

PRINCIPLE OF EXTENSIONALITY: *Two sets are equal iff they have the same elements.*

It follows from this principle that if two sets are to be distinct then there must be an object that is a member of one but not the other. Since empty collections have no elements they cannot be so distinguished and so the Extensionality Principle implies that there is only one empty set.

Subsets

The definition of equality of sets can alternatively be conveyed through the notion of *subsets*. A set A is a *subset* of set B, written $A \subseteq B$, if each member of A is also a member of B.

EXAMPLE 1. The set $\{0, 1, 2\}$ is a subset of $\{0, 1, 2, 3\}$, $\{0, 1, 2\} \subseteq \{0, 1, 2, 3\}$.

EXAMPLE 2. For any set A, we have $A \subseteq A$, since each member of A is a member of A.

EXAMPLE 3. For any set A, $\emptyset \subseteq A$, for if $\emptyset$ was not a subset of A, there would be an element of $\emptyset$ that did not belong to A. However $\emptyset$ has no elements at all.

Using this latest concept we can see that, for any sets A and B,

$$A = B \quad \text{iff} \quad A \subseteq B \quad \text{and} \quad B \subseteq A.$$

If $A \subseteq B$ but $A \neq B$, we may write $A \subset B$ (A is a *proper* subset of B).

Russell's Paradox

In stating and using the Comprehension Principle we gave no precise explanation of what a "condition pertaining to objects x" is, nor indeed what sort of entities the letter x is referring to. Do we intend the elements of our sets to be physical objects, like tables, people, or the Eiffel Tower, or are they to be abstract things, like numbers, or other sets themselves? What about the collection

$$V = \{x: x = x\}?$$

All things, being equal to themselves, satisfy the defining condition for this set. Is V then to include everything in the world (itself as well) or should it be restricted to a particular kind of object, a particular universe of discourse?

To demonstrate the significance of these questions we consider the condition "$x \notin x$". It is easy to think of sets that do not belong to themselves. For example the set $\{0, 1\}$ is distinct from its two elements 0 and 1. It is not so easy to think of a collection that includes itself amongst its members. One might contemplate something like "the set of all sets". A somewhat intriguing example derives from the condition

> "x is a set derived from the Comprehension Principle by a defining condition expressed in less than 22 words of English".

The sentence in quotation marks has less than 22 words, and so defines a set that satisfies its own defining condition.

Using the Comprehension Principle we form the so-called *Russell set*

$$R = \{x: x \notin x\}.$$

The crunch comes when we ask "Does R itself satisfy the condition $x \notin x$?" Now if $R \notin R$, it does satisfy the condition, so it belongs to the set defined by that condition, which is R, hence $R \in R$. Thus the assumption $R \notin R$ leads to the contradictory conclusion $R \in R$. We must therefore reject this assumption, and accept the alternative $R \in R$. But if $R \in R$, i.e. R is an element of R, it must satisfy the defining condition for R, which is $x \notin x$. Thus $R \notin R$. This time the assumption $R \in R$ has lead to contradiction, so it is rejected in favor of $R \notin R$. So now we have proven both $R \in R$ and $R \notin R$, i.e. R both is, and is not, an element of itself. This is hardly an acceptable situation.

The above argument, known as Russell's Paradox, was discovered by Bertrand Russell in 1901. Set theory itself began a few decades earlier with the work of George Cantor. Cantor's concern was initially with the analysis of the real number system, and his theory, while rapidly becoming of intrinsic interest, was largely intended to give insight into properties of infinite sets of real numbers (e.g. that the set of irrational numbers has "more" elements than the set of rational numbers). During this same period the logician Gottlob Frege made the first attempt to found a definition of "number" and a development of the laws of arithmetic on formal logic and set theory. Frege's system included the Comprehension Principle in a form much as we have given it, and so was shown to be inconsistent (contradictory) by Russell's paradox. The appearance of the later, along with other set-theoretical paradoxes, constituted a crisis in the development of a theoretical basis for mathematical knowledge. Mathematicians were faced with the problem of revising their intuitive ideas about sets and reformulating them in such a way as to avoid inconsistencies. This challenge provided one of the major sources for the burgeoning growth in this century of mathematical logic, a subject which, amongst other things, undertakes a detailed analysis of the axiomatic method itself.

NBG

Set theory now has a rigorous axiomatic formulation – in fact several of them, each offering a particular resolution of the paradoxes.

John von Neuman proposed a solution in the mid-1920's that was later refined and developed by Paul Bernays and Kurt Gödel. The outcome is a group of axioms known as the system NBG. Its central feature is a very simple and yet powerful conceptual distinction between *sets* and *classes*. All entities referred to in NBG are to be thought of as classes, which correspond to our intuitive notion of collections of objects. The word "set" is reserved for those classes that are themselves members of other classes. The statement "x is a set" is then short-hand for "there is a class y such that $x \in y$". Classes that are not sets are called *proper classes*. Intuitively we think of them as "very large" collections. The Comprehension Principle is modified by requiring the objects x referred to there to be sets. Thus from a condition $\varphi(x)$ we can form the *class* of all *sets* (elements of other classes) that satisfy $\varphi(x)$. This is denoted

$$\{x : x \text{ is a set and } \varphi(x)\}.$$

The definition of the Russell class must now be modified to read

$$R = \{x : x \text{ is a set and } x \notin x\}.$$

Looking back at the form of the paradox we see that we now have a way out. In order to derive $R \in R$ we would need the extra assumption that R is a set. If this were true the contradiction would obtain as before, and so we reject it as false. Thus the paradox disappears and the argument becomes nothing more than a proof that R is a *proper class* i.e. a large collection that is not an element of any other collection. In particular $R \notin R$.

Another example of a proper class is V, which we now take to be the class

$$\{x : x \text{ is a set and } x = x\}$$

whose elements are all the sets. In fact NBG has further axioms that imply that $V = R$, i.e. no set is a member of itself.

ZF

A somewhat different and historically prior approach to the paradoxes was proposed by Ernst Zermelo in 1908. This system was later extended by Abraham Fraenkel and is now known as ZF. It can be informally regarded as a theory of "set-building". There is only one kind of entity, the set. All sets are built up from certain simple ones (in fact one can start just with $\emptyset$) by operations like intersection $\cap$, union $\cup$, and complementation $-$. The axioms of ZF legislate as to when such operations can be effected. They can only be applied to sets that have already been constructed, and the result is always a set. Thus proper classes like R are never actually constructed within ZF.

The Comprehension Principle can now only be used relative to a given set, i.e. we cannot collect together all objects satisfying a certain condition, but only those we already know to be members of some previously defined set. In ZF this is known as the

Separation Principle. *Given a set A and a condition $\varphi(x)$ there exists a set whose elements are precisely those members of A that satisfy $\varphi(x)$.*

This set is denoted

$$\{x : x \in A \text{ and } \varphi(x)\}.$$

Again we can no longer form the Russell class per se, but only for each set A the set

$$R(A) = \{x : x \in A \text{ and } x \notin x\}.$$

To obtain a contradiction involving the statements $R(A) \in R(A)$ and $R(A) \notin R(A)$ we would need to know that $R(A) \in A$. Our conclusion then is simply that $R(A) \notin A$. In fact in ZF as in NBG no set is an element of itself, so $R(A) = A$. (Note the similarity of this argument to the resolution in NBG – replacing V everywhere by A makes the latter formally identical to the former.)

NBG and ZF then offer some answers to the questions posed earlier. In practical uses of set theory, members of collections may well be physicial objects. In axiomatic presentations of set theory however all objects have a conceptual rather than a material existence. The entities considered are "abstract" collections, whose members are themselves sets. NBG offers a "larger" ontology than ZF. Indeed ZF can be construed as a subsystem of NBG, consisting of the part of NBG that refers only to sets, (i.e. classes that are not proper). We still have not shed any real light on what we mean by a "condition pertaining to objects x" (since sets are never members of themselves, the "less than 22 words" condition mentioned earlier will not be admissible in ZF or NBG). Some clarification of this notion will come later when we consider formal languages and take a closer look at the details of the axioms for systems like ZF.

Consistency

The fact that a particular system avoids Russell's Paradox does not guarantee that it is consistent, i.e. entirely free of contradictions. It is known an inconsistency in either ZF or NBG would imply an inconsistency in the other, and so the two systems stand or fall together. They have been intensively and extensively studied in the last 60 or so years without any contradiction emerging. However there is a real conceptual barrier to the possibility of proving that no such contradiction will ever be found. This was demonstrated by Gödel, around 1930, who showed in effect that any proof of consistency would have to depend on principles whose own consistency was no more certain than that of ZF and NBG themselves. In the decade prior to Gödel's work a group of mathematicians lead by David Hilbert had attempted to establish the consistency of arithmetic and mathematics generally by using only so-called *finitary* methods. These methods are confined to the description of concrete, particular, directly perceivable objects, and principles whose truth is evident by direct inspection. Gödel showed that such methods could never establish the consistency of any system that was powerful enough to develop the arithmetic of ordinary whole numbers. This discovery is regarded as one of the major mathematical events of the 20th century. Its impact on Hilbert's program was devastating, but many people have found in it a source of encouragement, an affirmation of the essentially creative nature of mathematical thought, and evidence against the mechanistic thesis that the mind can be adequately modelled as a physical computing device. As Gödel himself has put it, "either mathematics is too big for the human mind, or the human mind is more than a machine." (cf. Bergamini [65]).

While it would seem there can be no absolute demonstration of the consistency of ZF, there is considerable justification, of an experiential and epistemological nature, for the belief that it contains no contradictions. Certainly if the opposite were the case then, in view of the central role of set theory in contemporary mathematics, a great deal more would be at stake than simply the adequacy of a particular set of postulates.

Which of ZF and NBG is a "better" treatment of set theory? The choice is largely a matter of philosophical taste, together with practical need. ZF seems to enjoy the widest popularity amongst mathematicians generally. Its principle of relativising constructions to particular sets closely reflects the way set theory is actually used in mathematics, where sets are specified within clearly given, mathematically defined contexts (universes). The collection of all sets has not been an object of concern for most working mathematicians. Indeed the sets that they need can generally be obtained within a small fragment of ZF. It is only very recently, with the advent of category theory that a genuine need has arisen amongst mathematicians (other than set-theorists) for a means of handling large collections. These needs are met in a more flexible way by the class-set dichotomy, and have offered a more significant role to NBG and even stronger systems.

The moral to be drawn from these observations is that there is no "correct" way to do set theory. The system a mathematician chooses to work with will depend on what he wishes to achieve.

1.2. Foundations of mathematics

The aim of Foundational studies is to produce a rigorous explication of the nature of mathematical reality. This involves a precise and formal definition, or representation of mathematical concepts, so that their interrelationships can be clarified and their properties better understood. Most approaches to foundations use the axiomatic method. The language to be used is first introduced, generally itself in a precise and formal description. This language then serves for the definition of mathematical notions and the statement of postulates, or axioms, concerning their properties.

The axioms codify ways we regard mathematical objects as actually behaving. The theory of these objects is then developed in the form of statements derived from the axioms by techniques of deduction that are themselves rendered explicit.

It would be somewhat misleading to infer from this that foundational systems act primarily as a basis out of which mathematics is actually created. The artificiality of that view is evident when one reflects that the essential content of mathematics is already there before the basis is made explicit, and does not depend on it for its existence. We may for example think of a real number as an infinite decimal expression, or a point on the number line. Alternatively it could be introduced as an element of a complete ordered field, an equivalence class of Cauchy sequences, or a Dedekind cut. None of these could be said to be *the* correct explanation of what a real number is. Each is an enbodiment of an intuitive notion and we evalute it, not in terms of its correctness, but rather in terms of its effectiveness in explicating the nature of the real number system.

Mathematical discovery is by no means a matter of systematic deductive procedure. It involves insight, imagination, and long explorations along paths that sometimes lead nowhere. Axiomatic presentations serve to describe and communicate the fruits of this activity, often in a different order to that in which they were arrived at. They lend a coherence and unity to their subject matter, an overview of its extent and limitations.

Having clarified our intuitions, the formal framework may then be used for further exploration. It is at this level that the axiomatic method does have a creative role. The systematisation of a particular theory may lead to new internal discoveries, or the recognition of similarities with other theories and their subsequent unification. This however belongs to the "doing" of mathematics. As far as Foundational studies are concerned the role of axiomatics is largely descriptive. A Foundational system serves not so much to prop up the house of mathematics as to clarify the principles and methods by which the house was built in the first place. "Foundations" is a discipline that can be seen as a branch of mathematics standing apart from the rest of the subject in order to describe the world in which the working mathematician lives.

1.3. Mathematics as set theory

The equation of mathematics with set theory can with some justification be seen as a summary of the direction that mathematics has taken in modern times. Many will have heard of the revolution in school curricula called the "New Math". This has largely revolved around the introduction of set theory into elementary education and indicates the preoccupation of the mathematical community with that subject. Of all the foundational frameworks that have been proposed, the set theories have enjoyed the widest acceptance and the most detailed attention. Systems like ZF and NBG provide an elegant formalisation and explanation of the basic notions that the mathematician uses. Paul Cohen, whose work on the independence of the Continuum Hypothesis in 1963 lead to a veritable explosion of set-theoretic activity, has said "by analysing mathematical arguments logicians became convinced that the notion of "set" is the most fundamental concept of mathematics."

Apart from, or perhaps because of, its central role in Foundations, set theory has also dominated the stage of mathematical practise. This is not intended to imply that mathematicians think in set-theoretical concepts, although that is very often the case. Rather the point is that set theory is the basic tool of communication and exposition. It has provided the vehicle for an enormous proliferation of mathematics, both in terms of quantity of knowledge and range of topics and applications. It would be hard to find a recent book on any pure mathematical subject, be it algebra, geometry, analysis, or probability theory, that used no set-theoretical symbolism.

The group of French mathematicians who work under the name of Nicolas Bourbaki undertook in 1935 the formidable task of producing a "fully axiomatised presentation of mathematics in entirety". The result, over 40 years, has been about that many volumes to date, ranging over algebra, analysis and topology. Book 1 of this influential work is devoted to the theory of sets, which provides the framework for the whole enterprise. Bourbaki has said (1949) "... all mathematical theories may be regarded as extensions of the general theory of sets ... on these foundations I state that I can build up the whole of the mathematics of the present day".

The point to be made in this book is that the emergence of category theory has changed the perspectives just described, and that Cohen's statement is no longer even prima facie acceptable. It may be the case that the objects of mathematical study *can* be thought of as sets, but it is not certain that in the future they *will* be so regarded. No doubt the basic language of set theory will continue to be an important tool whenever collections of things are to be dealt with. But the conception of the things themselves as sets has lost some of its prominence through the development of a natural and attractive alternative. It seems indeed very likely that the role of set theory as the lingua universalis for mathematical foundations will be a declining one in the years to come. In case the wrong impression should have been conveyed by the last quotation above, it should be noted that the French mathematicians have been amongst the first to recognise this. René Thom [71] has written that "the old hope of Bourbaki, to see mathematical structures arise naturally from a hierachy of sets, from their subsets, and from their combination, is, doubtless, only an illusion". And in an address given in 1961, Jean Dieudonné made the following prophetic statement:

"In the years between 1920 and 1940 there occurred, as you know, a complete reformation of the classification of different branches of mathematics, necessitated by a new conception of the essence of mathematical thinking itself, which originated from the works of Cantor and Hilbert. From the latter there sprang the *systematic axiomatization* of mathematical science in entirety and the fundamental concept of *mathematical structure.* What you may perhaps be unaware of is that mathematics is about to go through a second revolution at this very moment. This is the one which is in a way completing the work of the first revolution, namely, which is releasing mathematics from the far too narrow conditions by 'set'; it is the theory of *categories and functors*, for which estimation of its range or perception of its consequences is still too early . . . ". (Quoted from Fang [70].)

CHAPTER 2

WHAT CATEGORIES ARE

> "*. . . understanding consists in reducing one type of reality to another.*"
>
> Claude Levi-Strauss

2.1. Functions are sets?

A good illustration of the way in which set theory formalises an intuitive mathematical idea is provided by an examination of the notion of a *function.* A function is an association between objects, a correspondence that assigns to a given object one and only one other object. It may be thought of as a rule, or operation, which is applied to something to obtain its associated thing. A useful way of envisaging a function is as an input–output process, a kind of "black box" (see figure). For a given input the function produces a uniquely determined output. For example, the instruction "multiply by 6" determines a function which for input 2 gives output $6 \times 2 = 12$, which associates with the number 1 the number 6, which assigns 24 to 4, and so on. The inputs are called *arguments* of the function and the outputs *values,* or *images* of the inputs that they are produced by. If f denotes a function, and x an input, then the corresponding output, the image of x under f, is denoted $f(x)$. The above example may then be displayed as that function f given by the rule $f(x) = 6x$.

If A is the set of all appropriate inputs to function f (in our example A will include the number 2, but not the Eiffel Tower), and B is a set that includes all the f-images of the members of A (and possibly the Eiffel Tower as well), then we say that f is a function *from A to B*. This is symbolised as $f: A \to B$ or $A \xrightarrow{f} B$. A is called the *domain* or *source* of f and B is the *codomain* or *target.*

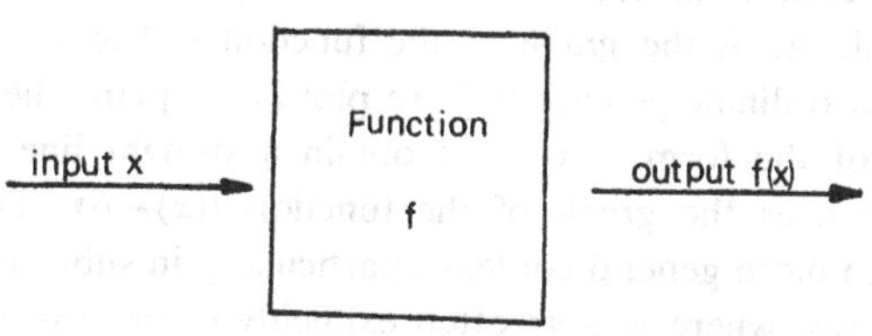

Fig. 2.1.

How does set theory deal with this notion? To begin with we introduce the notion of an *ordered* pair, as consisting of two objects with one designated as first, and the other as second. The notation $\langle x, y\rangle$ is used for the ordered pair having x as first element and y as second. The essential property of this notion is that $\langle x, y\rangle = \langle z, w\rangle$ if and only if $x = z$ and $y = w$.

We now define a (binary) *relation* as being a set whose elements are all ordered pairs. This formalises the intuitive idea of an association referred to earlier. If R is a relation (set of ordered pairs) and $\langle x, y\rangle \in R$ (sometimes written xRy) then we think of x being assigned to y by the association that R represents. For example the expression "is less than" establishes an association between numbers and determines the set

$$\{\langle x, y\rangle : x \text{ is less than } y\}.$$

Note that the pairs $\langle 1, 2\rangle$ and $\langle 1, 3\rangle$ both belong to this set, i.e. a relation may associate several objects to a given one.

From a function we obtain the relation

$$\hat{f} = \{\langle x, y\rangle : y \text{ is the } f\text{-image of } x\}.$$

To distinguish those relations that represent functions we have to incorporate the central feature of functions, namely that a given input produces one uniquely corresponding output. This means that each x can be the first element of only one of the ordered pairs in $\hat{f}$. That is

$$(*) \qquad \text{if} \quad \langle x, y\rangle \in \hat{f} \quad \text{and} \quad \langle x, z\rangle \in \hat{f}, \quad \text{then} \quad y = z.$$

This then is our set-theoretical characterisation of a function; as a set of ordered pairs satisfying the condition (*). What happens next is a ploy often used in mathematics – a formal representation becomes an actual definition. It is quite common, in books at all levels, to find near the beginning a statement to the effect that "a function *is* a set of ordered pairs such that . . .".

How successful is this set-theoretical formulation of the function concept? Technically it works very well and allows an easy development of the theory of functions. But there are a number of rejoinders that can be made on the conceptual level. Some would say that the set $\hat{f}$ is not a function at all, but is the *graph* of the function f. The word of course comes from co-ordinate geometry. If we plot in the plane the points with co-ordinates of the form $\langle x, 6x\rangle$ we obtain a straight line (see figure) which is known as the *graph* of the function $f(x) = 6x$. This usage is carried over to more general contexts, particularly in subjects like topology and analysis, where writers often explicitly distinguish the function $f: A \to B$ from the *graph of* f as the set $\{\langle x, f(x)\rangle : x \in A\}$. Conflation of the two notions can easily lead to confusion.

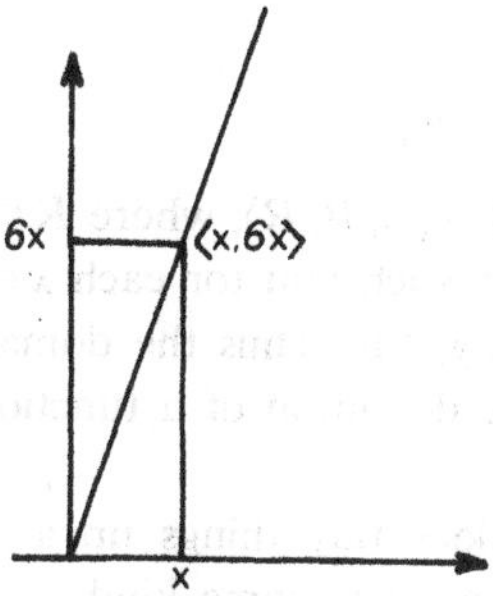

Fig. 2.2.

Another difficulty relates to the notion of codomain. Given a function f simply as a set of ordered pairs we can readily recover the domain (set of inputs) as the set

$$\operatorname{dom} f = \{x: \text{for some } y, \langle x, y\rangle \in f\}.$$

But what about the codomain of f? Recall that this can be any set that includes all the outputs of f. The outputs themselves form the so-called *range* or *image* of f, symbolically

$$\operatorname{Im} f = \{y: \text{for some } x, \langle x, y\rangle \in f\}.$$

In general f can be called a function from A to B whenever $A = \operatorname{dom} f$ and $\operatorname{Im} f \subseteq B$. Thus a function given simply as a set of ordered pairs does not have a uniquely determined codomain. This may seem a trifling point, but it leads to an interesting complication with the very important notion of *identity function*. This function is characterised by the rule $f(x) = x$, i.e. the output assigned to a given input is just that input itself. Each set A has its own identity function, called the *identity function on* A, denoted id_A, whose domain is the set A. Thus the image of id_A is also A, i.e. $\mathrm{id}_A : A \to A$. On the set-theoretic account, $\mathrm{id}_A = \{\langle x, x\rangle : x \in A\}$.

Now if A is a subset of a set B, then the rule $f(x) = x$ provides a function from A to B. In this case we talk of the *inclusion* function from A to B, for which we reserve the symbol $A \hookrightarrow B$. The use of a new word indicates a different intention. It conveys the sense of the function acting to include the elements of A amongst those of B. However even though the identity function on A and the inclusion map from A to B are conceptually quite different, as set-theoretical entities they are identical, i.e. exactly the same set of ordered pairs.

One way to cope with this point would be to modify the definition of function in the following way. Firstly for sets A and B we define the *product set* or *Cartesian product* of A and B to be the set of all ordered pairs whose first elements are in A and second elements in B. This is denoted $A \times B$, and so

$$A \times B = \{\langle x, y\rangle : x \in A \text{ and } y \in B\}.$$

A function is now defined as a triple $f = \langle A, B, R\rangle$, where $R \subseteq A \times B$ is a relation from A to B (the *graph* of f), such that for each $x \in A$ there is one and only one $y \in B$ for which $\langle x, y\rangle \in R$. Thus the domain ($A$) and codomain ($B$) are incorporated in the definition of a function from the outset.

Although the modified definition does tidy things up a little it still presents a function as being basically a set of some kind – a fixed, static object. It fails to convey the "operational" or "transitional" aspect of the concept. One talks of "applying" a function to an argument, of a function "acting" on a domain. There is a definite impression of action, even of motion, as evidenced by the use of the arrow symbol, the source-target terminology, and commonly used synonyms for "function" like "transformation" and "mapping". The impression is analogous to that of a physical force acting on an object to move it somewhere, or replace it by another object. Indeed in geometry, transformations (rotations, reflections, dilations etc.) are functions that quite literally describe motion, while in applied mathematics forces are actually modelled as functions. This dynamical quality that we have been describing is an essential part of the meaning of the word "function" as it is used in mathematics. The "ordered-pairs" definition does not convey this. It is a formal set-theoretic *model* of the intuitive idea of a function, a model that captures an aspect of the idea, but not its full significance.

2.2. Composition of functions

Given two functions $f: A \to B$ and $g: B \to C$, with the target of one being the source of the other, we can obtain a new function by the rule "apply f and then g". For $x \in A$, the output $f(x)$ is an element of B, and hence an input to g. Applying g gives the element $g(f(x))$ of C. The passage from x to $g(f(x))$ establishes a function with domain A and codomain C. It is called the *composite of f and g*, denoted $g \circ f$, and symbolically defined by the rule $g \circ f(x) = g(f(x))$.

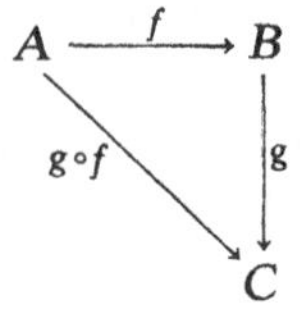

Now suppose we have three functions $f:A \to B$, $g:B \to C$, and $h:C \to D$ whose domains and codomains are so related that we can apply the three in succession to get a function from A to D. There are actually two ways to do this, since we can first form the composites $g \circ f : A \to C$ and $h \circ g : B \to D$. Then we follow either the rule "do f and then $h \circ g$", giving the function $(h \circ g) \circ f$, or the rule "do $g \circ f$ and then h", giving the composite $h \circ (g \circ f)$.

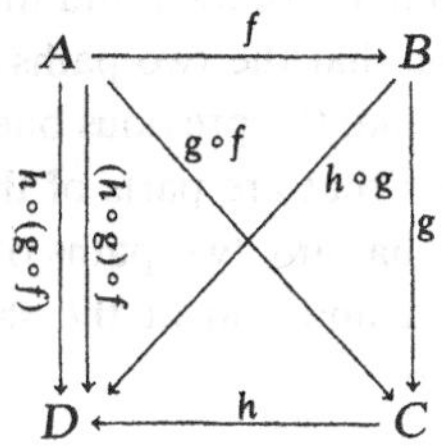

In fact these two functions are the same. When we examine their outputs we find that

$$[h \circ (g \circ f)](x) = h(g \circ f(x)) = h(g(f(x))),$$

while

$$[(h \circ g) \circ f](x) = h \circ g(f(x)) = h(g(f(x))).$$

Thus the two functions have the same domain and codomain, and they give the same output for the same input. They each amount to the rule "do f, and then g, and then h." In other words, they are the same function, and we have established the following.

Associative Law for Functional Composition. $h \circ (g \circ f) = (h \circ g) \circ f$.

This law allows us to drop brackets and simply write $h \circ g \circ f$ without ambiguity. Note that the law does not apply to *any* three functions – the equation only makes sense when they "follow a path", i.e. their sources and targets are arranged as described above.

The last figure is an example of the notion of *commutative diagram*, a very important aid to understanding used in category theory. By a diagram we simply mean a display of some objects, together with some arrows (here representing functions) linking the objects. The "triangle" of arrows f, g, h as shown is another diagram.

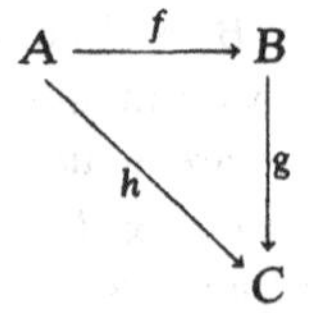

It will be said to *commute* if $h = g \circ f$. The point is that the diagram offers two paths from A to C, either by composing to follow f and then g, or by following h directly. Commutativity means that the two paths amount to the same thing. A more complex diagram, like the previous one, is said to be commutative when all possible triangles that are parts of the diagram are themselves commutative. This means that any two paths of arrows in the diagram that start at the same object and end at the same object compose to give the same overall function.

Composing with identities

What happens when we compose a function with an identity function? Given $f: A \to B$ we can follow f by id_B. Computing outputs we find, for $x \in A$, that

$$\mathrm{id}_B \circ f(x) = \mathrm{id}_B(f(x)) = f(x).$$

Similarly, given $g: B \to C$ we can precede g by id_B, in which case, for $x \in B$,

$$g \circ \mathrm{id}_B(x) = g(\mathrm{id}_B(x)) = g(x).$$

Since $\mathrm{id}_B \circ f$ and f have the same source and target, as do $g \circ \mathrm{id}_B$ and g, we have established the following.

IDENTITY LAW FOR FUNCTIONAL COMPOSITION. *For any* $f: A \to B$, $g: B \to C$, $\mathrm{id}_B \circ f = f$, *and* $g \circ \mathrm{id}_B = g$.

The Identity Law amounts to the assertion of the commutativity of the following diagram

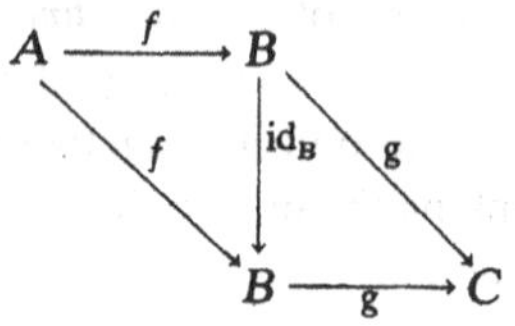

2.3. Categories: First examples

We have already stated that a category can initially be conceived as a universe of mathematical discourse, and that such a universe is determined by specifying a certain kind of object and a certain kind of "function" between objects. The less suggestive word "arrow" is used in place of "function" in the general theory of categories (the word "morphism" is also used). The following table lists some categories by specifying their objects and arrows.

CATEGORY	OBJECTS	ARROWS
Set	all sets	all functions between sets
Finset	all finite sets	all functions between finite sets
Nonset	all nonempty sets	all functions between nonempty sets
Top	all topological spaces	all continuous functions between topological spaces
Vect	vector spaces	linear transformations
Grp	groups	group homomorphisms
Mon	monoids	monoid homomorphisms
Met	metric spaces	contraction maps
Man	manifolds	smooth maps
Top Grp	topological groups	continuous homomorphisms
Pos	partially ordered sets	monotone functions

In each of these examples the objects are sets with, apart from the first three cases, some additional structure. The arrows are all set functions which in each appropriate case satisfy conditions relating to this structure. It is not in fact vital that the reader be familiar with all of these examples. What is important is that she or he understands what they all have in common – what it is that makes each of them a category. The key lies, not in the particular nature of the objects or arrows, but in the way the arrows behave. In each case the following things occur;

(a) each arrow has associated with it two special objects, its *domain* and its *codomain*,

(b) there is an operation of *composition* that can be performed on certain pairs $\langle g, f\rangle$ of arrows in the category (when domain of $g =$ codomain of f) to obtain a new arrow $g \circ f$, *which is also in the category.*

(A composite of group homomorphisms is a group homomorphism, a composite of continuous functions between topological spaces is itself continuous etc.) This operation of composition always obeys the Associative Law described in the last section,

(c) each object has associated with it a special arrow in the category, the *identity arrow* on that object. (The identity function on a group is a group homomorphism, on a topological space is continuous etc). Within the category the identity arrows satisfy the Identity Law described in §2.2.

There are other features common to our list of examples. But as categories it is the two properties of associative composition and existence of identities that we single out for particular attention in the

Axiomatic Definition of a Category. A *category* $\mathscr{C}$ comprises

(1) a collection of things called *$\mathscr{C}$-objects*;

(2) a collection of things called *$\mathscr{C}$-arrows*;

(3) operations assigning to each $\mathscr{C}$-arrow f a $\mathscr{C}$-object $\operatorname{dom} f$ (the "domain" of f) and a $\mathscr{C}$-object $\operatorname{cod} f$ (the "codomain" of f). If $a = \operatorname{dom} f$ and $b = \operatorname{cod} f$ we display this as

$$f : a \to b \quad \text{or} \quad a \xrightarrow{f} b;$$

(4) an operation assigning to each pair $\langle g, f\rangle$ of $\mathscr{C}$-arrows with $\operatorname{dom} g = \operatorname{cod} f$, a $\mathscr{C}$-arrow $g \circ f$, *the composite of f and g*, having $\operatorname{dom}(g \circ f) = \operatorname{dom} f$ and $\operatorname{cod}(g \circ f) = \operatorname{cod} g$, i.e. $g \circ f : \operatorname{dom} f \to \operatorname{cod} g$, and such that the following condition obtains:

Associative Law: Given the configuration

$$a \xrightarrow{f} b \xrightarrow{g} c \xrightarrow{h} d$$

of $\mathscr{C}$-objects and $\mathscr{C}$-arrows then $h \circ (g \circ f) = (h \circ g) \circ f$.

The associative law asserts that a diagram having the form

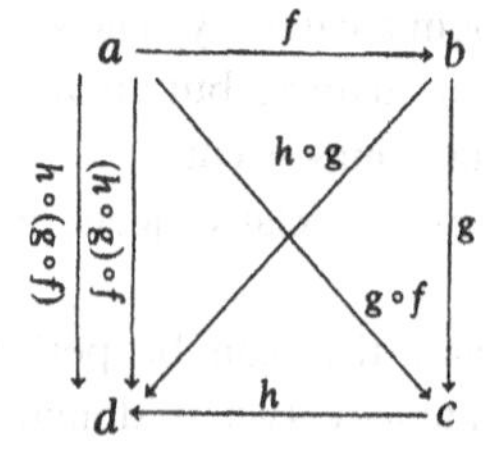

always commutes;

(5) an assignment to each $\mathscr{C}$-object b of a $\mathscr{C}$-arrow $1_b : b \to b$, called *the identity arrow on b*, such that

Identity Law: For any $\mathscr{C}$-arrows $f : a \to b$ and $g : b \to c$

$$1_b \circ f = f, \quad \text{and} \quad g \circ 1_b = g$$

i.e. the diagram

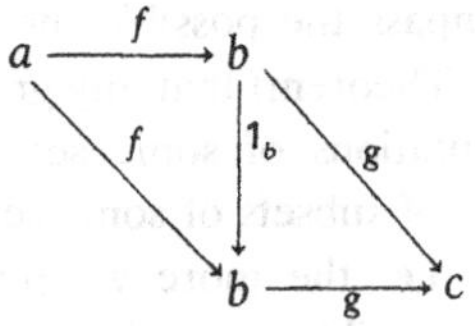

commutes.

2.4. The pathology of abstraction

The process we have just been through in identifying the notion of a category is one of the basic modi operandi of pure mathematics. It is called *abstraction*. It begins with the recognition, through experience and examination of a number of specific situations, that certain phenomena occur repeatedly, that there are a number of common features, that there are formal analogies in the behaviour of different entities. Then comes the actual *process* of abstraction, wherein these common features are singled out and presented in isolation; an axiomatic description of an "abstract" concept. This is precisely how we obtained our general definition of a category from an inspection of a list of particular categories. It is the same process by which all of the abstract structures that mathematics investigates (group, vector space, topological space etc) were arrived at.

Having obtained our abstract concept we then develop its general theory, and seek further instances of it. These instances are called *examples* of the concept or *models* of the axioms that define the concept. Any statement that belongs to the general theory of the concept (i.e. is derivable from the axioms) will hold true in all models. The search for new models is a process of specialisation, the reverse of abstraction. Progress in understanding comes as much from the recognition that a particular new structure is an instance of a more general phenomenon, as from the recognition that several different structures have a common core. Our knowledge of mathematical reality advances through the interplay of these two processes, through movement from the particular to the general and back again. The procedure is well illustrated, as we shall see, by the development of topos theory.

An important aspect of specialisation concerns so-called *representation* theorems. These are propositions to the effect that any model of the axioms for a certain abstract structure must be (equivalent to) one of a particular list of concrete models. They "measure" the extent to which the original motivating examples encompass the possible models of the general notion. Thus we know (Cayley's Theorem) that any group can be thought of as being a group of permutations of some set, while any Boolean algebra is essentially an algebra of subsets of some set. Roughly speaking, the stronger the abstraction, i.e. the more we put into the abstract concept, the fewer will be the possible examples. The extreme case is where there is only one model. A classic example of this is the axiomatically presented concept of a complete ordered field. There is in fact only one such field, viz the real number system.

The category axioms represent a very weak abstraction. There is no representation theorem in terms of our original list. We talked at the outset of "general universes of mathematical discourse". However we have picked out only the bare bones of our initial examples, and so little of the flesh that the axioms admit of all sorts of "pathological" cases that differ wildly in appearance from **Set**, **Top**, **Vect** etc. One readily finds categories that are not universes of discourse at all, in which the objects are not sets, the arrows look nothing like functions, and the operation $\circ$ has nothing to do with functional composition. The following list includes a number of such categories. The reader is urged to examine these closely, to fill out the details of their definition, and to check that in each case the Associative and Identity axioms are satisfied.

2.5. Basic examples

Example 1. **1**: This category is to have only one object, and one arrow. Having said that, we find that its structure is completely determined. Suppose we call the object a, and the arrow f. Then we must put $\text{dom}\, f = \text{cod}\, f = a$, as a is the only available object. Since f is the only arrow, we have to take it as the identity arrow on a, i.e. we put $1_a = f$. The only composable pair of arrows is $\langle f, f \rangle$, and we put $f \circ f = f$. This gives the identity law, as $1_a \circ f = f \circ 1_a = f \circ f = f$, and the associative law holds as $f \circ (f \circ f) = (f \circ f) \circ f = f$. Thus we have a category, which we

display diagramatically as

We did not actually say what a and f are. The point is that they can be anything you like. a might be a set, with f its identity function. But f might be a number, or a pair of numbers, or a banana, or the Eiffel tower, or Richard Nixon. Likewise for a. Just take any two things, call them a and f, make the above *definitions* of $\text{dom}\, f$, $\text{cod}\, f$, 1_a, and $f \circ f$, and you have produced a structure that satisfies the axioms for a category. Whatever a and f are, the category will look like the above diagram. In this sense there is "really" only one category that has one object and one arrow. We give it the name **1**. As a paradigm description of it we might as well take the object to be the number 0, and the arrow to be the ordered pair $\langle 0, 0\rangle$.

Example 2. **2**: This category has two objects, three arrows, and looks like

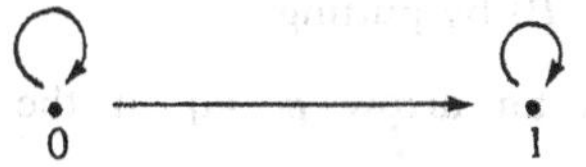

We take the two objects to be the numbers 0 and 1. For the three arrows we take the pairs $\langle 0, 0\rangle$, $\langle 0, 1\rangle$, and $\langle 1, 1\rangle$, putting

$$\langle 0,0\rangle : 0 \to 0$$
$$\langle 0,1\rangle : 0 \to 1$$
$$\langle 1,1\rangle : 1 \to 1$$

Thus we must have

$$\langle 0,0\rangle = 1_0 \quad \text{(the identity on 0)}$$

and

$$\langle 1,1\rangle = 1_1.$$

There is only one way to define composition for this set up:

$$1_0 \circ 1_0 = 1_0$$
$$\langle 0,1\rangle \circ 1_0 = \langle 0,1\rangle$$
$$1_1 \circ \langle 0,1\rangle = \langle 0,1\rangle$$

and

$$1_1 \circ 1_1 = 1_1$$

EXAMPLE 3. **3**: This category has three objects and six arrows, the three non-identity arrows being arranged in a triangle thus:

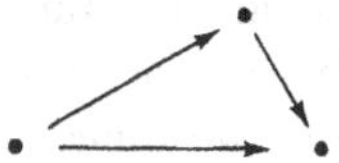

Again there is only one possible way to define composites.

EXAMPLE 4. *Preorders in general.* In each of our first three examples there is only one way that composites can be defined. This is because between any two objects there is never more than one arrow, so once the dom and cod are known, there is no choice about what the arrow is to be. In general a category with this property, that between any two objects p and q there is *at most one* arrow $p \to q$, is called a *pre-order.* If P is the collection of objects of a pre-order category then we may define a binary relation R on P (i.e. a set $R \subseteq P \times P$) by putting

> $\langle p, q\rangle \in R$ iff there is an arrow $p \to q$ in the pre-order category.

The relation R then has the following properties (writing "pRq" in place of "$\langle p, q\rangle \in R$"); it is

(i) *reflexive*, i.e. for each p we have pRp, and

(ii) *transitive*, i.e. whenever pRq and qRs, we have pRs.

(Condition (i) holds as there is always the identity arrow $p \to p$, for any p. For (ii), observe that an arrow from p to q composes with one from q to s to give an arrow from p to s).

A binary relation that is reflexive and transitive is commonly known as a *pre-ordering*. We have just seen that a pre-order category has a natural pre-ordering relation on its collection of objects (hence its name of course). Conversely if we start simply with a set P that is pre-ordered by a relation R (i.e. $R \subseteq P \times P$ is reflexive and transitive) then we can obtain a pre-order category as follows. The objects are the members p of P. The arrows are the pairs $\langle p, q\rangle$ for which pRq. $\langle p, q\rangle$ is to be an arrow from p to q. Given a composable pair

$$p \xrightarrow{\langle p,q\rangle} q \xrightarrow{\langle q,s\rangle} s,$$

we put

$$\langle q, s\rangle \circ \langle p, q\rangle = \langle p, s\rangle.$$

Note that if $\langle p, q\rangle$ and $\langle q, s\rangle$ are arrows then pRq and qRs, so pRs (transitivity) and hence $\langle p, s\rangle$ is an arrow. There is at most one arrow from p to q, depending on whether or not pRq, and by transitivity there is only one way to compose arrows. By reflexivity, $\langle p, p\rangle$ is always an arrow, for any p, and indeed $\langle p, p\rangle = 1_p$.

Examples 1–3 are pre-orders whose associated pre-ordering relation R satisfies a further condition, viz it is

(iii) *antisymmetric*, i.e. whenever pRq and qRp, we have

$$p = q.$$

An antisymmetric pre-ordering is called a *partial ordering*. The symbol "$\sqsubseteq$" will generally be used for this type of relation, i.e. we write $p \sqsubseteq q$ in place of pRq. A *poset* is a pair $\mathbf{P} = \langle P, \sqsubseteq\rangle$, where P is a set and $\sqsubseteq$ is a partial ordering on P. These structures will play a central role in our study of topoi.

The set $\{0\}$ becomes a poset when we put $0 \sqsubseteq 0$. The corresponding pre-order category is **1** (Example 1). The pre-order **2** corresponds to the partial ordering on the set $\{0, 1\}$ that has $0 \sqsubseteq 1$ (and of course $0 \sqsubseteq 0$ and $1 \sqsubseteq 1$). This is the usual numerical ordering, $\leqslant$, of the numbers 0 and 1 (where "$\leqslant$" means "less than or equal to"). The category **3** corresponds to the usual ordering on the three element set $\{0, 1, 2\}$. We could continue this process indefinitely, constructing a pre-order **4** from the usual ordering on $\{0, 1, 2, 3\}$, and in general for each natural number n, a pre-order $\mathbf{n}$ from the usual ordering on $\{0, 1, 2, \ldots, n-1\}$. Continuing even further we can consider the infinite collection

$$\omega = \{0, 1, 2, 3, \ldots\}$$

of all natural numbers under the usual ordering, to obtain a pre-order category which has the diagram

$$0 \to 1 \to 2 \to 3 \to \ldots$$

(composites and identities not shown).

A simple example of a pre-order that is not partially ordered would be a two-objects, four-arrows category

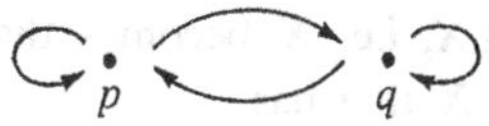

which has pRq and qRp, but $p \neq q$.

A categorial expression of the antisymmetry condition will be given in the next chapter, while the above numerical examples will be reconsidered in Example 9.

EXAMPLE 5. *Discrete categories.* If b is an object of a category $\mathscr{C}$, then the $\mathscr{C}$-arrow 1_b is uniquely determined by the property expressed in the Identity Law. For if $1' : b \to b$ has the property that

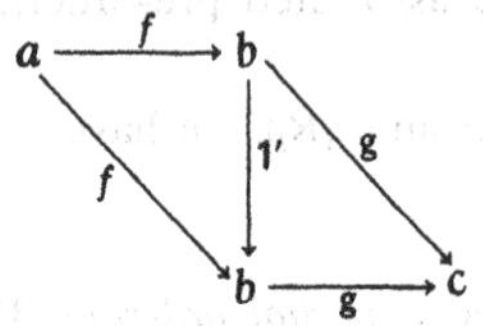

commutes for any $\mathscr{C}$-arrows f and g as shown, then in the particular case of $f = 1'$ and $g = 1_b$,

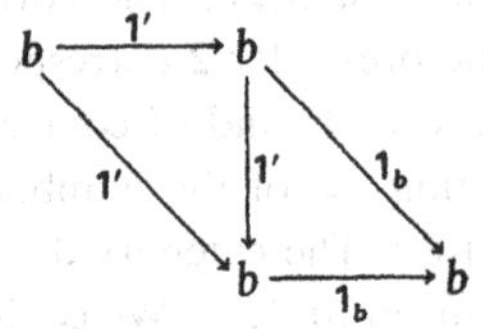

commutes giving $1_b = 1_b \circ 1'$ (right triangle). But by the Identity Law (with $f = 1'$), $1_b \circ 1' = 1'$, and so $1_b = 1'$.

Since 1_b is thus uniquely determined, the practice is sometimes adopted of identifying the object b with the arrow 1_b and writing $b : b \to b$, $b \circ f$ etc. Now the category axioms require that the $\mathscr{C}$-arrows include, at a minimum, an identity arrow for each $\mathscr{C}$-object (why must distinct objects have distinct identity arrows?). $\mathscr{C}$ is a *discrete* category if these are the only arrows, i.e. every arrow is the identity on some object. A discrete category is a pre-order since, as we have just seen, there can only be one identity arrow on a given object. Equating objects with identity arrows, we see that a discrete category is really nothing more than a collection of objects. Indeed, any set X can be made into a discrete category by adding an identity arrow $x \to x$ for each $x \in X$, i.e. X becomes the pre-order corresponding to the relation $R \subseteq X \times X$ that has

$$xRy \quad \text{iff} \quad x = y.$$

EXAMPLE 6. **N**: It is time we looked at some categories that have more than one arrow between given objects. The present example has only one object, which we shall call N, but an infinite collection of arrows from N to N. The arrows are, by definition, the natural numbers 0, 1, 2, 3, ... Each arrow has the same dom and cod, viz the unique object N. This means that all pairs of arrows are composable. The composite of two arrows (numbers) m and n is to be another number. The definition is

$$m \circ n = m + n.$$

Thus the diagram

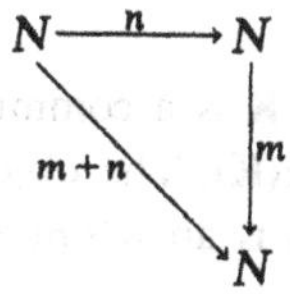

commutes by definition. The associative law is satisfied, since addition of numbers is an associative operation, i.e., $m+(n+k)=(m+n)+k$ is true for any numbers m, n and k.

The identity arrow 1_N on the object N is defined to be the number 0. The diagram

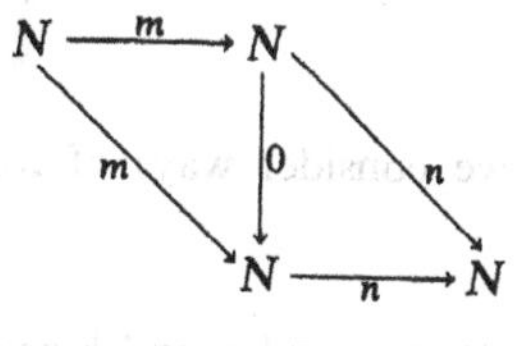

commutes because $0+m=m$ and $n+0=n$.

EXAMPLE 7. *Monoids*. The category **N** of the last example *is* a category because the structure $(N, +, 0)$ is an example of the abstract algebraic concept of a *monoid*.

A monoid is a triple $\mathbf{M}=(M, *, e)$ where

(i) M is a set

(ii) $*$ is a binary operation on M, i.e. a function from $M \times M$ to M assigning to each pair $\langle x, y\rangle \in M \times M$ an element $x * y$ of M, that is *associative*, i.e. satisfies $x * (y * z) = (x * y) * z$ for all $x, y, z \in M$.

(iii) e is a member of M, the monoid *identity*, that satisfies $e * x = x * e = x$, for all $x \in M$.

Any monoid **M** gives rise to a category with one object, exactly as in Example 6. We take the object to be M the arrows $M \to M$ to be the members of M, and put $e = 1_M$. Composition of arrows $x, y \in M$ is given by

$$x \circ y = x * y.$$

Conversely, if $\mathscr{C}$ is a category with only one object a, and M is its collection of arrows, then $(M, \circ, 1_a)$ is a monoid. All arrows have the same dom and cod and so all pairs are composable. Hence composition $\circ$ is a function from $M \times M$ to M, i.e. a binary operation on M, that is associative by the Associative Law for categories. 1_a is an identity for the monoid by the Identity Law for categories.

Example 8. **Matr(K)** (for linear algebraists). If **K** is a commutative ring then the matrices over **K** yield a category **Matr(K)**. The objects are the positive integers 1, 2, 3, ... An arrow $m \to n$ is an $n \times m$ matrix with entries in **K**. Given composable arrows

$$m \xrightarrow{\mathrm{B}} n \xrightarrow{\mathrm{A}} p,$$

i.e. A a $p \times n$ matrix and B $n \times m$, *we define* $A \circ B$ to be the matrix product AB *of* A and B (which is $p \times m$ and hence an arrow $m \to p$). The Associative Law is given by the associativity of matrix multiplication. 1_m is the identity matrix of order m.

In the remainder of this chapter we consider ways of forming new categories from given ones.

Example 9. *Subcategories.* If $\mathscr{C}$ is a category, and a and b are $\mathscr{C}$-objects, we introduce the symbol $\mathscr{C}(a, b)$ to denote the collection of all $\mathscr{C}$-arrows with dom $= a$ and cod $= b$, i.e.

$$\mathscr{C}(a, b) = \left\{f : f \text{ is a } \mathscr{C}\text{-arrow and } a \xrightarrow{f} b\right\}.$$

$\mathscr{C}$ is said to be a *subcategory* of category $\mathscr{D}$, denoted $\mathscr{C} \subseteq \mathscr{D}$, if

(i) every $\mathscr{C}$-object is a $\mathscr{D}$-object, and

(ii) if a and b are any two $\mathscr{C}$-objects, then $\mathscr{C}(a, b) \subseteq \mathscr{D}(a, b)$, i.e. all the $\mathscr{C}$-arrows $a \to b$ are present in $\mathscr{D}$.

For example, we have **Finset** $\subseteq$ **Set**, and **Nonset** $\subseteq$ **Set**, although neither of **Finset** and **Nonset** are subcategories of each other.

$\mathscr{C}$ is a *full* subcategory of $\mathscr{D}$ if $\mathscr{C} \subseteq \mathscr{D}$, and

(iii) for any $\mathscr{C}$-objects a and b, $\mathscr{C}(a, b) = \mathscr{D}(a, b)$, i.e. $\mathscr{D}$ has no arrows $a \to b$ other than the ones already in $\mathscr{C}$.

If $\mathscr{D}$ is a category and C is any collection of $\mathscr{D}$-objects we obtain a full subcategory $\mathscr{C}$ of $\mathscr{D}$ by taking as $\mathscr{C}$-arrows all the $\mathscr{D}$-arrows between members of C. Thus we see that **Finset** and **Nonset** are each full subcategories of **Set**.

An important full subcategory of **Finset** (and hence of **Set**) is the category **Finord** of all finite *ordinals*. The finite ordinals are sets that are used in set-theoretic foundations as representations of the natural numbers. We use the natural numbers as names for these sets and put

0 for $\emptyset$ (the empty set)
1 for $\{0\}$ $(=\{\emptyset\})$
2 for $\{0, 1\}$ $(=\{\emptyset, \{\emptyset\}\})$
3 for $\{0, 1, 2\}$ $(=\{\emptyset, \{\emptyset\}, \{\emptyset, \{\emptyset\}\}\})$
4 for $\{0, 1, 2, 3\}$

and so on.

Proceeding "inductively", where n is a natural number, we put

$$n \quad \text{for} \quad \{0, 1, 2, \ldots, n-1\}.$$

The sequence of finite sets thus generated are the finite ordinals. They form the objects of the category **Finord**, whose arrows are all the set functions between finite ordinals.

Of course it is ridiculous to suggest that the number 1 *is* the set $\{0\}$ whose only member is the null set. The point is that in axiomatic set theory, where we seek an explicit and precise account of mathematical entities and their intuitively understood properties, the finite ordinals provide such a paradigmatic representation of the natural numbers. They have an intricate and elegant structure that exhibits all the arithmetic and algebraic properties of the natural number system. They are related by set inclusion and set membership as follows:

$$0 \subseteq 1 \subseteq 2 \subseteq 3 \subseteq \ldots$$

$$0 \in 1 \in 2 \in 3 \in \ldots$$

In fact the following three statements are equivalent

(a) $n < m$ (the number n is numerically less than the number m)
(b) $n \subset m$ (the set n is a proper subset of set m)
(c) $n \in m$ (n is a member of set m)

Thus $n \leqslant m$ iff $n \subseteq m$.

So the ordinal (set) $n = \{0, 1, \ldots, n-1\}$ has the ordering $\leqslant$ built into its structure in a natural set-theoretic way. The corresponding pre-order category is none other than **n** of Example 4. Notice that if $n \leq m$, the pre-order **n** is a full subcategory of **m**.

Example 10. **Product categories.** The category $\mathbf{Set}^2$ of pairs of sets has as objects all pairs $\langle A, B \rangle$ of sets. An arrow in $\mathbf{Set}^2$ from $\langle A, B \rangle$ to $\langle C, D \rangle$ is a pair $\langle f, g \rangle$ of set functions such that $f: A \to C$ and $g: B \to D$. Composition is defined by $\langle f, g \rangle \circ \langle f', g' \rangle = \langle f \circ f', g \circ g' \rangle$, where $f \circ f'$ and $g \circ g'$ are the functional compositions. The identity arrow on $\langle A, B \rangle$ is $\langle id_A, id_B \rangle$.

This construction generalises: given any two categories $\mathscr{C}$ and $\mathscr{D}$, the product category $\mathscr{C} \times \mathscr{D}$ has objects the pairs $\langle a, b \rangle$ where a is a $\mathscr{C}$-object and b a $\mathscr{D}$-object. A $\mathscr{C} \times \mathscr{D}$-arrow $\langle a, b \rangle \to \langle c, d \rangle$ is a pair $\langle f, g \rangle$ where $f: a \to c$ is a $\mathscr{C}$-arrow and $g: b \to d$ a $\mathscr{D}$-arrow. Composition is defined "componentwise" with respect to composition in $\mathscr{C}$, and composition in $\mathscr{D}$.

Example 11. **Arrow categories.** The category $\mathbf{Set}^{\to}$ of functions has as objects the set functions $f: A \to B$. An arrow in $\mathbf{Set}^{\to}$ from the $\mathbf{Set}^{\to}$-object $f: A \to B$ to the $\mathbf{Set}^{\to}$-object $g: C \to D$ is a pair of functions $\langle h, k \rangle$ such that

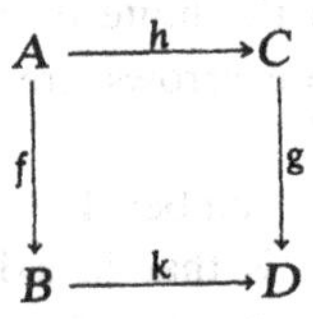

commutes, i.e. $g \circ h = k \circ f$.

For composition we put

$$\langle j, l \rangle \circ \langle h, k \rangle = \langle j \circ h, l \circ k \rangle$$

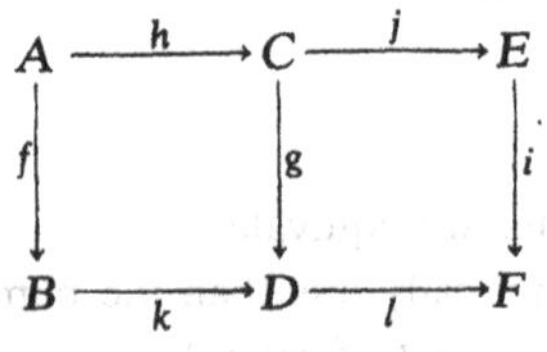

The identity arrow for the $\mathbf{Set}^{\to}$-object $f: A \to B$ is the function pair $\langle id_A, id_B \rangle$.

This construction can also be generalised to form, from any category $\mathscr{C}$, the *arrow category* $\mathscr{C}^{\to}$ whose objects are all the $\mathscr{C}$-arrows.

EXAMPLE 12. **Comma categories.** These can be thought of as specialisations of arrow categories, where we restrict attention to arrows with fixed domain or codomain.

Thus if $\mathbb{R}$ is the set of real numbers, we obtain the category $\mathbf{Set}\downarrow\mathbb{R}$ of *real valued functions*. The objects are all functions $f:A\to\mathbb{R}$ that have codomain $\mathbb{R}$. An arrow from $f:A\to\mathbb{R}$ to $g:B\to\mathbb{R}$ is a function $k:A\to B$ that makes the triangle

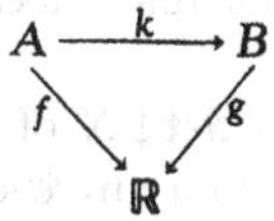

commute, i.e. has $g\circ k=f$.

It is sometimes convenient to think of $\mathbf{Set}\downarrow\mathbb{R}$-objects as pairs (A,f), where $f:A\to\mathbb{R}$. Then the $\mathbf{Set}\downarrow\mathbb{R}$ composite of

$$(A,f)\xrightarrow{k}(B,g)\xrightarrow{l}(C,h)$$

is defined as $l\circ k:(A,f)\to(C,h)$

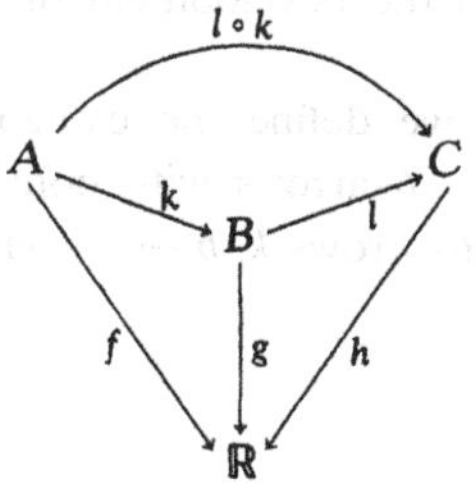

The identity arrow on the object $f:A\to\mathbb{R}$ is $\mathrm{id}_A:(A,f)\to(A,f)$. $\mathbf{Set}\downarrow\mathbb{R}$ is not as it stands a subcategory of $\mathbf{Set}^{\to}$ as the two have different sorts of arrows. However, we could equate the $\mathbf{Set}\downarrow\mathbb{R}$ arrow $k:(A,f)\to(B,g)$ with the $\mathbf{Set}^{\to}$ arrow $\langle k,\mathrm{id}_{\mathbb{R}}\rangle$, as

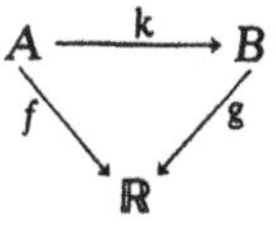

commutes iff

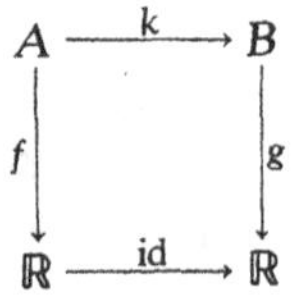

does.

In this way **Set**$\downarrow\mathbb{R}$ can be "construed" as a (not full) subcategory of $\mathbf{Set}^{\rightarrow}$.

Similarly for any set X we obtain the category **Set**$\downarrow X$ of "X-valued functions". More generally if $\mathscr{C}$ is any category, and a any $\mathscr{C}$-object then the category $\mathscr{C}\downarrow a$ of *objects over* a has the $\mathscr{C}$-arrows with codomain a as objects, and as arrows from $f:b\to a$, to $g:c\to a$ the $\mathscr{C}$-arrows $k:b\to c$ such that

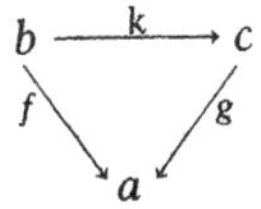

commutes, i.e. $g\circ k=f$.

Categories of this type are going to play an important role both in the provision of examples of topoi, and in the development of the general theory.

Turning our attention to domains, we define the category $\mathscr{C}\uparrow a$ of *objects under* a to have as objects the $\mathscr{C}$-arrows with dom$=a$ and as arrows from $f:a\to b$ to $g:a\to c$ the $\mathscr{C}$-arrows $k:b\to c$ such that

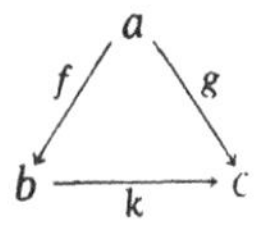

commutes, i.e. $k\circ f=g$.

Categories of the type $\mathscr{C}\downarrow a$ and $\mathscr{C}\uparrow a$ are known as *comma* categories.

CHAPTER 3

ARROWS INSTEAD OF EPSILON

> *"The world of ideas is not revealed to us in one stroke; we must both permanently and unceasingly recreate it in our consciousness".*
>
> René Thom

In this chapter we examine a number of standard set-theoretic constructions and reformulate them in the language of arrows. The general theme, as mentioned in the introduction, is that concepts defined by reference to the "internal" membership structure of a set are to be characterised "externally" by reference to connections with other sets, these connections being established by functions. The analysis will eventually lead us to the notions of *universal property* and *limit*, which encompass virtually all constructions within categories.

3.1. Monic arrows

A set function $f: A \to B$ is said to be *injective*, or *one-one* when no two distinct inputs give the same output, i.e. for inputs $x, y \in A$,

$$\text{if} \quad f(x) = f(y), \quad \text{then} \quad x = y.$$

Now let us take an injective $f: A \to B$ and two "parallel" functions $g, h: C \rightrightarrows A$ for which

$$\begin{array}{ccc} C & \xrightarrow{g} & A \\ {\scriptstyle h}\downarrow & & \downarrow{\scriptstyle f} \\ A & \xrightarrow{f} & B \end{array}$$

commutes, i.e. $f \circ g = f \circ h$.

Then for $x \in C$, we have $f \circ g(x) = f \circ h(x)$, i.e. $f(g(x)) = f(h(x))$. But as f is injective, this means that $g(x) = h(x)$. Hence g and h, giving the same output for every input, are the same function, and we have shown that an injective f is "left-cancellable", i.e.

$$\textit{whenever} \quad f \circ g = f \circ h, \quad \text{then} \quad g = h.$$

On the other hand, if f has this left-cancellation property, it must be injective. To see this, take x and y in A, with $f(x) = f(y)$.

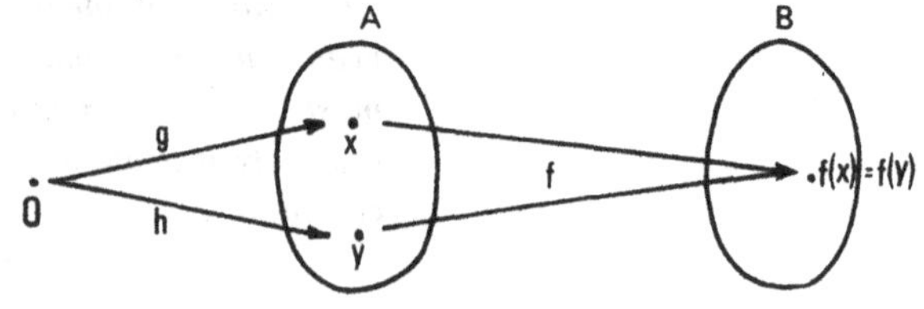

Fig. 3.1

The instructions "$g(0) = x$", "$h(0) = y$" establishes a pair of functions g, h from $\{0\}$ (i.e. the ordinal 1) to A for which we have $f \circ g = f \circ h$. By left cancellation, $g = h$, so $g(0) = h(0)$, i.e. $x = y$.

We thus see that the injective arrows in **Set** are precisely the ones that are left cancellable. The point of all this is that the latter property is formulated entirely by reference to arrows and leads to the following abstract definition:

An arrow $f: a \to b$ in a category $\mathscr{C}$ is *monic* in $\mathscr{C}$ if for any parallel pair $g, h: c \rightrightarrows a$ of $\mathscr{C}$-arrows, the equality $f \circ g = f \circ h$ implies that $g = h$. The symbolism $f: a \rightarrowtail b$ is used to indicate that f is monic. The name comes from the fact that an injective algebraic homomorphism (i.e. an arrow in a category like **Mon** or **Grp**) is called a "monomorphism".

Example 1. In the category **N** (Example 6, Chapter 2) *every arrow is monic.* Left-cancellation here means that

$$\text{if} \quad m + n = m + p, \quad \text{then} \quad n = p$$

which is certainly a true statement about addition of numbers.

Example 2. In a pre-order, every arrow is monic: given a pair $g, h: c \rightrightarrows a$, we must have $g = h$, as there is at most one arrow $c \to a$.

Example 3. In **Mon, Grp, Met, Top** the monics are those arrows that are injective as set functions (see e.g. Arbib and Manes [75]).

Example 4. In a comma category $\mathscr{C} \downarrow a$, an arrow k from (b, f) to (c, g),

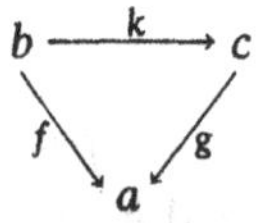

is monic in $\mathscr{C} \downarrow a$ iff k is monic in $\mathscr{C}$ as an arrow from b to c.

Exercises

In any category

(1). $g \circ f$ is monic if both f and g are monic.

(2) If $g \circ f$ is monic then so is f.

3.2. Epic arrows

A set function $f: A \to B$ is *onto*, or *surjective* if the codomain B is the range of f, i.e. for each $y \in B$ there is some $x \in A$ such that $y = f(x)$, i.e. every member of B is an output for f. The "arrows-only" definition of this concept comes from the definition of "monic" by simply reversing the arrows. Formally:

An arrow $f: a \to b$ is *epic* (right-cancellable) in a category $\mathscr{C}$ if for any pair of $\mathscr{C}$-arrows $g, h: b \rightrightarrows c$, the equality $g \circ f = h \circ f$ implies that $g = h$, i.e. whenever a diagram

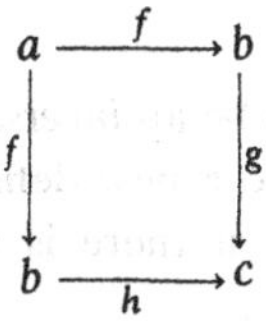

commutes, then $g = h$. The notation $f: a \twoheadrightarrow b$ is used for epic arrows.

In **Set**, the epic arrows are precisely the surjective functions (exercise for the reader, or Arbib and Manes, p. 2). A surjective homomorphism is known as an *epimorphism*.

In the category **N**, every arrow is epic, as $n + m = p + m$ implies that $n = p$. In any pre-order, all arrows are epic.

In the categories of our original list, where arrows are functions, the arrows that are surjective as functions are always epic. The converse is true in **Grp**, but not in **Mon**. The inclusion of the natural numbers into the integers is a monoid homomorphism (with respect to +), that is certainly not onto, but nevertheless is right cancellable in **Mon**. (Arbib and Manes p. 57).

3.3. Iso arrows

A function that is both injective and surjective is called *bijective*. If $f: A \rightarrowtail\!\!\!\!\twoheadrightarrow B$ is bijective then the passage from A to B under f can be reversed or "inverted". We can think of f as being simply a "relabelling" of A. Any $b \in B$ is the image $f(a)$ of some $a \in A$ (surjective property) and in fact is the image of only one such a (injective property). Thus the rule which assigns to b this unique a, i.e. has

$$g(b) = a \quad \text{iff} \quad f(a) = b$$

establishes a function $B \to A$ which has

$$g(f(a)) = a, \quad \text{all} \quad a \in A$$

and

$$f(g(b)) = b, \quad \text{all} \quad b \in B.$$

Hence

$$g \circ f = id_A$$

and

$$f \circ g = id_B.$$

A function that is related to f in this way is said to be an *inverse* of f. This is an essentially arrow-theoretic idea, and leads to a new definition.

A $\mathscr{C}$-arrow $f: a \to b$ is *iso*, or *invertible*, in $\mathscr{C}$ if there is a $\mathscr{C}$-arrow $g: b \to a$, such that $g \circ f = 1_a$ and $f \circ g = 1_b$.

There can in fact be at most one such g, for if $g' \circ f = 1_a$, and $f \circ g' = 1_b$, then $g' = 1_a \circ g' = (g \circ f) \circ g' = g \circ (f \circ g') = g \circ 1_b = g$. So this g, when it exists, is called *the inverse of f*, and denoted by $f^{-1}: b \to a$. It is defined by the conditions $f^{-1} \circ f = 1_a$, $f \circ f^{-1} = 1_b$. The notation $f: a \cong b$ is used for iso's.

An iso arrow is always monic. For if $f \circ g = f \circ h$, and f^{-1} exists, then $g = 1_a \circ g = (f^{-1} \circ f) \circ g = f^{-1} \circ (f \circ g) = f^{-1} \circ (f \circ h) = 1_a \circ h = h$, and so f is left-cancellable. An analogous argument shows that iso's are always epic.

Now in **Set** a function that is epic and monic has an inverse, as we saw at the beginning of this section. So in **Set**, "iso" is synonymous with "monic and epic". The same, we shall learn, goes for any topos, but is certainly not so in all categories.

In the category **N** we already know that *every* arrow is both monic and epic. But the only iso is $0: N \to N$. For if m has inverse n, $m \circ n = 1_N$, i.e. $m + n = 0$. Since m and n are both natural numbers, hence both non-negative, this can only happen if $m = n = 0$.

The inclusion map mentioned at the end of the last section is in fact epic and monic, but cannot be iso, since if it had an inverse it would, as a set function, be bijective.

In a poset category $\mathbf{P}=(P, \sqsubseteq)$, if $f: p \to q$ has an inverse $f^{-1}: q \to p$, then $p \sqsubseteq q$ and $q \sqsubseteq p$, whence by antisymmetry, $p = q$. But then f must be the unique arrow 1_p from p to p. Thus in a poset, every arrow is monic and epic, but the only iso's are the identities.

Groups

A *group* is a monoid $(M, *, e)$ in which for each $x \in M$ there is a $y \in M$ satisfying $x * y = e = y * x$. There can in fact be only one such y for a given x. It is called the inverse of x, and denoted x^{-1}. Thinking of a monoid as a category with one object, the terminology and notation is tied to its above usage: a group is essentially the same thing as a one-object category in which every arrow is iso.

EXERCISE 1. Every identity arrow is iso.

EXERCISE 2. If f is iso, so is f^{-1}.

EXERCISE 3. $f \circ g$ is iso if f, g are, with $(f \circ g)^{-1} = g^{-1} \circ f^{-1}$.

3.4. Isomorphic objects

Objects a and b are *isomorphic* in $\mathscr{C}$, denoted $a \cong b$, if there is a $\mathscr{C}$-arrow $f: a \to b$ that is iso in $\mathscr{C}$, i.e. $f: a \cong b$.

In **Set**, $A \cong B$ when there is a bijection between A and B, in which case each set can be thought of as being a "relabelling" of the other. As a specific example take a set A and put

$$B = A \times \{0\} = \{\langle x, 0\rangle : x \in A\}.$$

In effect B is just A with the label "0" attached to each of its elements. The rule $f(x) = \langle x, 0\rangle$ gives the bijection $f: A \to B$ making $A \cong B$.

In **Grp**, two groups are isomorphic if there is a group homomorphism (function that "preserves" group structure) from one to the other whose set-theoretic inverse exists and is a group homomorphism (hence is present in **Grp** as an inverse). Such an arrow is called a *group isomorphism.*

In **Top**, isomorphic topological spaces are usually called *homeomorphic.* This means there is a homeomorphism between them, i.e. a continuous bijection whose inverse is also continuous.

In these examples, isomorphic objects "look the same". One can pass freely from one to the other by an iso arrow and its inverse. Moreover these arrows, which establish a "one-one correspondence" or "matching" between the elements of the two objects, preserve any relevant structure. This means that we can replace some or all of the members of one object by their counterparts in the other object without making any difference to the structure of the object, to its appearance. Thus isomorphic groups look exactly the same, *as groups;* homeomorphic topological spaces are indistinguishable by any topological property, and so on. Within any mathematical theory, isomorphic objects are indistinguishable in terms of that theory. The aim of that theory is to identify and study constructions and properties that are "invariant" under the isomorphisms of the theory (thus topology studies properties that are not altered or destroyed when a space is replaced by another one homeomorphic to it). An object will be said to be "unique up to isomorphism" in possession of a particular attribute if the only other objects possessing that attribute are isomorphic to it. A concept will be "defined up to isomorphism" if its description specifies a particular entity, not uniquely, but only uniquely up to isomorphism.

Category theory then is the subject that provides an abstract formulation of the idea of mathematical isomorphism and studies notions that are invariant under all forms of isomorphism. In category theory, "is isomorphic to" is virtually synonymous with "is". Indeed most of the basic definitions and constructions that one can perform in a category do not specify things uniquely at all, but only, as we shall see, "up to isomorphism".

Skeletal categories

A skeletal category is one in which "isomorphic" does actually mean the same as "is", i.e. in which whenever $a \cong b$, then $a = b$. We saw in the last section that in a poset, the only iso arrows are the identities. This then gives us a categorial account of antisymmetry in pre-orders. A poset is precisely a *skeletal pre-order category*.

EXERCISE 1. For any $\mathscr{C}$-objects

(i) $a \cong a$;

(ii) if $a \cong b$ then $b \cong a$;

(iii) If $a \cong b$ and $b \cong c$, then $a \cong c$.

EXERCISE 2. **Finord** is a skeletal category.

3.5. Initial objects

What arrow properties distinguish $\emptyset$, the null set, in **Set**? Given a set A, can we find any function $\emptyset \to A$? Recalling our formulation of a function as a triple $\langle A, B, X\rangle$ with $X \subseteq A \times B$ (§2.1), we find by checking the details of that definition that $f = \langle \emptyset, A, \emptyset\rangle$ is a function from $\emptyset \to A$. The graph of f is empty, and f is known as the *empty function* for A. Since $\emptyset \times A$ is empty, $\emptyset$ is the only subset of $\emptyset \times A$, and hence f is the only function from $\emptyset$ to A. This observation leads us to the following:

DEFINITION. An object 0 is *initial* in category $\mathscr{C}$ if for every $\mathscr{C}$-object a there is one and only one arrow from 0 to a in $\mathscr{C}$.

Any two initial $\mathscr{C}$*-objects must be isomorphic in* $\mathscr{C}$. For if $0, 0'$ are such objects there are unique arrows $f: 0' \to 0$, $g: 0 \to 0'$. But then $f \circ g: 0 \to 0$ must be 1_0, as 1_0 is the *only* arrow $0 \to 0$, 0 being initial. Similarly, as $0'$ is initial, $g \circ f: 0' \to 0'$ is $1_{0'}$. Thus f has an inverse (g), and $f: 0' \cong 0$.

The symbol 0 of course is used because in **Set** it is a name for $\emptyset$, and $\emptyset$ is initial in **Set**. In fact $\emptyset$ is the only initial object in **Set**, so whereas the initial $\mathscr{C}$-object may only be "unique up to isomorphism", when $\mathscr{C} = \textbf{Set}$ it is actually unique.

In a pre-order $(P, \sqsubseteq)$ an initial object is an element $0 \in P$ with $0 \sqsubseteq p$ for all $p \in P$ (i.e. a *minimal* element). In a poset, where "isomorphic" means "equal", then there can be at most one initial object (the *minimum*, or *zero* element). Thus in the poset $\{0, \ldots, n-1\}$, 0 is *the* initial object, whereas in the two-object category with diagram

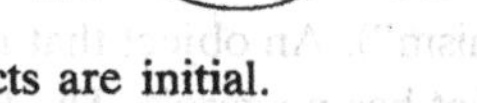

both objects are initial.

In **Grp**, and **Mon**, an initial object is any one element algebra $(M, *, e)$, i.e. $M = \{e\}$, and $e * e = e$. Each of these categories has infinitely many initial objects.

In $\textbf{Set}^2$, the category of pairs of sets, the initial object is $\langle \emptyset, \emptyset\rangle$, while in $\textbf{Set}^{\to}$, the category of functions, it is $\langle \emptyset, \emptyset, \emptyset\rangle$, the empty function from $\emptyset$ to $\emptyset$. In $\textbf{Set}\downarrow\mathbb{R}$, the category of real valued functions, it is $f = \langle \emptyset, \mathbb{R}, \emptyset\rangle$. Given $g: A \to \mathbb{R}$, the only way to make the diagram

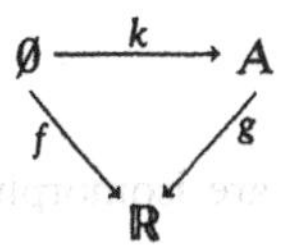

commute is to put $k = \langle \emptyset, A, \emptyset \rangle$, the empty map from $\emptyset$ to A.

NOTATION. The exclamation mark "!" is often used to denote a uniquely existing arrow. We put $! : 0 \to a$ for the unique arrow from 0 to a. It is also denoted 0_a, i.e. $0_a : 0 \to a$.

3.6. Terminal objects

By reversing the direction of the arrows in the definition of initial object, we have the following idea:

DEFINITION. An object 1 is *terminal* in a category $\mathscr{C}$ if for every $\mathscr{C}$-object a there is one and only one arrow from a to 1 in $\mathscr{C}$.

In **Set**, the terminal objects are the singletons, i.e. the one-element sets $\{e\}$. Given set A, the rule $f(x) = e$ gives a function $f : A \to \{e\}$. Since e is the only possible output, this is the only possible such function. Thus **Set** has many terminal objects. They are all isomorphic (terminal objects in any category are isomorphic) and the paradigm is the ordinal $1 = \{0\}$, whence the notation.

Again we may write $! : a \to 1$ to denote the unique arrow from a to 1, or alternatively $!_a : a \to 1$.

In a pre-order a terminal object satisfies $p \sqsubseteq 1$, all p (a maximal element). In a poset, 1 is unique (the *maximum*), when it exists, and is also called the *unit* of **P**.

In **Grp** and **Mon**, terminal objects are again the one element monoids. Hence the initial objects are the same as the terminal ones (and so the equation $0 = 1$ is "true up to isomorphism"). An object that is both initial and terminal is called a *zero* object. **Set** has no zero's. The fact that **Grp** and **Mon** have zeros precludes them, as we shall see, from being topoi.

In **Set**$\downarrow\mathbb{R}$, $(\mathbb{R}, \mathrm{id}_{\mathbb{R}})$ is a terminal object. Given (A, f), the only way to make

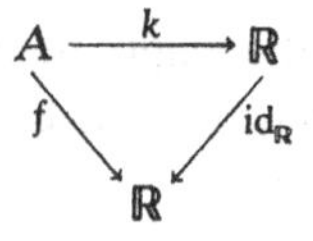

commute is to put $k = f$.

EXERCISE 1. Prove that all terminal $\mathscr{C}$-objects are isomorphic.

EXERCISE 2. Find terminals in $\mathbf{Set}^2$, $\mathbf{Set}^{\rightarrow}$, and the poset $\mathbf{n}$.

EXERCISE 3. Show that an arrow $1 \rightarrow a$ whose domain is a terminal object must be monic.

3.7. Duality

We have already observed that the notion of epic arrow arises from that of monic by "reversing the arrows". The same applies to the concepts of terminal and initial objects. These are two examples of the notion of *duality* in category theory, which we will now describe a little more precisely.

If Σ is a statement in the basic language of categories, the *dual* of Σ, Σ^{op}, is the statement obtained by replacing "dom" by "cod", "cod" by "dom", and "$h = g \circ f$" by "$h = f \circ g$". Thus all arrows and composites referred to by Σ are reversed in Σ^{op}. The notion or construction described by Σ^{op} is said to be *dual* to that described by Σ. Thus the notion of epic arrow is dual to that of monic arrow. The dual of "initial object" is "terminal object", and so on.

From a given category $\mathscr{C}$ we construct its *dual* or *opposite* category $\mathscr{C}^{\mathrm{op}}$ as follows:

$\mathscr{C}$ and $\mathscr{C}^{\mathrm{op}}$ have the same objects. For each $\mathscr{C}$-arrow $f : a \rightarrow b$ we introduce an arrow $f^{\mathrm{op}} : b \rightarrow a$ in $\mathscr{C}^{\mathrm{op}}$, these being all and only the arrows in $\mathscr{C}^{\mathrm{op}}$. The composite $f^{\mathrm{op}} \circ g^{\mathrm{op}}$ is defined precisely when $g \circ f$ is defined in $\mathscr{C}$ and has

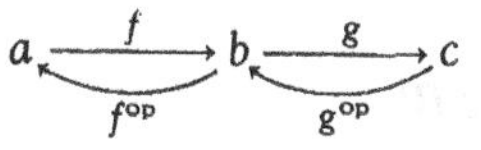

$f^{\mathrm{op}} \circ g^{\mathrm{op}} = (g \circ f)^{\mathrm{op}}$. Note that $\mathrm{dom}\, f^{\mathrm{op}} = \mathrm{cod}\, f$, and $\mathrm{cod}(f^{\mathrm{op}}) = \mathrm{dom}\, f$.

EXAMPLE 1. If $\mathscr{C}$ is discrete, $\mathscr{C}^{\mathrm{op}} = \mathscr{C}$.

EXAMPLE 2. If $\mathscr{C}$ is a pre-order (P, R), with $R \subseteq P \times P$, then $\mathscr{C}^{\mathrm{op}}$ is the pre-order (P, R^{-1}), where $pR^{-1}q$ iff qRp, i.e. R^{-1} is the inverse relation to R.

EXAMPLE 3. For any $\mathscr{C}$, $(\mathscr{C}^{\mathrm{op}})^{\mathrm{op}} = \mathscr{C}$.

The dual of a construction expressed by Σ can be interpreted as the original construction applied to the opposite category. If Σ is true of $\mathscr{C}$, Σ^{op} will be true of $\mathscr{C}^{op}$. Thus the initial object $\emptyset$ in **Set** is the terminal object of **Set**^{op}. Now if Σ is a theorem of category theory, i.e. derivable from the category axioms, then Σ will be true in all categories. Hence Σ^{op} will hold in all categories of the form $\mathscr{C}^{op}$. But any category $\mathscr{D}$ has this form (put $\mathscr{C} = \mathscr{D}^{op}$), and so Σ^{op} holds in all categories. Thus from any true statement Σ of category theory we immediately obtain another true statement Σ^{op} by this *Duality Principle*.

The Duality Principle cuts the number of things to be proven in half. For example, we note first that the concept of iso arrow is self-dual. The dual of an invertible arrow is again an invertible arrow – indeed $(f^{op})^{-1} = (f^{-1})^{op}$. So having proven

any two initial $\mathscr{C}$-objects are isomorphic

we can conclude without further ado, the dual fact that

any two terminal $\mathscr{C}$-objects are isomorphic.

The Duality Principle comes from the domain of logic. It is discussed in a more rigorous fashion in Hatcher [68] §8.2.

3.8. Products

We come now to the problem of giving a characterisation, using arrows, of the product set

$$A \times B = \{\langle x, y\rangle : x \in A \text{ and } y \in B\}$$

of two sets A and B. The uninitiated may find it hard to believe that this can be achieved without any reference to ordered pairs. But in fact it can be, up to isomorphism, and the way it is done will lead us to a general description of what a "construction" in a category is.

Associated with $A \times B$ are two special maps, the *projections*

$$p_A : A \times B \to A$$

and

$$p_B : A \times B \to B$$

given by the rules

$$p_A(\langle x, y\rangle) = x$$

$$p_B(\langle x, y\rangle) = y.$$

Now suppose we are given some other set C with a pair of maps $f: C \to A$, $g: C \to B$, Then we define $p: C \to A \times B$

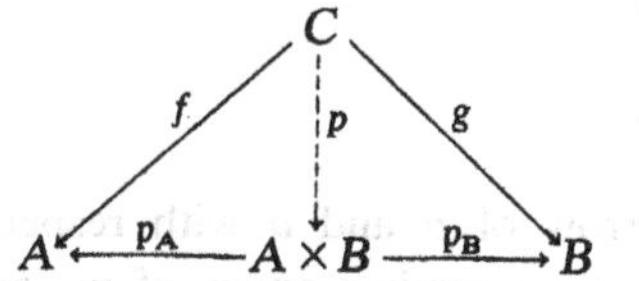

by the rule $p(x) = \langle f(x), g(x)\rangle$. Then we have $p_A(p(x)) = f(x)$, and $p_B(p(x)) = g(x)$ for all $x \in C$, so $p_A \circ p = f$ and $p_B \circ p = g$, i.e. the above diagram commutes. Moreover, p as defined is the only arrow that can make the diagram commute. For if $p(x) = \langle y, z\rangle$ then simply knowing that $p_A \circ p = f$ tells us that $p_A(p(x)) = f(x)$, i.e. $y = f(x)$. Similarly if $p_B \circ p = g$, we must have $z = g(x)$.

The map p associated with f and g is usually denoted $\langle f, g\rangle$, the *product map* of f and g. Its definition in **Set** is $\langle f, g\rangle(x) = \langle f(x), g(x)\rangle$.

The observations just made motivate the following:

DEFINITION. A *product* in a category $\mathscr{C}$ of two objects a and b is a $\mathscr{C}$-object $a \times b$ together with a pair $(pr_a: a \times b \to a,\ pr_b: a \times b \to b)$ of $\mathscr{C}$-arrows such that for *any* pair of $\mathscr{C}$-arrows of the form $(f: c \to a,$ $g: c \to b)$ there is exactly one arrow $\langle f, g\rangle: c \to a \times b$ making

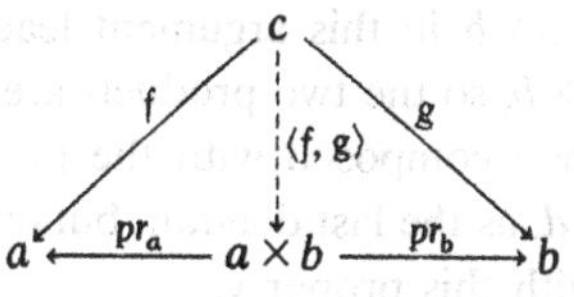

commute, i.e. such that $pr_a \circ \langle f, g\rangle = f$ and $pr_b \circ \langle f, g\rangle = g$. $\langle f, g\rangle$ is the *product arrow of f and g with* respect to the *projections* pr_a, pr_b.

Notice that we said *a* product of a and b, not *the product*. This is because $a \times b$ *is only defined up to isomorphism*. For suppose $(p: d \to a, q: d \to b)$ also satisfies the definition of "a product of $a \times b$" and

consider the diagram

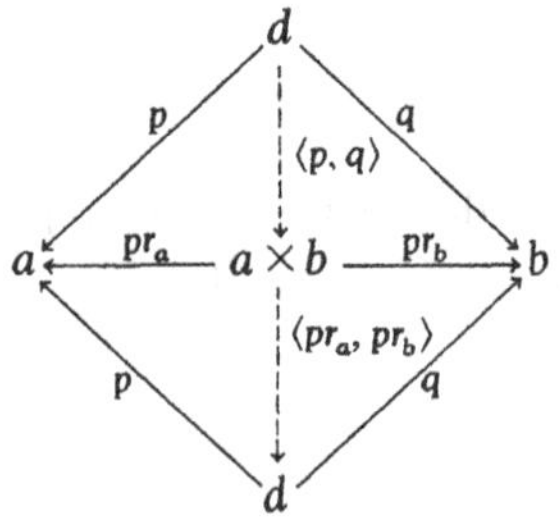

$\langle p, q\rangle$ is the unique product arrow of p and q with respect to "the" product $a \times b$. $\langle pr_a, pr_b\rangle$ is the unique product arrow of pr_a and pr_b with respect to "the" product d.

Now, since d is a product of a and b there can be only one arrow $s: d \to d$ such that

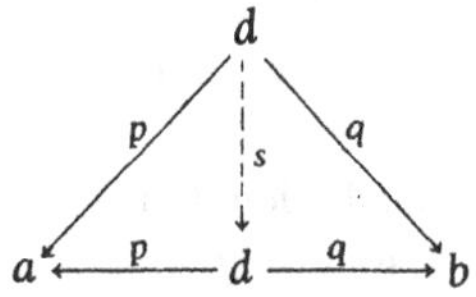

commutes. But putting $s = 1_d$ makes this diagram commute, while the commutativity of the previous diagram implies that putting $s = \langle pr_a, pr_b\rangle \circ \langle p, q\rangle$ also works (more fully – $p \circ \langle pr_a, pr_b\rangle \circ \langle p, q\rangle = pr_a \circ \langle p, q\rangle = p$ etc.). By the uniqueness of s we must conclude

$$\langle pr_a, pr_b\rangle \circ \langle p, q\rangle = 1_d.$$

Interchanging the roles of d and $a \times b$ in this argument leads to $\langle p, q\rangle \circ \langle pr_a, pr_b\rangle = 1_{a\times b}$. Thus $\langle p, q\rangle: d \cong a \times b$, so the two products are isomorphic and furthermore the iso $\langle p, q\rangle$ when composed with the projections for $a \times b$ produces the projections for d, as the last diagram but one indicates. Indeed, $\langle p, q\rangle$ is the only arrow with this property.

In summary then our definition characterises the product of a and b "uniquely up to a unique commuting isomorphism", which is enough from the categorial viewpoint.

Example 1. In **Set**, **Finset**, **Nonset**, the product of A and B is the Cartesian product set $A \times B$.

EXAMPLE 2. In **Grp** the product of two objects is the standard direct product of groups, with the binary operation defined "component-wise" on the product set of the two groups.

EXAMPLE 3. In **Top**, the product is the standard notion of product space.

EXAMPLE 4. In a pre-order $(P, \sqsubseteq)$ a product of p and q when it exists is defined by the properties

(i) $p \times q \sqsubseteq p$, $p \times q \sqsubseteq q$, i.e. $p \times q$ is a "lower bound" of p and q;

(ii) if $c \sqsubseteq p$ and $c \sqsubseteq q$, then $c \sqsubseteq p \times q$, i.e. $p \times q$ is "greater" than any other lower bound of p and q.

In other words $p \times q$ is a *greatest lower bound* (g.l.b.) of p and q. In a poset, being skeletal, the g.l.b. is unique, when it exists, and will be denoted $p \sqcap q$. A poset in which every two elements have a g.l.b. is called a *lower semilattice*. Categorially a lower semilattice is a skeletal pre-order category in which any two objects have a product.

EXAMPLE 5. If A and B are finite sets, with say m and n elements respectively, then the product set $A \times B$ has $m \times n$ elements (where the last "$\times$" denotes multiplication). This has an interesting manifestation in the skeletal category **Finord**. There the product of the ordinal numbers m and n exists and is quite literally the ordinal $m \times n$.

EXERCISE 1. $\langle pr_a, pr_b \rangle = 1_{a \times b}$

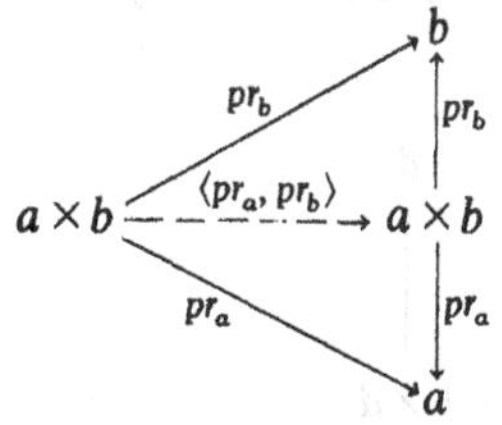

EXERCISE 2. If $\langle f, g \rangle = \langle k, h \rangle$, then $f = k$ and $g = h$.

EXERCISE 3. $\langle f\circ h, g\circ h\rangle = \langle f, g\rangle\circ h$

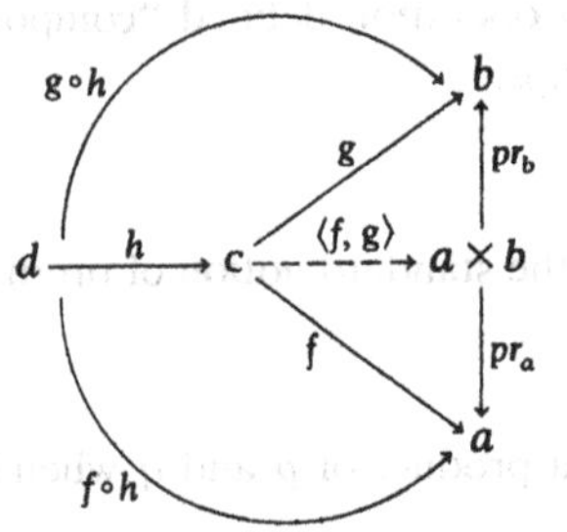

EXERCISE 4. We saw earlier that in **Set**, $A \cong A\times\{0\}$. Show that if category $\mathscr{C}$ has a terminal object 1 and products, then for any $\mathscr{C}$-object a, $a \cong a\times 1$ and indeed $\langle 1_a, !_a\rangle$ is iso

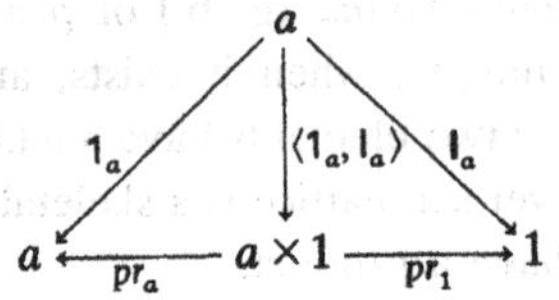

Product maps

Given set functions $f:A\to B$, $g:C\to D$ we obtain a function from $A\times C$ to $B\times D$ that outputs $\langle f(x), g(y)\rangle$ for input $\langle x, y\rangle$. This map is denoted $f\times g$, and we have

$$f\times g(\langle x, y\rangle) = \langle f(x), g(y)\rangle,$$

It is not hard to see that $f\times g$ is just the product map of the two composites $f\circ p_A : A\times C\to A\to B$ and $g\circ p_C : A\times C\to C\to D$, so we can define the following.

DEFINITION If $f:a\to b$ and $g:c\to d$ are $\mathscr{C}$-arrows then $f\times g: a\times b\to c\times d$ is the $\mathscr{C}$-arrow $\langle f\circ pr_a, g\circ pr_b\rangle$

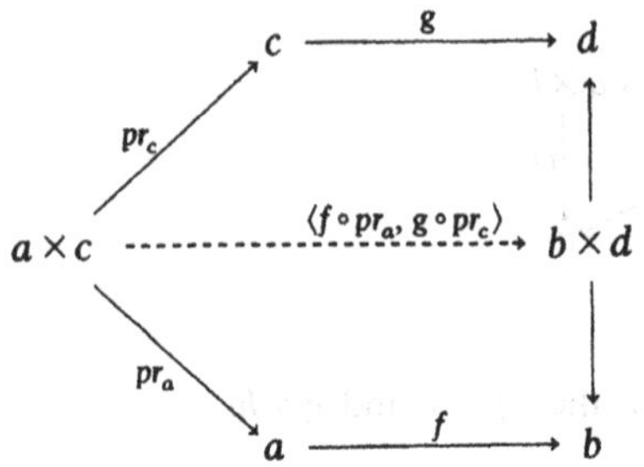

(Of course $f \times g$ is only defined when $a \times c$ and $b \times d$ exist in $\mathscr{C}$).

EXERCISE 5. $1_a \times 1_b = 1_{a \times b}$

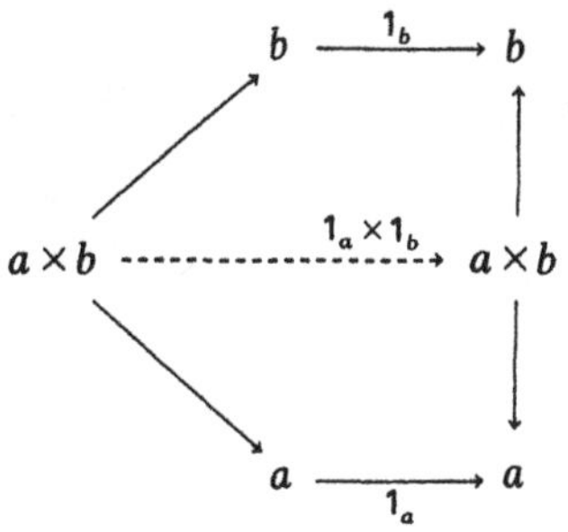

EXERCISE 6. $a \times b \cong b \times a$.

EXERCISE 7. Show that $(a \times b) \times c \cong a \times (b \times c)$

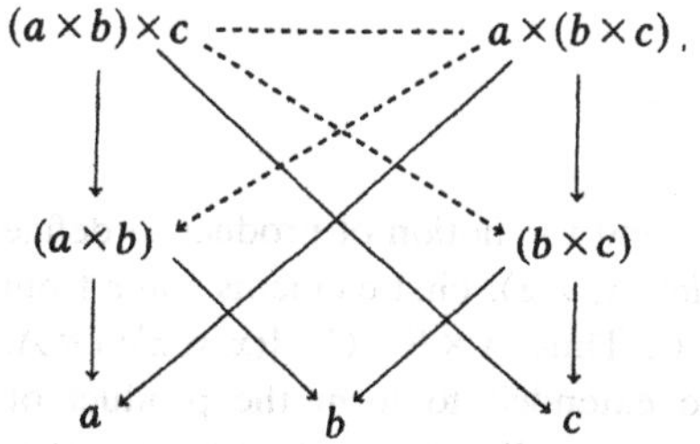

EXERCISE 8. Show that (i)

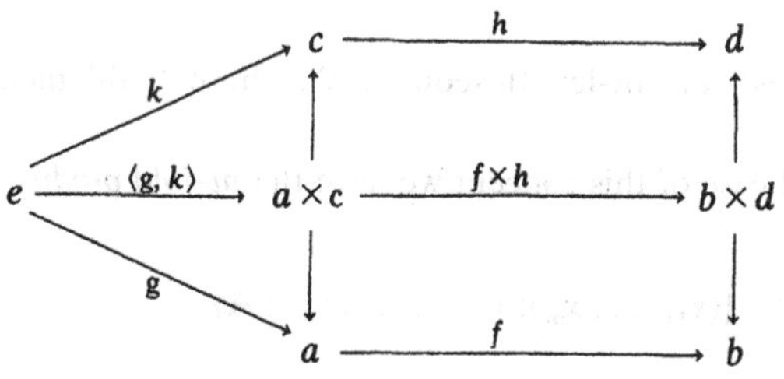

$(f \times h) \circ \langle g, k \rangle = \langle f \circ g, h \circ k \rangle$ and

(ii)

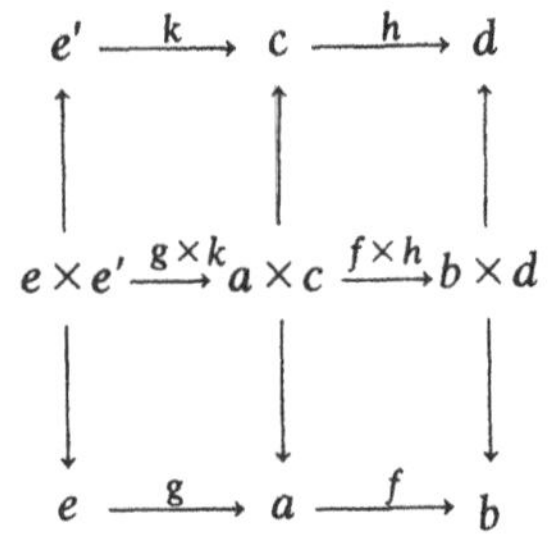

$(f\times h)\circ(g\times k)=(f\circ g)\times(h\circ k).$ □

The use we have been making of the broken arrow symbol ⇢ is a standard one in category theory. When present in any diagram it indicates that there is one and only one arrow that can occupy that position and allow the diagram to commute.

Finite products

Given sets A, B, C we extend the notion of product to define $A\times B\times C$ as the set of ordered *triples* $\langle x, y, z\rangle$. First elements come from A, second from B, and third from C. Thus $A\times B\times C=\{\langle x, y, z\rangle: x\in A, y\in B$, and $z\in C\}$. This idea can be extended to form the product of any finite sequence of sets $A_1, A_2, \ldots, A_m$. We define $A_1\times A_2\times\ldots\times A_m$ to be the set

$$\{\langle x_1, \ldots, x_m\rangle: x_1\in A_1, x_2\in A_2, \ldots, x_m\in A_m\}$$

of all "m-tuples", or "m-length sequences", whose "i-th" members come from A_i.

As a special case of this concept we have the *m-fold product* of a set A, as the set

$$A^m=\{\langle x_1, \ldots, x_m\rangle: x_1, x_2, \ldots, x_m\in A\}$$

of all m-tuples whose members all come from A. Associated with A^m are m different projection maps $pr_1^m, pr_2^m, \ldots, pr_m^m$ from A^m to A, given by the rules

$$pr_1^m(\langle x_1, \ldots, x_m \rangle) = x_1$$

$$pr_2^m(\langle x_1, \ldots, x_m \rangle) = x_2$$

$$\vdots$$

$$pr_m^m(\langle x_1, \ldots, x_m \rangle) = x_m$$

Given a set C and m maps $f_1 : C \to A, \ldots, f_m : C \to A$, we can then form a product map $\langle f_1, \ldots, f_m \rangle$ from C to A^m by stipulating, for input $c \in C$, that

$$\langle f_1, \ldots, f_m \rangle(c) = \langle f_1(c), f_2(c), \ldots, f_m(c) \rangle.$$

The construction just outlined can be developed in any category $\mathscr{C}$ that has products of any two $\mathscr{C}$-objects. For a given $\mathscr{C}$-object a, we define the m-fold product of a (with itself) to be

$$a^m = \underbrace{a \times a \times \ldots \times a}_{m\text{-copies}}$$

There is an ambiguity here. Should, for example, a^3 be taken as $(a \times a) \times a$ or $a \times (a \times a)$? However, Exercise 7 above allows us to gloss over this point, since these last two objects are isomorphic.

By applying the definition of products of pairs objects to the formation of a^m we may show that a^m has associated with it m projection arrows $pr_1^m : a^m \to a, \ldots, pr_m^m : a^m \to a$, with the universal property that for any $\mathscr{C}$-arrows $f_1 : c \to a, \ldots, f_m : c \to a$ with common domain, there is exactly one (product) arrow $\langle f_1, \ldots, f_m \rangle : c \to a^m$ making

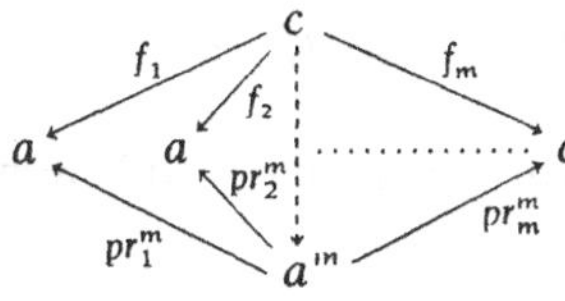

commute. For $m = 1$, we take a^1 to be just a, and $pr_1 : a \to a$ to be 1_a.

Finite products will play an important role in the "first-order" semantics of Chapter 11.

EXERCISE 9. Analyse in detail the formation of the projection arrows $pr_1^m, \ldots, pr_m^m$, and verify all assertions relating to the last diagram. Show that for any product arrow

$$c \xrightarrow{\langle f_1, \ldots, f_m \rangle} a^m,$$

we have $pr_j^m \circ \langle f_1, \ldots, f_m \rangle = f_j$, for $1 \leqslant j \leqslant m$.

EXERCISE 10. Develop the notion of the product $a_1 \times a_2 \times \ldots \times a_m$ of m objects (possibly different) and the product $f_1 \times f_2 \times \ldots \times f_m$ of m arrows (possibly different).

3.9. Co-products

The dual notion to "product" is the *co-product*, or *sum*, of objects, which by the duality principle we directly define as follows.

DEFINITION A *co-product* of $\mathscr{C}$-objects a and b is a $\mathscr{C}$-object $a+b$ together with a pair $i_a : a \to a+b$, $i_b : b \to a+b)$ of $\mathscr{C}$-arrows such that for any pair of $\mathscr{C}$-arrows of the form $(f : a \to c,\ g : b \to c)$ there is exactly one arrow $[f, g] : a+b \to c$ making

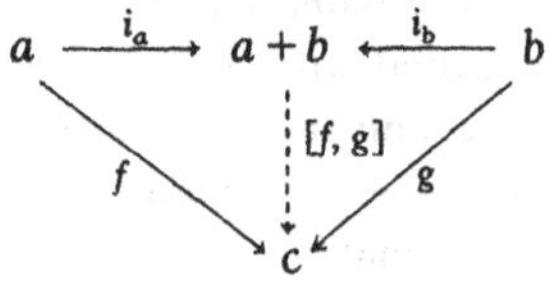

commute, i.e. such that $[f, g] \circ i_a = f$ and $[f, g] \circ i_b = g$.

$[f, g]$ is called the *co-product* arrow of f and g with respect to the *injections* i_a and i_b.

In **Set**, the co-product of A and B is their *disjoint union*, $A + B$. This is the union of two sets that look the same as (i.e. are isomorphic to) A and B but are *disjoint* (have no elements in common). We put

$$A' = \{\langle a, 0 \rangle : a \in A\} = A \times \{0\}$$

and

$$B' = \{\langle b, 1 \rangle : b \in B\} = B \times \{1\}$$

(why does $A' \cap B' = \emptyset$?) and then define

$$A+B = A' \cup B'.$$

The injection $i_A : A \to A+B$ is given by the rule

$$i_A(a) = \langle a, 0 \rangle,$$

while $i_B : B \to A+B$ has $i_B(b) = \langle b, 1 \rangle$.

EXERCISE 1. Show that $A+B$, i_A, i_B as just defined satisfy the co-product definition. (First you will have to determine the rule for the function $[f, g]$ in this case.)

EXERCISE 2. If $A \cap B = \emptyset$, show $A \cup B \cong A+B$. □

In a pre-order $(P, \sqsubseteq)$, $p+q$ is defined by the properties

(i) $p \sqsubseteq p+q$, $q \sqsubseteq p+q$ (i.e. $p+q$ is an "upper bound" of p and q);

(ii) if $p \sqsubseteq c$ and $q \sqsubseteq c$, then $p+q \sqsubseteq c$, i.e. $p+q$ is "less than" any other upper bound of p and q.

In other words $p+q$ is a *least upper bound* (l.u.b.) of p and q. In a poset the l.u.b. is unique when it exists, and will be denoted $p \sqcup q$. A poset in which any two elements have a l.u.b. *and a g.l.b.* (§3.8) is called a *lattice*.

Categorially then a lattice is a skeletal pre-order having a product and a co-product for any two of its elements.

The disjoint union of two finite sets, with say m and n elements respectively is a set with (m plus n) elements. Indeed in **Finord**, the co-product of m and n is the ordinal number $m+n$ (where "+" means "plus" quite literally). With regard to the ordinals $1 = \{0\}$ and $2 = \{0, 1\}$ it is true then in the skeletal category **Finord** that

$$1+1=2,$$

while in **Finset**, or **Set** it would be more accurate to say

$$1+1 \cong 2$$

(Co-products being defined only up to isomorphism.)

Later in §5.4 we shall see that there are categories in which this last statement, under an appropriate interpretation, is false.

EXERCISE 3. Define the co-product arrow $f+g : a+b \to c+d$ of arrows $f : a \to c$ and $g : b \to d$ and dualise all of the Exercises in §3.8.

3.10. Equalisers

Given a pair $f,g : A \rightrightarrows B$ of parallel functions in **Set**, let E be the subset of A on which f and g agree, i.e.

$$E = \{x : x \in A \text{ and } f(x) = g(x)\}$$

Then the inclusion function $i : E \hookrightarrow A$ is called the *equaliser* of f and g. The reason for the name is that under composition with i we find that $f \circ i = g \circ i$, i.e. the two functions are "equalised" by i. Moreover, i is a "canonical" equaliser of f and g – if $h : C \to A$ is any other such equaliser of f and g, i.e. $f \circ h = g \circ h$,

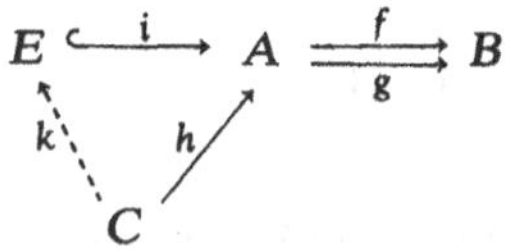

then h "factors" uniquely through $i : E \hookrightarrow A$, i.e. there is exactly one function $k : C \to E$ such that $i \circ k = h$. In other words, given h, there is only one way to fill in the broken arrow to make the above diagram commute. That there can be at most one way is clear – if $i \circ k$ is to be the same as h, then for $c \in C$ we must have $i(k(c)) = h(c)$, i.e. $k(c) = h(c)$ (i being the inclusion). But this *does* work, for $f(h(c)) = g(h(c))$, and so $h(c) \in E$.

The situation just considered is now abstracted and applied to categories in general.

An arrow $i : e \to a$ in $\mathscr{C}$ is an *equaliser* of a pair $f,g : a \to b$ of $\mathscr{C}$-arrows if

(i) $f \circ i = g \circ i$, and

(ii) Whenever $h : c \to a$ has $f \circ h = g \circ h$ in $\mathscr{C}$ there is exactly one $\mathscr{C}$-arrow $k : c \to e$ such that $i \circ k = h$

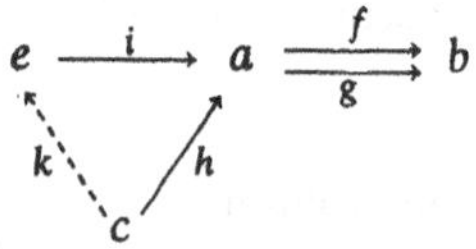

An arrow will simply be called an *equaliser* in $\mathscr{C}$ if there are a pair of $\mathscr{C}$-arrows of which it is an equaliser.

THEOREM 1. *Every equaliser is monic.*

PROOF. Suppose i equalises f and g. To show i monic (left cancellable), let $i \circ j = i \circ l$, where $j, l : c \rightrightarrows e$. Then in the above diagram let $h : c \to a$ be the arrow $i \circ j$. We have $f \circ h = f \circ (i \circ j) = (f \circ i) \circ j = (g \circ i) \circ j = g \circ h$, and so there is a unique k with $i \circ k = h$. But $i \circ j = h$ (by definition), so k must be j. However, $i \circ l = i \circ j = h$, so $k = l$. Hence $j = l$. □

The converse of Theorem 1 does not hold in all categories. For instance in the category **N**, 1 is monic (all arrows are), but cannot equalise any pair (m, n) of arrows. If it did, we would have $m \circ 1 = n \circ 1$, i.e. $m + 1 = n + 1$, hence $m = n$. But then $m + 0 = n + 0$, which would imply that 0 factors uniquely through 1, i.e. there is a unique k having $1 + k = 0$. But of course there is no such *natural number* k.

Recalling that in **N** every arrow is epic, while 0 is the only iso, the next theorem gives a somewhat deeper explanation of the situation just described.

THEOREM 2. *In any category, an epic equaliser is iso.*

PROOF. If i equalises f and g, then $f \circ i = g \circ i$, so if i is epic, $f = g$. Then in the equaliser diagram, put $c = a$, and $h = 1_a$. We have

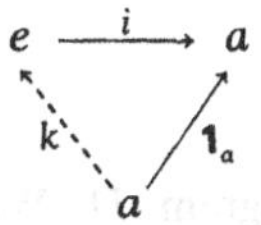

$f \circ 1_a = g \circ 1_a = f$, so there is a unique k with $i \circ k = 1_a$. Then $i \circ k \circ i = 1_a \circ i = i = i \circ 1_b$. But i is an equaliser, therefore left-cancellable, (Theorem 1), so $k \circ i = 1_b$. This gives k as an inverse to i, so i is iso. □

While monics may not be equalisers in all categories, they are certainly so in **Set** (and in fact in any topos). For if $f : E \rightarrowtail A$ is injective, define $h : A \to \{0, 1\}$ by the rule $h(x) = 1$, all $x \in A$, and $g : A \to \{0, 1\}$ by the rule

$$g(x) = \begin{cases} 1 & \text{if} \quad x \in \operatorname{Im} i \\ 0 & \text{if} \quad x \notin \operatorname{Im} i \end{cases}$$

Then f equalises g and h.

EXERCISE 1. Prove the last assertion.

EXERCISE 2. Show that in a poset, the only equalisers are the identity arrows.

3.11 Limits and co-limits

The definitions of the product of two objects and the equaliser of two arrows have the same basic form. In each case the entity in question has a certain property "canonically", in that any other object with that property "factors through" it in the manner indicated above. In the case of an equaliser the property is that of "equalising" the two original arrows. In the case of the product of a and b the property is that of being the domain of a pair of arrows whose codomains are a and b. This sort of situation is called a *universal construction*. The entity in question is *universal* amongst the things that have a certain property.

We can make this idea a little more precise (without being too pedantic, hopefully) by considering diagrams. By a diagram D in a category $\mathscr{C}$ we simply mean a collection of $\mathscr{C}$-objects $d_i, d_j, \ldots$, together with some $\mathscr{C}$-arrows $g: d_i \to d_j$ between certain of the objects in the diagram. (Possibly more than one arrow between a given pair of objects, possibly none.)

A *cone* for diagram D consists of a $\mathscr{C}$-object c together with a $\mathscr{C}$-arrow $f_i : c \to d_i$ for each object d_i in D, such that

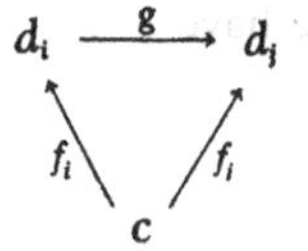

commutes, whenever g is an arrow in the diagram D. We use the symbolism $\{f_i : c \to d_i\}$ to denote a cone for D.

A *limit* for a diagram D is a D-cone $\{f_i : c \to d_i\}$ with the property that for any other D-cone $\{f_i' : c' \to d_i\}$ there is exactly one arrow $f: c' \to c$ such

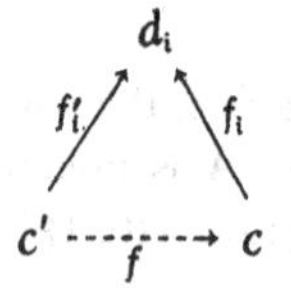

commutes for every object d_i in D.

This limiting cone, when it exists, is said to have the *universal property* with respect to D-cones. It is universal amongst such cones – any other D-cone factors uniquely through it as in the last diagram. A limit for diagram D is unique up to isomorphism:- if $\{f_i : c \to d_i\}$ and $\{f_i' : c' \to d_i\}$ are both limits for D, then the unique commuting arrow $f: c' \dashrightarrow c$ above is iso (its inverse is the unique commuting arrow $c \dashrightarrow c'$ whose existence follows from the fact that $\{f_i' : c' \to d_i\}$ is a limit).

EXAMPLE 1. Given $\mathscr{C}$-objects a and b let D be the arrow-less diagram

$$a \qquad b$$

A D-cone is then an object c, together with two arrows f, and g of the form

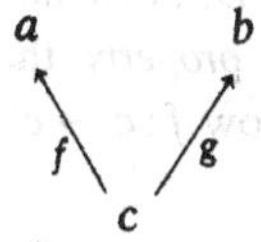

A limiting D-cone, one through which all such cones factor, is none other than a product of a and b in $\mathscr{C}$.

EXAMPLE 2. Let D be the diagram

$$a \underset{g}{\overset{f}{\rightrightarrows}} b$$

A D-cone is a pair $h: c \to a$, $j: c \to b$ such that

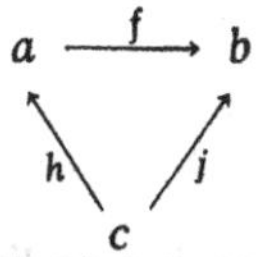

and

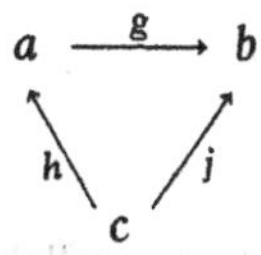

commute. But this requires that $j = f \circ h = g \circ h$, so we can simply say that a D-cone in this case is an arrow $h: c \to a$ such that

$$c \xrightarrow{h} a \underset{g}{\overset{f}{\rightrightarrows}} b$$

commutes, i.e. $f \circ h = g \circ h$. We then see that a D-limit is an *equaliser* of f and g.

EXAMPLE 3. Let D be the *empty* diagram

i.e. no objects and no arrows. A D-cone is then simply a $\mathscr{C}$-object c (there are no f_i's as D has no d_i's). A limiting cone is then an object c

such that for any other $\mathscr{C}$-object (D-cone) c', there is exactly one arrow $c' \dashrightarrow c$. In other words, a limit for the empty diagram is a *terminal object!*

□

By *duality* we define a *co-cone* $\{f_i : d_i \to c\}$ for diagram D to consist of an object c, and arrows $f_i : d_i \to c$ for each object d_i in D. A *co-limit* for D is then a co-cone $\{f_i : d_i \to c\}$ with the *co-universal property* that for any other co-cone $\{f_i' : d_i \to c'\}$ there is exactly one arrow $f : c \to c'$ such

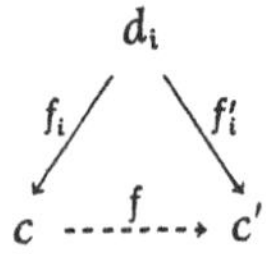

commutes for every d_i in D.

A co-limit for the diagram of Example 1 is a co-product of a and b, while a co-limit for the empty diagram is a category $\mathscr{C}$ is an *initial object* for $\mathscr{C}$.

3.12. Co-equalisers

The co-equaliser of a pair (f, g) of parallel $\mathscr{C}$-arrows is a co-limit for the diagram

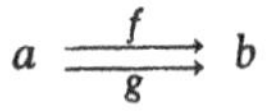

It can be described as a $\mathscr{C}$-arrow $q : b \to e$ such that

(i) $q \circ f = q \circ g$, and

(ii) whenever $h : b \to c$ has $h \circ f = h \circ g$ in $\mathscr{C}$ there is exactly one $\mathscr{C}$-arrow $k : e \to c$ such that

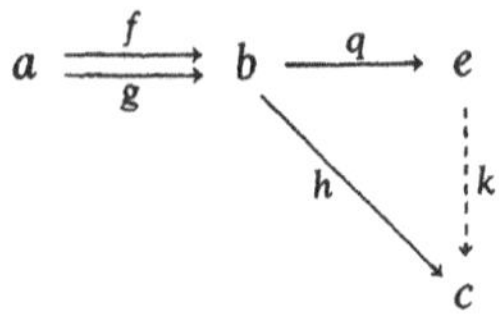

commutes. The results of §3.10 immediately dualise to tell us that co-equalisers are epic, that the converse is true in **Set**, and that a monic co-equaliser is iso.

In **Set** an "$\in$-related" description of the co-equaliser comes through the very important notion of *equivalence relation.* An equivalence relation on a set A is, by definition, a relation $R \subseteq A \times A$ that is

(a) *reflexive,* i.e. aRa, for every $a \in A$;

(b) *transitive,* i.e. whenever aRb and bRc, then aRc; and

(c) *symmetric,* i.e. whenever aRb, then bRa.

Equivalence relations arise throughout mathematics (and elsewhere) in situations where one wishes to identify different things that are 'equivalent'. Typically one may be concerned with some particular property (properties) with respect to which different things may be indistinguishable. The relation that holds between two things when they are thus indistinguishable will then be an equivalence relation.

We have in fact already met this idea in the discussion in §3.4 of isomorphism. Two objects in a category that are isomorphic might just as well be the same object, as far as categorial properties are concerned, and indeed

$$\{\langle a, b\rangle : a \cong b \text{ in } \mathscr{C}\}$$

is a relation on $\mathscr{C}$-objects that is reflexive, transitive, and symmetric. (Exercise 3.4.1).

The process of "identifying equivalent things" is rendered explicit by lumping together all things that are related to each other and treating the resulting collection as a single entity. Formally, for $a \in A$ we define the *R-equivalence class* of a to be the set

$$[a] = \{b : aRb\}$$

of all members of A to which a is R-related. Different elements may have the same subset of A as their equivalence class, and the situation in general is as follows:

(1) $[a] = [b]$ iff aRb

(2) if $[a] \neq [b]$ then $[a] \cap [b] = \emptyset$

(3) $a \in [a]$

(the proof of these depends on properties (a), (b), (c) above). Statement (1) tells us that equivalent elements are related to precisely the same elements, and conversely (2) says if two equivalence classes are not the same, then they have no elements in common at all. This, together with

(3) (which holds by (a)), implies that each $a \in A$ is a member of one and only one R-equivalence class.

The actual identification process consists in passing from the original set to a new set whose elements are the R-equivalence classes, i.e. we shift from A to the set

$$A/R = \{[a] : a \in A\}$$

The transfer is effected by the *natural map* $f_R : A \to A/R$, where $f_R(a) = [a]$, for $a \in A$.

Thus, by (1), when aRb we have $f_R(a) = f_R(b)$, and so R-equivalent elements are identified by the application of f_R.

What has all this to do with co-equalisers? Well the point is that f_R is the co-equaliser of the pair $f,g : R \rightrightarrows A$ of projection functions from R to A, i.e. the functions

$$f(\langle a, b\rangle) = a$$

and

$$g(\langle a, b\rangle) = b.$$

The last paragraph explained in effect why $f_R \circ f = f_R \circ g$. To see why the diagram

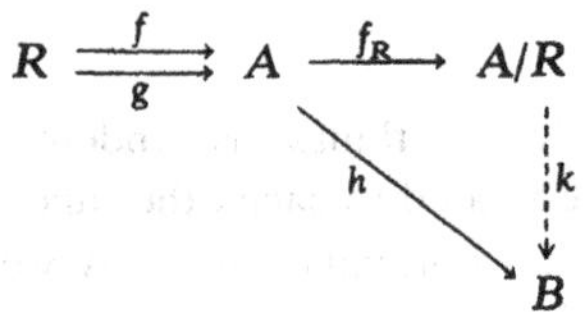

can be "filled in" by only one k, given $h \circ f = h \circ g$, we suppose we have a k such that $k \circ f_R = h$. Then for $[a] \in A/R$ we must have $k([a]) = k(f_R(a)) = k \circ f_R(a) = h(a)$. So the only thing we can do is define k to be the function that for input $[a]$ gives output $h(a)$. There is a problem here about whether k is a *well-defined* function, for if $[a] = [b]$, our rule also tells us to output $h(b)$ for input $[a] = [b]$. In order for there to be a unique output for a given input, we would need to know in this case that $h(a) = h(b)$. But in fact if $[a] = [b]$ then $\langle a, b\rangle \in R$ and our desideratum follows, because $h \circ f = h \circ g$.

The question of "well-definedness" just dealt with occurs repeatedly in working with so called "quotient" sets of the form A/R. Operations on, and properties of an R-equivalence class are defined by reference to some selected member of the equivalence class, called its *representative*.

One must always check that the definition does not depend on which representative is chosen. In other words a well defined concept is one that is *stable* or *invariant* under R, i.e. is not altered or destroyed when certain things are replaced by others to which they are R-equivalent.

Equivalence relations can be used to construct the co-equaliser in **Set** of any pair $f,g: A \rightrightarrows B$ of parallel functions. To co-equalise f and g we have to identify $f(x)$ with $g(x)$, for $x \in A$. So we consider the relation

$$S = \{\langle f(x), g(x)\rangle : x \in A\} \subseteq B \times B.$$

S may not be an equivalence relation on B. However, it is possible to build up S until it becomes an equivalence relation, and to do this in a "minimal" way. There is an equivalence relation R on B such that

(i) $S \subseteq R$, and

(ii) if T is any other equivalence on B such that T contains S, then $R \subseteq T$

(i.e. R is the "smallest" equivalence relation on B that contains S). The co-equaliser of f and g is then the natural map $f_R : B \to B/R$. (See Arbib and Manes, p. 19, for the details of how to construct this R).

3.13. The pullback

A *pullback* of a pair $a \xrightarrow{f} c \xleftarrow{g} b$ of $\mathscr{C}$-arrows with a common codomain is a limit in $\mathscr{C}$ for the diagram

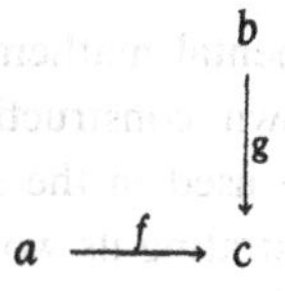

A cone for this diagram consists of three arrows f', h, g', such that

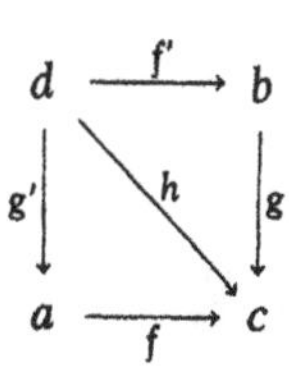

commutes. But this requires that $h = g \circ f' = f \circ g'$, so we may simply say that a cone is a pair $a \xleftarrow{g'} d \xrightarrow{f'} b$ of $\mathscr{C}$-arrows such that the "square"

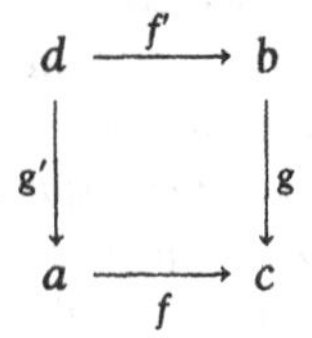

commutes, i.e. $f \circ g' = g \circ f'$.

Thus we have, by the definition of universal cone, that a pullback of the pair $a \xrightarrow{f} c \xleftarrow{g} b$ in $\mathscr{C}$ is a pair of $\mathscr{C}$-arrows $a \xleftarrow{g'} d \xrightarrow{f'} b$ such that

(i) $f \circ g' = g \circ f'$, and

(ii) whenever $a \xleftarrow{h} e \xrightarrow{j} b$ are such that $f \circ h = g \circ j$, then

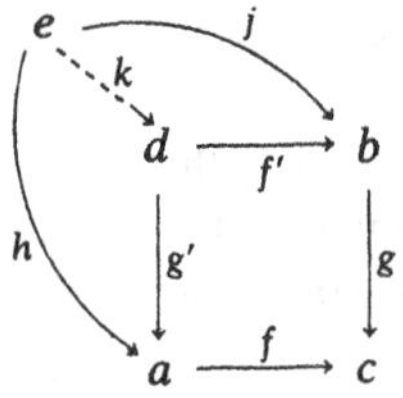

there is exactly one $\mathscr{C}$-arrow $k : e \dashrightarrow d$ such that $h = g' \circ k$ and $j = f' \circ k$. In other words when h and j are such that the outer "square", or "boundary" of the above diagram commutes, then there is only one way to fill in the broken arrow to make the whole diagram commute.

The inner square (f, g, f', g') of the diagram is called a *pullback square*, or *Cartesian square*. We also say that f' arises by *pulling back f along g*, and g' arises by *pulling back g along f*.

The pullback is a very important and fundamental mathematical notion, that incorporates a number of well known constructions. It is certainly the most important limit concept to be used in the study (and definition) of topoi. The following examples, illustrating its workings and generality, are commended as worthy of detailed examination.

EXAMPLE 1. In **Set**, the pullback

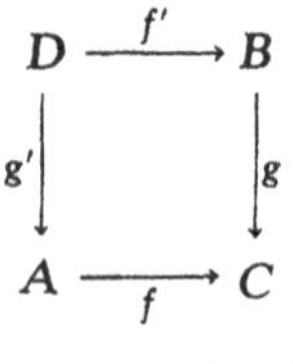

of two set function f and g is defined by putting

$$D = \{\langle x, y \rangle : x \in A,\ y \in B, \text{ and } f(x) = g(y)\}$$

with f' and g' as the projections:

$$f'(\langle x, y\rangle) = y$$

$$g'(\langle x, y\rangle) = x.$$

D is then a subset of the product set $A \times B$. It is sometimes denoted $A \underset{C}{\times} B$, the product of A and B *over* C. Pullbacks are also known as "fibred products" (the use of the word "fibred" is explained in Chapter 4).

EXAMPLE 2. **Inverse images.** If $f: A \to B$ is a function, and C a subset of B, then the *inverse image of C under f*, denoted $f^{-1}(C)$, is that subset of A consisting of all the f-inputs whose corresponding outputs lie in C, i.e.

$$f^{-1}(C) = \{x : x \in A \text{ and } f(x) \in C\}$$

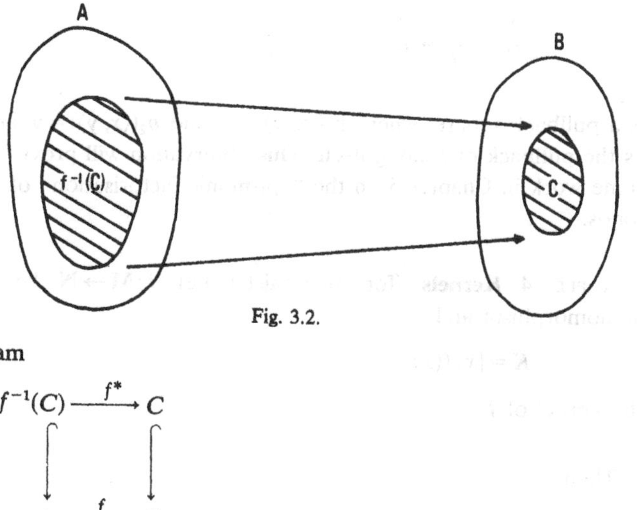

Fig. 3.2.

The diagram

$$\begin{array}{ccc} f^{-1}(C) & \xrightarrow{f^*} & C \\ \downarrow & & \downarrow \\ A & \xrightarrow{f} & B \end{array}$$

is a pullback square in **Set**, where the arrows with curved tails denote inclusions as usual, and $f^*(x) = f(x)$ for $x \in f^{-1}(C)$ (i.e. f^* is the *restriction* of f to $f^{-1}(C)$). Thus *the inverse image of C under f arises by pulling C back along f.*

The dynamical quality inherent in the notion of function (cf. §2.1) is quite forcefully present in this example of "pulling back". It would be quite unconvincing to suggest we were just dealing with sets of ordered pairs.

EXAMPLE 3. **Kernel relation.** Associated with any function $f: A \to B$ is a special equivalence relation on A, denoted R_f, and called the *kernel relation* of f (the kernel "congruence" in universal algebra, where it lies at the heart of the First Isomorphism Theorem). As a set of ordered pairs we have

$$R_f = \{\langle x, y\rangle : x \in A \text{ and } y \in A \text{ and } f(x) = f(y)\}$$

or

$$xR_f y \quad \text{iff} \quad f(x) = f(y).$$

In the light of our first example we see that

$$\begin{array}{ccc} R_f & \xrightarrow{p_2} & A \\ {\scriptstyle p_1}\downarrow & & \downarrow{\scriptstyle f} \\ A & \xrightarrow[f]{} & B \end{array}$$

is a pullback square, where $p_1(\langle x, y\rangle) = x$ and $p_2(\langle x, y\rangle) = y$, i.e. R_f arises as the pullback of f along itself. This observation will provide the key to some work in Chapter 5 on the "epi-monic factorisation" of arrows in a topos.

EXAMPLE 4. **Kernels** (for algebraists). Let $f: \mathbf{M} \to \mathbf{N}$ be a monoid homomorphism and

$$K = \{x : f(x) = e\}$$

the kernel of f.

Then

$$\begin{array}{ccc} \mathbf{K} & \hookrightarrow & \mathbf{M} \\ {\scriptstyle !}\downarrow & & \downarrow{\scriptstyle f} \\ \mathbf{O} & \xrightarrow{!} & \mathbf{N} \end{array}$$

is a pullback square in **Mon**, where **O** is the one-element monoid (which is initial and terminal).

This characterisation of kernels applies also to the categories **Grp** and **Vect**.

EXAMPLE 5. In a pre-order $(P, \sqsubseteq)$,

$$\begin{array}{ccc} s & \longrightarrow & q \\ \downarrow & & \downarrow \\ p & \longrightarrow & r \end{array}$$

is a pullback square iff s is a product of p and q.

EXAMPLE 6. In any category with a terminal object, if

$$\begin{array}{ccc} d & \xrightarrow{g} & b \\ {\scriptstyle f}\downarrow & & \downarrow{\scriptstyle !} \\ a & \xrightarrow{!} & 1 \end{array}$$

is a pullback, then $\langle f, g\rangle$ is a product (g.l.b.) of a and b.

EXAMPLE 7. In any category, if

$$\begin{array}{ccc} e & \xrightarrow{i} & a \\ {\scriptstyle i}\downarrow & & \downarrow{\scriptstyle g} \\ a & \xrightarrow[f]{} & b \end{array}$$

is a pullback, then i is an equaliser of f and g.

EXAMPLE 8. THE PULLBACK LEMMA (PBL). *If a diagram of the form*

$$\begin{array}{ccccc} \bullet & \longrightarrow & \bullet & \longrightarrow & \bullet \\ \downarrow & & \downarrow & & \downarrow \\ \bullet & \longrightarrow & \bullet & \longrightarrow & \bullet \end{array}$$

commutes, then

(i) *if the two small squares are pullbacks, then the outer "rectangle"* (with top and bottom edges the evident composites) *is a pullback;*

(ii) *if the outer rectangle and the right hand square are pullbacks then so is the left hand square.*

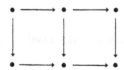

The PBL is a key fact, and will be used repeatedly in what follows. Its proof, though rather tedious, will certainly familiarise the reader with how a pullback works.

The PBL will often be used for a diagram of the form,

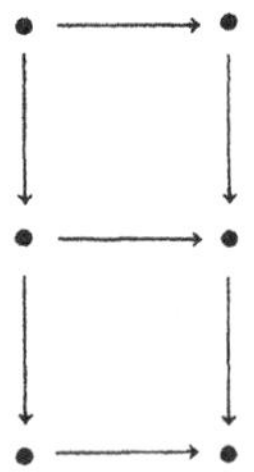

in which case when the outer rectangle and *bottom* square are pullbacks, we will conclude that the *top* square is a pullback.

Example 9. In any category, an arrow $f: a \to b$ is monic iff

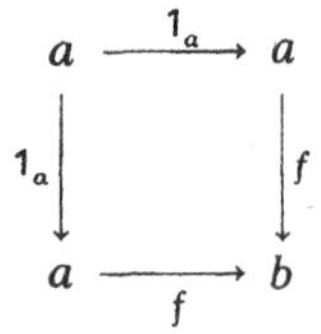

is a pullback square.

Exercise. Show that if

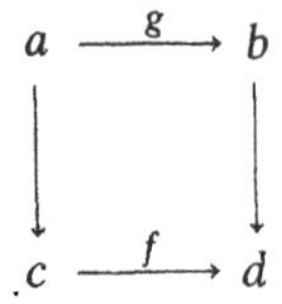

is a pullback square, and f is monic, then g is also monic.

3.14. Pushouts

The dual of a pullback of a pair of arrows with common codomain is a *pushout* of the two arrows with common domain:

a pushout of $b \xleftarrow{f} a \xrightarrow{g} c$ is a co-limit for the diagram

$$\begin{array}{ccc} a & \xrightarrow{g} & c \\ {\scriptstyle f}\downarrow & & \\ b & & \end{array}$$

In **Set** it obtained by forming the disjoint union $b+c$ and then identifying $f(x)$ with $g(x)$, for each $x \in a$ (by a co-equaliser).

Exercise. Dualise §3.13.

3.15. Completeness

A category $\mathscr{C}$ is *complete* if every diagram in $\mathscr{C}$ has a limit in $\mathscr{C}$. Dually $\mathscr{C}$ is *co-complete* when every $\mathscr{C}$-diagram has a co-limit. A *bi-complete* category is one that is complete and co-complete.

A *finite* diagram is one that has a finite number of objects, and a finite number of arrows between them.

A category is *finitely complete* if it has a limit for every *finite* diagram. Finite co-completeness and finite bi-completeness are defined similarly.

Theorem 1. *If $\mathscr{C}$ has a terminal object, and a pullback for each pair of $\mathscr{C}$-arrows with common codomain, then $\mathscr{C}$ is finitely complete.* □

A proof of this theorem is beyond our present scope (and outside our major concerns). The details may be found in Herrlich and Strecker [73], Theorem 23.7, along with a number of other characterisations of finite completeness.

To illustrate the Theorem, we observe that

(A) given a terminal object and pullbacks, the product of a and b is got from the pullback of $a \to 1 \leftarrow b$ (cf. §3.13, Example 6);

(B) given pullbacks and products, from a parallel pair $f,g : a \rightrightarrows b$ we first form the product arrows

$$a \xrightarrow{\langle 1_a, f\rangle} a \times b \quad \text{and} \quad a \xrightarrow{\langle 1_a, g\rangle} a \times b$$

and then their pullback

$$\begin{array}{ccc} d & \xrightarrow{q} & a \\ {\scriptstyle p}\downarrow & & \downarrow{\scriptstyle \langle 1_a, g\rangle} \\ a & \xrightarrow[\langle 1_a, f\rangle]{} & a \times b \end{array}$$

It follows readily (§3.8) that $p = q$, and that this arrow is an equaliser of f and g.

Exercises

(1) Verify (B), and consider the details of that construction in **Set**.
(2) Show how to construct pullbacks from products and equalisers. A hint is given by the description (Example 1, §3.13) of pullbacks in **Set**. A co-hint appears in §3.14.
(3) Dualise the Theorem of this section.

3.16. Exponentiation

Given sets A and B we can form in **Set** the collection B^A of all functions that have domain A and codomain B, i.e.

$$B^A = \{f : f \text{ is a function from } A \text{ to } B\}$$

To characterise B^A by arrows we observe that associated with B^A is a special arrow

$$ev : B^A \times A \to B,$$

given by the rule

$$ev(\langle f, x\rangle) = f(x).$$

ev is the *evaluation* function. Its inputs are pairs of the form $\langle f, x\rangle$ where $f : A \to B$ and $x \in A$. The action of *ev* for such as input is to apply f to x, to evaluate f at x, yielding the output $f(x) \in B$. The categorial description of B^A comes from the fact that *ev* enjoys a universal property amongst all set functions of the form

$$C \times A \xrightarrow{g} B.$$

Given any such g, there is one and only one function $\hat{g}:C \to B^A$ such that

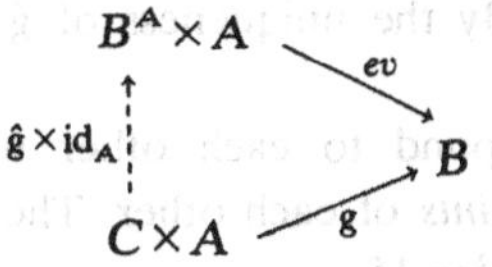

commutes where $\hat{g}\times \mathrm{id}_A$ is the product function described in §3.8. For input $\langle c, a\rangle \in C\times A$ it gives output $\langle \hat{g}(c), \mathrm{id}_A(a)\rangle = \langle \hat{g}(c), a\rangle$.

The idea behind the definition of $\hat{g}$ is that the action of g causes any particular c to determine a function $A \to B$ by fixing the first elements of arguments of g at c, and allowing the second elements to range over A. In other words for a given $c \in C$ we define $g_c : A \to B$ by the rule

$$g_c(a) = g(\langle c, a\rangle), \quad \text{for each} \quad a \in A.$$

$\hat{g}:C \to B^A$ can now be defined by $\hat{g}(c) = g_c$, all $c \in C$. For any $\langle c, a\rangle \in C\times A$ we then get

$$ev(\langle \hat{g}(c), a\rangle) = g_c(a) = g(\langle c, a\rangle)$$

and so the above diagram commutes. But the requirement that the diagram commutes, i.e. that $ev(\langle \hat{g}(c), a\rangle) = g(\langle c, a\rangle)$, means that $\hat{g}(c)$ must be the function that for input a gives output $g(\langle c, a\rangle)$, i.e. $\hat{g}(c)$ must be g_c as above.

By abstraction then we say that a category $\mathscr{C}$ has *exponentiation* if it has a product for any two $\mathscr{C}$-objects, and if for any given $\mathscr{C}$-objects a and b there is a $\mathscr{C}$-object b^a and a $\mathscr{C}$-arrow $ev : b^a \times a \to b$, called an *evaluation* arrow, such that for any $\mathscr{C}$-object c and $\mathscr{C}$-arrow $g : c \times a \to b$, there is a unique $\mathscr{C}$-arrow $\hat{g} : c \to b^a$ making

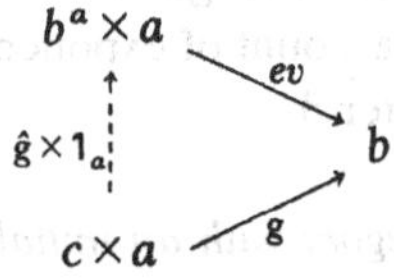

commute, i.e. a unique $\hat{g}$ such that $ev \circ (\hat{g} \times 1_a) = g$. The assignment of $\hat{g}$ to g establishes a bijection

$$\mathscr{C}(c \times a, b) \cong \mathscr{C}(c, b^a)$$

between the collection of $\mathscr{C}$-arrows from $c \times a$ to b, and the collection of those from c to b^a. For if $\hat{g} = \hat{h}$, then $ev \circ (\hat{g} \times 1_a) = ev \circ (\hat{h} \times 1_a)$, i.e. $g = h$, and so the assignment is injective. To see that it is surjective, take $h : c \to b^a$ and *define* $g = ev \circ (h \times 1_a)$. By the uniqueness of $\hat{g}$ we must have $h = \hat{g}$.

Two arrows (g and $\hat{g}$) that correspond to each other under this bijection will be called *exponential adjoints* of each other. The origin of this terminology may be found in Chapter 15.

A finitely complete category with exponentiation is said to be *Cartesian closed.*

EXAMPLE 1. If A and B are finite sets with say m and n elements, then B^A is finite and has n^m ("n to the power m") elements. In the expression n^m, the "m" is called an *exponent*, hence the above terminology. **Finord** is Cartesian closed, and indeed the exponential is literally the number n^m.

EXAMPLE 2. A *chain* is a poset $\mathbf{P} = (P, \sqsubseteq)$ that is linearly ordered, i.e. has $p \sqsubseteq q$ or $q \sqsubseteq p$ for any $p, q \in P$. If $\mathbf{P}$ is a chain with a terminal object 1, then we put

$$q^p = \begin{cases} 1 & \text{if } \; p \sqsubseteq q \\ q & \text{if } \; q \sqsubset p \quad (\text{i.e. } q \sqsubseteq p \; \text{ and } \; q \neq p) \end{cases}$$

A chain always has products:

$$p \times q = \text{g.l.b. of } p \text{ and } q = \begin{cases} p & \text{if } \; p \sqsubseteq q \\ q & \text{if } \; q \sqsubseteq p. \end{cases}$$

We thus have two cases to consider for ev.

(i) $p \sqsubseteq q$. Then $q^p \times p = 1 \times p = p \sqsubseteq q$;

(ii) $q \sqsubseteq p$. Then $q^p \times p = q \times p = q$.

In either case $q^p \times p \sqsubseteq q$ and so ev is the unique arrow $q^p \times p \to q$ in $\mathbf{P}$. We leave it to the reader to verify that this definition gives $\mathbf{P}$ exponentiation. An explanation of why it works, and an account of exponentiation in posets in general will be forthcoming in Chapter 8. □

THEOREM 1. *Let $\mathscr{C}$ be a Cartesian closed category with an initial object* 0. *Then in $\mathscr{C}$,*

(1) $0 \cong 0 \times a$, *for any object a;*

(2) *if there exists an arrow $a \to 0$, then $a \cong 0$;*

(3) *if $0 \cong 1$, then the category $\mathscr{C}$ is degenerate, i.e. all $\mathscr{C}$-objects are isomorphic;*

(4) *any arrow* $0 \to a$ *with dom* 0 *is monic;*
(5) $a^1 \cong a$, $a^0 \cong 1$, $1^a \cong 1$.

PROOF. (1) For any $\mathscr{C}$-object b, $\mathscr{C}(0, b^a)$ has only one member (as 0 is initial). By definition of exponentiation, $\mathscr{C}(0, b^a) \cong \mathscr{C}(0 \times a, b)$. Hence the latter collection has only one member. Thus there is only one arrow $0 \times a \to b$, for any b. Hence $0 \times a$ is an initial $\mathscr{C}$-object, and since the latter are unique up to isomorphism, $0 \cong 0 \times a$.

(2) Given $f : a \to 0$, we show that $a \cong 0 \times a$, and hence by (1), $a \cong 0$. From the universal definition of product

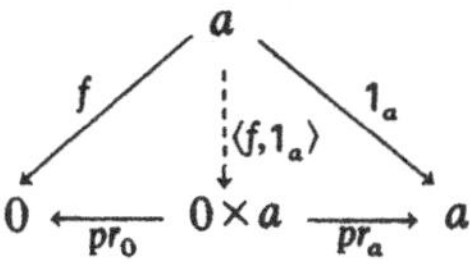

$pr_a \circ \langle f, 1_a \rangle = 1_a$. But $\langle f, 1_a \rangle \circ pr_a$ is an arrow from $0 \times a$ to $0 \times a$, and there is only one such, $0 \times a$ being initial. thus $\langle f, 1_a \rangle \circ pr_a = 1_{0 \times a}$, giving $\langle f, 1_a \rangle = pr_a^{-1}$ and $pr_a : 0 \times a \cong a$.

(3) If $0 \cong 1$, then for any a, since there is an arrow from a to 1, there will be one from a to 0 whence, by (2), $a \cong 0$. Thus all objects are isomorphic to 0. Ergo they are all isomorphic to each other.

(4) Given $f : 0 \to a$, suppose $f \circ g = f \circ h$, i.e.

$$b \underset{h}{\overset{g}{\rightrightarrows}} 0 \xrightarrow{\ f\ } a$$

commutes. But then by (2), $b \cong 0$, so b is an initial object and there is only one arrow $b \to 0$. Thus $g = h$, and f is left-cancellable. □

EXERCISE. Prove part (5) of the Theorem, and interpret (1)–(5) as they apply to **Set**. □

Having reached the end of this chapter, we can look back on an extensive catalogue of categorial versions of mathematical concepts and constructions. We now have some idea of how category theory has recreated the world of mathematical ideas, and indeed expanded the horizons of mathematical thought. And we have seen a number of features that distinguish **Set** from other categories. In **Set**, monic epics are iso, a property not enjoyed by **Mon**. It is however, enjoyed by **Grp** – but then **Grp** is not Cartesian closed (this follows from the above Theorem – **Grp** is not degenerate, but does have $0 \cong 1$). On the other hand the Cartesian-closed categories are not all "**Set**-like". The poset $\mathbf{n} = \{0, \ldots, n-1\}$ is Cartesian-closed (being a chain with terminal object), but has monic epics that are not iso. It would appear then that to develop a categorial set theory we will have to work in categories that have some other special features in common with **Set**, something at least that is not possessed by **Mon**, **n**, etc. In fact what we need is one more construction, a conceptually straightforward but very powerful one whose nature will be revealed in the next chapter.

CHAPTER 4

INTRODUCING TOPOI

> *"This is the development on the basis of elementary (first-order) axioms of a theory of "toposes" just good enough to be applicable not only to sheaf theory, algebraic spaces, global spectrum, etc. as originally envisaged by Grothendieck, Giraud, Verdier, and Hakim but also to Kripke semantics, abstract proof theory, and the Cohen–Scott–Solovay method for obtaining independence results in set theory."*
>
> F. W. Lawvere

4.1. Subobjects

If A is a subset of B, then the inclusion function $A \hookrightarrow B$ is injective, hence monic. On the other hand any monic function $f: C \rightarrowtail B$ determines a subset of B, viz $\operatorname{Im} f = \{f(x) : x \in C\}$. It is easy to see that f induces a bijection between C and $\operatorname{Im} f$, so $C \cong \operatorname{Im} f$.

Thus the domain of a monic function is isomorphic to a subset of the codomain. Up to isomorphism, the domain is a subset of the codomain. This leads us to the categorial versions of subsets, which are known as *subobjects*:

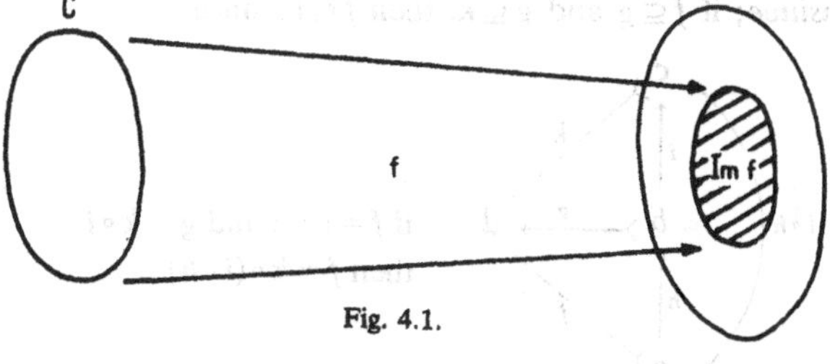

Fig. 4.1.

a *subobject* of a $\mathscr{C}$-object d is a monic $\mathscr{C}$-arrow $f: a \rightarrowtail d$ with codomain d.

Now if D is a set, then the collection of all subsets of D is known as the *powerset of* D, denoted $\mathscr{P}(D)$. Thus

$$\mathscr{P}(D) = \{A : A \text{ is a subset of } D\}.$$

The relation of set inclusion is a partial ordering on the power set $\mathscr{P}(D)$, i.e. $(\mathscr{P}(D), \subseteq)$ is a poset, and becomes a category in which there is an arrow $A \to B$ iff $A \subseteq B$. When there is such an arrow, the diagram

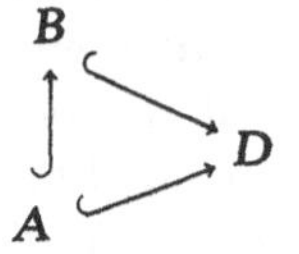

commutes. This suggests a way of defining an "inclusion" relation between subobjects of d. Given $f: a \rightarrowtail d$ and $g: b \rightarrowtail d$, we put $f \subseteq g$ iff there is a $\mathscr{C}$-arrow $h: a \to b$ such that

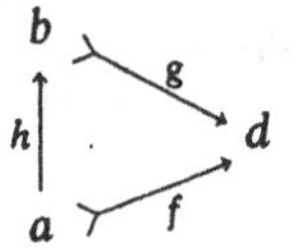

commutes, i.e. $f = g \circ h$. (such an h will always be monic, by Exercise 3.1.2, so h will be a subobject of b, enhancing the analogy with the **Set** case). Thus $f \subseteq g$ precisely when f factors through g.

The inclusion relation on subobjects is

(i) *reflexive*; $f \subseteq f$, since

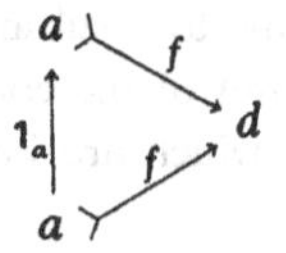

$f = f \circ 1_a$,

and

(ii) *transitive*; if $f \subseteq g$ and $g \subseteq k$, then $f \subseteq k$, since

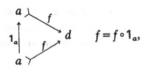

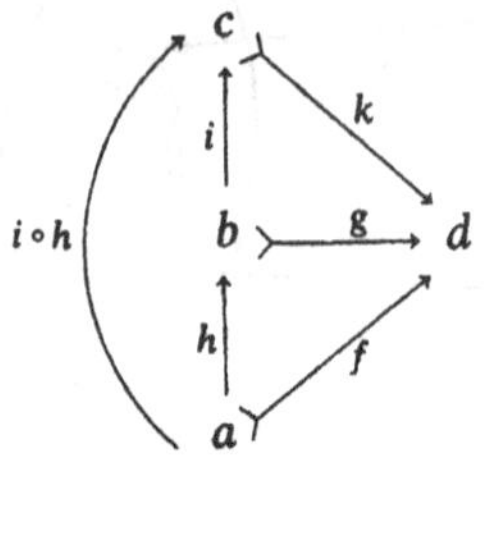

if $f = g \circ h$ and $g = k \circ i$ then $f = k \circ (i \circ h)$.

Now if $f \subseteq g$ and $g \subseteq f$, then f and g each factor through each other, as in

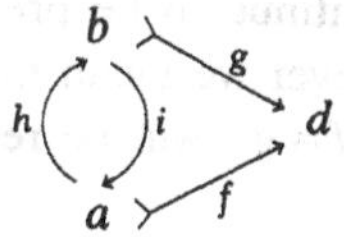

$$f = g \circ h$$
$$g = f \circ i.$$

In that case, $h : a \to b$ is iso, with inverse i (exercise for the reader). Thus when $f \subseteq g$ and $g \subseteq f$, they have isomorphic domains, and so we call them *isomorphic subobjects* and write $f \simeq g$. Now in order for $\subseteq$ to be antisymmetric, we require that when $f \simeq g$, then $f = g$. This may not in fact be so, indeed we may have $a \neq b$. So $\subseteq$ will in general be a preordering on the subobjects of d as defined, and not a partial ordering. If we left things there, we would run into difficulties later. We really do want to be able to think of $\subseteq$ as being antisymmetric. The machinery that allows this was set up in §3.12. The relation $\simeq$ is an equivalence relation (exercise – use (i), (ii) above). Each $f : a \rightarrowtail d$ determines an equivalence class

$$[f] = \{g : f \simeq g\},$$

and we form the collection

$$\mathrm{Sub}(d) = \{[f] : f \text{ is a monic with } \mathrm{cod}\, f = d\}.$$

We are now going to refer to the members of Sub(d) as the subobjects, i.e. we *redefine* a subobject of d to be an equivalence class of monics with codomain d. To obtain an inclusion notion for these entities, we put (using the same symbol as before)

$$[f] \subseteq [g] \quad \text{iff} \quad f \subseteq g.$$

Here we come up against the question mentioned in §3.12. Is the definition, given via representatives of equivalence classes, independent of the choice of representative? The answer is yes. If $[f] = [f']$ and $[g] = [g']$, then $f \subseteq g$ iff $f' \subseteq g'$, i.e. $\subseteq$ is stable under $\simeq$ (exercise).

The point of this construction was to make $\subseteq$ antisymmetric. But when $[f] \subseteq [g]$ and $[g] \subseteq [f]$, then $f \subseteq g$ and $g \subseteq f$, so $f \simeq g$ and hence $[f] = [g]$. Thus the subobjects of d, as now defined, form a poset (Sub(d), $\subseteq$).

This lengthy piece of methodology is not done with yet. It now starts to bite its own tail as we blur the distinction between equivalence class and representative. We shall usually say "the subobject f" when we mean "the subobject $[f]$", and "$f \subseteq g$" when strictly speaking "$[f] \subseteq [g]$" is intended, etc. All properties and constructions of subobjects used will however be stable under $\simeq$ (indeed being categorial they will only be *defined* up to isomorphism anyway). So this *abus de langage* is technically justifiable and has great advantages in terms of conceptual and notational clarity. The only point on which we shall continue to be precise is the matter of identity. "$f \simeq g$" will be used whenever we mean that f and g are the same subobject, i.e. $[f]=[g]$, while "$f=g$" will be reserved for when they are the same actual arrow.

EXERCISE 1. In **Set**, $\mathrm{Sub}(D) \cong \mathscr{P}(D)$. □

Elements

Having described subsets categorially, we turn to actual elements of sets. A member x of set A, $(x \in A)$, can be identified with the "*singleton*" subset $\{x\}$ of A, and hence with the arrow $\{x\} \hookrightarrow A$, from the terminal object $\{x\}$ to A. In the converse direction, a function $f: 1 \to A$ in **Set** determines an element of A, viz the f-image of the only member of the terminal object 1. Thus; if category $\mathscr{C}$ has a terminal object 1, then an *element* of a $\mathscr{C}$-object a is defined to be a $\mathscr{C}$-arrow $x: 1 \to a$. (Note that $x: 1 \to a$ is always monic – Exercise 3.6.3.)

Of course the question is – does this notion in general reflect the behaviour of elements in **Set**? Must a non-initial $\mathscr{C}$-object have elements? Can two different $\mathscr{C}$-objects have the same elements? Can we characterise monic and epic arrows in terms of elements of their dom and cod? These matters will be taken up in due course.

Naming arrows

A function $f: A \to B$ from set A to set B is an *element* of the set B^A, i.e. $f \in B^A$, and so determines a function $\ulcorner f \urcorner : \{0\} \to B^A$, with $\ulcorner f \urcorner(0) = f$. Then if x is an element of A, we have a categorial "element" $\bar{x}: \{0\} \to A$, with $\bar{x}(0) = x$. Since $ev(\langle f, x \rangle) = f(x)$ we find that $ev \circ \langle \ulcorner f \urcorner, \bar{x} \rangle(0) = ev(\ulcorner f \urcorner(0), \bar{x}(0)) = f(x) = f(\bar{x}(0))$, and hence we have an equality of functions:

$$ev \circ \langle \ulcorner f \urcorner, \bar{x} \rangle = f \circ \bar{x}.$$

This situation can be lifted to any category $\mathscr{C}$ that has exponentials. Given a $\mathscr{C}$-arrow $f: a \to b$, let $f \circ pr_a : 1 \times a \to b$ be the composite $f \circ pr_a : 1 \times a \to a \to b$. Then the *name of* f is, by definition, the arrow $\ulcorner f \urcorner : 1 \to b^a$ that is the exponential adjoint of $f \circ pr_a$. Thus $\ulcorner f \urcorner$ is the unique arrow making

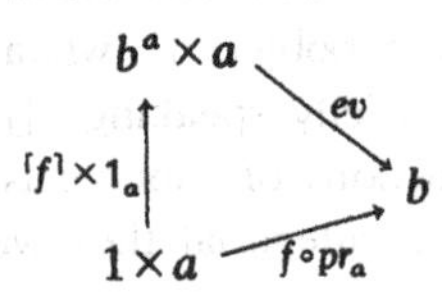

commute. Then we have that for any $\mathscr{C}$-element $x:1 \rightarrow a$ of a,

$$ev \circ \langle \ulcorner f \urcorner, x \rangle = f \circ x.$$

EXERCISE 2. Prove this last statement.

4.2. Classifying subobjects

In set theory, the powerset $\mathscr{P}(D)$ is often denoted 2^D. The later symbol, according to our earlier definition, in fact denotes the collection of all functions from D to $2 = \{0, 1\}$. The justification for the usage is that $\mathscr{P}(D) \cong 2^D$, i.e. there is a bijective correspondence between subsets of D and functions $D \rightarrow 2$. This isomorphism is established as follows: given a subset $A \subseteq D$, we define the function $\chi_A : D \rightarrow 2$, called the *characteristic function of* A, by the rule "for those elements of D in A, give output 1 and for those not in A, give output 0". i.e.

$$\chi_A(x) = \begin{cases} 1 & \text{if} \quad x \in A \\ 0 & \text{if} \quad x \notin A \end{cases}$$

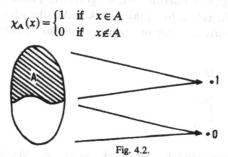

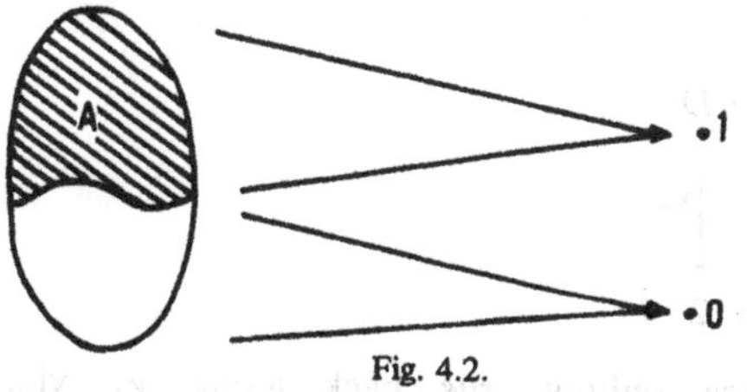

Fig. 4.2.

The assignment of χ_A to A is injective from $\mathscr{P}(D)$ to 2^D, i.e. if $\chi_A = \chi_B$ then $A = B$ (why?). It is also surjective, for if $f \in 2^D$, then $f = \chi_{A_f}$, where

$$A_f = \{x : x \in D \text{ and } f(x) = 1\}.$$

This correspondence between subset and characteristic function can be "captured" by a pullback diagram. The set A_f just defined is the inverse image under f of the subset $\{1\}$ of $\{0, 1\}$, i.e.

$$A_f = f^{-1}(\{1\}),$$

and so according to §3.13

$$\begin{array}{ccc} A_f & \hookrightarrow & D \\ {\scriptstyle !}\downarrow & & \downarrow{\scriptstyle f} \\ \{1\} & \hookrightarrow & 2 \end{array}$$

is a pullback square, i.e. A_f arises by pulling back $\{1\} \hookrightarrow 2$ along f. We are going to modify this picture slightly. The bottom arrow, which outputs the element 1 of $\{0, 1\}$ is replaced by the function from $1 = \{0\}$ to $2 = \{0, 1\}$ that outputs 1. We give this function the name *true*, for reasons that will emerge in Chapter 6. It has the rule; *true* $(0) = 1$. Then the inner square of

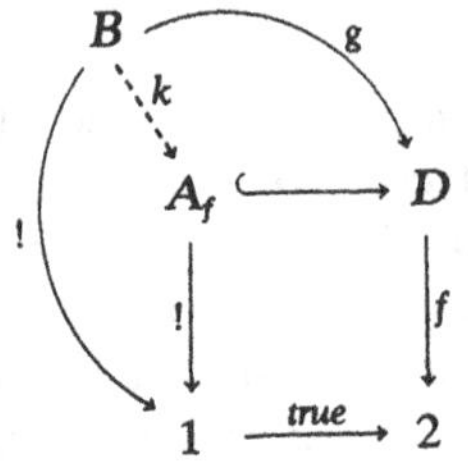

is a pullback. To see this, suppose the "outer square" commutes for some g. Then if $b \in B$, $f(g(b)) = true(!(b)) = 1$, so $g(b) \in A_f$. Hence $k: B \to A_f$ can be defined by the rule $k(b) = g(b)$. This k makes the whole diagram commute, and is clearly the only one that could do so. It follows that if $A \subseteq D$, then

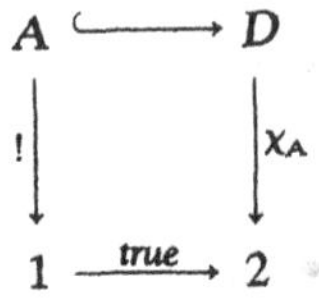

is a pullback, since pulling *true* back along χ_A yields the set $\{x: \chi_A(x) = 1\}$, which is just A. But more than this follows – χ_A can be identified as the *one and only* function from D to 2 that makes the above diagram a pullback, i.e. the only function along which *true* pulls back to yield A. If, for some f, the inner square of

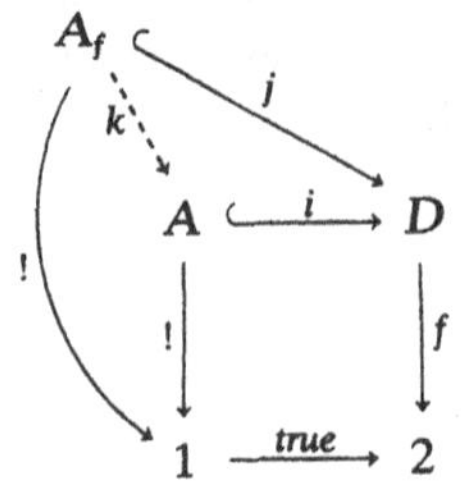

is a pullback, then for $x \in A$, $f(x) = 1$, so $x \in A_f$. Hence $A \subseteq A_f$. But the outer square commutes – indeed it is a pullback as we saw above – and so the unique k exists with $i \circ k = j$. Since i and j are inclusions, k must be as well. Thus $A_f \subseteq A$, and altogether $A = A_f$. But f is the characteristic function of A_f, and so, $f = \chi_A$.

So the set 2 together with the function $true : 1 \to 2$ play a special role in the transfer from subset to characteristic function, a role that has been cast in the language of categories, in such a way as to lead to an abstract definition:

DEFINITION. If $\mathscr{C}$ is a category with a terminal object 1, then a *subobject classifier* for $\mathscr{C}$ is a $\mathscr{C}$-object Ω together with a $\mathscr{C}$-arrow $true : 1 \to \Omega$ that satisfies the following axiom.

Ω-AXIOM. *For each monic $f : a \rightarrowtail d$ there is one and only one $\mathscr{C}$-arrow $\chi_f : d \to \Omega$ such that*

$$\begin{array}{ccc} a & \overset{f}{\rightarrowtail} & d \\ {\scriptstyle !}\downarrow & & \downarrow{\scriptstyle \chi_f} \\ 1 & \underset{true}{\longrightarrow} & \Omega \end{array}$$

is a pullback square.

The arrow χ_f is called the *characteristic arrow*, or the *character*, of the monic f (subobject of d). The arrow *true* will often be denoted by the letter "$\top$".

A *subobject classifier*, when it exists in a category, *is unique up to isomorphism*. If $\top : 1 \to \Omega$ and $\top' : 1 \to \Omega'$ are both subobject classifiers we have the diagram

$$\begin{array}{ccc} 1 & \overset{\top}{\rightarrowtail} & \Omega \\ \downarrow & & \downarrow{\scriptstyle \chi'_\top} \\ 1 & \overset{\top'}{\rightarrowtail} & \Omega' \\ \downarrow & & \downarrow{\scriptstyle \chi_{\top'}} \\ 1 & \overset{\top}{\rightarrowtail} & \Omega \end{array}$$

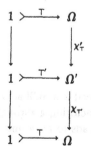

The top square is the pullback that gives the character $\chi'_\top$ of $\top$ using $\top'$ as classifier (remember any arrow with dom = 1 is monic). The bottom square is the pullback that gives the character of $\top'$, when $\top$ is used as classifier.

Hence by the PBL (§3.13, Example 8) the outer rectangle

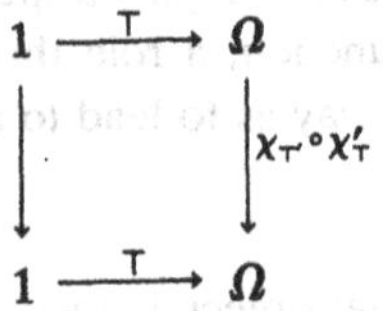

is a pullback. But by the Ω-axiom there is only one arrow $\Omega \rightarrow \Omega$ making this square a pullback, and 1_Ω would do that job (why?) Thus $\chi_{\top'} \circ \chi'_{\top} = 1_\Omega$. Interchanging $\top$ and $\top'$ in this argument gives

$$\chi'_{\top} \circ \chi_{\top'} = 1_{\Omega'},$$

and so $\chi_{\top'}: \Omega' \cong \Omega$.

Since $\top' = \chi'_{\top} \circ \top$ we have that any two subobject classifiers may be obtained from each other by composing with an iso arrow between their codomains.

The assignment of χ_f to f establishes a one-one correspondence between subobjects of an object d, and arrows $d \rightarrow \Omega$, as shown by:

THEOREM. *For $f: a \rightarrowtail d$ and $g: b \rightarrowtail d$,*

$$f \simeq g \quad \text{iff} \quad \chi_f = \chi_g.$$

PROOF. Suppose first that $\chi_f = \chi_g$. Consider

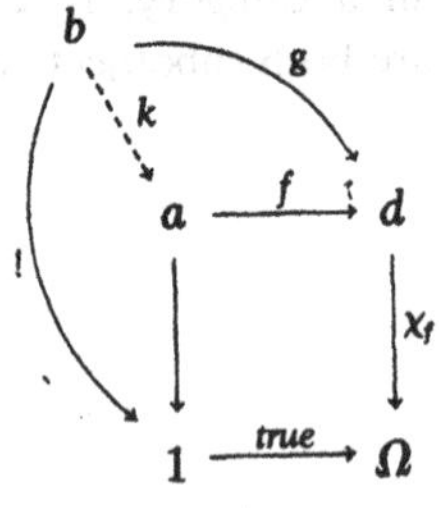

Since $\chi_f = \chi_g$, the outer square commutes (indeed is a pullback) and so as the inner square is a pullback there exists k factoring g through f, hence $g \subseteq f$. Interchanging f and g on the diagram leads to $f \subseteq g$ and altogether $f \simeq g$.

Conversely if $f \simeq g$, then the arrow k in the above diagram does exist and is iso with an inverse $k^{-1}: a \cong b$. Using this one can show that the outer square is a pullback, which can only be so if χ_f is the unique character of g, $\chi_f = \chi_g$.   □

Thus the assignment of χ_f to f (more exactly to $[f]$) injects Sub(d) into $\mathscr{C}(d, \Omega)$. But given any $h : d \to \Omega$, if we pull *true* back along h,

$$
\begin{array}{ccc}
a & \xrightarrow{\ f\ } & d \\
\downarrow & & \downarrow h \\
1 & \xrightarrow{\ true\ } & \Omega
\end{array}
$$

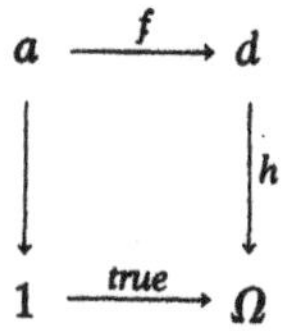

the resulting arrow f will be monic (since *true* is monic and the pullback of a monic is always itself monic – Exercise, §3.13). Hence h must be χ_f. So in a category where these constructions are possible we get

$$\text{Sub}(d) \cong \mathscr{C}(d, \Omega).$$

NOTATION. For any $\mathscr{C}$-object a, the composite *true* $\circ !_a$, of arrows $! : a \to 1$ and *true*, will be denoted *true*$_a$, or T_a, or sometimes *true*!

$$
\begin{array}{ccc}
a & \longrightarrow & 1 \\
 & \searrow{\scriptstyle true_a} & \downarrow{\scriptstyle true} \\
 & & \Omega
\end{array}
$$

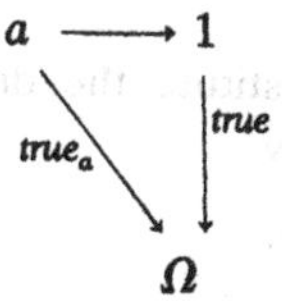

EXERCISE 1. Show that the character of *true* : $1 \rightarrowtail \Omega$ is 1_Ω

$$
\begin{array}{ccc}
1 & \xrightarrow{\ true\ } & \Omega \\
\downarrow & & \downarrow 1_\Omega \\
1 & \xrightarrow{\ true\ } & \Omega
\end{array}
$$

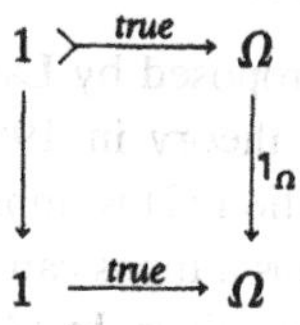

i.e. $\chi_{true} = 1_\Omega$.

EXERCISE 2. Show that $\chi_{1_\Omega} = true_\Omega = true \circ !_\Omega$.

$$
\begin{array}{ccc}
\Omega & \xrightarrow{\ 1_\Omega\ } & \Omega \\
\downarrow{\scriptstyle !_\Omega} & & \downarrow{\scriptstyle true \circ !_\Omega} \\
1 & \xrightarrow{\ true\ } & \Omega
\end{array}
$$

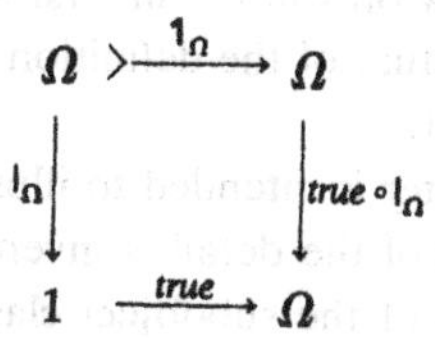

EXERCISE 3. Show that for any $f: a \to b$,

$$\begin{array}{ccc} a & \xrightarrow{f} & b \\ & {\scriptstyle true_a}\searrow \quad \swarrow{\scriptstyle true_b} & \\ & \Omega & \end{array}$$

$true_b \circ f = true_a$. □

4.3. Definition of topos

DEFINITION. An *elementary topos* is a category $\mathscr{E}$ such that

(1) $\mathscr{E}$ is finitely complete,
(2) $\mathscr{E}$ is finitely co-complete,
(3) $\mathscr{E}$ has exponentiation,
(4) $\mathscr{E}$ has a subobject classifier.

As observed in Chapter 3, (1) and (3) constitute the definition of "Cartesian closed", while (1) can be replaced by

(1′) $\mathscr{E}$ has a terminal object and pullbacks,

and dually (2) replaced by

(2′) $\mathscr{E}$ has an initial object 0, and pushouts.

The definition just given is the one originally proposed by Lawvere and Tierney, in terms of which they started topos theory in 1969. Subsequently C. Juul Mikkelsen discovered that condition (2) is implied by the combination of (1), (3) and (4) (cf. Paré [74]). Thus a topos can be defined as a Cartesian closed category with a subobject classifier. In §4.7 we shall consider a different definition, based on a categorial characterisation of power sets.

The word "elementary" (which from now on will be understood) has a special technical meaning to do with the nature of the definition of topos. This usage will be explained in Chapter 11.

The list of topoi that follows in this chapter is intended to illustrate the generality of the concept. By no means all of the detail is given – for the most part we concentrate on the structure of the subobject classifier.

4.4. First examples

EXAMPLE 1. **Set** is a topos – the prime example and the motivation for the concept in the first place.

EXAMPLE 2. **Finset** is a topos, with limits, exponentials, and $\top : 1 \to \Omega$ exactly as in **Set.**

EXAMPLE 3. **Finord** is a topos. Every finite set is isomorphic to some finite ordinal ($A \cong n$ if A has n elements). Hence all categorial constructions in **Finset** "transfer" into **Finord** (as we have already observed for product, exponentials). The subobject classifier in **Finord** is the same function $true : \{0\} \to \{0, 1\}$ as in **Finset** and **Set**.

EXAMPLE 4. **Set**^{2}, the category of pairs of sets is a topos. All constructions are obtained by "doubling up" the corresponding constructions in **Set** (cf. Example 10, §2.5).

A terminal object is a pair $\langle\{0\}, \{0\}\rangle$ of singleton sets. Given two arrows $\langle f, g\rangle : \langle A, B\rangle \to \langle E, F\rangle$, $\langle h, k\rangle : \langle C, D\rangle \to \langle E, F\rangle$ with common codomain in **Set**^{2}, form the pullbacks

$$\begin{array}{ccc} P & \xrightarrow{j} & C \\ {\scriptstyle i}\downarrow & & \downarrow{\scriptstyle h} \\ A & \xrightarrow{f} & E \end{array} \qquad \begin{array}{ccc} Q & \xrightarrow{v} & D \\ {\scriptstyle u}\downarrow & & \downarrow{\scriptstyle k} \\ B & \xrightarrow[g]{} & F \end{array}$$

in **Set**. Then

$$\begin{array}{ccc} \langle P, Q\rangle & \xrightarrow{\langle j, v\rangle} & \langle C, D\rangle \\ {\scriptstyle \langle i, u\rangle}\downarrow & & \downarrow{\scriptstyle \langle h, k\rangle} \\ \langle A, B\rangle & \xrightarrow[\langle f, g\rangle]{} & \langle E, F\rangle \end{array}$$

will be a pullback in **Set**^{2}.

The exponential has

$$\langle C, D\rangle^{\langle A, B\rangle} = \langle C^A, D^B\rangle$$

with evaluation arrow from

$$\langle C, D\rangle^{\langle A, B\rangle} \times \langle A, B\rangle = \langle C^A \times A, D^B \times B\rangle$$

to $\langle C, D\rangle$ as the pair $\langle e, f\rangle$ where $e: C^A \times A \to C$ and $f: D^B \times B \to D$ are the appropriate evaluation arrows in **Set**.

The subobject classifier is $\langle \top, \top\rangle: \langle\{0\}, \{0\}\rangle \to \langle 2, 2\rangle$. The category **Set** plays no special role here. If $\mathscr{E}_1$ and $\mathscr{E}_2$ are any topoi, then the product category $\mathscr{E}_1 \times \mathscr{E}_2$ is a topos.

EXAMPLE 5. $\mathbf{Set}^{\to}$, the category of functions. The terminal object is the identity function $\mathrm{id}_{\{0\}}$ from $\{0\}$ to $\{0\}$.

Pullback: Consider the "cube"

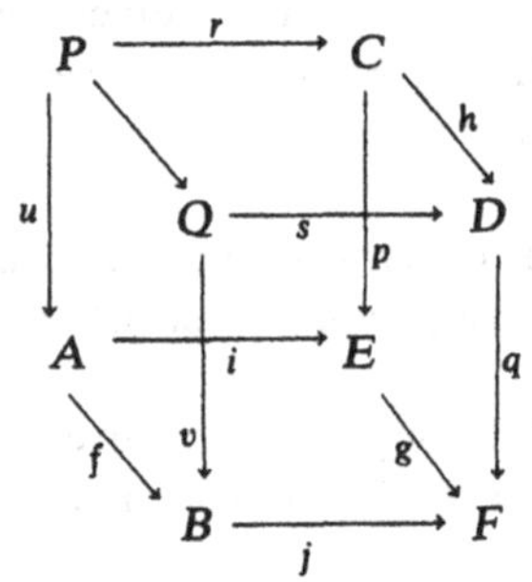

f, g, h are given as $\mathbf{Set}^{\to}$-objects with $\langle i, j\rangle$ an arrow from f to g, $\langle p, q\rangle$ an arrow from h to g. The rest of the diagram obtains by forming the pullbacks

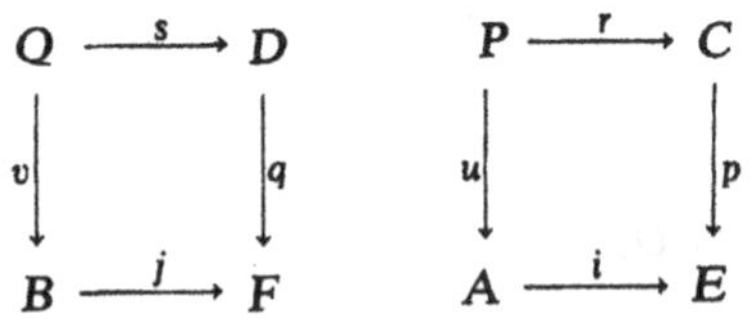

in **Set**. The arrow k exists by the universal property of the pullback of j and q. Then in $\mathbf{Set}^{\to}$ the arrows $\langle u, v\rangle$ and $\langle r, s\rangle$ are the pullbacks of $\langle i, j\rangle$ and $\langle p, q\rangle$.

Classifier: If $f: A \to B$ is a subobject of $g: C \to D$ in $\mathbf{Set}^{\to}$ then there is a commutative **Set** diagram

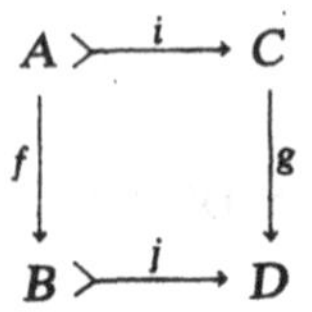

We will take the monics to be actual inclusions, so that $A \subseteq C$, $B \subseteq D$ and f is the restriction of g, i.e. $f(x) = g(x)$ for $x \in A$. The picture is

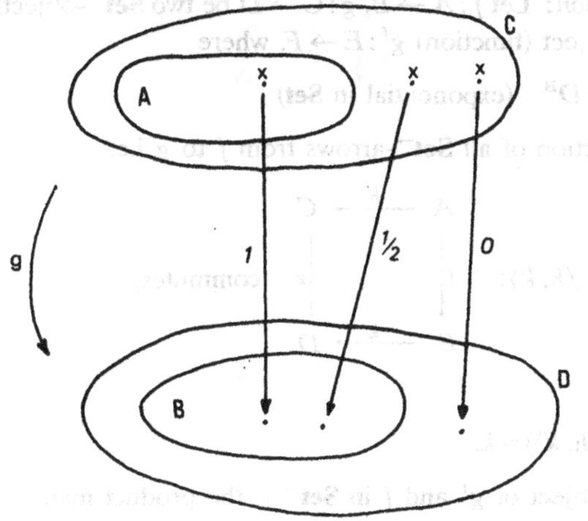

Fig. 4.3.

An element x of C can be classified now in three ways. Either

(i) $x \in A$, or
(ii) $x \notin A$, but $g(x) \in B$, or
(iii) $x \notin A$, and $g(x) \notin B$.

So we introduce a 3-element set $\{0, \frac{1}{2}, 1\}$ and define $\Psi : C \to \{0, \frac{1}{2}, 1\}$ by

$$\psi(x) = \begin{cases} 1 & \text{if} \quad \text{(i) holds} \\ \frac{1}{2} & \text{if} \quad \text{(ii) holds} \\ 0 & \text{if} \quad \text{(iii) holds} \end{cases}$$

We can now form the cube

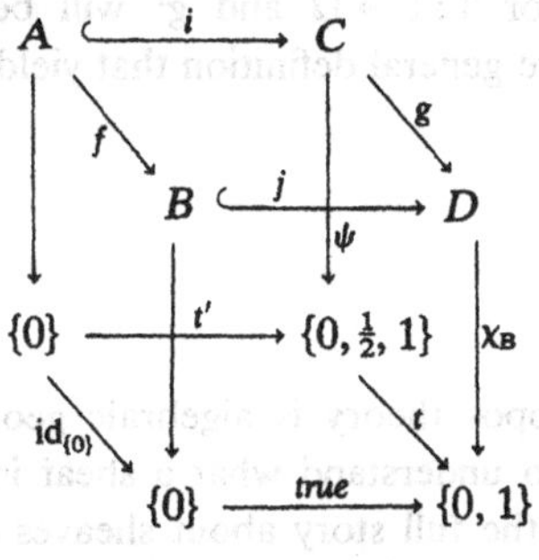

where $true(0) = t'(0) = 1$, $t : \{0, \frac{1}{2}, 1\} \to \{0, 1\}$ has $t(0) = 0$, and $t(1) = t(\frac{1}{2}) = 1$. χ_B is the characteristic function of B.

The base of the cube displays the subobject classifier $\top : 1 \to \Omega$ for $\mathbf{Set}^{\to}$. $\top$ is the pair $\langle t', true \rangle$ from $1 = \mathrm{id}_{\{0\}}$ to $\Omega = t : \{0, \frac{1}{2}, 1\} \to \{0, 1\}$.

The front and back faces of the cube are each pullbacks in **Set**. The whole diagram exhibits $\langle \psi, \chi_B \rangle$ as the character in $\mathbf{Set}^{\rightarrow}$ of the monic $\langle i, j \rangle$.

Exponentiation: Let $f: A \to B$, $g: C \to D$ be two $\mathbf{Set}^{\rightarrow}$-objects. Then g^f is the $\mathbf{Set}^{\rightarrow}$-object (function) $g^f: E \to F$, where

$$F = D^B \quad \text{(exponential in } \mathbf{Set}\text{)}$$

E is the collection of all $\mathbf{Set}^{\rightarrow}$-arrows from f to g i.e.

$$E = \left\{\langle h, k \rangle : \begin{array}{ccc} A & \xrightarrow{h} & C \\ {\scriptstyle f}\downarrow & & \downarrow{\scriptstyle g} \\ B & \xrightarrow{k} & D \end{array} \text{ commutes}\right\}$$

and

$$g^f(\langle h, k \rangle) = k.$$

The product object of g^f and f in $\mathbf{Set}^{\rightarrow}$ is the product map

$$g^f \times f : E \times A \to F \times B \qquad \text{(cf. §3.8)}$$

and the evaluation arrow from $g^f \times f$ to g is the pair $\langle u, v \rangle$

$$\begin{array}{ccc} E \times A & \xrightarrow{u} & C \\ {\scriptstyle g^f \times f}\downarrow & & \downarrow{\scriptstyle g} \\ F \times B & \xrightarrow{v} & D \end{array}$$

where v is the usual evaluation arrow in **Set**, and u takes input $\langle \langle h, k \rangle, x \rangle$ to output $h(x)$.

The constructions just given for $\top: 1 \to \Omega$ and g^f will be seen in Chapter 9 to be instances of a more general definition that yields a whole family of topoi.

4.5. Bundles and sheaves

One of the primary sources of topos theory is algebraic geometry, in particular the study of *sheaves*. To understand what a sheaf is requires some knowledge of topology and the full story about sheaves and their relation to topoi would take us beyond our present scope. The idea is closely tied up with models of intuitionistic logic, but is much more general than that. Indeed, sheaf theory constitutes a whole conceptual framework and language of its own, and to ignore it completely, even at this stage, would be to distort the overall significance and point of view of topos theory.

For the benefit of the reader unfamiliar with topology we shall delay its introduction and first consider the underlying set-theoretic structure of the sheaf concept, to be called a *bundle.*

Let us assume we have a collection $\mathscr{A}$ of sets, no two of which have any elements in common. That is, any two members of $\mathscr{A}$ are sets that are disjoint. We need a convenient notation for referring to these sets so we presume we have a set I of *labels,* or *indices,* for them. For each index $i \in I$, there is a set A_i that belongs to our collection, and each member of $\mathscr{A}$ is labelled in this way, so we write $\mathscr{A}$ as the collection of all these A_i's,

$$\mathscr{A} = \{A_i : i \in I\}.$$

The fact that the members of $\mathscr{A}$ are pairwise disjoint is expressed by saying that for *distinct* indices $i, j \in I$

$$A_i \cap A_j = \emptyset$$

We visualise the A_i's as "sitting over" the index set I thus:

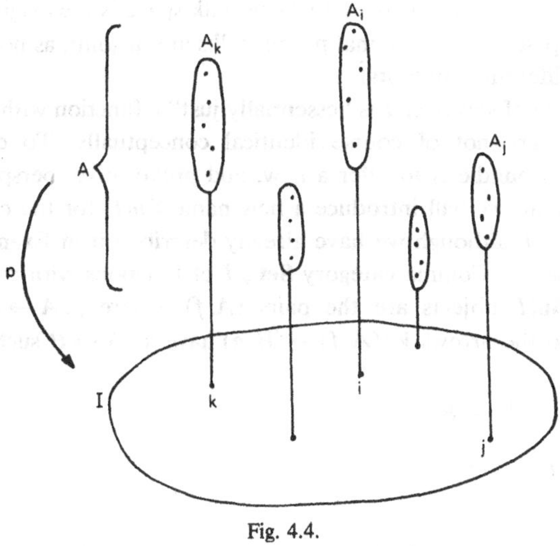

Fig. 4.4.

If we let A be the union of all the A_i's, i.e.

$$A = \{x : \text{for some } i, x \in A_i\}$$

then there is an obvious map $p : A \to I$. If $x \in A$ then there is exactly one A_i such that $x \in A_i$, by the disjointness condition. We put $p(x) = i$. Thus all the members of A_i get mapped to i, all the members of A_j to j, etc. We can then re-capture A_i as the inverse image under p of $\{i\}$, for

$$p^{-1}(\{i\}) = \{x : p(x) = i\} = A_i.$$

The set A_i is called the *stalk*, or *fibre* over i. The members of A_i are called the *germs* at i. The whole structure is called a *bundle* of sets over the *base* space I. The set A is called the *stalk space* (l'espace étalé) of the bundle. The reason for the botanical terminology is evident – what we have is a bundle of stalks, each with its own head of germs (think of a bunch of asparagus spears).

This construction looks rather special, but it is to be found whenever there are functions. We have just seen that a bundle has an associated map p from its stalk space to the base. (If in fact every stalk is nonempty then p will be surjective, but in general we will allow the possibility that $A_i = \emptyset$). Conversely, if $p: A \to I$ is an arbitrary function from some set A to I, then we can *define* A_i to be $p^{-1}(\{i\})$, for each $i \in I$, and *define*

$$\mathcal{A} = \{p^{-1}(\{i\}) : i \in I\} = \{A_i : i \in I\}.$$

Then $\mathcal{A}$ is a bundle of sets over I whose stalk space is the original A, and induced map $A \to I$ the original p (the stalks are disjoint, as no $x \in A$ can have two different p-outputs).

So a bundle of sets over I is "essentially just" a function with codomain I. The two are not of course identical conceptually. To construe a function as a bundle is to offer a new, and provocative, perspective. To emphasise that, we will introduce a new name **Bn**(I) for the category of bundles over I, although we have already described it in Example 12 of Chapter 2 as the Comma category **Set** $\downarrow I$ of functions with codomain I. Thus the **Bn**(I)-objects are the pairs (A, f), where $f: A \to I$ is a set function and the arrows $k: (A, f) \to (B, g)$ have $k: A \to B$ such that

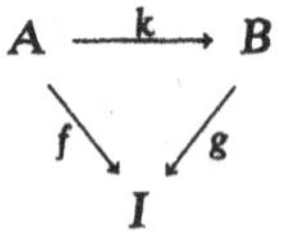

commutes, i.e. $g \circ k = f$. This means that if $f(x) = i$, for $x \in A$, then $g(k(x)) = i$, i.e. if $x \in A_i$, then $k(x) \in B_i$. Thus k maps germs at i in (A, f) to germs at i in (B, g).

Now a topos is to be thought of as a generalisation of the category **Set**. An object in a topos is a "generalised set". A "set" in the topos **Bn**(I) is a bundle of ordinary sets. Many categorial notions when applied to **Bn**(I)

prove to be bundles of the corresponding entities in **Set**, as we shall now see.

The terminal object 1 for **Bn**(I) is $\mathrm{id}_I : I \to I$, and for any bundle (A, f), the unique arrow $(A, f) \to (I, \mathrm{id}_I)$ is $f : A \to I$ itself (cf. §3.6). Now the stalk of id_I over i is $\mathrm{id}^{-1}(\{i\}) = \{i\}$, which is terminal in **Set**. Thus the **Bn**(I) terminal is a bundle of **Set**-terminals over I, and the unique arrow $f : (A, f) \to (I, \mathrm{id}_I)$ can be construed as a bundle

$$\{f_i : i \in I\}$$

of unique **Set**-arrows, where

$$f_i = \,! : f^{-1}(\{i\}) \to \{i\}.$$

Pullback: Given **Bn**(I)-arrows $k : \langle A, f\rangle \to \langle C, h\rangle$ and $l : \langle B, g\rangle \to \langle C, h\rangle$, so that

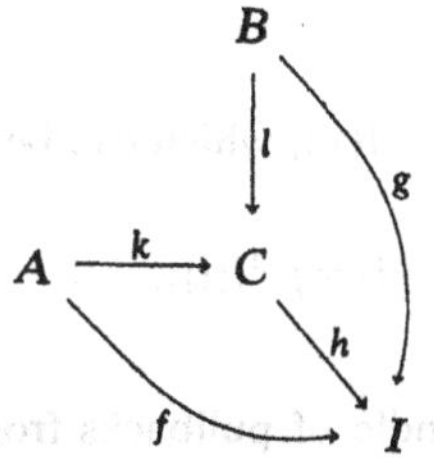

commutes, form the pullback

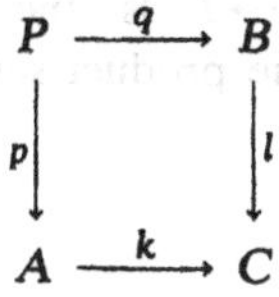

in **Set** of k and l. Then

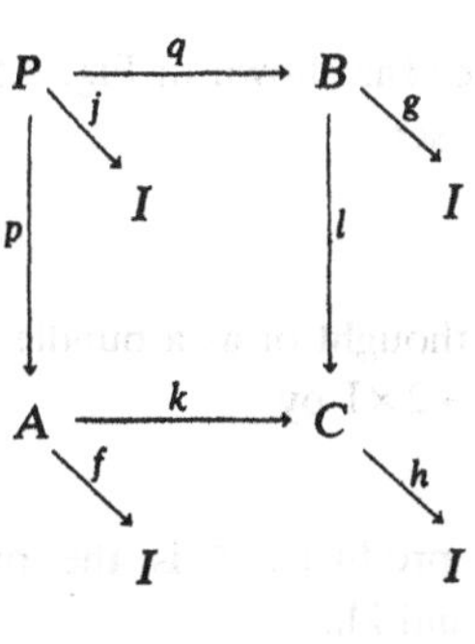

is a pullback of k and l in $\mathbf{Bn}(I)$, where $j = f \circ p = h \circ k \circ p = h \circ l \circ q = g \circ q$. The diagram is probably more usefully given as the commutative

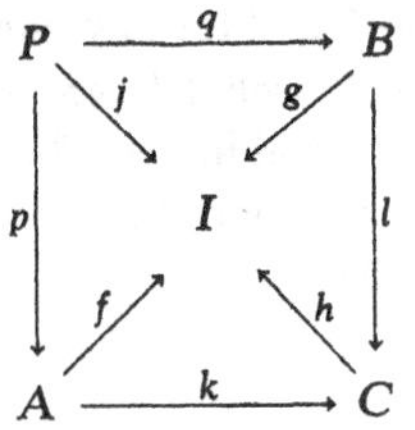

Now if A_i, B_i, C_i are the stalks over i for the bundles f, g, h, then the pullback of

$$\begin{array}{ccc} & & B_i \\ & & \downarrow{\scriptstyle l^*} \\ A_i & \xrightarrow{k^*} & C_i \end{array}$$

has domain $\{\langle x, y\rangle : x \in A_i,\ y \in B_i, \text{ and } k(x) = l(y)\}$ which can be seen to be the same as

$$\{\langle x, y\rangle : x \in A,\ y \in B \text{ and } j\langle x, y\rangle = i\} = j^{-1}(\{i\}),$$

which is the stalk over i of $j: P \to I$.

Thus the pullback object (P, j) is a bundle of pullbacks from **Set**.

Subobject classifier: The classifier for $\mathbf{Bn}(I)$ is a bundle of two-element sets, i.e. a bundle of **Set**-classifiers.

We define $\Omega = (2 \times I, p_I)$, where $p_I : 2 \times I \to I$ is the projection $p_I(\langle x, y\rangle) = y$ onto the "second factor". Now the product set $2 \times I$ is in fact the (disjoint) union of the sets

$$\{0\} \times I = \{\langle 0, i\rangle : i \in I\}$$

and

$$\{1\} \times I = \{\langle 1, i\rangle : i \in I\},$$

each isomorphic to I, and we visualise Ω as shown in Fig. 4.5. The stalk over a particular i is the two-element set

$$\Omega_i = \{\langle 0, i\rangle, \langle 1, i\rangle\} = 2 \times \{i\}.$$

The classifier arrow $\top : 1 \to \Omega$ can be thought of as a bundle of copies of the set function *true*. We define $\top : I \to 2 \times I$ by

$$\top(i) = \langle 1, i\rangle.$$

In terms of the limit approach to products, $\top$ is the product map $\langle true!, \mathrm{id}_I\rangle$ of $true \circ ! : I \to \{0\} \to \{0, 1\}$ and id_I.

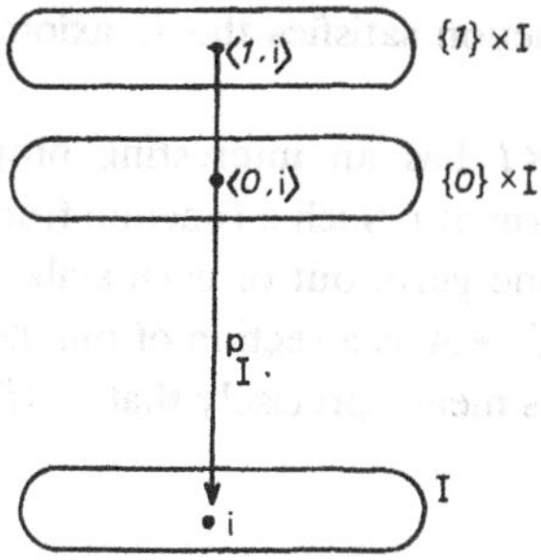

Fig. 4.5.

To see how ⊤ classifies subobjects we take a monic $k:(A, f) \rightarrowtail (B, g)$ in **Bn**(I), and in fact suppose that k is an inclusion, i.e. $A \subseteq B$ and $f(x) = g(x)$, all $x \in A$. We wish to define the character $\chi_k : \langle B, g\rangle \to \Omega = (2\times I, p_I)$ so that

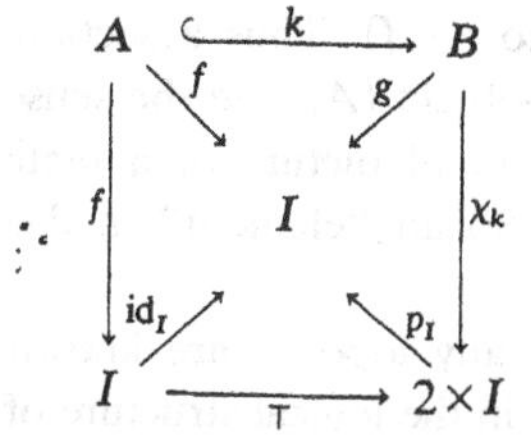

commutes and gives a pullback in **Bn**(I). Now any $x \in B$ is classified according to whether $x \in A$ or $x \notin A$.

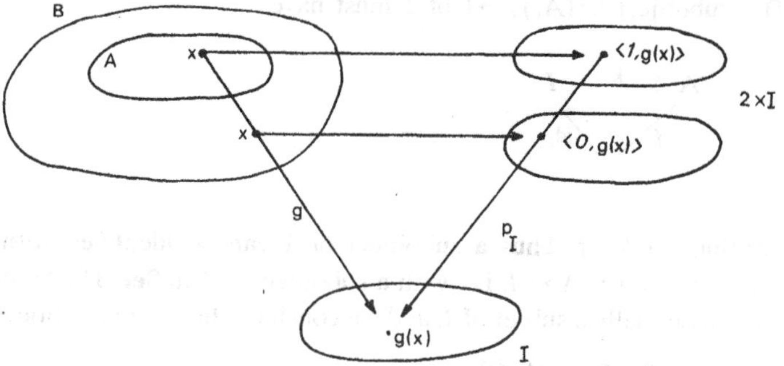

Fig. 4.6.

We make χ_k assign as "1" or "0" accordingly, and also make these choices in the right stalks, so that $p_I \circ \chi_k = g$. Formally, $\chi_k : B \to 2\times I$ is the product map $\langle \chi_A, g\rangle : B \to 2\times I$, where $\chi_A : B \to 2$ is the usual characteristic function of A, i.e.

$$\chi_k(x) = \begin{cases} \langle 1, g(x)\rangle & \text{if} \quad x \in A \\ \langle 0, g(x)\rangle & \text{if} \quad x \notin A. \end{cases}$$

EXERCISE 1. Verify that this construction satisfies the Ω-axiom. □

Sections: The function $\top : I \to 2 \times I$ has an interesting property – for input i the output $\top(i) = \langle 1, i \rangle$ is a germ at i. Such a function from the base set I to the stalk space that picks one germ out of each stalk is called a *section* of the bundle. In general $s : I \to A$ is a section of bundle $f : A \to I$ if $s(i) \in A_i = f^{-1}(\{i\})$, for all $i \in I$. This means precisely that $f(s(i)) = i$, all i, and hence that

$$\begin{array}{ccc} I & \xrightarrow{s} & A \\ & {\scriptstyle \mathrm{id}_I} \searrow \quad \swarrow {\scriptstyle f} & \\ & I & \end{array}$$

commutes. So another way of looking at a section is to say that it is a $\mathbf{Bn}(I)$-arrow from the terminal (I, id_I) to (A, f). Thus a section of the bundle (A, f) is an *element* of the $\mathbf{Bn}(I)$-object (A, f) in the sense of the definition at the end of §4.1. But our initial picture of a section is a bundle of germs, one from each stalk. So an "element" in $\mathbf{Bn}(I)$ is a bundle of ordinary elements.

Elements of Ω, i.e. arrows $1 \to \Omega$, in any topos $\mathscr{E}$ are known as the *truth-values* of $\mathscr{E}$, and have a special role in the logical structure of $\mathscr{E}$ (See Chapter 6). We know (§4.2) that there is a bijective correspondence $\mathrm{Sub}(1) \cong \mathscr{E}(1, \Omega)$ between elements of Ω and subobjects of 1. Now in $\mathbf{Bn}(I)$ a subobject $k : (A, f) \rightarrowtail 1$ of 1 must have

$$\begin{array}{ccc} A & \overset{k}{\rightarrowtail} & I \\ & {\scriptstyle f} \searrow \quad \swarrow {\scriptstyle \mathrm{id}_I} & \\ & I & \end{array}$$

commuting, so $k = f$. Thus a subobject of 1 can be identified with an injective function $f : A \rightarrowtail I$, i.e. with a subobject of I in $\mathbf{Set}$. The latter of course is essentially a subset of I, and we conclude that there is a bijection

$$\mathscr{P}(I) \cong \mathbf{Bn}(I)(1, \Omega)$$

i.e. we may identify truth-values (elements of Ω) in $\mathbf{Bn}(I)$ with subsets of I. It is instructive to spell this out fully:

Given $A \subseteq I$, let $S_A : I \to 2 \times I$ be the product map $\langle \chi_A, \mathrm{id}_I \rangle$, i.e.

$$S_A(i) = \begin{cases} \langle 1, i \rangle & \text{if } i \in A \\ \langle 0, i \rangle & \text{if } i \notin A \end{cases}$$

then S_A is a section of Ω, whose image is shown shaded in the picture.

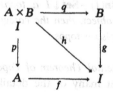

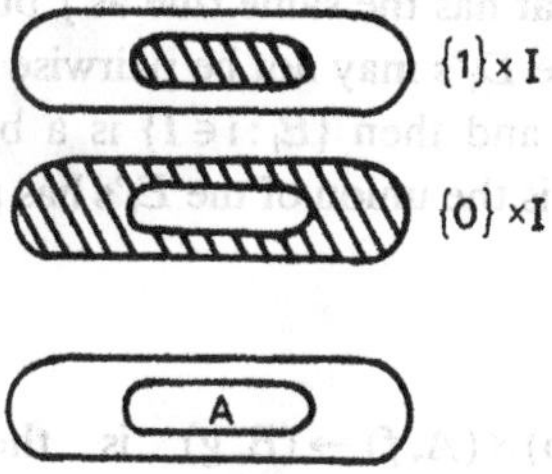

Fig. 4.7.

The assignment of $S_A : 1 \to \Omega$ to A is injective (exercise). Moreover if $S : 1 \to \Omega$ is any section, and $A = \{i : S(i) = \langle 1, i\rangle\}$, then $S = S_A$, so the assignment is also surjective.

Note that whereas **Set** has two truth values, $\mathcal{P}(I)$ may well be infinite (it certainly will be if I is infinite).

EXERCISE 2. What are the truth-values in $\mathbf{Set}^2$ and in $\mathbf{Set}^{\to}$? □

Products. Let (A, f) and (B, g) be bundles over I and form the pullback

$$\begin{array}{ccc} A\times_I B & \xrightarrow{q} & B \\ {\scriptstyle p}\downarrow & \searrow^{h} & \downarrow{\scriptstyle g} \\ A & \xrightarrow[f]{} & I \end{array}$$

Then $(A \times_I B, h)$ is the product of (A, f) and (B, g) in $\mathbf{Bn}(I)$, where $h = f \circ p = g \circ q$, and has projection arrows p and q. Note that the stalk (fibre) over i is

$$\{\langle x, y\rangle : f(x) = g(y) = i\} = A_i \times B_i,$$

the product of the fibres over i in (A, f) and (B, g). Hence the name "fibred product" that is sometimes used for "pullback".

Exponentials. Given bundles $f : A \to I$ and $g : B \to I$ we form their exponential as a bundle of the exponentials $B_i^{A_i}$ of the stalks of A and B. More precisely let D_i be the collection of functions $k : A_i \to B$ such that

$$\begin{array}{ccc} A_i & \xrightarrow{k} & B \\ & {\scriptstyle f^*}\searrow \quad \swarrow{\scriptstyle g} & \\ & I & \end{array}$$

commutes and so k carries A_i into the stalk B_i of g over i (where, as previously, f^* denotes a function that has the same rule as f but may vary as to domain or codomain). Now the D_i's may not be pairwise disjoint, so we define $E_i = \{i\} \times D_i$, for each i, and then $\{E_i : i \in I\}$ is a bundle. The induced function $p : E \to I$ where E is the union of the E_i's has $p(\langle i, k \rangle) = i$. (E, p) is the exponential

$$(B, g)^{(A,f)}.$$

The evaluation arrow $ev : (E, p) \times (A, f) \to (B, g)$ is the function $ev : E \times_I A \to B$, where

$$ev(\langle\langle i, k\rangle, x\rangle) = k(x).$$

The reader who has the patience to wade through the details of checking that this construction is well defined and satisfies the definition of exponentiation will no doubt get his reward in heaven. For the present he will perhaps appreciate the advantages of the categorial viewpoint, wherein all we need to say about the exponential, to know what it is, is that it satisfies the universal property described in §3.16. (We shall return to this example in Chapter 15).

FUNDAMENTAL THEOREM. *Not only is* $\mathbf{Bn}(I) = \mathbf{Set}\downarrow I$ *a topos, but more generally if* $\mathscr{E}$ *is any topos and* a *an* $\mathscr{E}$*-object, then the category* $\mathscr{E}\downarrow a$ *of* $\mathscr{E}$*-arrows over* a (§2.5, Example 12) *is also a topos.*

This fact has been called the *Fundamental Theorem of Topoi* by Freyd [72]. The reader can probably sort out many of the details from the above, e.g. if $\top : 1 \to \Omega$ is the classifier in $\mathscr{E}$, then in $\mathscr{E}\downarrow a$ it is $\langle \top_a, 1_a \rangle$, i.e.

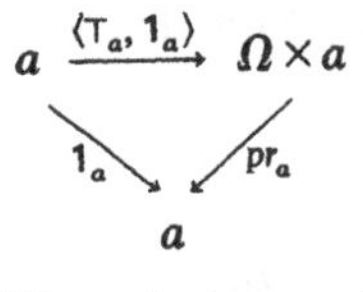

The definition of exponentials in $\mathscr{E}\downarrow a$ would carry us too far afield at present. It requires the development of a categorial theory of "partial functions" and their classification, which will be considered in Chapters 11 and 15.

Sheaves

A *sheaf* is a bundle with some additional topological structure. Let I be a topological space, with Θ its collection of open sets. A *sheaf* over I is a pair (A, p) where A is a topological space and $p: A \to I$ is a continuous map that is a local homeomorphism. This means that each point $x \in A$ has an open neighbourhood U in A that is mapped homeomorphically by p onto $p(U) = \{p(y): y \in U\}$, and the latter is open in I. The category **Top**(I) of sheaves over I has such pairs (A, p) as objects, and as arrows $k: (A, p) \to (B, q)$ the *continuous* maps $k: A \to B$ such that

$$\begin{array}{ccc} A & \xrightarrow{k} & B \\ & {\scriptstyle p}\searrow \quad \swarrow{\scriptstyle q} & \\ & I & \end{array}$$

commutes. Such a k is in fact an open map (as is a local homeomorphism) and in particular $\text{Im } k = k(A)$ will be an open subset of B.

Top(I) is a topos, known as a *spatial topos*. The terminal object is $\text{id}_I: I \to I$. The subobject classifier is the *sheaf of germs of open sets in I*. Its construction illustrates a common method of building a bundle over I. There will be some ambient set X and each point $i \in I$ will determine an equivalence relation $\sim_i$ on X. The stalk over i will then be defined as the quotient set $X/\sim_i$ of equivalence classes of X under $\sim_i$.

In the present case X is the collection Θ of open sets in I. At $i \in I$, we define $\sim_i$ by declaring, for $U, V \in \Theta$

$$U \sim_i V \quad \text{iff there is some open set } W \text{ such that } i \in W \text{ and } U \cap W = V \cap W$$

Then $\sim_i$ is an equivalence relation. The intuitive idea is that $U \sim_i V$ when the points in U that are close to i are the same as those that are in V and close to i, i.e. "locally" around i, U and V look the same, i.e. the statement "$U = V$" is "locally true" at i.

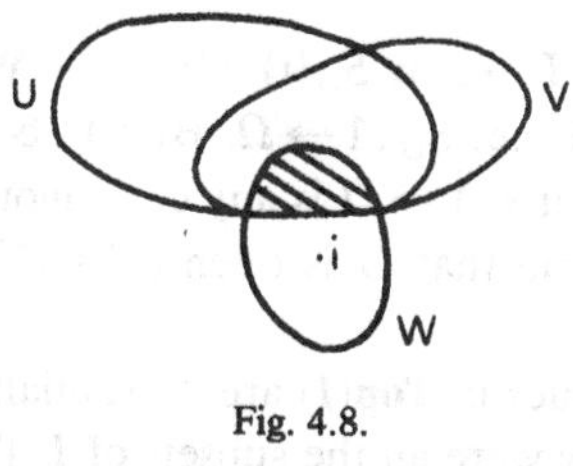

Fig. 4.8.

The equivalence class

$$[U]_i = \{V: U \sim_i V\}$$

is called the *germ of U at i*. Intuitively it "represents" the collection of points in U that are "close" to i.

We then take as the stalk over i,

$$\Omega_i = \{\langle i, [U]_i\rangle : U \text{ open in } I\}.$$

Then Ω is the corresponding function $p : \hat{I} \to I$, where $\hat{I}$ is the union of the stalks Ω_i, and p gives output i for inputs from Ω_i. The topology on $\hat{I}$ has as base all sets of the form

$$[U, V] = \{\langle i, [U]_i\rangle : i \in V\}$$

where V is open and $U \subseteq V$. This makes p a local homeomorphism, and also makes each stalk a discrete space under the relative topology.

If we denote by Θ_i the collection of open neighbourhoods of i then we have the following facts about germs of open sets:

(i) $[U]_i = [I]_i$ iff $i \in U$

(ii) $[I]_i = \Theta_i$

(iii) $[U]_i = [\emptyset]_i$ iff i is separated from U (i.e. there exists $V \in \Theta_i$ such that $U \cap V = \emptyset$)

[The reader familiar with lattices may care to note that the open sets in I form a distributive lattice $(\Theta, \cap, \cup)$ in which Θ_i is a (prime) filter. The stalk Ω_i is essentially the quotient lattice Θ/Θ_i, i.e. $\sim_i$ is the standard definition of the lattice congruence determined by Θ_i.]

Before examining Ω as a subobject classifier we will look at truth-values $s : 1 \to \Omega$. Such an arrow is a *continuous* section of Ω, generally called a *global* section of the sheaf. (We may also consider *local sections* $s : U \to \hat{I}$ of $\hat{I}$ defined on (open) subsets U of I).

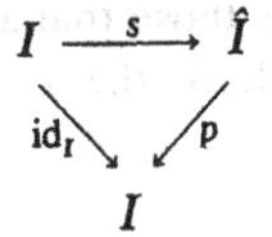

Now if U is open in I, define $S_U : I \to \hat{I}$ by $S_U(i) = \langle i, [U]_i\rangle$. We then find S_U is a continuous global section, i.e. $S_U : 1 \to \Omega$. By (i) above we note that $S_U(i) = \langle i, [I]_i\rangle$ iff $i \in U$. Then if $s : 1 \to \Omega$ is any continuous section of Ω and $U = \{i : s(i) = \langle i, [I]_i\rangle\}$ we find that U is open ($U = s^{-1}([I, I])$) and $S_U = s$.

We thus have that the truth values in **Top**(I) are "essentially" the *open* subsets of I, whereas in **Bn**(I) they were all the subsets of I. This will be a continuing theme. We shall later see other constructions that have a set-theoretic and a topological version, and find that the latter arise from the form by replacing "subset" by "open subset".

The arrow $\top : 1 \to \Omega$ is the continuous section $\top : I \to \hat{I}$ that has $\top(i) = \langle i, [I]_i \rangle$, all $i \in I$. Now if k is monic, where

$$\begin{array}{ccc} A & \overset{k}{\hookrightarrow} & B \\ & {\scriptstyle p}\searrow \quad \swarrow{\scriptstyle q} & \\ & I & \end{array}$$

commutes, and A is an open subset of B, we obtain the character $\chi_k : (B, q) \to \Omega$ as follows.

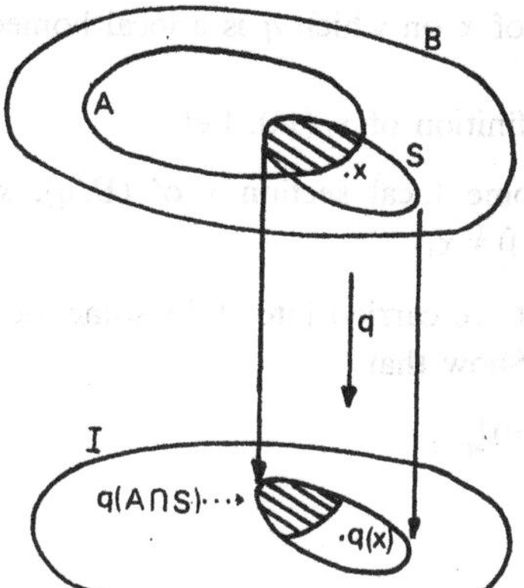

Fig. 4.9.

If $x \in B$, choose a neighbourhood S of x on which q is a local homeomorphism. Then $\chi_k : B \to \hat{I}$ takes x to the germ of $q(A \cap S)$ at $q(x)$, i.e.

$$\chi_k(x) = \langle q(x), [q(A \cap S)]_{q(x)} \rangle$$

Intuitively, the germ of $q(A \cap S)$ at $q(x)$ represents in I, under the local homemorphism q, the set of points in A close to x. It provides a measure of the *extent* to which x is in A. Whereas in set theory classification admits of only two possibilities – either $x \in A$ or $x \notin A$ – in a topological context we may make more subtle distinctions by classifying according to how close x is to A. We use the germs at $q(x)$ as a system of entities for measuring proximity of x to open subsets of B. A partial ordering on $\Omega_{q(x)}$ is given by

$$[U]_{q(x)} \sqsubseteq [V]_{q(x)} \quad \text{iff there is some open set } W \text{ such that } q(x) \in W \text{ and } U \cap W \subseteq V \cap W,$$

i.e. iff the statement "$U \subseteq V$" is locally true at $q(x)$.

Then the "larger" the germ of $q(A \cap S)$ is in terms of this ordering, the closer will x be to A. If in fact $x \in A$, then $q(x) \in q(A \cap S)$ and so by (i) above, the germ of $q(A \cap S)$ is as large as it could be, i.e. $[q(A \cap S)]_{q(x)} = [I]_{q(x)}$. At the other extreme, if x is separated from A, then the germ of $q(A \cap S)$ is as small as it could be, i.e. $[q(A \cap S)] = [\emptyset]$. Otherwise, when x is on the boundary of A, $[q(A \cap S)]$ is strictly between the germs of $\emptyset$ and I, $[\emptyset] \sqsubset [q(A \cap S)] \sqsubset [I]$.

EXERCISE 1. Verify that the definition of $\chi_k(x)$ does not depend on the choice of neighbourhood S of x on which q is a local homeomorphism.

EXERCISE 2. (Alternative definition of $\chi_k(x)$). Let

$$U_x = \{i \in I: \text{ for some local section } s \text{ of } (B, q),\ s(i) \in A \text{ and } s(q(x)) = x\}$$

be the set of points in I that are carried into A by some local section of (B, q) that takes $q(x)$ to x. Show that

$$[U_x]_{q(x)} = [q(A \cap S)]_{q(x)},$$

where S is as above. □

4.6. Monoid actions

Let $\mathbf{M} = (M, *, e)$ be a monoid (cf. §2.5). Then any given $m \in M$ determines a function $\lambda_m : M \to M$, called *left-multiplication by m*, and defined by the rule $\lambda_m(n) = m * n$, for all $n \in M$. We thus obtain a family $\{\lambda_m : m \in M\}$ of functions, indexed by M, which satisfies

(i) $\lambda_e = \mathrm{id}_M$, since $\lambda_e(m) = e * m = m$, and

(ii) $\lambda_m \circ \lambda_p = \lambda_{m*p}$, since $\lambda_m(\lambda_p(n)) = m * (p * n) = (m * p) * n$.

Condition (ii) in fact says that the collection of λ_m's is closed under functional composition. Indeed, it forms a monoid under this operation with identity λ_e.

The notion just described can be generalised. Suppose we have a set X and a collection $\{\lambda_m : X \to X : m \in M\}$ of functions λ_m from X to X, the collection being indexed by the elements of our original monoid, and satisfying

$$\lambda_e = \mathrm{id}_X$$

$$\lambda_m \circ \lambda_p = \lambda_{m*p}.$$

The collection of λ_m's is called an *action* of $\mathbf{M}$ on the set X, and can be replaced by a single function $\lambda : M \times X \to X$, defined by

$$\lambda(m, x) = \lambda_m(x), \quad \text{all} \quad m \in M, \qquad x \in X.$$

The above two conditions become

$$\lambda(e, x) = x$$

and

$$\lambda(m, \lambda(p, x)) = \lambda(m * p, x).$$

An **M**-set is defined to be a pair (X, λ), where $\lambda : M \times X \to X$ is such an action of **M** on X.

EXAMPLE 1. **M** is the monoid $(N, +, 0)$ of natural numbers under addition. X is the set of real numbers. λ is *addition*:— $\lambda(m, r) = m + r$.

EXAMPLE 2. X is the set of vectors of a vector space, **M** the multiplicative monoid of its scalars, λ is scalar multiplication of vectors.

EXAMPLE 3. X is the set of points in the Euclidean plane. **M** is the group of Euclidean transformations (rotations, reflections, translations) with $*$ as function composition. $\lambda(m, x)$ is $m(x)$, i.e. the result of applying transformation m to point x.

EXAMPLE 4. X is the set of states of a computing device. **M** is the set of input words (strings) with $*$ the operation of concatenation or juxtaposition of strings. $\lambda(m, x)$ is the state the machine goes into in response to being fed input m while in state x. □

For a given monoid **M**, the **M**-sets are the objects of a category **M-Set**, which is a topos. An arrow $f : (X, \lambda) \to (Y, \mu)$ is an *equivariant*, or *action-preserving* function $f : X \to Y$, i.e. one such that

$$\begin{array}{ccc} X & \xrightarrow{f} & Y \\ {\scriptstyle \lambda_m}\downarrow & \cdot & \downarrow{\scriptstyle \mu_m} \\ X & \xrightarrow{f} & Y \end{array}$$

commutes for each $m \in M$. In other words, $f(\lambda(m, x)) = \mu(m, f(x))$, all m and x. Composition of arrows is functional composition.

The terminal object is a singleton **M**-set. We take $1=(\{0\},\lambda_0)$ where $\lambda_0(m,0)=0$, all m.

The product of (X,λ) and (Y,μ) is $(X\times Y,\delta)$, where δ_m is $\lambda_m\times\mu_m : X\times Y\to X\times Y$. The pullback of

$$\begin{array}{ccc} & & (Y,\mu) \\ & & \big\downarrow{\scriptstyle g} \\ (X,\lambda) & \xrightarrow{\ f\ } & (Z,\gamma) \end{array}$$

is $(X\times_Z Y,\delta)$ with δ as above.

Now a set $B\subseteq M$ is called a *left ideal* of **M** if it is closed under left-multiplication, i.e. if $m*b\in B$ whenever $b\in B$ and m is any element of M. For example, M and $\emptyset$ are left ideals of **M**. We put $\Omega=(L_M,\omega)$ where L_M is the set of left ideals in **M**, and $\omega: M\times L_M\to L_M$ has $\omega(m,B)=\{n: n*m\in B\}$. $\top:1\to\Omega$ is the function $\top:\{0\}\to L_M$ with $\top(0)=M$. Thus $\top$ picks out the largest left-ideal M of **M**.

To illustrate the workings of the subobject classifier, suppose $k:(X,\lambda)\rightarrowtail(Y,\mu)$ is in fact the inclusion $X\hookrightarrow Y$ (since k is *equivariant* this means $\mu(m,x)=\lambda(m,x)$, all $x\in X$). The character $\chi_k:(Y,\mu)\to\Omega$ of k is $\chi_k: Y\to L_M$ defined by

$$\chi_k(y)=\{m: \mu(m,y)\in X\},\quad \text{all}\quad y\in Y.$$

EXERCISE 1. Check all the details – that ω is an action of **M** on L_M, that $\chi_k(y)$ is a left-ideal, and that χ_k satisfies the Ω-axiom. □

Exponentiation

Our initial motivation showed that $*: M\times M\to M$ is itself an action of **M** on M, i.e. that $(M,*)$ is an **M**-set. Given (X,λ) and (Y,μ) we define the exponential

$$(Y,\mu)^{(X,\lambda)}=(E,\sigma)$$

where E is the set of equivariant maps f of the form $f:(M,*)\times(X,\lambda)\to(Y,\mu)$ and $\sigma_m: E\to E$ takes such an f to the function $g=\sigma_m(f): M\times X\to Y$ given by

$$g(n,x)=f(m*n,x)$$

The evaluation arrow

$$ev:(E,\sigma)\times(X,\lambda)\to(Y,\mu)$$

has

$$ev(f, x) = f(e, x).$$

Then given an arrow $f:(X, \lambda)\times(Y, \mu)\to(Z, \nu)$, the exponential adjoint $\hat{f}:(X, \lambda)\to(Z, \nu)^{(Y,\mu)}$ takes $x\in X$ to the equivariant map $\hat{f}_x: M\times Y\to Z$ having

$$\hat{f}_x(m, y) = f(\lambda_m(x), y).$$

Categories of the form **M-Set** provide a rich source of examples, particularly of topoi that have "non-classical" properties. They will be "re-created" from a different perspective in Chapter 9.

EXERCISE 2. Describe all the left-ideals in $(N, +, 0)$.

EXERCISE 3. Show that M is a group iff M and $\emptyset$ are the only left-ideals of **M**, i.e. iff $L_M = \{M, \emptyset\}$. □

4.7. Power objects

The exponential Ω^a in a topos is the analogue of 2^A in **Set**. Since $2^A \cong \mathcal{P}(A)$ it is natural to wonder whether the object Ω^a behaves like the "powerset" of the "set" a. In fact it does, as we shall see by first developing an independent categorial description of $\mathcal{P}(A)$ in **Set**.

Now given sets A and B there is a bijective correspondence between the functions from B to $\mathcal{P}(A)$ and the relations from B to A. Given function $f: B\to\mathcal{P}(A)$ define relation $R_f \subseteq B\times A$ by stipulating xR_fy iff $y\in f(x)$, for $x\in B$, $y\in A$. Conversely, given $R\subseteq B\times A$, define $f_R: B\to \mathcal{P}(A)$ by $f_R(x) = \{y: y\in A \text{ and } xRy\}$.

It is not hard to see that the assignments of f_R to R and R_f to f are inverse to each other and establish the asserted isomorphism.

In order to capture this correspondence in terms of arrows we examine a special relation $\in_A$ from $\mathcal{P}(A)$ to A. $\in_A$ is the membership relation and contains all the information about which subsets of A contain which elements of A. Precisely

$$\in_A = \{\langle U, x\rangle: U\subseteq A, x\in A, \text{and } x\in U\}.$$

Passing from $\mathcal{P}(A)$ to 2^A, the condition "$x\in U$" becomes "$\chi_U(x) = 1$", and we see that $\in_A$ is isomorphic to the set

$$\in'_A = \{\langle \chi_U, x\rangle: U\subseteq A, x\in A, \text{and } \chi_U(x) = 1\}\subseteq 2^A\times A$$

What is the characteristic function of $\in'_A$ as a subset of $2^A \times A$? Well it is none other than the evaluation arrow $ev: 2^A \times A \to 2$, since $ev(\chi_U, x) = \chi_U(x)$. Thus we are lead to a characterisation of $\in'_A$ (and hence $\in_A$ up to isomorphism) by the pullback square

$$\begin{array}{ccc} \in'_A & \hookrightarrow & 2^A \times A \\ {\scriptstyle !}\downarrow & & \downarrow{\scriptstyle ev} \\ 1 & \xrightarrow{\ true\ } & 2 \end{array}$$

Now given a relation $R \subseteq B \times A$, we have $\langle x, y\rangle \in R$ iff $y \in f_R(x)$ iff $\langle f_R(x), y\rangle \in \in_A$, and so R is the inverse image of $\in_A$ under the map $f_R \times 1_A$, that takes $\langle x, y\rangle$ to $\langle f_R(x), y\rangle$.

So we see that (§3.13) the diagram

$$\begin{array}{ccc} R & \hookrightarrow & B \times A \\ {\scriptstyle g}\downarrow & & \downarrow{\scriptstyle f_R \times \mathrm{id}_A} \\ \in_A & \hookrightarrow & \mathcal{P}(A) \times A \end{array}$$

is a pullback, where g is the restriction of $f_R \times \mathrm{id}_A$ to R. But something stronger than this can be said – given R, then without considering what g is, f_R is the *only* function $B \to \mathcal{P}(A)$ that will give a pullback of the form of the diagram.

EXERCISE 1. Prove this last assertion. □

We are therefore lead to the following definition:

DEFINITION. A category $\mathscr{C}$ with products is said to have *power objects* if to each $\mathscr{C}$-object a there are $\mathscr{C}$-objects $\mathcal{P}(a)$ and $\in_a$, and a monic $\in : \in_a \rightarrowtail \mathcal{P}(a) \times a$, such that for any $\mathscr{C}$-object b, and "relation", $r: R \rightarrowtail b \times a$ there is exactly one $\mathscr{C}$-arrow $f_r: b \rightarrowtail \mathcal{P}(a)$ for which there is a pullback in $\mathscr{C}$ of the form

$$\begin{array}{ccc} R & \overset{r}{\rightarrowtail} & b \times a \\ \downarrow & & \downarrow{\scriptstyle f_r \times 1_a} \\ \in_a & \overset{\in}{\rightarrowtail} & \mathcal{P}(a) \times a \end{array}$$

THEOREM 1. *Any topos $\mathscr{E}$ has power objects.*

PROOF. For given $\mathscr{E}$-object a, let $\mathscr{P}(a) = \Omega^a$ and let $\in : \in_a \rightarrowtail \Omega^a \times a$ be the subobject of $\Omega^a \times a$ whose character is $ev_a : \Omega^a \times a \to \Omega$, i.e.

$$
\begin{array}{ccc}
\in_a & \xrightarrow{\ \in\ } & \Omega^a \times a \\
\downarrow{\scriptstyle !} & & \downarrow{\scriptstyle ev_a} \\
1 & \xrightarrow{\ \top\ } & \Omega
\end{array}
$$

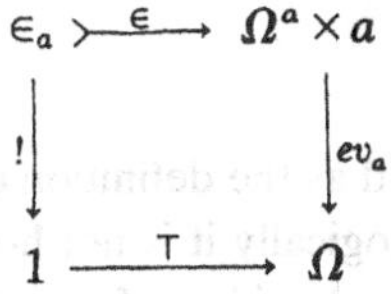

is a pullback, where ev_a is the evaluation arrow from $\Omega^a \times a$ to Ω. To show that this construction gives power objects take any monic $r : R \rightarrowtail b \times a$ and let $\chi_r : b \times a \to \Omega$ be its character. Then let $f_r : b \to \Omega^a$ be the exponential adjoint to χ_r, i.e. the unique arrow that makes

$$
\begin{array}{ccc}
\Omega^a \times a & & \\
\downarrow{\scriptstyle f_r \times 1_a} & \searrow^{ev_a} & \\
& & \Omega \\
b \times a & \nearrow_{\chi_r} &
\end{array}
$$

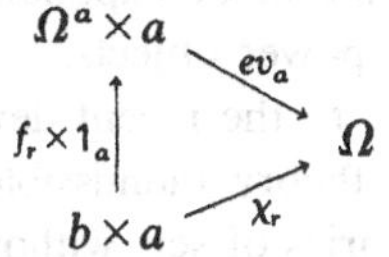

commute. Now consider the diagram

$$
\begin{array}{ccc}
R & \xrightarrow{\ r\ } & b \times a \\
\vdots & & \downarrow{\scriptstyle f_r \times 1_a} \\
\in_a & \xrightarrow{\ \in\ } & \Omega^a \times a \\
\downarrow & & \downarrow{\scriptstyle ev_a} \\
1 & \xrightarrow{\ \top\ } & \Omega
\end{array}
$$

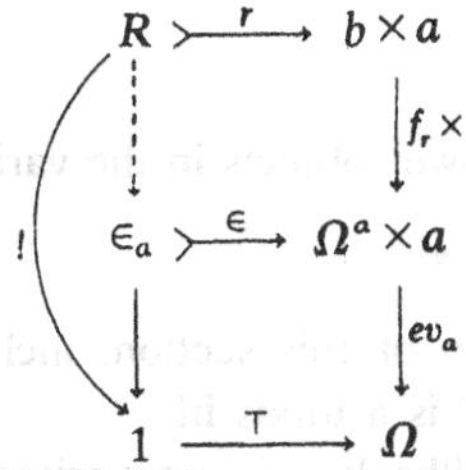

Since $ev_a \circ (f_r \times 1_a) = \chi_r$, the "perimeter" of this diagram is a pullback, by the Ω-axiom. In particular it commutes, so as the bottom square is a pullback, the unique arrow $R \dashrightarrow \in_a$ does exist to make the whole diagram commute. But then by the PBL the top square is a pullback, as required by the definition of power objects. Moreover simply knowing that f_r is some arrow making the top square a pullback gives both squares as pullbacks and hence (PBL) the outer rectangle is a pullback. The Ω-axiom then implies that $ev_a \circ (f_r \times 1_a) = \chi_r$ and thus from the previous diagram f_r is uniquely determined as *the* exponential adjoint of χ_r. $\square$

Now given power objects we can recover Ω, as $\Omega \cong \Omega^1 = \mathscr{P}(1)$. The monic $\in_1 \rightarrowtail \Omega^1 \times 1 \cong \Omega^1$ proves to be a subobject classifier. Anders Kock and C. Juul Mikkelsen have shown that power objects can also be used to construct exponentials, and that

> *a category $\mathscr{C}$ is a topos iff $\mathscr{C}$ is finitely complete and has power objects*

(for details consult Wraith [75]).

Currently this characterisation is being used as the definition of a topos, it being the best in terms of brevity. Paedogogically it is not however the best, for a number of reasons. Historically the idea of an elementary topos arose through examination of subobject classifiers, and this path provides the most suitable motivation. As will be evident it is the Ω-axiom that is the key to the basic structure of a topos and it would have to be introduced anyway for the theory to get off the ground. Moreover each of the Ω-axiom, and the notion of exponentiation, is conceptually simpler than the description of power objects.

There is another more remote matter, due to the recent development of weak set theories relating to recursion theory (admissible sets – cf. Barwise [75]). These theories produce categories of sets without general powerset formation. It therefore becomes of interest to study the ramifications of the Ω-axiom without having to relate it to the notion of power-object.

Exercise 2. Examine the structure of power objects in the various topoi described in this chapter.

Exercise 3. Deduce from the discussion of this section, including the proof of the Theorem, that a category $\mathscr{C}$ is a topos iff

(i) $\mathscr{C}$ has a terminal object and pullbacks of appropriate pairs of arrows,

(ii) $\mathscr{C}$ has a subobject classifier $true : 1 \to \Omega$

(iii) For each $\mathscr{C}$-object a there is a $\mathscr{C}$-object Ω^a and an arrow $ev_a : \Omega^a \times a \to \Omega$ such that for each $\mathscr{C}$-object b and "relation" $r : R \rightarrowtail b \times a$ there is exactly one $\mathscr{C}$-arrow $f_r : b \to \Omega^a$ making

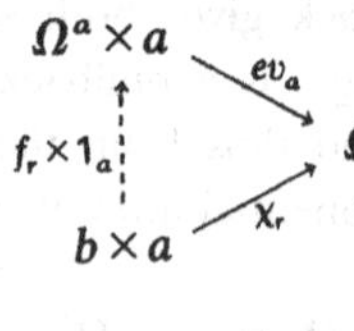

commute.

Exercise 4. Show that the unique arrow $\Omega^a \to \Omega^a$ corresponding to the relation $\in_a \rightarrowtail \Omega^a \times a$ is 1_{Ω^a}. □

4.8. Ω and comprehension

In Lawvere [72] it is suggested that the Ω-axiom is a form of the ZF Comprehension principle. To see this, suppose that B is a set and φ a property that applies to members of B. We represent φ in **Set** as a function $\varphi : B \to 2$ given by

$$\varphi(x) = \begin{cases} 1 & \text{if } x \text{ has property } \varphi \\ 0 & \text{otherwise.} \end{cases}$$

Now the comprehension (seperation) principle allows us to form the subset $\{x : x \in B \text{ and } \varphi(x)\}$ of all elements of B satisfying φ. This set is determined by φ qua function as what we earlier called $A_\varphi = \{x : \varphi(x) = 1\}$. We have $y \in \{x : \varphi(x)\}$ iff $\varphi(y) = 1$, and

$$\begin{array}{ccc} A_\varphi & \hookrightarrow & B \\ {\scriptstyle !}\downarrow & & \downarrow{\scriptstyle \varphi} \\ 1 & \xrightarrow{\ true\ } & 2 \end{array}$$

is a pullback. By analogy, in a topos $\mathscr{E}$, if $\varphi : b \to \Omega$ is an arrow with $\mathrm{cod} = \Omega$, we let $\{x : \varphi\} : a \rightarrowtail b$ be the subobject of b obtained by pulling *true* back along φ, as in

$$\begin{array}{ccc} a & \xrightarrow{\ \{x:\varphi\}\ } & b \\ {\scriptstyle !}\downarrow & & \downarrow{\scriptstyle \varphi} \\ 1 & \xrightarrow{\ true\ } & \Omega \end{array}$$

Now in a general category, if $x : 1 \to b$ is an element of object b, and $f : a \rightarrowtail b$ a subobject, we define x to be a member of f, $x \in f$, when x factors through f, i.e. there exists $k : 1 \to a$ making

$$\begin{array}{ccc} & 1 & \\ {\scriptstyle k}\swarrow & & \searrow{\scriptstyle x} \\ a & \xrightarrow{\ f\ } & b \end{array}$$

commute. This naturally generalises the situation in **Set**.

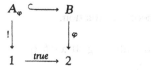

Applying this notion of membership to the above pullback we see that if $y:1 \to b$ is a b-element then

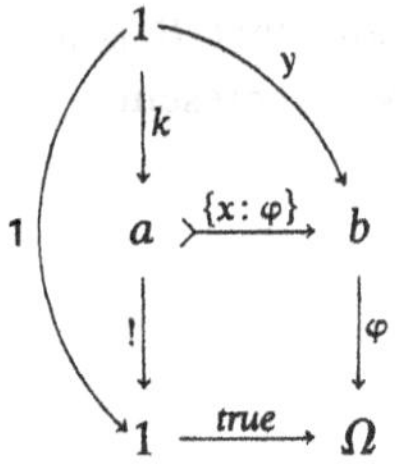

$y \in \{x\colon \varphi\}$ iff the arrow k exists to make the whole diagram commute. But as the inner square is a pullback, k will exist (uniquely) iff the perimeter of the diagram commutes. Hence

$$y \in \{x\colon \varphi\} \quad \text{iff} \quad \varphi \circ y = \mathit{true},$$

giving us an analogue of the set-theoretic situation.

Exercise 1. Take $f: a \rightarrowtail b$, $g: c \rightarrowtail b$ with $f \subseteq g$. If $x \in b$ (i.e. $x: 1 \to b$, or $x \in 1_b$ as above) has $x \in f$, show $x \in g$.

Exercise 2. For any $f: a \rightarrowtail d$ and $x: 1 \to d$, $x \in f$ iff $\chi_f \circ x = \mathit{true}$.

CHAPTER 5

TOPOS STRUCTURE: FIRST STEPS

> *"The development of elementary topoi by Lawvere and Tierney strikes this writer as the most important event in the history of categorical algebra since its creation... It is not just that they proved these things, its that they dared to believe them provable."*
>
> Peter Freyd

5.1. Monics equalise

In §3.10 it was stated that an injective function $f: A \rightarrowtail B$ is an equaliser for a pair of functions g and h. We now see that g is $\chi_{\mathrm{Im} f}: B \rightarrow 2$ and h is the composite of $!: B \rightarrow \{0\}$ and $true: \{0\} \rightarrow \{0, 1\}$. This situation generalises directly:—

THEOREM 1: *If $f: a \rightarrowtail b$ is a monic $\mathscr{E}$-arrow ($\mathscr{E}$ any topos) then f is an equaliser of χ_f and $true_b = true \circ !_b$.*

PROOF: Since the pullback square of

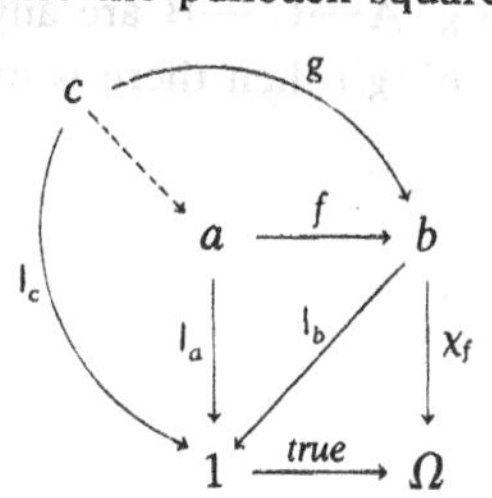

commutes, and $!_a = !_b \circ f$, we have $\chi_f \circ f = true_b \circ f$. But if $\chi_f \circ g = true_b \circ g$

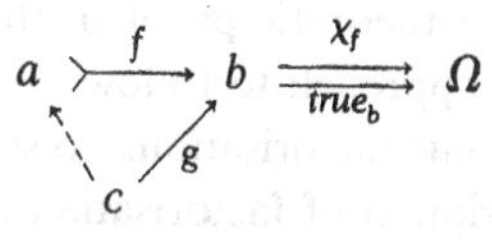

then the perimeter of the first diagram must commute, since $1_b \circ g = 1_c$. So, by the universal property of pullbacks, g factors uniquely through f as required. □

Corollary: *In any topos, an arrow is iso iff it is both epic and monic.*

Proof. In any category, an iso is monic and epic (§3.3). On the other hand, in a topos an epic monic is, by the Theorem, an epic equaliser. Such a thing is always iso (§3.10). □

Exercise. $true: 1 \to \Omega$ equalises $\mathbf{1}_\Omega: \Omega \to \Omega$ and $true_\Omega: \Omega \to \Omega$. □

5.2. Images of arrows

Any set function $f: A \to B$ can be factored into a surjection, followed by an injection. We have the commutative diagram

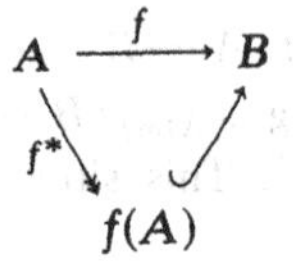

where $f(A) = \operatorname{Im} f = \{f(x): x \in A\}$, and $f^*(x) = f(x)$, all $x \in A$.

This "epi-monic" factorisation of f is unique up to a *unique commuting isomorphism* as shown in the

Exercise 1. If $h \circ g: A \twoheadrightarrow C \rightarrowtail B$ and $h' \circ g': A \twoheadrightarrow C' \rightarrowtail B$ are any two epi-monic factorisations of f (i.e. $f = h \circ g = h' \circ g'$) then there is exactly one $k: C \to C'$ such that

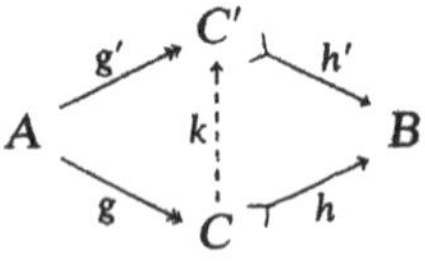

commutes, and furthermore k is iso in **Set** (a bijection). □

The reader may care to develop a set-theoretic proof of this exercise and contrast it with the "arrows-only" approach to follow.

In all topoi, each arrow has an epi-monic factorisation. To see how this works, we turn first to a different description of factorisation in **Set**, one that has a categorial formulation. Given $f: A \to B$ we define, as in §3.13, the Kernel equivalence relation $R_f \subseteq A \times A$ by

$$xR_fy \quad \text{iff} \quad f(x)=f(y).$$

Now a map $h: A/R_f \to B$ is well-defined by $h([x])=f(x)$. Moreover h is injective and

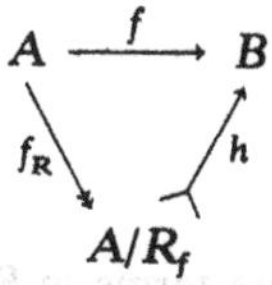

commutes, where $f_{\mathbf{R}}$ is the surjective natural map $f_{\mathbf{R}}(x)=[x]$.

Now as observed in §3.13, R_f as a set of ordered pairs yields a pullback

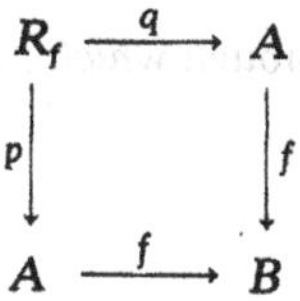

where p and q, the projections, are the *kernel pair* of f. The considerations of §3.12 then show that $f_{\mathbf{R}}$ co-equalises the kernel pair (p, q) and that h is the unique arrow making

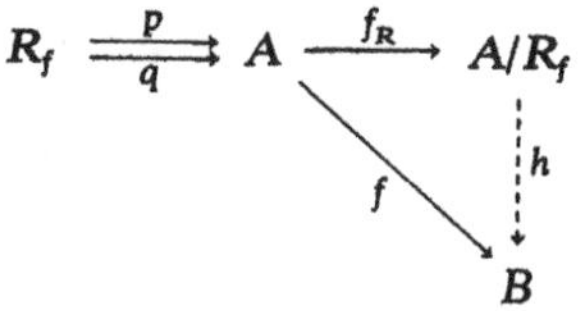

commute. This suggests that in a more general category we attempt to factor an arrow by co-equalising its pullback along itself. However, for technical reasons (the availability of the results of the last section) it is simpler now to dualise the construction, i.e. to equalise the pushout of the arrow with itself.

So, let $\mathscr{E}$ be any topos, and $f: a \to b$ any $\mathscr{E}$-arrow. We form the pushout

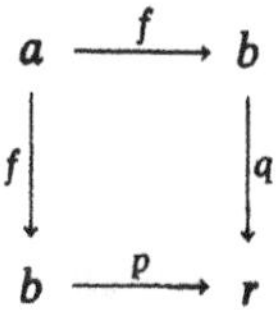

of f with f, and let $\operatorname{im} f : f(a) \rightarrowtail b$ be the equaliser of p and q ($\operatorname{im} f$ is monic by Theorem 3.10.1). Since $q \circ f = p \circ f$, there is a unique arrow $f^* : a \to f(a)$ making

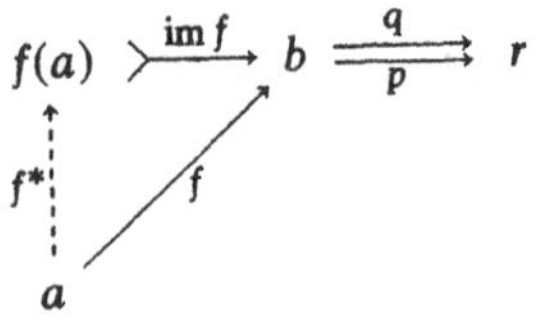

commute.

EXERCISE 2. Analyse this construction in concrete terms in **Set**.

EXERCISE 3. If $p = q$, then f is epic.   □

THEOREM 1. $\operatorname{im} f$ is the smallest subobject of b through which f factors. That is, if

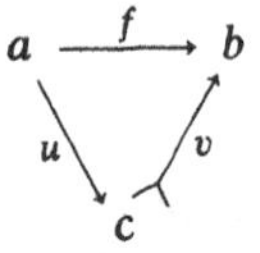

commutes, for any u and monic v as shown, then there is a (unique) $k : f(a) \to c$ making

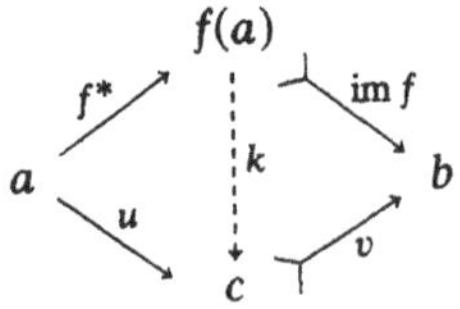

commute, and hence $\operatorname{im} f \subseteq v$.

PROOF. Being monic, v equalises a pair $s, t : b \rightrightarrows d$ of $\mathscr{E}$-arrows (§5.1). Thus $s \circ f = s \circ v \circ u = t \circ v \circ u = t \circ f$, so

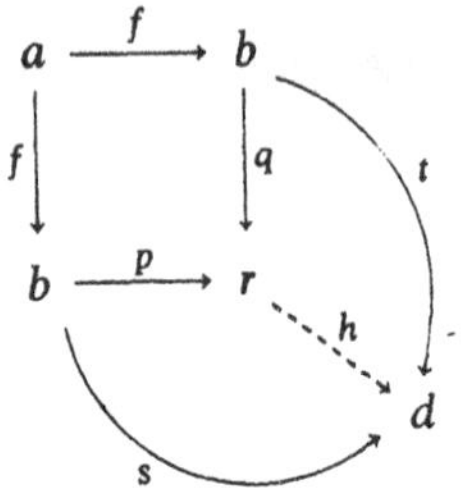

there is a unique $h:r \to d$ such that $h \circ p = s$ and $h \circ q = t$. But then

$$\begin{aligned} s \circ \operatorname{im} f &= h \circ p \circ \operatorname{im} f \\ &= h \circ q \circ \operatorname{im} f \\ &= t \circ \operatorname{im} f, \end{aligned}$$

so, as v equalises s and t

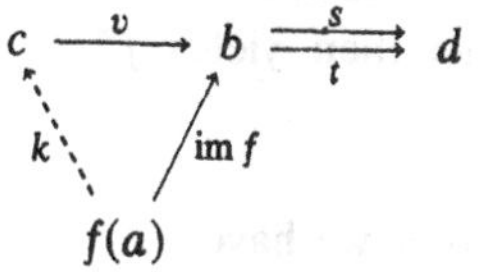

we get a unique arrow k that has $v \circ k = \operatorname{im} f$. This k is the unique arrow making the right-hand triangle in the diagram in the statement of the theorem commute. But then $v \circ k \circ f^* = \operatorname{im} f \circ f^* = f = v \circ u$, and v is monic (left-cancellable), so $k \circ f^* = u$. Thus k makes the left-hand triangle commute as well. □

Corollary. *$f^*: a \to f(a)$ is epic.*

Proof. Apply the image construction to f^* itself, giving the commuting diagram

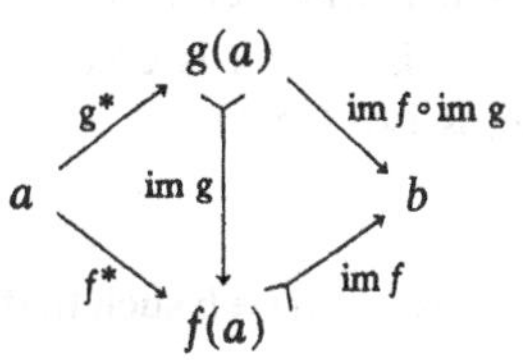

where $g = f^*$.

But $\operatorname{im} f \circ \operatorname{im} g$ is monic, being a product of monics, and so, as $\operatorname{im} f$ is left cancellable, we must have $\operatorname{im} g$ as the unique arrow making $\operatorname{im} f \circ \operatorname{im} g \subseteq \operatorname{im} f$. But also, applying the Theorem to $\operatorname{im} f$ we must have $\operatorname{im} f \subseteq \operatorname{im} f \circ \operatorname{im} g$, and so $\operatorname{im} f \simeq \operatorname{im} f \circ \operatorname{im} g$ in $\operatorname{Sub}(b)$, hence $g(a) \cong f(a)$. Thus the unique arrow $\operatorname{im} g$ must be iso.

But $\operatorname{im} g$ is, by definition, the equaliser

$$g(a) \xrightarrow{\operatorname{im} g} f(a) \underset{q}{\overset{p}{\rightrightarrows}} r,$$

where p and q, the *cokernel pair* of $g = f^*$, form a pushout thus:

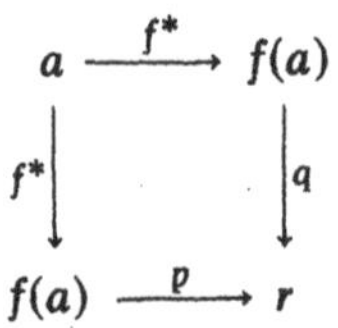

Since $p \circ \text{im } g = q \circ \text{im } g$, and im g is iso, hence epic, we cancel to get $p = q$. The co-universal property of pushouts then yields f^* as epic (as in Exercise 3, above). □

Bringing the work of this section together we have

THEOREM 2. $\text{im } f \circ f^* : a \twoheadrightarrow f(a) \rightarrowtail b$ *is an epi-monic factorisation of f that is unique up to a unique commuting isomorphism. That is, if* $v \circ u : a \twoheadrightarrow c \rightarrowtail b$ *has* $v \circ u = f$, *then there is exactly one arrow* $k : f(a) \to c$ *such that*

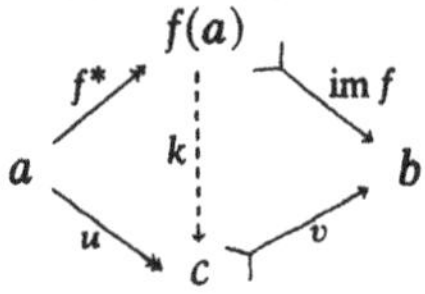

commutes, and k is iso.

PROOF. The unique k exists by Theorem 1. But then $v \circ k = \text{im } f$ is monic, so k is monic by Exercise 2, §3.1. Also $k \circ f^* = u$ is epic, so dually k is epic. Hence k, being epic and monic, is iso. (§5.1). □

EXERCISE 4. $f : a \to b$ is epic iff there exists $g : f(a) \cong b$ such that $g \circ f^* = f$. □

5.3. Fundamental facts

If $\mathscr{E}$ is a topos then the comma category $\mathscr{E} \downarrow a$ of objects over a is also a topos. As mentioned in Chapter 4, this is (part of) a result known as the Fundamental Theorem of Topoi. The proof of this theorem involves a construction too advanced for our present stage of development, but yielding some important information that we shall need now. We therefore record these consequences of the Fundamental Theorem without proof:

FACT 1. *Pullbacks preserve epics. If*

$$\begin{array}{ccc} a & \longrightarrow & b \\ {\scriptstyle g}\downarrow & & \downarrow{\scriptstyle f} \\ c & \longrightarrow & d \end{array}$$

is a pullback square in a topos, and f is epic, then g, the pullback of f, is also epic.

FACT 2. *Coproducts preserve pullbacks. If*

$$\begin{array}{ccc} a & \xrightarrow{f} & d \\ {\scriptstyle g}\downarrow & & \downarrow{\scriptstyle k} \\ b & \xrightarrow{h} & e \end{array} \quad \textit{and} \quad \begin{array}{ccc} a' & \xrightarrow{f'} & d \\ {\scriptstyle g'}\downarrow & & \downarrow{\scriptstyle k} \\ b' & \xrightarrow{h'} & e \end{array}$$

are pullbacks in a topos, then so is

$$\begin{array}{ccc} a+a' & \xrightarrow{[f,f']} & d \\ {\scriptstyle g+g'}\downarrow & & \downarrow{\scriptstyle k} \\ b+b' & \xrightarrow{[h,h']} & e \end{array}$$

Proofs of these results may be found in Kock and Wraith [71], Freyd [72], and Brook [74].

5.4. Extensionality and bivalence

Since a general topos $\mathscr{E}$ is supposed to be "**Set**-like", its initial object 0 ought to behave like the null set $\emptyset$, and have no elements. This in fact obtains, except in one case. If there is an arrow $x: 1 \to 0$, then by the work in §3.16 on Cartesian closed categories, $\mathscr{E}$ is degenerate, i.e. all $\mathscr{E}$-objects are isomorphic. This happens for example in the category **1** with one object and one arrow – **1** is a degenerate topos. So in a non-degenerate topos, 0 has no elements.

Now if we call an object *a non-zero* if it is not isomorphic to 0, $a \not\cong 0$, and *non-empty* if there is at least one $\mathscr{E}$-arrow $1 \to a$, then when $\mathscr{E} = \mathbf{Set}$, "non-zero" and "non-empty" are co-extensive. But when $\mathscr{E} = \mathbf{Set}^2$, the topos of pairs of sets, the situation is different. The object $\langle \emptyset, \{0\} \rangle$ is not isomorphic to the initial object $\langle\emptyset,\emptyset\rangle$, hence is *non-zero*. But an element $\langle f,g\rangle:\langle\{0\},\{0\}\rangle\to\langle\emptyset,\{0\}\rangle$ of $\langle\emptyset,\{0\}\rangle$ would require f to be a set function $\{0\}\to\emptyset$, of which there is no such thing. Thus $\langle\emptyset,\{0\}\rangle$ is non-zero but empty.

EXERCISE 1. Are there any other non-zero empty objects in **Set**^{2}? What about non-empty zero objects?

EXERCISE 2. Are there non-zero empty objects in **Set**$^{\to}$? In **Bn**(I)? □

The question of the existence of elements of objects relates to the notion of *extensionality*, the principle that sets with the same elements are identical. For functions, this principle takes the following form (which we have used repeatedly): two parallel functions $f, g: A \rightrightarrows B$ are equal if they give the same output for the same input, i.e. if for each $x\in A$, $f(x)=g(x)$. Categorially this takes the form of the:

EXTENSIONALITY PRINCIPLE FOR ARROWS. *If $f, g: a \rightrightarrows b$ are a pair of* distinct *parallel arrows, then there is an element $x:1\to a$ of a such that $f\circ x\neq g\circ x$.*

(Category-theorist will recognise this as the statement "1 is a generator".) This principle holds in **Set**, but not in **Set**^{2}. It is easy to see that in the latter there are two distinct arrows from $\langle\emptyset,\{0\}\rangle$ to $\langle\emptyset,2\rangle$. But $\langle\emptyset,\{0\}\rangle$ has no elements at all to distinguish them.

A non-degenerate topos that satisfies the extensionality principle for arrows is called *well-pointed*. The purpose of this section is to examine the properties of such categories.

THEOREM 1. *If $\mathscr{E}$ is well-pointed, then every non-zero $\mathscr{E}$-object is non-empty.*

PROOF. If a is non-zero then $0_a:0\rightarrowtail a$ and $1_a:a\rightarrowtail a$ have different domains, and so are distinct. Hence $\chi_{0_a}:a\to\Omega$ and $\chi_{1_a}:a\to\Omega$ are distinct (otherwise $0_a\simeq 1_a$, hence $0\cong a$). By extensionality it follows that there is some $x:1\to a$ such that $\chi_{0_a}\circ x\neq\chi_{1_a}\circ x$. In particular a has an element, so is non-empty. □

False

In **Set** there are exactly two arrows from $1=\{0\}$ to $\Omega=\{0,1\}$. One of course is the map *true*, with *true*$(0)=1$. The other we call *false*, and is defined by $false(0) = 0$. This map, having codomain Ω is the characteristic function of

$$\{x : false(x) = 1\} = \emptyset, \quad \text{the null set,}$$

so in **Set** we have a pullback

$$\begin{array}{ccc} \emptyset & \overset{!}{\hookrightarrow} & 1 \\ {\scriptstyle !}\downarrow & & \downarrow{\scriptstyle false} \\ 1 & \xrightarrow{true} & \Omega \end{array}$$

Abstracting this, we define in any topos $\mathscr{E}$, $false : 1 \to \Omega$ to be the unique $\mathscr{E}$-arrow such that

$$\begin{array}{ccc} 0 & \xrightarrow{0_1} & 1 \\ {\scriptstyle !}\downarrow & & \downarrow{\scriptstyle false} \\ 1 & \xrightarrow{\top} & \Omega \end{array}$$

is a pullback in $\mathscr{E}$. Thus $false = \chi_{0_1}$. We will also use the symbol "$\perp$" for this arrow.

Example 1. In $\mathbf{Set}^2$, $\perp : 1 \to \Omega$ is $\langle false, false\rangle : \langle\{0\},\{0\}\rangle \to \langle 2, 2\rangle$.

Example 2. In $\mathbf{Bn}(I)$, $\perp : 1 \to \Omega$ *is* $\perp : I \to 2 \times I$ where $\perp(i) = \langle 0, i\rangle$, all $i \in I$.

Example 3. In $\mathbf{Top}(I)$, $\perp : I \to \hat{I}$ has $\perp(i) = \langle i, [\emptyset]_i\rangle$, the germ of $\emptyset$ at i.

Example 4. In **M-Set**, $0 = (\emptyset, \emptyset)$, with $\emptyset : M \times \emptyset \to \emptyset$, *the "empty action"*. $\perp : \{0\} \to L_M$ has $\perp(0) = \{m : \lambda_0(m, 0) \in \emptyset\} = \emptyset$. □

Exercise 3. For any $\mathscr{E}$-object a,

$$\begin{array}{ccc} 0 & \xrightarrow{0_a} & a \\ {\scriptstyle !}\downarrow & & \downarrow{\scriptstyle \perp \circ !_a} \\ 1 & \xrightarrow{\top} & \Omega \end{array}$$

is a pullback, i.e. $\chi_{0_a} = \perp \circ !_a (= \perp_a = false_a)$.

(Hint: you may need the PBL)

EXERCISE 4. In a non-degenerate topos, *true* $\neq$ *false*. □

A non-degenerate topos $\mathscr{E}$ is called *bivalent* (two-valued) if *true* and *false* are its only truth-values (elements of Ω).

THEOREM 2. *If $\mathscr{E}$ is well-pointed, then $\mathscr{E}$ is bivalent.*

PROOF. Let $f:1\to\Omega$ be any element of Ω and form the pullback

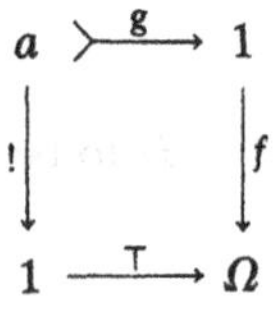

of f and $\top$.

Case 1: If $a\cong 0$, then a is an initial object, with $g\simeq 0_1$. Then $f=\chi_g=\chi_{0_1}=$ *false*.

Case 2: If not $a\cong 0$, then as $\mathscr{E}$ is well-pointed, a has an element $x:1\to a$ (Theorem 1). We use this to show that g is epic. For, if $h,k:1\rightrightarrows b$ have $h\circ g=k\circ g$, then $h\circ g\circ x=k\circ g\circ x$. But $g\circ x:1\to 1$ can only be 1_1 (1 is terminal) so $h=k$. Thus g is right cancellable. Hence g is both epic and monic (being the pullback of a monic), giving $g:a\cong 1$. So a is terminal, yielding $g\simeq 1_1$, hence $f=\chi_g=\chi_{1_1}=$ *true*.

Altogether then we have shown that an element of Ω must be either *true* or *false*. □

Now in **Set**, the co-product $1+1$ is a two-element set and hence isomorphic to $\Omega=2$ (this was observed in §3.9). In fact the isomorphism is given by the co-product arrow $[\top,\bot]:1+1\to\Omega$

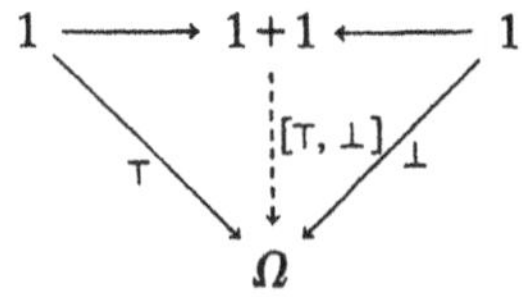

But any topos $\mathscr{E}$ has co-products, and so the arrow $[\top,\bot]$ is certainly *defined*. If $[\top,\bot]$ is an *iso* $\mathscr{E}$-arrow we will say that $\mathscr{E}$ is a *classical* topos. Shortly we shall see that there are non-classical topoi. However we do have

THEOREM 3. *In any topos, $[\top,\bot]$ is monic.*

To prove this we need to do some preliminary work with co-product arrows. If $f:a \to b$ and $g:c \to b$ are $\mathscr{E}$-arrows, we say that f and g are *disjoint* if their pullback is 0, i.e. if

$$\begin{array}{ccc} 0 & \xrightarrow{!} & c \\ {\scriptstyle !}\downarrow & & \downarrow{\scriptstyle g} \\ a & \xrightarrow[f]{} & b \end{array}$$

is a pullback square in $\mathscr{E}$. (In **Set** this means precisely that $\operatorname{Im} f \cap \operatorname{Im} g = \emptyset$.)

Lemma. *If* $f:a \rightarrowtail b$ and $g:c \rightarrowtail b$ *are disjoint* monics *in* $\mathscr{E}$, *then* $[f,g]:a+c \to b$ *is monic.*

Proof. g being monic means

$$\begin{array}{ccc} c & \xrightarrow{1_c} & c \\ {\scriptstyle 1_c}\downarrow & & \downarrow{\scriptstyle g} \\ c & \xrightarrow[g]{} & b \end{array}$$

is a pullback. This, with the previous diagram, and Fact 2 of §5.3, gives the pullback

$$\begin{array}{ccc} 0+c & \xrightarrow{[0_c,\,1_c]} & c \\ {\scriptstyle 0_a+1_c}\downarrow & & \downarrow{\scriptstyle g} \\ a+c & \xrightarrow[{[f,g]}]{} & b \end{array}$$

Now $[0_c, 1_c]: 0+c \cong c$ (dual of Exercise 3.8.4), from which it can be shown that

$$\begin{array}{ccc} c & \xrightarrow{1_c} & c \\ {\scriptstyle i_c}\downarrow & & \downarrow{\scriptstyle g} \\ a+c & \xrightarrow[{[f,g]}]{} & b \end{array}$$

is a pullback (i_c being the injection associated with $a+c$).

Analogously we get

$$\begin{array}{ccc} a & \xrightarrow{1_a} & a \\ {\scriptstyle i_a}\big\downarrow & & \big\downarrow{\scriptstyle f} \\ a+c & \xrightarrow[{[f,\,g]}]{} & b \end{array}$$

as a pullback. These last two diagrams (suitably rotated and reflected), with Fact 2 again, give

$$\begin{array}{ccc} a+c & \xrightarrow{[i_a,\,i_c]} & a+c \\ {\scriptstyle 1_a+1_c}\big\downarrow & & \big\downarrow{\scriptstyle [f,\,g]} \\ a+c & \xrightarrow[{[f,\,g]}]{} & b \end{array}$$

as a pullback. But $[i_a, i_c] = 1_{a+c} = 1_a + 1_c$ (dual of Exercises 1, 4, §3.8), and from this it follows that $[f, g]$ is monic (cf. Example 9, §3.13). □

Now, for the proof of Theorem 3 we observed that

$$\begin{array}{ccc} 0 & \xrightarrow{!} & 1 \\ {\scriptstyle !}\big\downarrow & & \big\downarrow{\scriptstyle \bot} \\ 1 & \xrightarrow[\top]{} & \Omega \end{array}$$

is a pullback, indeed this diagram gives the definition of $\bot$. Thus $\top$ and $\bot$ are *disjoint* monics, and so by the Lemma, $[\top, \bot]: 1+1 \to \Omega$ is monic.

□

THEOREM 4. *If* $\mathscr{E}$ *is well-pointed, then* $[\top, \bot]$: $1+1 \cong \Omega$, i.e. $\mathscr{E}$ *is classical.*

PROOF. In view of Theorem 3, we need only establish that $[\top, \bot]$ is epic, when $\mathscr{E}$ is well-pointed. So, suppose $f \circ [\top, \bot] = g \circ [\top, \bot]$.

$$\begin{array}{ccccc} 1 & \xrightarrow{i} & 1+1 & \xleftarrow{j} & 1 \\ & {\scriptstyle \top}\searrow & \big\downarrow{\scriptstyle [\top,\,\bot]} & \swarrow{\scriptstyle \bot} & \\ & & \Omega & & \\ & & f\big\downarrow\big\downarrow g & & \\ & & a & & \end{array}$$

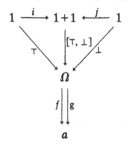

Then

$$f \circ \top = f \circ [\top, \bot] \circ i$$
$$= g \circ [\top, \bot] \circ i$$
$$= g \circ \top$$

and similarly, (using j), $f \circ \bot = g \circ \bot$. Since $\top$ and $\bot$ are the only elements of Ω (Theorem 2), and neither of them distinguish f and g, the extensionality principle for arrows implies that $f = g$. Thus $[\top, \bot]$ *is right-cancellable.* □

The major link between the concepts of this section is:

THEOREM 5. *A topos $\mathscr{E}$ is well-pointed iff it is classical and every non-zero $\mathscr{E}$-object is non-empty in $\mathscr{E}$.*

The "only if" part of this theorem is given by Theorems 4 and 1. The proof of the "if" part requires some notions to be introduced in subsequent chapters, and will be held in abeyance until §7.6.

The category $\mathbf{Set}^2$ is classical, but not bivalent (it has four truth-values – what are they?) The category $\mathbf{Set}^{\rightarrow}$ of functions on the other hand is neither bivalent (having three truth-values) nor classical (cf. Chapter 10). To construct an example of a non-classical but bivalent topos we use the following interesting fact:

THEOREM 6. *If* **M** *is a monoid, then the category* **M-Set** *is classical iff* **M** *is a group.*

PROOF. In **M-Set**, $1 = (\{0\}, \lambda_0)$ is the one-element **M**-set. $1+1$ can be described as the disjoint union of 1 with itself, i.e. two copies of 1 acting independently. To be specific we put $1+1 = (\{0, 1\}, \gamma)$, where $\gamma(m, 0) = 0$ and $\gamma(m, 1) = 1$, all $m \in M$. We then have the co-product diagram

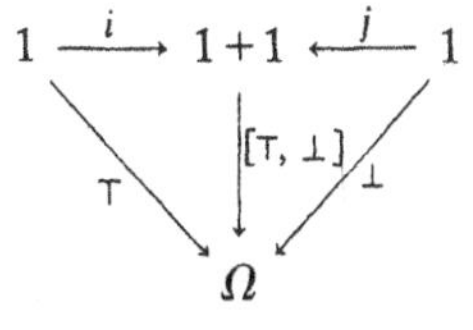

where the injections are $i(0) = 0$ and $j(0) = 1$, with $[\top, \bot]$ mapping 0 to M and 1 to $\emptyset$ in $\Omega = (L_M, \omega)$. Now if $[\top, \bot]$ is iso, it is a bijection of sets, and so L_M has only two elements. Hence $L_M = \{M, \emptyset\}$. Conversely if $L_M = \{M, \emptyset\}$ then as $\omega(m, M) = M$ and $\omega(m, \emptyset) = \emptyset$, $[\top, \bot]$ is an equivariant bijection, i.e. an iso arrow in **M-Set**. Thus **M-Set** is classical iff $L_M = \{M, \emptyset\}$. But this last condition holds precisely when **M** is a group, (Exercise 4.6.3). □

So to construct a non-classical topos we need only select a monoid that is not a group. The natural thing to do is pick the smallest one. This is a two element algebra which can be described simply as consisting of the numbers 0 and 1 under multiplication. Formally it is the structure $\mathbf{M}_2 = (2, \cdot, 1)$ where $2 = \{0, 1\}$ and $\cdot$ is defined by

$$1 \cdot 1 = 1, \qquad 1 \cdot 0 = 0 \cdot 1 = 0 \cdot 0 = 0,$$

or in a table

	1	0
1	1	0
0	0	0

$\mathbf{M}_2$ is a monoid with identity 1, in which 0 has no inverse. The category of $\mathbf{M}_2$-sets is a kind of "universal counterexample" that will prove extremely useful for illustrative purposes. We will call it simply "the topos $\mathbf{M}_2$".

The set L_2 of left ideals of $\mathbf{M}_2$ has three elements, 2, $\emptyset$, and $\{0\}$ (why is $\{1\}$ not a left ideal?). Thus in $\mathbf{M}_2$, $\Omega = (L_2, \omega)$, where the action

$$\omega : 2 \times L_2 \to L_2,$$

defined by

$$\omega(m, B) = \{n : n \in 2 \text{ and } n \cdot m \in B\},$$

can be presented by the table

ω	2	$\{0\}$	$\emptyset$
1	2	$\{0\}$	$\emptyset$
0	2	2	$\emptyset$

Now the map $[\top, \bot]$ as considered in Theorem 6 is not iso. To show explicitly that it is not epic, consider $f_\Omega : L_2 \to L_2$ defined by

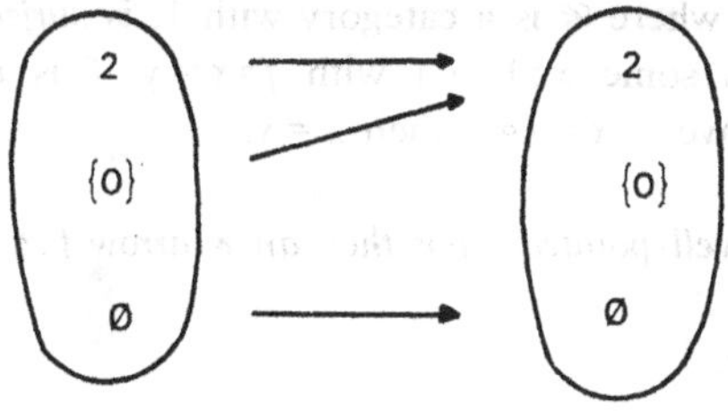

Fig. 5.1.

$$f_\Omega(2) = f_\Omega(\{0\}) = 2$$
$$f_\Omega(\emptyset) = \emptyset$$

By the table for ω, f_Ω is equivariant, so is an arrow $f_\Omega : \Omega \to \Omega$ in $\mathbf{M_2}$. But $f_\Omega \circ [\top, \bot] = 1_\Omega \circ [\top, \bot]$, while $f_\Omega \neq 1_\Omega$, hence $[\top, \bot]$ is not right-cancellable. Though $\mathbf{M_2}$ is non-classical, it is bivalent. For if $h : 1 \to \Omega$ is an $\mathbf{M_2}$-arrow, then $h : \{0\} \to L_2$ is an equivariant map, so $\omega(0, h(0)) = h(\lambda_0(0, 0)) = h(0)$. Since $\omega(0, \{0\}) = 2 \neq \{0\}$, we cannot have $h(0) = \{0\}$. Thus either $h(0) = 2$, whence $h = \top$, or $h(0) = \emptyset$, whence $h = \bot$. So $\mathbf{M_2}$ has only two truth-values.

By Theorem 4, $\mathbf{M_2}$ is not well-pointed. To see this explicitly, observe that $f_\Omega \neq 1_\Omega$ (f_Ω as above), but $f_\Omega \circ \top = 1_\Omega \circ \top$ (both output 2) while $f_\Omega \circ \bot = 1_\Omega \circ \bot$ (both output $\emptyset$). Thus no element of Ω distinguishes the distinct arrows $f_\Omega, 1_\Omega : \Omega \rightrightarrows \Omega$.

Exercise 5. Show that if $a = (X, \lambda)$ is an object in **M-Set** (**M** any monoid) then an element $x : 1 \to a$ of a in **M-Set** can be identified with a *fixed point* of a, i.e. an element $y \in X$ such that $\lambda(m, y) = y$, all $m \in M$. □

In the light of this exercise we can show that the converse of Theorem 1 above is false. If $a = (X, \lambda)$ is a non-zero object in $\mathbf{M_2}$, then $X \neq \emptyset$. Take some $x \in X$, and put $y = \lambda(0, x)$. Then y is a fixed point of a, since $\lambda(m, y) = \lambda(m \cdot 0, x) = \lambda(0, x) = y$. In this way we see that every non-zero object in $\mathbf{M_2}$ is non-empty, even though $\mathbf{M_2}$ is not well-pointed.

5.5. Monics and epics by elements

Using our notion of elements as arrows of the form $1 \to a$ we can give categorial definitions of "injective" and "surjective".

A $\mathscr{C}$-arrow $f:a \to b$, where $\mathscr{C}$ is a category with 1, is *surjective* if for each $y:1 \to b$ there is some $x:1 \to a$ with $f \circ x = y$. f is *injective* if whenever $x, y:1 \rightrightarrows a$ have $f \circ x = f \circ y$, then $x = y$.

THEOREM 1. *If $\mathscr{E}$ is a well-pointed topos then an $\mathscr{E}$-arrow $f:a \to b$ is*
 (i) *surjective iff epic*
 (ii) *injective iff monic.*

PROOF. (i) Suppose f surjective. Let $g,h:b \rightrightarrows c$ be such that $g \circ f = h \circ f$. If $g \neq h$ then there is some $y:1 \to b$ such that $g \circ y \neq h \circ y$. But as f is surjective, $y = f \circ x$ for some $x:1 \to a$. Then $g \circ y = g \circ f \circ x = h \circ f \circ x = h \circ y$, a contradiction. So we must conclude that $g = h$, and that f cancels on the right.

Conversely assume f epic. Given $y:1 \to b$, form the pullback

$$\begin{array}{ccc} c & \xrightarrow{p} & 1 \\ {\scriptstyle q}\downarrow & & \downarrow{\scriptstyle y} \\ a & \xrightarrow{f} & b \end{array}$$

Now p is epic, by Fact 1 of §5.3, so if $c \cong 0$, then p would be monic (Theorem 3.16.1), hence iso, making $0 \cong 1$ and $\mathscr{E}$ degenerate. So c must be non-zero, ergo (Theorem 1) there exists $z:1 \to c$. Then putting $x = q \circ z$ we get $x:1 \to a$ and $f \circ x = y$ (details?).

EXERCISE 1. Prove Part (ii) of the Theorem.

EXERCISE 2. Show that in $\mathbf{M}_2$, f_Ω is surjective, although not epic, and similarly for $[\top, \bot]$.

EXERCISE 3. Show that f_Ω is not monic, but is injective. □

We will return to the subject of well-pointed topoi and extensionality in Chapters 7 and 12.

CHAPTER 6

LOGIC CLASSICALLY CONCEIVED

"It is not easy, and perhaps not even useful, to explain briefly what logic is."

E. J. Lemmon

6.1. Motivating topos logic

In any systematic development of set theory one of the first topics to be examined is the so-called *algebra of classes*. This is concerned with ways of defining new sets, and when relativised to the subsets of a given set D focuses on the operations of

Intersection: $A \cap B = \{x: x \in A \text{ and } x \in B\}$
Union: $A \cup B = \{x: x \in A \text{ or } x \in B\}$
Complement: $-A = \{x: x \in D \text{ and not } x \in A\}$

The power set $\mathcal{P}(D)$ together with the operations $\cap$, $\cup$, $-$ exhibit the structure of what is known as a *Boolean algebra*. These algebras, to be defined shortly, are intimately connected with the classical account of logical truth.

Now the operations $\cap$, $\cup$, $-$ can be characterised by universal properties, and hence defined in any topos, yielding an "algebra of subobjects". It turns out that in some cases, this algebra does not satisfy the laws of Boolean algebra, indicating that the "logic" of the topos is not the same as classical logic. The proper perspective, it would seem, is that the algebra of subobjects is non-Boolean *because* the topos logic is non-classical, rather than the other way round. In defining $\cap$, $\cup$, $-$ we used the words "and", "or", and "not", and so the properties of the set operations are determined by the meaning, the logical behaviour, of these words. It is the rules of classical logic that dictate that $\mathcal{P}(D)$ should be a Boolean algebra.

The classical rules of logic are representable in **Set** by operations on the set $2 = \{0, 1\}$, and can then be developed in any topos $\mathscr{E}$ by using Ω in place of 2. This gives the "logic" of $\mathscr{E}$, which proves to characterise the behaviour of subobjects in $\mathscr{E}$. It is precisely when this logic fails to reflect all the principles of classical logic (i.e. the logic of **Set**) that the algebra of subobjects in $\mathscr{E}$ fails to be Boolean.

In this chapter we will briefly (in spite of Lemmon's caveat) outline the basics of classical logic and show how it generalises to the topos setting. Later chapters will deal with non-classical logic, and its philosophical motivation, leading eventually to a full account of what the logic of the general topos looks like.

6.2. Propositions and truth-values

A *proposition*, or *statement*, or *sentence*, is simply an expression that is either true or false. Thus

"2+2=4"

and

"2 plus 2 equals 5"

are to count as propositions, while

"Is 2+2 equal to 4?"

and

"Add 2 and 2!"

are not.

Thus each sentence has one of two *truth-values*. It is either *true*, which we indicate by assigning it the number 1, or *false*, indicated by the assignment of 0. The set of truth-values is $2=\{0,1\}$ (hence the terminology used earlier for arrows $1 \to \Omega$).

We may construct compound sentences from given ones by the use of the *logical connectives* "and", "or", and "not", i.e. given sentences α and β we form the new sentences

"α and β" symbolised "$\alpha \wedge \beta$"

"α or β" symbolised "$\alpha \vee \beta$"

"not α" symbolised "$\sim\alpha$".

These are said to be obtained by *conjunction*, *disjunction*, and *negation*, respectively.

The truth-value of a compound sentence can be computed from the truth-values of its components, using some simple rules that we now describe.

Negation

The sentence $\sim\alpha$ is to be true (assigned 1) when α is false (assigned 0), and false (0) when α is true (1).

We present this rule in the form of a table

α	$\sim\alpha$
1	0
0	1

called the *truth-table* for negation. Alternatively we can regard it as determining a function $\neg$ from 2 to 2 that outputs 0 (resp. 1) for input 1 (resp. 0). This $\neg: 2 \to 2$, defined by $\neg 1 = 0$, $\neg 0 = 1$, is called the *negation truth-function.*

Conjunction

In order for $\alpha \wedge \beta$ to be true, both of α and β must be true. Otherwise $\alpha \wedge \beta$ is false.

Now, given two sentences α, and β, there are four ways their possible truth-values can be combined, as in the four rows of the truth-table

α	β	$\alpha \wedge \beta$
1	1	1
1	0	0
0	1	0
0	0	0

for conjunction. The corresponding truth-value for $\alpha \wedge \beta$ in each row is determined according to the above rule.

The table provides a function $\cap$ from pairs of truth-values to truth-values, i.e. $\cap: 2 \times 2 \to 2$, defined by $1 \cap 1 = 1$, $1 \cap 0 = 0 \cap 1 = 0 \cap 0 = 0$. This is called the *conjunction truth-function*, which can also be presented in a tabular display as

$\cap$	1	0
1	1	0
0	0	0

Disjunction

$\alpha \vee \beta$ is true provided at least one of α and β are true, and is false only if both of α and β are false.

From this rule we obtain the disjunction truth-table

α	β	$\alpha \vee \beta$
1	1	1
1	0	1
0	1	1
0	0	0

and the corresponding *disjunction truth-function* $\cup : 2 \times 2 \to 2$, which has $1 \cup 1 = 1 \cup 0 = 0 \cup 1 = 1$, $0 \cup 0 = 0$, i.e.

$\cup$	1	0
1	1	1
0	1	0

Implication

The implication connective allows us to form the sentence "α implies β" symbolised "$\alpha \supset \beta$".

(synonyms: "if α then β", "α only if β".)

The classical interpretation of the connective "implies" is that $\alpha \supset \beta$ cannot be a true implication if it allows us to infer something false from something true. So we make $\alpha \supset \beta$ false if α is true while β is false. In all other cases $\alpha \supset \beta$ counts as true. The truth-table is

α	β	$\alpha \supset \beta$
1	1	1
1	0	0
0	1	1
0	0	1

The *implication truth function* $\Rightarrow : 2 \times 2 \to 2$ has $1 \Rightarrow 0 = 0$, $1 \Rightarrow 1 = 0 \Rightarrow 1 = 0 \Rightarrow 0 = 1$, or

$\Rightarrow$	1	0
1	1	0
0	1	1

Tautologies

By successive applications of the rules just given we can construct a truth-table for any compound sentence. For example

α	$\sim\alpha$	$\alpha\wedge\sim\alpha$	$\alpha\vee\sim\alpha$	$\alpha\wedge\alpha$	$\alpha\supset(\alpha\wedge\alpha)$
1	0	0	1	1	1
0	1	0	1	0	1

α	β	$\alpha\supset\beta$	$\beta\supset(\alpha\supset\beta)$	$\alpha\vee\beta$	$\alpha\supset(\alpha\vee\beta)$
1	1	1	1	1	1
1	0	0	1	1	1
0	1	1	1	1	1
0	0	0	1	0	1

A *tautology* is by definition a sentence whose truth-table contains only 1's. Thus $\alpha\vee\sim\alpha$, $\alpha\supset(\alpha\wedge\alpha)$, $\beta\supset(\alpha\supset\beta)$, $\alpha\supset(\alpha\vee\beta)$, are all tautologies. Such sentences are true no matter what truth-values their component parts have. The truth of $\alpha\vee\sim\alpha$ comes not from the truth or falsity of α, but from the logical "shape" of the sentence, the way its logical connectives are arranged. A tautology then expresses a logical law, a statement that is true for purely logical reasons, and not because of any facts about the world that happen to be the case.

6.3. The propositional calculus

In order to further our study of logic we need to give a somewhat more precise rendering of our description of propositions and truth-values. This is done by the device of a *formal language*. Such a language is presented as an alphabet (list of basic symbols) together with a set of *formation rules* that allow us to make *formulae* or *sentences* out of the alphabet symbols. The language we shall use, called PL, has the following ingredients:

Alphabet for PL

(i) an infinite list $\pi_0, \pi_1, \pi_2, \ldots$ of symbols, to be called *propositional variables*, or *sentence letters*;

(ii) the symbols $\sim, \wedge, \vee, \supset$;

(iii) the bracket symbols), (.

Formation Rules for PL-sentences

(1) Each sentence letter π_i is a sentence;
(2) If α is a sentence, so is $\sim\alpha$;
(3) If α and β are sentences, then so are $(\alpha\wedge\beta)$, $(\alpha\vee\beta)$, $(\alpha\supset\beta)$.

Notice that we are using the letters α and β as general names for sentences. Thus α might stand for a letter, like π_{24}, or something more complex, like $(\sim(\pi_2\wedge\pi_{11})\vee(\pi_0\supset\pi_0))$. The collection of sentence letters is denoted Φ_0, while Φ denotes the set of all sentences, i.e.

$$\Phi_0=\{\pi_0, \pi_1, \pi_2, \ldots\}$$

$$\Phi=\{\alpha: \alpha \text{ is a PL-sentence}\}.$$

To develop a theory of meaning, or *semantics*, for PL we use the truth-functions defined in §6.2. By a *value assignment* we shall understand any function V from Φ_0 to $\{0, 1\}$. Such a $V:\Phi_0\to 2$ assigns a truth-value $V(\pi_i)$ to each sentence letter, and so provides a "meaning" or "interpretation" to the members of Φ_0. This interpretation can then be systematically extended to all sentences, so that V extends to a function from Φ to 2. This is done by "induction over the formation rules", through successive application of the rules

(a) $V(\sim\alpha)=\neg V(\alpha)$
(b) $V(\alpha\wedge\beta)=V(\alpha)\cap V(\beta)$
(c) $V(\alpha\vee\beta)=V(\alpha)\cup V(\beta)$
(d) $V(\alpha\supset\beta)=V(\alpha)\Rightarrow V(\beta)$

Example. If $V(\pi_0)=V(\pi_1)=1$, and $V(\pi_2)=0$, then

$$V(\sim\pi_1)=\neg V(\pi_1)=\neg 1=0$$

$$V(\sim\pi_1\wedge\pi_2)=V(\sim\pi_1)\cap V(\pi_2)=0\cap 0=0$$

$$V(\pi_0\supset(\sim\pi_1\wedge\pi_2))=V(\pi_0)\Rightarrow V(\sim\pi_1\wedge\pi_2)=1\Rightarrow 0=0$$

etc. □

In this way any $V:\Phi_0\to 2$ is "lifted" in a unique way to become a function $V:\Phi\to 2$.

A sentence $\alpha\in\Phi$ is then defined to be a *tautology*, or *classically valid*, if it receives the value "true" from every assignment whatsoever. Thus α is a tautology, denoted $\vDash\alpha$, iff for each value-assignment V, $V(\alpha)=1$.

Axiomatics

The semantics for PL allows us to single out a special class of sentences – the tautologies. There is another way of characterising this class, namely by the use of an axiom system. Axiomatics are concerned with methods of generating new sentences from given ones, through the application of *rules of inference.* These rules, allowing us to "infer", or "derive", certain sentences, embodying principles of deduction and techniques of reasoning.

The basic ingredients of an axiom system then are

(i) a collection of sentences, called *axioms* of the system;

(ii) a collection of *rules of inference* which prescribe operations to be performed on sentences, to derive new ones.

Sentences derivable from the axioms are called *theorems.* To specify these a little more precisely we introduce the notion of a *proof sequence* as a finite sequence of sentences, each of which is either

(i) an axiom, or

(ii) derivable from earlier members of the sequence by one of the system's inferential rules.

A theorem can then be defined as a sentence which is the last member of some proof sequence. The set of theorems of an axiom system is said to be *axiomatised* by that system.

There are several known systems that axiomatise the classically valid sentences, i.e. whose theorems are precisely the tautologies of PL. The one we shall deal with will be called CL (for Classical Logic).

The axioms for CL comprise all sentences that are instances of one of the following twelve forms (α, β, and γ denote arbitrary sentences).

I $\alpha \supset (\alpha \wedge \alpha)$

II $(\alpha \wedge \beta) \supset (\beta \wedge \alpha)$

III $(\alpha \supset \beta) \supset ((\alpha \wedge \gamma) \supset (\beta \wedge \gamma))$

IV $((\alpha \supset \beta) \wedge (\beta \supset \gamma)) \supset (\alpha \supset \gamma)$

V $\beta \supset (\alpha \supset \beta)$

VI $(\alpha \wedge (\alpha \supset \beta)) \supset \beta$

VII $\alpha \supset (\alpha \vee \beta)$

VIII $(\alpha \vee \beta) \supset (\beta \vee \alpha)$

IX $((\alpha \supset \gamma) \wedge (\beta \supset \gamma)) \supset ((\alpha \vee \beta) \supset \gamma)$

X $\sim\alpha \supset (\alpha \supset \beta)$

XI $((\alpha \supset \beta) \wedge (\alpha \supset \sim\beta)) \supset \sim\alpha$

XII $\alpha \vee \sim\alpha$

The system CL has a single rule of inference;

RULE OF DETACHMENT. *From α and $\alpha \supset \beta$, the sentence β may be derived.*

This rule is known also by its medieval name, *modus ponens*, more correctly modus ponendo ponens. It operates on a pair of theorems, an implication and its *antecedent*, to "detach" the *consequent* as a new theorem.

By writing "$\vdash_{CL} \alpha$" to indicate that α is a CL-theorem the rule of detachment can be expressed as

$$\textit{if} \vdash_{CL} \alpha \quad \textit{and} \quad \vdash_{CL} (\alpha \supset \beta), \quad \textit{then} \quad \vdash_{CL} \beta.$$

The demonstration that the CL-theorems are precisely the tautologies falls into two parts:

SOUNDNESS THEOREM. *If* $\vdash_{CL} \alpha$, *then* α *is classically valid.*

COMPLETENESS THEOREM. *If* α *is classically valid, then* $\vdash_{CL} \alpha$.

In general a "soundness" theorem for an axiom system is a result to the effect that only sentences of a certain kind are derivable as theorems, while a "completeness" theorem states that all sentences of a certain kind are derivable. Together they give an exact characterisation of a particular type of sentence in terms of derivability. Thus the results just quoted state that theoremhood in CL characterises classical validity.

To prove the Soundness theorem is easy, in the sense that a computer could do it. First one shows that all of the axioms are tautologies (the truth-tables in §6.2 show that the axioms of the forms I, V, VII, and XII are tautologies). Then one shows that *detachment* "preserves" validity, i.e. if α and $\alpha \supset \beta$ are tautologies, then β is also a tautology. This implies that a proof sequence can consist only of valid sentences, hence every theorem of CL is valid.

The Completeness theorem requires more than a mechanical procedure for its verification. The first result of this kind for classical logic was established in 1921 by Emil Post, who proved that all tautologies were derivable in the system used by Russell and Whitehead in *Principia Mathematica*. Since then a number of methods have been developed for proving completeness of various axiomatisations of classical logic. A survey of these may be found in a paper by Surma [73].

6.4 Boolean algebra

The set 2, together with the truth-functions $\neg$, $\cap$, $\cup$ forms a Boolean algebra, a structure that we have mentioned several times and now at last are going to define. The definition proceeds in several stages.

Recall from Chapter 3 that a lattice is a poset $\mathbf{P} = (P, \sqsubseteq)$ in which any two elements $x, y \in P$ have

(i) a greatest lower bound (g.l.b.), $x \sqcap y$; and
(ii) a least upper bound (l.u.b.), $x \sqcup y$.

$x \sqcap y$ is also known as the lattice *meet* of x and y, while $x \sqcup y$ is the *join* of x and y. As observed in §§3.8, 3.9, when $\mathbf{P}$ is considered as a category, meets are products and joins are co-products.

Recall from §§3.5, 3.6 that a *zero* or *minimum* for a lattice is an element 0 having $0 \sqsubseteq x$, all $x \in P$, while a *unit* or *maximum* is an element 1 having $x \sqsubseteq 1$, all $x \in P$. A lattice is said to be *bounded* if it has a unit and a zero. Categorially, 0 is initial and 1 is terminal. Now a lattice always has pullbacks and pushouts (§3.13, Example 5 and its dual), so a bounded lattice is precisely (§3.15) a *finitely bicomplete skeletal pre-order category*.

EXAMPLE 1. $(\mathscr{P}(D), \subseteq)$ is a bounded lattice. The unit is D, the zero $\emptyset$, the meet of A and B is their intersection $A \cap B$, and the join is their union $A \cup B$.

EXAMPLE 2. The set $2 = \{0, 1\}$ has the natural ordering $0 \leqslant 1$ which makes it into the ordinal pre-order $\mathbf{2}$ (Example 2, Chapter 2)

0 is the zero, and 1 the unit in this poset. $x \cap y$ is both the lattice meet and the result of applying the conjunction truth function to $\langle x, y\rangle \in 2 \times 2$. Likewise $x \cup y$ is both the join of x and y and their disjunction.

EXAMPLE 3. If I is a topological space with Θ its collection of open sets, then $(\Theta, \subseteq)$ is a poset exactly as in Example 1 – joins and meets are unions and intersections, the zero is $\emptyset$, and the unit is I.

EXAMPLE 4. $(L_M, \subseteq)$ is a bounded lattice, where L_M is the set of left ideals of monoid M. Joins and meets are as in Examples 1 and 3. □

A lattice is said to be *distributive* if it satisfies the following laws (each of which implies the other in any lattice):

(a) $x \sqcap (y \sqcup z) = (x \sqcap y) \sqcup (x \sqcap z)$

(b) $x \sqcup (y \sqcap z) = (x \sqcup y) \sqcap (x \sqcup z)$ all x, y, z.

EXAMPLE 5. All four examples above are distributive. □

To complete our description of a Boolean algebra we need one further notion – a lattice version of complementation.

In a bounded lattice, y is said to be a *complement* of x if

$$x \sqcup y = 1$$

and

$$x \sqcap y = 0.$$

A bounded lattice is *complemented* if each of its elements has a complement in the lattice.

EXAMPLE 6. $(\mathscr{P}(D), \subseteq)$ is complemented. The lattice complement of A is its set complement $-A$.

EXAMPLE 7. $(2, \leqslant)$ is complemented. The complement of x is its negation $\neg x$ (cf. truth-tables for $\alpha \vee \sim\alpha$, $\alpha \wedge \sim\alpha$).

EXAMPLE 8. In $(\Theta, \subseteq)$ the only candidate for the complement of $U \in \Theta$ is its set complement. But $-U \notin \Theta$ unless U is closed. Thus $(\Theta, \subseteq)$ will only be complemented in the event that every open set is also closed.

EXAMPLE 9. If $\mathbf{M}$ is the monoid $\mathbf{M}_2 = (2, \cdot, 1)$ then in $(L_M, \subseteq)$, $\{0\}$ has no lattice complement, as $\{1\} \notin L_M$. □

EXERCISE 1. In a *distributive* lattice each element has *at most* one complement, i.e. if $x \sqcap y = x \sqcap z = 0$ and $x \sqcup y = x \sqcup z = 1$, then $y = z$. □

A *Boolean algebra* (**BA**) is, by definition, a *complemented distributive lattice.*

EXAMPLE. $(\mathscr{P}(D), \subseteq)$ and $\mathbf{2} = (2, \leqslant)$. □

If $\mathbf{B} = (B, \sqsubseteq)$ is a **BA** then each $x \in B$ has, by the above exercise, exactly one complement. We denote it in general by x'.

EXERCISE 2. In any **BA** we have: (1) $(x')' = x$; (2) $x \sqcap y = 0$ iff $y \sqsubseteq x'$; (3) $x \sqsubseteq y$ iff $y' \sqsubseteq x'$; (4) $(x \sqcap y)' = x' \sqcup y'$; (5) $(x \sqcup y)' = x' \sqcap y'$. □

Boolean algebras are named after George Boole (1815–1864) who first described the laws they satisfy in his work, *The Mathematical Analysis of Logic* (1847).

6.5. Algebraic semantics

Each **BA** $\mathbf{B} = (B, \sqsubseteq)$ has operations $\sqcap$ (meet), $\sqcup$ (join), and $'$ (complement) corresponding to the conjunction, disjunction, and negation truth functions on 2. It also has an operation corresponding to implication. The sentence $\alpha \supset \beta$ has exactly the same truth-table as the sentence $\sim\alpha \vee \beta$, and hence on the classical account the two sentences have the same meaning. So for $x, y \in B$ we *define*

$$x \Rightarrow y = x' \sqcup y.$$

EXERCISE 1. Verify that $\alpha \supset \beta$ and $\sim\alpha \vee \beta$ have the same truth-table, and hence that the definition just reproduces the implication truth-function on 2. □

The operations on **B** can be used to generalise the semantics of 6.3.

A **B**-valuation is a function $V: \Phi_0 \to B$. This is extended to a function $V: \Phi \to B$ by the rules

(a) $V(\sim\alpha) = V(\alpha)'$
(b) $V(\alpha \wedge \beta) = V(\alpha) \sqcap V(\beta)$
(c) $V(\alpha \vee \beta) = V(\alpha) \sqcup V(\beta)$
(d) $V(\alpha \supset \beta) = V(\alpha)' \sqcup V(\beta) = V(\alpha) \Rightarrow V(\beta)$.

Then a sentence α is **B**-*valid*, $\mathbf{B} \models \alpha$, iff for every **B**-valuation V, $V(\alpha) = 1$ (where 1 is the unit of **B**). Notice that a **2**-valuation is what we earlier called a value-assignment, and that $\mathbf{2} \models \alpha$ iff α is a tautology.

SOUNDNESS THEOREM FOR **B**-VALIDITY: *If* $\vdash_{CL} \alpha$ *then* $\mathbf{B} \models \alpha$.

The proof of this is as for **2**-validity. One shows that all the CL-axioms are **B**-valid, and that *Detachment* preserves this property.

Now the zero and unit of **B** provide an "isomorphic copy" of **2** within **B**. (**2** is a subobject of **B** in the category of **BA**'s). In this way any **2**-valuation can be construed as a **B**-valuation, hence $\mathbf{B} \models \alpha$ only if $\mathbf{2} \models \alpha$.

A sentence will be called **BA**-*valid* if it is valid in every **BA** (and hence in particular is **2**-valid).

All of these notions of validity are connected by the observation that the following four statements are equivalent to each other:

$\vdash_{\overline{\mathrm{CL}}} \alpha$
α is a tautology
*α is **B**-valid, for some particular **B***
*α is **BA**-valid.*

EXERCISE 2. (*The Lindenbaum Algebra*). Define a relation $\sim_c$ on Φ by

$$\alpha \sim_c \beta \quad \text{iff} \quad \vdash_{\overline{\mathrm{CL}}} \alpha \supset \beta \quad \text{and} \quad \vdash_{\overline{\mathrm{CL}}} \beta \supset \alpha$$

Show that $\sim_c$ is an equivalence relation on Φ and that a partial ordering is well defined on the quotient set $\Phi/{\sim_c}$ by

$$[\alpha] \sqsubseteq [\beta] \quad \text{iff} \quad \vdash_{\overline{\mathrm{CL}}} \alpha \supset \beta$$

The poset $\mathbf{B}_c = (\Phi/{\sim_c}, \sqsubseteq)$ is called the Lindenbaum Algebra of CL. Show that it is a **BA**, in which

$$[\alpha] \sqcap [\beta] = [\alpha \wedge \beta]$$

$$[\alpha] \sqcup [\beta] = [\alpha \vee \beta]$$

$$[\alpha] = [\sim\alpha]$$

$$[\alpha] = 1 \quad \text{iff} \quad \vdash_{\overline{\mathrm{CL}}} \alpha.$$

Define a $\mathbf{B}_c$-valuation V_c by $V_c(\pi_i) = [\pi_i]$, and prove that $V_c(\alpha) = [\alpha]$, all sentences α. Hence show

$$\vdash_{\overline{\mathrm{CL}}} \alpha \quad \text{iff} \quad \mathbf{B}_c \models \alpha. \qquad \square$$

The algebra $\mathbf{B}_c$ can be used to develop a proof that all tautologies are CL-theorems. The details of this can be found in Rasiowa and Sikorski [63], or Bell and Slomson [69].

6.6. Truth-functions as arrows

Each of the classical truth-functions has codomain 2, and so is the characteristic function of some subset of its domain. This observation will lead us to an arrows-only definition of the truth-functions that makes sense in any topos, through the Ω-axiom.

Negation

$\neg : 2 \to 2$ is the characteristic function of the set

$$\{x : \neg x = 1\} = \{0\} \subseteq 2.$$

But the inclusion function $\{0\} \hookrightarrow 2$ is the function we called *false* in §5.4. Hence in **Set** we have the pullback

$$\begin{array}{ccc} 1 & \xrightarrow{\ false\ } & 2 \\ {\scriptstyle !}\downarrow & & \downarrow{\scriptstyle \neg} \\ 1 & \xrightarrow{\ true\ } & 2 \end{array}$$

(recall that *false* is the characteristic function of $\emptyset \subseteq 1$).

Conjunction

The only input to $\cap : 2\times 2 \to 2$ that gives output 1 is $\langle 1, 1\rangle$. Hence $\cap = \chi_A$ where

$$A = \{\langle 1, 1\rangle\}$$

Now A being a one-element set can be identified with an arrow $1 \to 2\times 2$. We see that this arrow is the product map $\langle true, true\rangle$, which takes 0 to $\langle true\ (0),\ true\ (0)\rangle$, and hence

$$\begin{array}{ccc} 1 & \xrightarrow{\ \langle true,\, true\rangle\ } & 2\times 2 \\ \downarrow & & \downarrow{\scriptstyle \cap} \\ 1 & \xrightarrow{\ true\ } & 2 \end{array}$$

is a pullback.

Implication

$\Rightarrow : 2\times 2 \to 2$ is the characteristic function of

$$\circledleq = \{\langle 0, 0\rangle, \langle 0, 1\rangle, \langle 1, 1\rangle\},$$

and so

$$\begin{array}{ccc} \circledleq & \hookrightarrow & 2\times 2 \\ {\scriptstyle !}\downarrow & & \downarrow{\scriptstyle \Rightarrow} \\ 1 & \xrightarrow{\ true\ } & 2 \end{array}$$

is a pullback. Now $\circledleq$ is so named because, as a relation on 2, it is none other than the natural partial ordering on the ordinal **2**, i.e.

$$\circledleq = \{\langle x, y\rangle : x \leqslant y \text{ in } \mathbf{2}\}$$

But in any lattice we in fact have

$$x \sqsubseteq y \quad \text{iff} \quad x \sqcap y = x$$

(why?) so

$$\textcircled{\leqslant} = \{\langle x, y\rangle : x \cap y = x\}$$

and so according to §3.10, $\textcircled{\leqslant} \hookrightarrow 2\times 2$ is the equaliser of

$$2\times 2 \underset{pr_1}{\overset{\cap}{\rightrightarrows}} 2$$

where pr_1 is the projection $pr_1(\langle x, y\rangle) = x$.

Disjunction

$\cup : 2\times 2 \to 2$ is χ_D, where

$$D = \{\langle 1, 1\rangle \langle 1, 0\rangle \langle 0, 1\rangle\}.$$

The description of D by arrows is a little more complex than in the other cases.

Notice first that $D = A \cup B$, where

$$A = \{\langle 1, 1\rangle, \langle 1, 0\rangle\}, \quad \text{and} \quad B = \{\langle 1, 1\rangle, \langle 0, 1\rangle\}.$$

Now $A \subseteq 2\times 2$ can be identified with the monic product map $\langle true_2, 1_2\rangle : 2 \to 2\times 2$ which takes 1 to $\langle 1, 1\rangle$ and 0 to $\langle 1, 0\rangle$. Similarly B is identifiable with $\langle 1_2, true_2\rangle$. We then form the co-product

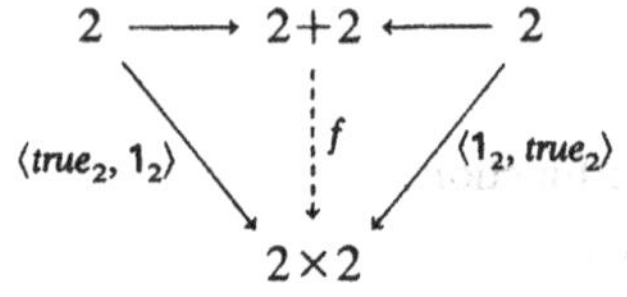

i.e. $f = [\langle true_2, \mathbf{1}_2\rangle, \langle \mathbf{1}_2, true_2\rangle]$ and find that $\operatorname{Im} f = D$. Thus we have an epi-monic factorisation

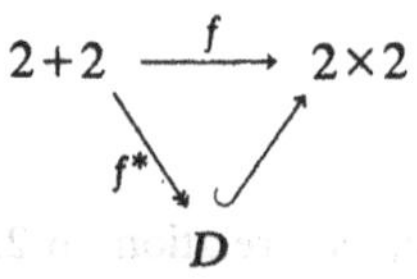

This specifies D uniquely up to isomorphism by properties that can all be expressed in the language of categories, and so we can now define the

Truth-arrows in a topos

If $\mathscr{E}$ is a topos with classifier $\top : 1 \to \Omega$;

(1) $\neg : \Omega \to \Omega$ is the unique $\mathscr{E}$-arrow such that

$$\begin{array}{ccc} 1 & \xrightarrow{\ \perp\ } & \Omega \\ \Big\downarrow & & \Big\downarrow \neg \\ 1 & \xrightarrow{\ \top\ } & \Omega \end{array}$$

is a pullback in $\mathscr{E}$. Thus $\neg = \chi_\perp$, where $\perp$ itself is the character of $! : 0 \to 1$.

(2) $\cap : \Omega \times \Omega \to \Omega$ is the character in $\mathscr{E}$ of the product arrow $\langle \top, \top \rangle : 1 \to \Omega \times \Omega$.

(3) $\cup : \Omega \times \Omega \to \Omega$ is defined to be the character of the image of the $\mathscr{E}$-arrow

$$[\langle \top_\Omega, 1_\Omega \rangle, \langle 1_\Omega, \top_\Omega \rangle] : \Omega + \Omega \to \Omega \times \Omega$$

(4) $\Rightarrow : \Omega \times \Omega \to \Omega$ is the character of

$$e : \textcircled{\scriptsize$\leq$} \rightarrowtail \Omega \times \Omega,$$

where the latter is the equaliser of

$$\Omega \times \Omega \underset{pr_1}{\overset{\cap}{\rightrightarrows}} \Omega,$$

$\cap$ being the conjunction truth arrow, and pr_1 the first projection arrow of the product $\Omega \times \Omega$.

Example 1. In **Set**, and **Finset** the truth arrows are the classical truth functions.

Example 2. In **Bn**(I), where $\Omega = (2 \times I, p_I)$, the stalk Ω_i over i is $2 \times \{i\}$, a "copy" of 2. The truth arrows in **Bn**(I) are essentially bundles of truth-functions, i.e. they consist of "copies" of the corresponding truth-functions acting on each stalk. Thus $\neg : \Omega \to \Omega$ is the function from $2 \times I$ to $2 \times I$ that takes $\langle 1, i \rangle$ to $\langle 0, i \rangle$ and $\langle 0, i \rangle$ to $\langle 1, i \rangle$. $\cap : \Omega \times \Omega \to \Omega$ takes a pair consisting of $\langle x, i \rangle$ and $\langle y, i \rangle$ to $\langle x \cap y, i \rangle$ (recall that $\Omega \times \Omega$ in **Bn**(I) consists only of those pairs that belong to the same stalk in Ω). The reader can readily define the other truth arrows in **Bn**(I).

Thus, whereas in **Set** Ω is the two-element **BA**, in **Bn**(I) Ω is a bundle of two-element **BA**'s, indexed by I.

EXAMPLE 3. In **M-Set**, where $\Omega = (L_M, \omega)$, the negation truth-arrow $\neg: L_M \to L_M$ is defined by

$$\begin{aligned}\neg(B) &= \{m: m \in M \quad \text{and } \omega_m(B) = \emptyset\}\\ &= \{m: \text{for all } n, n * m \notin B\}.\end{aligned}$$

The conjunction arrow is given by set intersection, i.e. it is that function from $L_M \times L_M$ to L_M that takes $\langle B, C\rangle$ to $B \cap C$.

The disjunction arrow is given by set union.

Implication $\Rightarrow: L_M \times L_M \to L_M$ has the description

$$B \Rightarrow C = \{m: \omega_m(B) \subseteq \omega_m(C)\},$$

and $\circledleq$ is the set inclusion relation on L_M.

EXAMPLE 4. In the particular case of our canonical (counter) example $\mathbf{M_2}$, the above definitions show the truth arrows to be given by the tables

	$\neg$
2	$\emptyset$
$\{0\}$	$\emptyset$
$\emptyset$	2

$\cap$	2	$\{0\}$	$\emptyset$
2	2	$\{0\}$	$\emptyset$
$\{0\}$	$\{0\}$	$\{0\}$	$\emptyset$
$\emptyset$	$\emptyset$	$\emptyset$	$\emptyset$

$\cup$	2	$\{0\}$	$\emptyset$
2	2	2	2
$\{0\}$	2	$\{0\}$	$\{0\}$
$\emptyset$	2	$\{0\}$	$\emptyset$

$\Rightarrow$	2	$\{0\}$	$\emptyset$
2	2	$\{0\}$	$\emptyset$
$\{0\}$	2	2	$\emptyset$
$\emptyset$	2	2	2

EXAMPLE 5. The description of truth-arrows in **Top**(I), which in itself gives further indication of the unification achieved by the present theory, will be delayed till Chapter 8. □

EXERCISE 1. Describe the truth-arrows in $\mathbf{Set}^2$.

EXERCISE 2. Describe Ω and the truth-arrows in $\mathbf{Z_2}$-**Set**, where $\mathbf{Z_2} = \langle 2, +, 0\rangle$ is the monoid of the numbers 0 and 1 under addition. □

6.7. $\mathscr{E}$-semantics

We are now able to do propositional logic in any topos $\mathscr{E}$. Recall that a truth value in $\mathscr{E}$ is an arrow $1 \to \Omega$ and that $\mathscr{E}(1, \Omega)$ denotes the collection of such $\mathscr{E}$-arrows.

An $\mathscr{E}$-*valuation* is a function $V: \Phi_0 \to \mathscr{E}(1, \Omega)$ assigning to each sentence letter π_i a truth value $V(\pi_i): 1 \to \Omega$. This function is extended to the whole of Φ by the rules

(a) $V(\sim\alpha) = \neg \circ V(\alpha)$

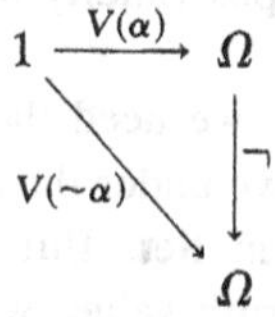

(b) $V(\alpha \wedge \beta) = \cap \circ \langle V(\alpha), V(\beta) \rangle$

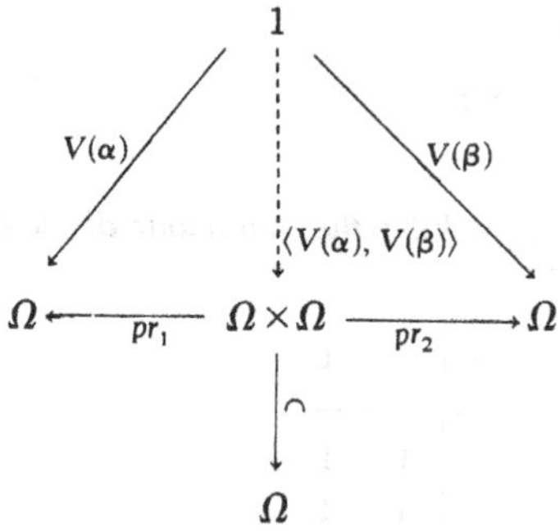

(c) $V(\alpha \vee \beta) = \cup \circ \langle V(\alpha), V(\beta) \rangle$

(d) $V(\alpha \supset \beta) = \Rightarrow \circ \langle V(\alpha), V(\beta) \rangle$.

In this way we extend V so that every sentence is assigned an $\mathscr{E}$-arrow $V(\alpha): 1 \to \Omega$.

We shall say that α is $\mathscr{E}$-*valid*, denoted $\mathscr{E} \vDash \alpha$, iff for every $\mathscr{E}$-valuation V, $V(\alpha) = \top : 1 \to \Omega$.

EXERCISE 1. **Set**$\vDash \alpha$ iff **Finset**$\vDash \alpha$ iff **Finord**$\vDash \alpha$ iff α is a tautology iff $\vdash_{\overline{\mathrm{CL}}} \alpha$.

EXERCISE 2. $\mathbf{Bn}(I) \vDash \alpha$ iff $(\mathscr{P}(I), \subseteq) \vDash \alpha$, i.e. topos-validity in $\mathbf{Bn}(I)$ is equivalent to Boolean-algebra-validity in $(\mathscr{P}(I), \subseteq)$. Hence

$\mathbf{Bn}(I) \vDash \alpha$ iff α is a tautology. □

In the topoi of these exercises, the system CL axiomatises the valid sentences. The natural question is – does this always happen? We are about to see that CL is complete for $\mathscr{E}$-validity, i.e. that any $\mathscr{E}$-valid sentence (whatever $\mathscr{E}$ is) is a CL-theorem. The question then reduces to – "is CL sound for $\mathscr{E}$-validity?" The short answer is – no! A slightly more revealing answer is that axioms I–XI of CL are $\mathscr{E}$-valid, but there are topoi in which the "law of excluded middle", $\alpha \vee {\sim}\alpha$, is not valid. An example is $\mathbf{Set}^{\to}$, the category of set functions, for reasons that will emerge in Chapter 10, where the full story on topos validity will be told, at least for propositional logic.

To show that $\mathscr{E}$-valid sentences are tautologies we need the following result, which shows that the arrows $\top$ and $\perp$ behave under the application of the truth-arrows in $\mathscr{E}$ exactly as they do in **Set**. But first some terminology. If $\langle f, g\rangle : 1 \to \Omega \times \Omega$ is a "pair" of truth-values we write

$$\begin{array}{lll} f \cap g & \text{for} & \cap \circ \langle f, g\rangle : 1 \to \Omega \\ f \cup g & \text{for} & \cup \circ \langle f, g\rangle \\ f \Rightarrow g & \text{for} & \Rightarrow \circ \langle f, g\rangle \quad \text{etc.} \end{array}$$

THEOREM 1. *In any* $\mathscr{E}$, $\top$ *and* $\perp$ *exhibit the behaviour displayed in the tables*

x	$\neg \circ x$
$\top$	$\perp$
$\perp$	$\top$

$\cap$	$\top$	$\perp$
$\top$	$\top$	$\perp$
$\perp$	$\perp$	$\perp$

(*i.e.* $\top \cap \top = \top, \top \cap \perp = \perp$ **etc.**)

$\cup$	$\top$	$\perp$
$\top$	$\top$	$\top$
$\perp$	$\top$	$\perp$

$\Rightarrow$	$\top$	$\perp$
$\top$	$\top$	$\perp$
$\perp$	$\top$	$\top$

PROOF. That $\neg \circ \perp = \top$ follows by commutativity of the pullback that *defines* $\neg$ (cf. §6.6). To see why $\neg \circ \top = \perp$, consider

$$\begin{array}{ccc} 0 & \xrightarrow{\;!\;} & 1 \\ {\scriptstyle !}\big\downarrow & & \big\downarrow{\scriptstyle \top} \\ 1 & \xrightarrow{\;\perp\;} & \Omega \\ \big\downarrow & & \big\downarrow{\scriptstyle \neg} \\ 1 & \xrightarrow{\;\top\;} & \Omega \end{array}$$

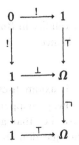

The bottom square is the pullback defining $\neg$. The top square is the pullback (inverted) defining $\perp$ as the character of $!:0 \to 1$. Hence by the PBL, the outer rectangle is a pullback showing $\neg \circ \top$ to be the character of $!:0 \to 1$.

It would be possible to derive the other tables from the relevant definitions, but in Chapter 7 some much deeper facts will be established which yield these tables as a rather easy corollary. So we will leave the details till then (cf. §7.6). □

Now suppose that $V: \Phi_0 \to 2$ is a classical value-assignment. We use V to define an $\mathscr{E}$-valuation $V': \Phi_0 \to \mathscr{E}(1, \Omega)$ by putting

$$V'(\pi_i) = \begin{cases} \top & \text{if} \quad V(\pi_i) = 1 \\ \perp & \text{if} \quad V(\pi_i) = 0. \end{cases}$$

LEMMA. *For any sentence* $\alpha \in \Phi$,

(a) *either* $V'(\alpha) = \top$ *or* $V'(\alpha) = \perp$

(b) $V'(\alpha) = \top$ *iff* $V(\alpha) = 1$.

PROOF. The statement of the Lemma is true when $\alpha = \pi_i$ by definition. The proof itself is by induction over the formation rules for sentences. One proves the statement is true when $\alpha = {\sim}\beta$ on the inductive assumption that it is true for β, is true when $\alpha = \beta \wedge \gamma$ assuming it is true for β and for γ etc. In view of the exact correspondence of the tables of Theorem 1 to the classical truth-tables it should be clear why the Lemma works, and the details are left as an exercise. □

THEOREM 2. *For any topos* $\mathscr{E}$,

$$\text{if} \quad \mathscr{E} \models \alpha \quad \text{then} \quad \vdash_{CL} \alpha$$

PROOF. Let V be any classical valuation and V' its associated $\mathscr{E}$-valuation, as above. Since $\mathscr{E} \models \alpha$, $V'(\alpha) = \top$ and so by the Lemma, $V(\alpha) = 1$. Hence α is assigned 1 by every classical valuation, so is a tautology, whence $\vdash_{CL} \alpha$.

□

THEOREM 3. *If* $\mathscr{E}$ *is bivalent, then*

$$\mathscr{E} \models \alpha \quad \text{iff} \quad \vdash_{CL} \alpha$$

PROOF. Theorem 2 gives the "only if" part. Conversely, suppose $\vdash_{CL} \alpha$, i.e. α is a tautology. If V' is *any* $\mathscr{E}$-valuation, define a classical valuation by $V(\pi_i) = 1$ or 0 according as $V'(\pi_i) = \top$ or $\perp$. Since $\mathscr{E}$ is bivalent, $\top$ and

$\perp$ are its only truth-values, so this definition is legitimate. But then V' and V are related as in the Lemma, so as $V(\alpha)=1$, we get $V'(\alpha)=\top$. □

This last result suggests perhaps that bivalent topoi look more like **Set** than ones with more than two truth-values. However, our example $\mathbf{M_2}$ is bivalent and yet differs from **Set** in other ways, e.g. is non-classical in having $1+1$ not isomorphic to Ω. On the other hand the topos $\mathbf{Set}^2$ is not bivalent, but is classical, and does have its valid sentences axiomatised by CL. We could then conclude that bivalence does not of itself lead to a categorial axiomatisation of classical set theory. Or should we perhaps conclude that our definition of topos validity is not the right generalisation of the notion of logical truth in **Set**? Read on.

Appendix

Sentences α and β are *logically equivalent* when they have the same truth-table, i.e. when $V(\alpha)=V(\beta)$ for every classical valuation V. As was mentioned above, $\alpha \supset \beta$ is logically equivalent to $\sim\alpha \vee \beta$, and because of this some presentations of CL introduce $\supset$, not as a basic symbol of the alphabet, but as a *definitional abbreviation* for a combination involving $\sim$ and $\vee$. Since $\alpha \wedge \beta$ is logically equivalent to $\sim(\sim\alpha \vee \sim\beta)$, $\wedge$ may also be introduced in this way. Alternatively we can start with $\sim$ and $\wedge$ and define $\vee$ and $\supset$, and there are still other approaches.

The definability of $\supset$ from $\sim$ and $\vee$ is reflected by the fact that in 2, $x \Rightarrow y = \neg x \cup y$. In arrow-language this means that

$$\Rightarrow \;=\; \cup \circ (\neg \times id_2)$$

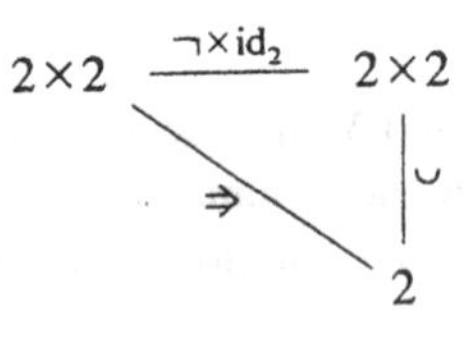

Now there are topoi in which the generalised truth-arrows do not satisfy this equation. So the question must be faced as to why the approach of this chapter is appropriate and why we do not simply define $\Rightarrow$ in $\mathscr{E}$ via $\neg$ and $\cup$ as above.

The point is that the connectives $\sim, \wedge, \vee, \supset$ were introduced separately, as they are all conceptually quite different, and each has its own intrinsic meaning. The construction of the truth-table was motivated independently in each case. That they prove to be inter-definable is *after the fact*. It is simply a feature of classical logic, a *consequence* of the classical account of truth and validity. Accordingly we defined the connectives independently, described them independently through the Ω-axiom, and lifted this description to the general topos. In so doing we find (in some cases) that the interdefinability is left behind. Later (Chapter 8) we shall see a different theory of propositional semantics in which the connectives are not inter-definable but in which they have exactly the same categorial description that they do in **Set**.

CHAPTER 7

ALGEBRA OF SUBOBJECTS

"Since new paradigms are born from old ones, they ordinarily incorporate much of the vocabulary and apparatus, both conceptual and manipulative, that the traditional paradigm had previously employed. But they seldom employ these borrowed elements in quite the traditional way."

Thomas Kuhn

7.1. Complement, intersection, union

At the beginning of Chapter 6 it was asserted that the structure of $(\mathcal{P}(D), \subseteq)$ as **BA** depends on the rules of classical logic, through the properties of the connectives "and", "or", and "not". This can be made quite explicit by the consideration of characteristic functions. We see from the following result just how set operations depend on truth-functions.

THEOREM 1. *If A and B are subsets of D, with characters* $\chi_A : D \to 2$, $\chi_B : D \to 2$, *then*

(i) $\chi_{-A} = \neg \circ \chi_A$

(ii) $\chi_{A \cap B} = \chi_A \frown \chi_B \qquad (= \frown \circ \langle \chi_A, \chi_B \rangle)$

(iii) $\chi_{A \cup B} = \chi_A \smile \chi_B$.

PROOF. If $\chi_{-A}(x) = 1$, for $x \in D$, then $x \in -A$, so $x \notin A$, whence $\chi_A(x) = 0$, so $\neg\chi_A(x) = 1$. But if $\chi_{-A}(x) = 0$, then $x \notin -A$, so $x \in A$, whence $\chi_A(x) = 1$ and $\neg\chi_A(x) = 0$. Thus χ_{-A} and $\neg \circ \chi_A$ give the same output for the same input, and are identical. The proofs of (ii) and (iii) follow similar lines, using the definitions of $\cap$, $\frown$, $\cup$, $\smile$. □

Theorem 1 suggests a generalisation – the result in one context becomes the definition in another, as follows.

Let $\mathscr{E}$ be a topos, and d an $\mathscr{E}$-object. We define operations on the collection Sub(d) of subobjects of d in $\mathscr{E}$ thus:

(1) *Complements:* Given $f: a \rightarrowtail d$, the complement of f (relative to d) is the subobject $-f: -a \rightarrowtail d$ whose character is $\neg \circ \chi_f$. Thus $-f$ is defined to be the pullback

$$\begin{array}{ccc} -a & \overset{-f}{\rightarrowtail} & d \\ \downarrow & & \downarrow{\scriptstyle \neg\circ\chi_f} \\ 1 & \xrightarrow{\top} & \Omega \end{array}$$

of $\top$ along $\neg \circ \chi_f$, yielding $\chi_{-f} = \neg \circ \chi_f$, by definition.

(2) *Intersections:* The intersection of $f: a \rightarrowtail d$ and $g: b \rightarrowtail d$ is the subobject $f \cap g: a \cap b \rightarrowtail d$ obtained by pulling $\top$ back along $\chi_f \cap \chi_g = \cap \circ \langle \chi_f, \chi_g \rangle$.

$$\begin{array}{ccc} a \cap b & \overset{f \cap g}{\rightarrowtail} & d \\ \downarrow & & \downarrow{\scriptstyle \chi_f \cap \chi_g} \\ 1 & \xrightarrow{\top} & \Omega \end{array}$$

Hence $\chi_{f \cap g} = \chi_f \cap \chi_g$.

(3) *Unions:* $f \cup g: a \cup b \rightarrowtail d$ is the pullback of $\top$ along $\chi_f \cup \chi_g = \cup \circ \langle \chi_f, \chi_g \rangle$,

$$\begin{array}{ccc} a \cup b & \xrightarrow{f \cup g} & d \\ {\scriptstyle !}\downarrow & & \downarrow{\scriptstyle \chi_f \cup \chi_g} \\ 1 & \xrightarrow{\top} & \Omega \end{array}$$

and so $\chi_{f \cup g} = \chi_f \cup \chi_g$. □

There is in fact a completely different approach available to the description of intersections and unions in **Set**.

(a) *Intersection:* The diagram

$$\begin{array}{ccc} A \cap B & \hookrightarrow & B \\ \downarrow & & \downarrow \\ A & \hookrightarrow & D \end{array}$$

is a pullback. Now in the poset $(\mathcal{P}(D), \subseteq)$, $A \cap B$ is the g.l.b. of A and B, hence their product, and indeed pullback. But we are saying something stronger than this, namely that the diagram is a pullback, not just in $\mathcal{P}(D)$, but in **Set** itself, as the reader may verify.

(b) *Unions:* In $\mathcal{P}(D)$, $A \cup B$ is the co-product of A and B. This description cannot be generalised as we do not yet know if $\text{Sub}(d)$ has co-products, and moreover in **Set** itself the co-product $A + B$ is the *disjoint* union of A and B, so $A + B \neq A \cup B$ unless A and B are disjoint.

However, $A \cup B$ can be described as the union of the images of the inclusions $f : A \hookrightarrow D$ and $g : B \hookrightarrow D$, and in §6.6, in defining the disjunction arrow $\cup$, we gave a general construction for the union of two images. We form the co-product arrow $[f, g] : A + B \to D$, and then $A \cup B$ obtains as the image of $A + B$ under $[f, g]$, i.e.

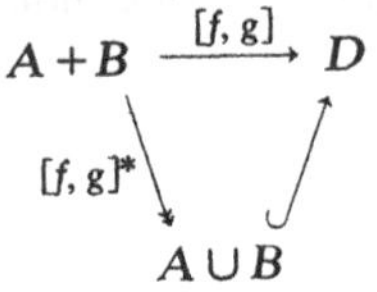

commutes as an epi-monic factorisation of $[f, g]$.

Although we have two descriptions of $\cap$ and $\cup$ in **Set** we are about to see that they present us with no choice in $\mathcal{E}$, i.e. that they lead to the same operations on $\text{Sub}(d)$ (topoi really are the right generalisations of **Set**). The full proof is somewhat lengthy and intricate, and so we shall confine ourselves to outlining the basic strategy and leave the details to the reader who has developed a penchant for "arrow-chasing".

THEOREM 2. *In any topos $\mathcal{E}$, if $f : a \rightarrowtail d$ and $g : b \rightarrowtail d$ have pullback*

$$
\begin{array}{ccc}
c & \xrightarrow{\ f'\ } & b \\
{\scriptstyle g'}\downarrow & & \downarrow{\scriptstyle g} \\
a & \xrightarrow[\ f\ ]{} & d
\end{array}
$$

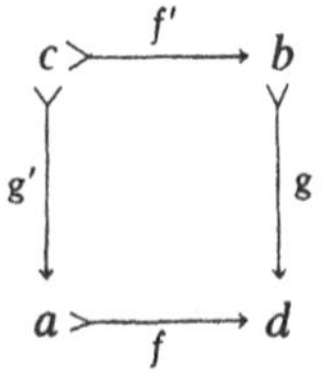

then $\alpha : c \rightarrowtail d$, where $\alpha = g \circ f' = f \circ g'$ has character $\chi_f \cap \chi_g$. Thus $\chi_\alpha = \chi_{f \cap g}$, so $\alpha \cong f \cap g$ and there is a pullback of the form

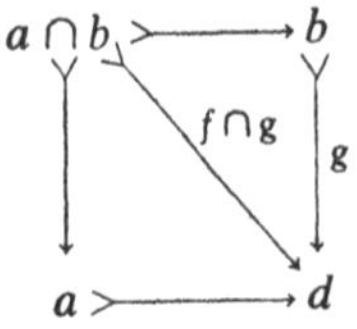

STRATEGY OF PROOF. The heart of the matter is to show that the top square of

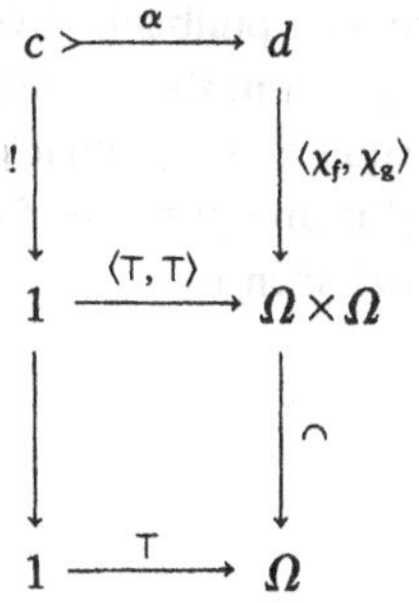

is a pullback. The bottom square is a pullback, by definition of $\cap$, so by the PBL the outer rectangle is a pullback, which by the Ω-axiom leads to the desired result that $\chi_\alpha = \cap \circ \langle \chi_f, \chi_g \rangle$. □

The analogous result for unions needs a preliminary

LEMMA. *In any* $\mathscr{E}$, *if*

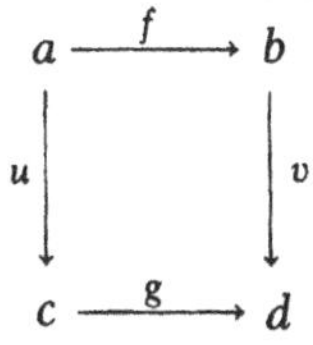

is a pullback, then there is an arrow $h : f(a) \to g(c)$ *that makes the right hand square of*

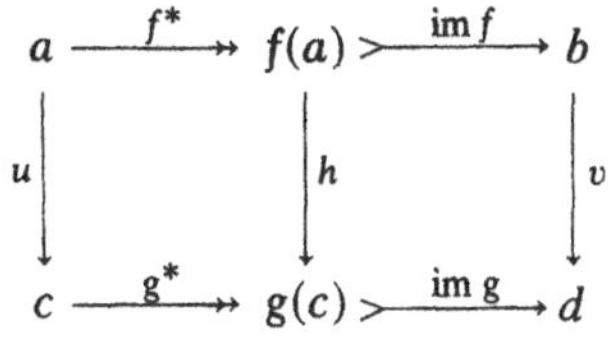

a pullback.

PROOF. Consider

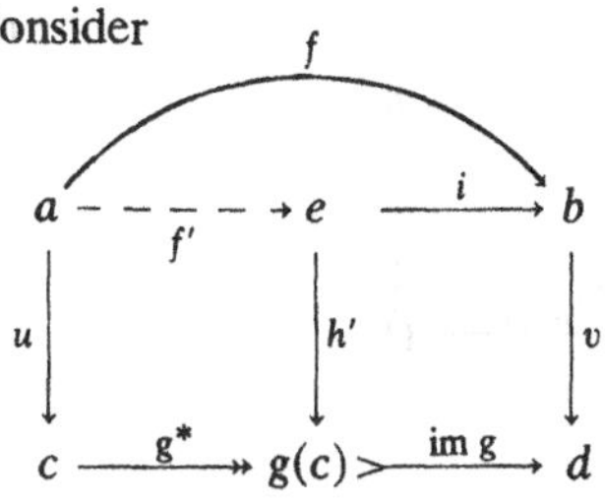

The right hand square obtains by pulling back im g along v, so i is monic. The existence of f', making the whole diagram commute follows from the universal property of the right hand square as a pullback, given that the "boundary" of the diagram is the pullback given in the hypothesis of the Lemma. The PBL then gives the left hand square as a pullback, and since the latter preserve epics (Fact 1, §5.3), $i \circ f'$ is an epi-monic factorisation of f. Hence there is a unique iso $k : e \to f(a)$ such that

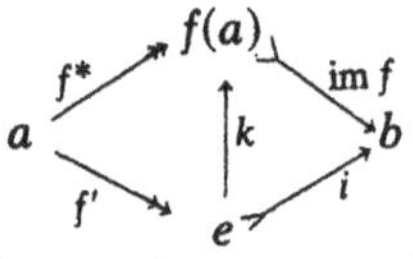

commutes. Then $h = h' \circ k^{-1}$ is the arrow required for the conclusion of the Lemma. □

THEOREM 3. *Given* $f : a \rightarrowtail d$ *and* $g : b \rightarrowtail d$ *in a topos* $\mathscr{E}$, *then the* $\mathscr{E}$-*arrow* $\alpha : c \rightarrowtail d$ *which is the image arrow of* $[f, g] : a + b \to d$,

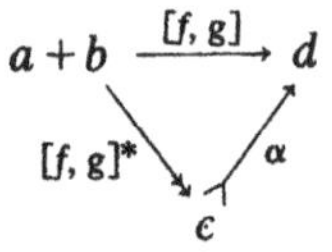

has character $\chi_f \cup \chi_g$.

Thus $\chi_\alpha = \chi_{f \cup g}$, *so* $\alpha \simeq f \cup g$ *and there is an epi-monic factorisation*

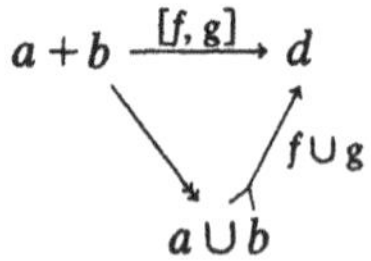

STRATEGY OF PROOF. The idea is to show that the two smaller squares of

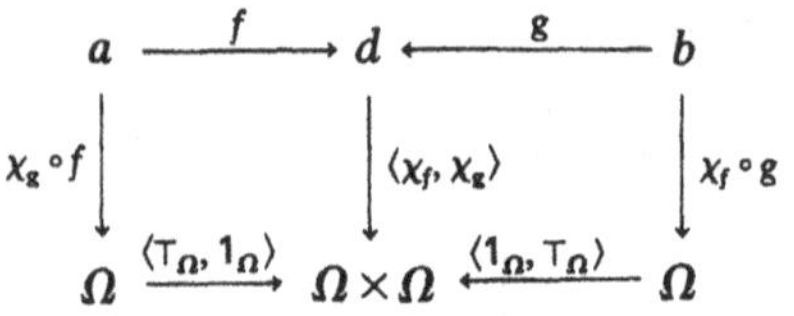

are pullbacks. Since co-products preserve pullbacks (Fact 2, §5.3) we then get a pullback of the form

$$\begin{array}{ccc} a+b & \xrightarrow{[f,g]} & d \\ \big\downarrow & & \big\downarrow \langle \chi_f, \chi_g \rangle \\ \Omega+\Omega & \xrightarrow{[\langle \top_\Omega, 1_\Omega\rangle \langle 1_\Omega, \top_\Omega\rangle]} & \Omega\times\Omega \end{array}$$

The Lemma then yields a pullback of the form

$$\begin{array}{ccc} c & \xrightarrow{\alpha} & d \\ \big\downarrow & & \big\downarrow \langle \chi_f, \chi_g \rangle \\ e & \xrightarrow{i} & \Omega\times\Omega \end{array}$$

where i is the image arrow of $[\langle \top_\Omega, 1_\Omega\rangle, \langle 1_\Omega, \top_\Omega\rangle]$; But i is the arrow whose character is $\cup : \Omega\times\Omega \to \Omega$, i.e.

$$\begin{array}{ccc} e & \xrightarrow{i} & \Omega\times\Omega \\ \big\downarrow & & \big\downarrow \cup \\ 1 & \xrightarrow{\top} & \Omega \end{array}$$

is by definition a pullback. Putting these last two diagrams together and invoking the PBL shows that $\chi_\alpha = \cup \circ \langle \chi_f, \chi_g \rangle$. □

In view of Theorem 3 we can now describe the disjunction truth arrow $\cup$ as the character of

$$\Omega \cup \Omega \xrightarrow{\langle \top_\Omega, 1_\Omega\rangle \cup \langle 1_\Omega, \top_\Omega\rangle} \Omega\times\Omega$$

7.2. Sub(d) as a lattice

THEOREM 1. (Sub(d), $\subseteq$) *is a lattice in which*
(1) $f \cap g$ *is the g.l.b. (lattice meet) of* f *and* g;
(2) $f \cup g$ *is the l.u.b.(join) of* f *and* g.

PROOF. (1) The characterisation of $f \cap g$ as a pullback of f and g makes it relatively easy to see why $f \cap g$ is the g.l.b. of f and g. The details are left to the reader.

(2) The characterisation of $f \cup g$ in Theorem 3 and the co-universal property of $[f, g]$ shows that

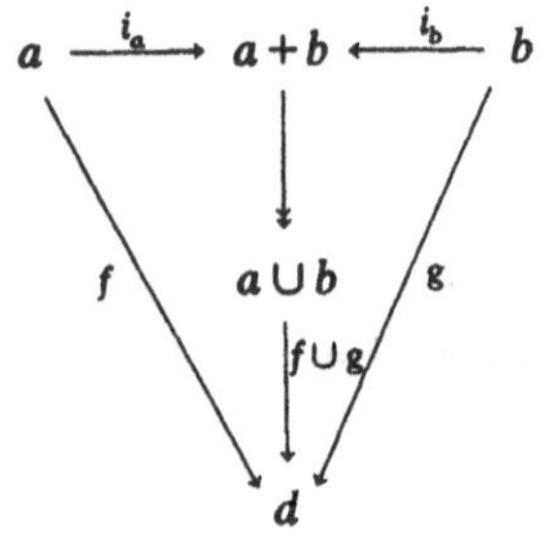

commutes and so each of f and g factors through $f \cup g$. Thus $f \subseteq f \cup g$, $g \subseteq f \cup g$, and $f \cup g$ is an upper bound of f and g. To show it is the least such, suppose $f \subseteq h$ and $g \subseteq h$. Then f and g each factor through h, so there are h_a, h_b making

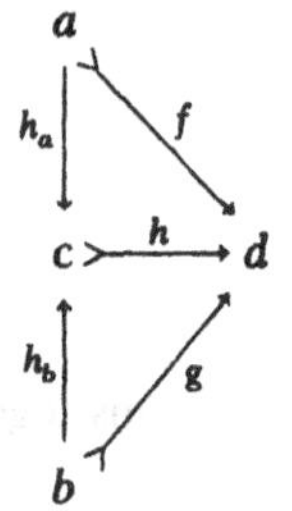

commute. Then

$$\begin{aligned}[f, g] &= [h \circ h_a, h \circ h_b] \\ &= h \circ [h_a, h_b] \qquad \text{(dual of Exercise 3.8.3)}\end{aligned}$$

and so $[f, g]$ is the composite of

$$[h_a, h_b]: a + b \to c \quad \text{and} \quad h: c \rightarrowtail d.$$

Replacing $[h_a, h_b]$ by its epi-monic factorisation we get $[f, g]$ as the composite of

$$a + b \overset{j}{\twoheadrightarrow} e \overset{k}{\rightarrowtail} c \overset{h}{\to} d$$

for some j and k. But then j followed by $h \circ k$ is an epi-monic factorisation of $[f, g]$. By the uniqueness, up to isomorphism, of such things there

is an iso u such that

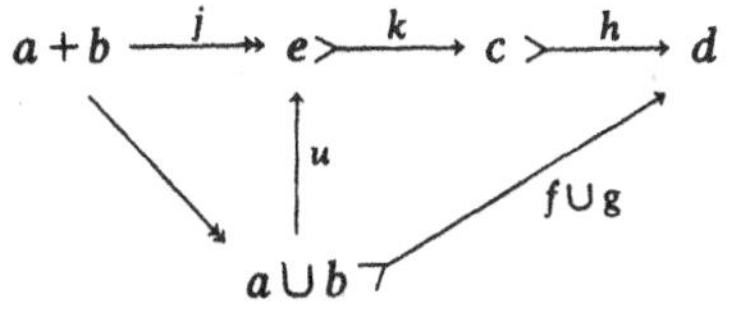

commutes. Then $k \circ u$ factors $f \cup g$ through h, yielding $f \cup g \subseteq h$ as required. □

COROLLARY. (1) *$f \subseteq g$ iff $f \cap g \simeq f$ iff $f \cup g \simeq g$.*

(2) *$f \subseteq g$ iff $\langle \chi_f, \chi_g \rangle$ factors (uniquely) through the equaliser*

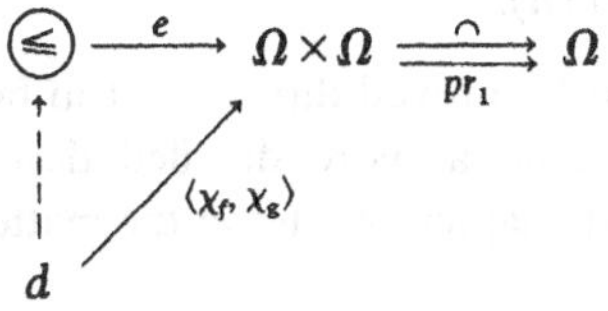

of $\cap$ and pr_1.

PROOF. (1) In any lattice, $x \sqsubseteq y$ iff $x \sqcap y = x$ iff $x \sqcup y = y$.

(2) $f \subseteq g$ iff $f \cap g \simeq f$

iff $\chi_{f \cap g} = \chi_f$

iff $\cap \circ \langle \chi_f, \chi_g \rangle = pr_1 \circ \langle \chi_f, \chi_g \rangle$

and the result follows by the universal property of equalisers. □

Part (2) of this Corollary is an analogue of the fact that in **Set** we have $A \subseteq B$ iff $\chi_A \leqslant \chi_B$ (the latter meaning $\chi_A(x) \leqslant \chi_B(x)$, all $x \in D$).

THEOREM 2. *(Sub(d), $\subseteq$) is a bounded lattice with unit 1_d and zero 0_d.*

PROOF. Given any $f: a \rightarrowtail d$, the commutativity of

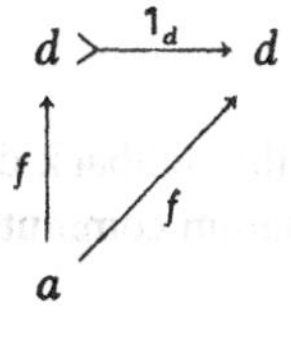

and of

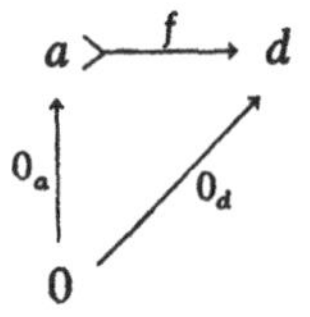

shows that $0_d \subseteq f$ and $f \subseteq \mathbf{1}_d$. □

EXERCISE 1. In Sub(d), $f \simeq \mathbf{1}_d$ iff f is iso, i.e. $f: a \cong d$. □

Sub(d) is in fact a distributive lattice, i.e. satisfies

$$f \cap (g \cup h) \simeq (f \cap g) \cup (f \cap h).$$

Again this is something that could be proved directly but in fact follows from some deeper results – this time a more detailed description of Sub(d) to be developed in the next chapter. We leave the matter till then (cf. §8.3).

What about complements? To date we have not used the definition $\chi_{-f} = \neg \circ \chi_f$. The first thing we shall prove in this connection is

THEOREM 3. *For $f: a \rightarrowtail d$, we have*

$$f \cap -f \simeq 0_d.$$

PROOF. The boundary of

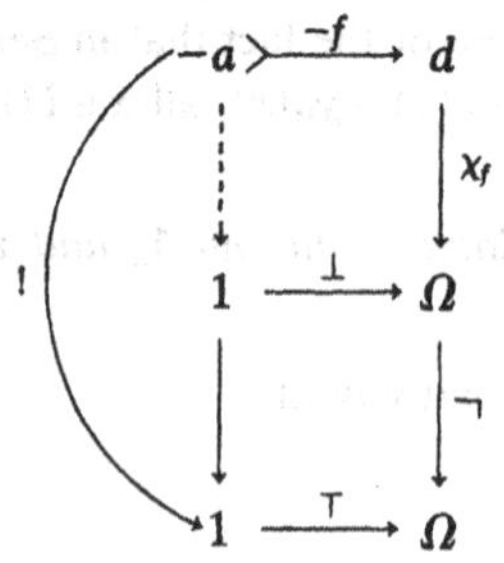

is the pullback defining $-f$, the bottom square is the pullback defining $\neg$, so the unique arrow $-a \rightarrow 1$ makes the whole diagram commute, and the top square a pullback.

Then each square of

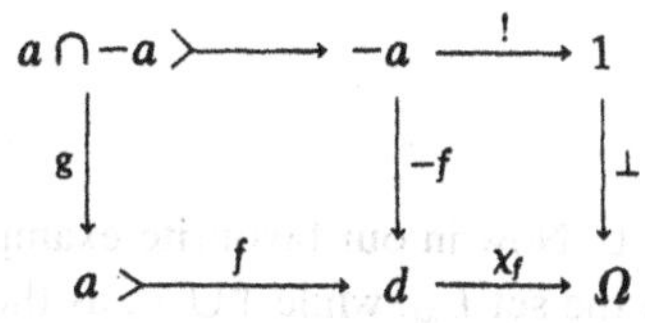

commutes (the left hand one is the pullback giving $f \cap -f$), so we get $\perp \circ ! = \chi_f \circ f \circ g$. But $\chi_f \circ f = true_a$ (Ω-axiom), so $\chi_f \circ f \circ g = true_a \circ g = true_{a \cap -a}$ (4.2.3). Hence the outer square of

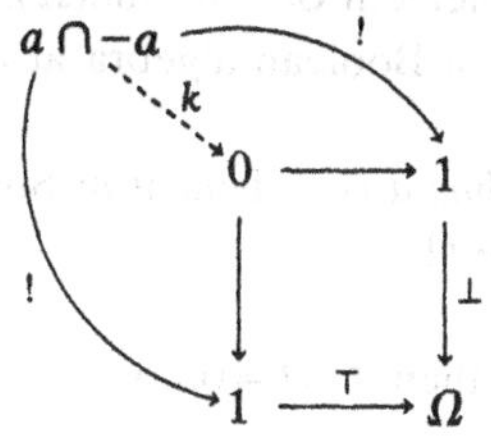

commutes. But the inner square is a pullback, so the arrow $k: a \cap -a \to 0$ does exist. But then $a \cap -a \cong 0$ (§3.16), so $a \cap -a$ is an initial object and

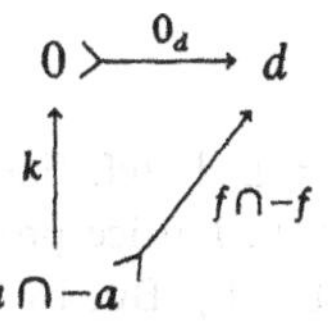

must commute. Thus $f \cap -f \subseteq 0_d$, and since 0_d is the minimum element of Sub(d), the result follows. □

We seem to be well on the way to a proof that Sub(d) is a Boolean algebra, and hence complete the analogy with $\mathcal{P}(D)$ in **Set**. We know it to be a bounded distributive lattice, with $f \cap -f$ always the zero. It remains only to show that $f \cup -f$ is the unit. But we cannot do this! There are topoi in which it is false. To give an example we need

THEOREM 4. *In* Sub(Ω), (*for any topos*),

$$\perp \simeq -\top.$$

PROOF. $\chi_\perp = \neg$ (definition of $\neg$)

$$= \neg \circ 1_\Omega$$
$$= \neg \circ \chi_\top$$
$$= \chi_{-\top}. \qquad \square$$

So in any topos, $\top \cup -\top = \top \cup \perp$. Now in our favourite example $\mathbf{M_2}$, 1_Ω in Sub(Ω) can be identified with the set L_M, while $\top \cup \perp$, as the image of $[\top, \perp]$ (recall the description of the latter in Theorem 5.4.6), can be identified with the set $\{M_2, \emptyset\} \neq L_2$. Hence

$$\top \cup \perp \not\simeq 1_\Omega,$$

and so $-\top$ $(=\perp)$ is not the lattice complement of $\top$ in Sub(Ω). But then, as the next result shows, Sub(Ω) is not a Boolean algebra at all.

THEOREM 5. *In any topos, if* $\top : 1 \to \Omega$ *has a complement in* Sub(Ω), *then this complement is the subobject* $\perp : 1 \to \Omega$.

PROOF. If $\top$ has a complement, f say, then $\top \cap f \simeq 0_\Omega$, so

$$\begin{array}{ccc} 0 & \xrightarrow{0_a} & a \\ \downarrow & & \downarrow f \\ 1 & \xrightarrow{\top} & \Omega \end{array}$$

is a pullback. The Ω-axiom then gives $f = \chi_{0_a} = \perp \circ !_a$ (cf. Exercise 5.4.3). But $\perp \circ !_a$ obviously factors through $\perp$, so $f \subseteq \perp$. Lattice properties then give $\top \cup f \subseteq \top \cup \perp$, and since $\top \cup f \simeq 1_\Omega$, $\top \cup \perp \simeq 1_\Omega$. But by Theorems 3 and 4 above, $\top \cap \perp = 0_\Omega$, and so $\perp$ is a complement of $\top$. But in a distributive lattice, complements are unique, hence $f \simeq \perp$. $\square$

7.3. Boolean topoi

A topos $\mathscr{E}$ will be called *Boolean* if for *every* $\mathscr{E}$-object d, (Sub(d), $\subseteq$) is a Boolean algebra.

THEOREM 1. *For any topos* $\mathscr{E}$, *the following statements are equivalent:*

(1) $\mathscr{E}$ *is Boolean*

(2) Sub(Ω) *is a* **BA**

(3) $\top : 1 \to \Omega$ *has a complement in* Sub(Ω)

(4) $\bot : 1 \to \Omega$ *is the complement of* $\top$ *in* Sub(Ω)
(5) $\top \cup \bot \simeq 1_\Omega$ *in* Sub(Ω)
(6) $\mathscr{E}$ *is classical, i.e.* $[\top, \bot] : 1+1 \to \Omega$ *is iso*
(7) $i_1 : 1 \to 1+1$ *is a subobject classifier.*

PROOF. (1) implies (2): definition of "Boolean"
(2) implies (3): definition of "**BA**"
(3) implies (4): Theorem 7.2.5
(4) implies (5): definition of "complement"
(5) implies (6): $[\top, \bot]$ is always monic, so

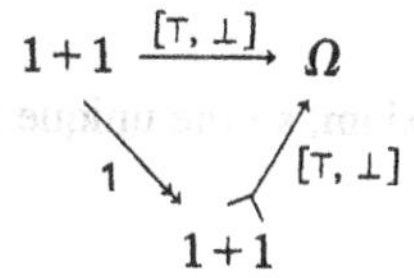

is an epi-monic factorisation of $[\top, \bot]$, i.e. in Sub(Ω), $\top \cup \bot \simeq [\top, \bot]$. Then if $\top \cup \bot \simeq 1_\Omega$, we get $[\top, \bot] \simeq 1_\Omega$, making $[\top, \bot]$ iso by Exercise 7.2.1.

(6) implies (7): Exercise – the essential point being that anything isomorphic to a classifier will be one itself.

(7) implies (1): Given $f : a \rightarrowtail d$, we wish to show that $f \cup -f \simeq 1_d$, and so by the work of §7.2 $-f$ will be a complement for f, and Sub(d) will be a **BA**.

The basic strategy can be seen in the diagram

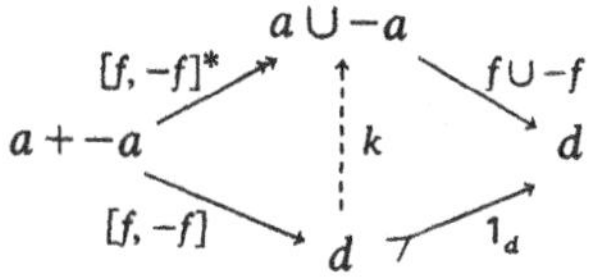

If we can show that $[f, -f]$ is epic, then the iso k as shown will exist to factor 1_d through $f \cup -f$ to make $f \cup -f \simeq 1_d$. We need first the following:

LEMMA. *In any topos,*

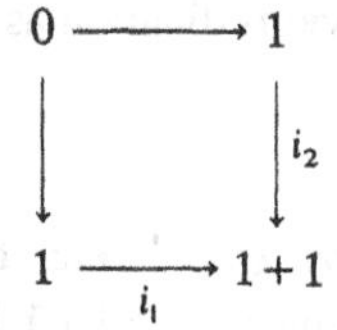

is a pullback, where i_1, i_2 *are the two injections for the co-product* $1+1$.

PROOF. The square commutes as 0 is initial. It is also a pushout by the co-universal property of the pair (i_1, i_2). But the outer square of

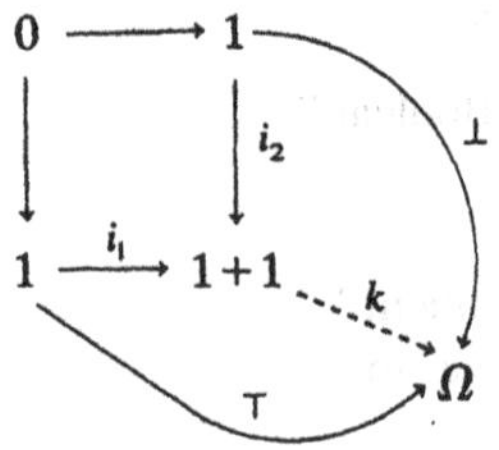

commutes, indeed is a pullback by the Ω-axiom, so the unique k exists as shown to make the diagram commute.

Then if the outer square of

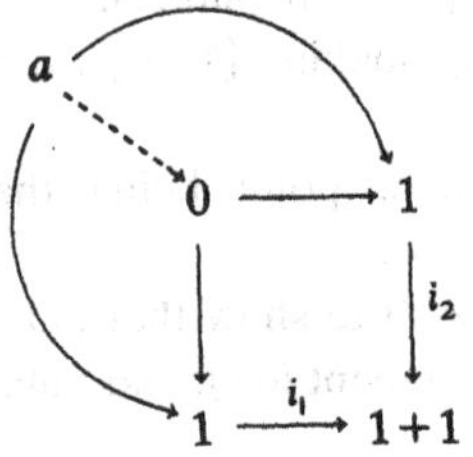

commutes, k can be used to show the outer square of

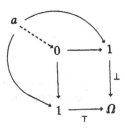

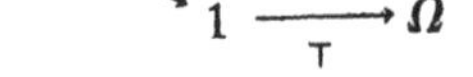

commutes, giving the unique $a \to 0$ for the previous diagram as required. □

To finish our Theorem we shall denote by χ'_f, $\perp'$ etc. the arrows defined in the same way as χ_f, $\perp$, etc., but using $i_1 : 1 \to 1+1$ in place of $\top : 1 \to \Omega$. Now the Lemma tells us that $i_2 = \perp'$, so by the argument at the beginning of Theorem 3 of §7.2,

$$\begin{array}{ccc} -a & \xrightarrow{-f} & d \\ \downarrow & & \downarrow \chi'_f \\ 1 & \xrightarrow{i_2} & 1+1 \end{array}$$

is a pullback. But so is

$$\begin{array}{ccc} a & \xrightarrow{f} & d \\ \downarrow & & \downarrow \chi'_f \\ 1 & \xrightarrow{i_1} & 1+1 \end{array}$$

and co-products preserve pullbacks, so

$$\begin{array}{ccc} a+-a & \xrightarrow{[f,-f]} & d \\ !+! \downarrow & & \downarrow \chi'_f \\ 1+1 & \xrightarrow{[i_1, i_2]} & 1+1 \end{array}$$

is a pullback. But $[i_1, i_2] = 1_{1+1}$ is epic, whence $[f, -f]$ is the pullback of an epic, i.e. an epic itself. □

7.4. Internal vs. External

THEOREM 1. *If $\mathscr{E}$ is Boolean, then $\mathscr{E} \models \alpha \vee \sim\alpha$, for any sentence α.*

PROOF. Let V be an $\mathscr{E}$-valuation. Form the pullback

$$\begin{array}{ccc} a & \xrightarrow{f} & 1 \\ \downarrow & & \downarrow V(\alpha) \\ 1 & \xrightarrow{\top} & \Omega \end{array}$$

of $\top$ along $V(\alpha)$, so that $\chi_f = V(\alpha)$.

Now if $\mathscr{E}$ is Boolean, Sub(1) is a **BA**, so $f \cup -f \simeq 1_1$, whence $\chi_{f\cup -f} = \chi_{1_1} = \top$. But

$$\begin{aligned}\chi_{f\cup -f} &= \chi_f \cup \neg\circ\chi_f \\ &= V(\alpha) \cup \neg\circ V(\alpha) \\ &= V(\alpha \vee {\sim}\alpha).\end{aligned}$$

Hence $V(\alpha \vee {\sim}\alpha) = \top$. □

One might think that if our theory was working well then the converse of Theorem 1 should hold. However our example $\mathbf{M}_2$ is non-Boolean, since in it Sub(Ω) is not a **BA**, and yet $\mathbf{M}_2 \models \alpha \vee {\sim}\alpha$, as observed at the end of Chapter 6. The proof of Theorem 1 in fact only required that Sub(1) be a **BA**. That *this* is the relevant condition is shown by

THEOREM 2. *In any topos $\mathscr{E}$, the following are equivalent:*

(1) $\mathscr{E} \models \alpha$ *iff* $\vdash_{CL} \alpha$, *all sentences* α

(2) $\mathscr{E} \models \alpha \vee {\sim}\alpha$, *all* α

(3) Sub(1) *is a* **BA**.

PROOF. Clearly (1) implies (2). Assuming (2) we take a subobject $f: a \rightarrowtail 1$ in Sub(1) and observe that χ_f is a truth value $1 \to \Omega$. Taking an $\mathscr{E}$-valuation that has $V(\pi_0) = \chi_f$, we have $\chi_{f\cup -f} = \chi_f \cup \neg\chi_f = V(\pi_0) \cup \neg V(\pi_0) = V(\pi_0 \vee {\sim}\pi_0) = \top = \chi_{1_1}$. Hence $f \cup -f \simeq 1_1$. This means that Sub(1) is a **BA**.

Finally assume (3), in order to derive (1). The "only if" part of (1) holds in any topos. The "if" part requires a proof that the CL-axioms are $\mathscr{E}$-valid and that detachment preserves $\mathscr{E}$-validity. We shall explain later why axioms I–XI are valid in any topos, and why *Detachment* is always validity preserving. For the present we note only that the proof of Theorem 1 shows that if Sub(1) is a **BA**, then axiom XII is $\mathscr{E}$-valid. □

COROLLARY. "Sub(1) *is a* **BA**" *does not imply that* $\mathscr{E}$ *is Boolean.*

The situation seems at first sight anomolous (at least it did to the author). In **Set** the logic is based on the **BA** 2, and in the general topos it seems to be intimately related to Sub(1). In **Set**, Sub(1) $\cong \mathscr{P}(1) \cong 2$ – so far so good. But the work of the previous sections shows that the properties of the "generalised power-sets" Sub(d) are determined by Sub(Ω), whereas in **Set**, Sub(Ω) is a four-element set that has played no special role to date.

Some clarification of this situation is afforded by the observation that Sub(d) is a collection of subobjects of d and may well not be itself an actual $\mathscr{E}$-object. Thinking of $\mathscr{E}$ as a "general universe of mathematical discourse" then a person living in that universe, i.e. one who uses only the individuals that exist in that universe, does not "see" Sub(d) at all as a single entity. Sub(d) is *external* to $\mathscr{E}$. What the topos-dweller does see is the power object Ω^d, which is the "object of subsets" of the object d. Ω^d is an individual in the universe $\mathscr{E}$, and is the *internal* version of the notion of power set, while Sub(d) is the external version.

Now the Law of Excluded Middle does have an internal version. The validity of $\alpha \vee \sim\alpha$ in **Set** corresponds to the truth of the equation

$$x \cup \neg x = 1, \quad \text{for} \quad x \in 2.$$

The truth of this equation is equivalent to the commutativity of

$$\begin{array}{ccc} 2 & \xrightarrow{\langle \mathrm{id}_2, \neg \rangle} & 2\times 2 \\ {\scriptstyle !}\downarrow & & \downarrow{\scriptstyle \cup} \\ 1 & \xrightarrow{\ \ true\ \ } & 2 \end{array}$$

(since $\langle \mathrm{id}_2, \neg \rangle(x) = \langle x, \neg x \rangle$).

Now this diagram has an analogue in any topos $\mathscr{E}$, and we have the interesting

THEOREM 3. *Sub(Ω) is a* **BA** *iff the diagram*

$$\begin{array}{ccc} \Omega & \xrightarrow{\langle 1_\Omega, \neg \rangle} & \Omega\times \Omega \\ \downarrow & & \downarrow{\scriptstyle \cup} \\ 1 & \xrightarrow{\ \ \top\ \ } & \Omega \end{array} \qquad \text{(EM)}$$

commutes.

PROOF. EM commutes when

$$\cup \circ \langle 1_\Omega, \neg \rangle = \top_\Omega$$

i.e.

$$1_\Omega \cup \neg = \top_\Omega$$

But we know that $1_\Omega = \chi_\top$, $\neg = \chi_\bot$, and $\top_\Omega = \chi_{1_\Omega}$, so

$$\begin{aligned} \mathrm{Sub}(\Omega) \text{ is a } \mathbf{BA} \quad &\text{iff} \quad \top \cup \bot \simeq 1_\Omega \qquad (\S 7.3) \\ &\text{iff} \quad \chi_{\top \cup \bot} = \chi_{1_\Omega} \\ &\text{iff} \quad \chi_\top \cup \chi_\bot = \chi_{1_\Omega} \\ &\text{iff} \quad 1_\Omega \cup \neg = \top_\Omega . \end{aligned}$$ □

EXERCISE 1. Show explicitly why EM does not commute in $\mathbf{M}_2$. □

Now in our theory of topos semantics we use the collection $\mathscr{E}(1, \Omega)$ of truth-values. This again is an external thing – the internal version of the collection of arrows from 1 to Ω would be the *object of truth-values* $\Omega^1 \cong \Omega$. Also a valuation $V: \Phi \to \mathscr{E}(1, \Omega)$ is external, i.e. is not an actual $\mathscr{E}$-arrow.

Thus the semantical theory we have developed is an external one, and this is why there can be topoi like $\mathbf{M}_2$ that look classical "from the outside" and yet can have non-classical properties (curiously, $\mathbf{M}_2$ is internally bivalent while "from the outside" Ω has three elements). We now see that a topos also has an internal logic, in the form of commuting diagrams like EM (cf. Exercise 2 below). It is precisely when this internal logic is classical that the topos is Boolean.

From the viewpoint that topoi offer a complete alternative to the category **Set** as a context for doing mathematics it is finally the internal structure that is important. Nonetheless the present external theory is very useful for elucidating the logical properties of topoi, and as we shall see, for describing the link between topoi and intuitionistic logic.

EXERCISE 2. Describe the validity of the CL-axioms I–XI in terms of commutativity of diagrams involving truth-arrows. (All of them commute in any topos – can you prove some of them?)

7.5. Implication and its implications

In the same way that we used the truth arrows $\cap$, $\cup$, $\neg$ to define operations $\cap$, $\cup$, $-$ on Sub(d) we can use implication $\Rightarrow$ to define the following operation: if $f: a \rightarrowtail d$ and $g: b \rightarrowtail d$ are subobjects of d, then $f \mapsto g: (a \mapsto b) \rightarrowtail d$ is the subobject obtained by pulling $\top$ back along

$\chi_f \Rightarrow \chi_g = \Rightarrow \circ \langle \chi_f, \chi_g \rangle$. Thus

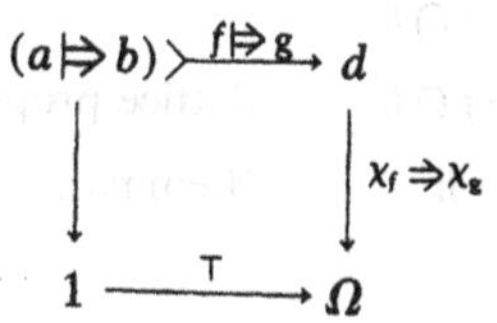

is a pullback, i.e. $\chi_{f \Rrightarrow g} = \chi_f \Rightarrow \chi_g$.

In order to study the properties of this new operation we need some technical results.

LEMMA 1. *If f, g, and h are subobjects of d (in any topos), then*

(1) $f \cap h \simeq g \cap h$ *iff* $\chi_f \circ h = \chi_g \circ h$,

and hence

(2) $\chi_f \cap \chi_h = \chi_g \cap \chi_h$ *iff* $\chi_f \circ h = \chi_g \circ h$.

PROOF. (1) Consider

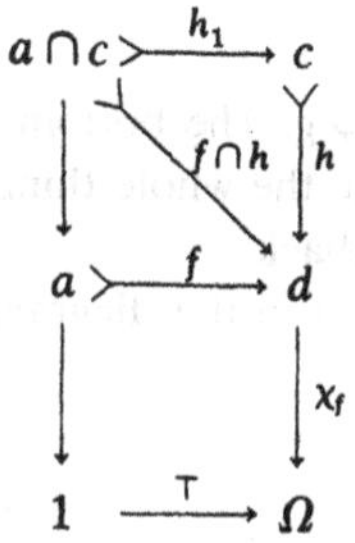

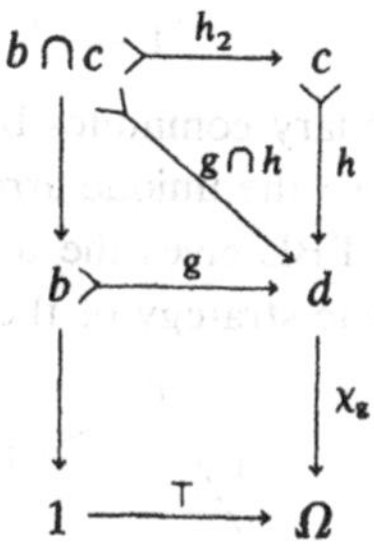

In each diagram the bottom squares are pullbacks by the Ω-axiom, and the top squares are pullbacks by the characterisation of intersections. So by the PBL, $\chi_f \circ h = \chi_{h_1}$ and $\chi_g \circ h = \chi_{h_2}$. Thus $\chi_f \circ h = \chi_g \circ h$ iff $h_1 \simeq h_2$. But this last condition holds only if there is an iso k giving $h_1 \circ k = h_2$, and so $h \circ h_1 \circ k = h \circ h_2$, i.e. $(f \cap h) \circ k = g \cap h$, and so $f \cap h \simeq g \cap h$. The argument reverses to show $f \cap h \simeq g \cap h$ only if $h_1 \simeq h_2$. Part (2) is immediate from (1). □

COROLLARY

$$f \cap h \subseteq g \quad \textit{iff} \quad \chi_{f \cap g} \circ h = \chi_f \circ h.$$

PROOF

$$
\begin{aligned}
f \cap h \subseteq g \quad & \text{iff} \quad (f \cap h) \cap g \simeq f \cap h \\
& \text{iff} \quad (f \cap g) \cap h \simeq f \cap h \qquad \text{(lattice properties)} \\
& \text{iff} \quad \chi_{f \cap g} \circ h = \chi_f \circ h \qquad \text{(Lemma).}
\end{aligned}
$$

□

THEOREM 1. *In* Sub(d) *we have:*

(1) $h \subseteq f \mapsto g$ *iff* $f \cap h \subseteq g$

(2) $f \subseteq g$ *iff* $f \mapsto g \simeq 1_d$

(3) $f \subseteq g$ *iff* $\chi_f \Rightarrow \chi_g = true_d$.

PROOF. (1) First consider

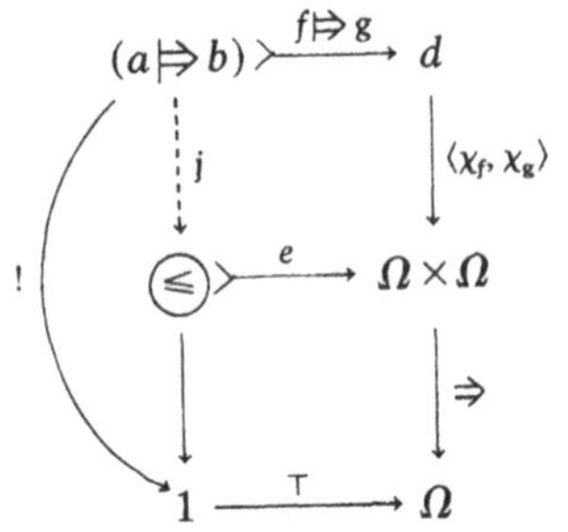

The boundary commutes by definition of $f \mapsto g$. The bottom square is a pullback, so the unique arrow j exists to make the whole thing commute. Then the PBL gives the top square as a pullback.

The basic strategy of the main proof is seen in the diagram

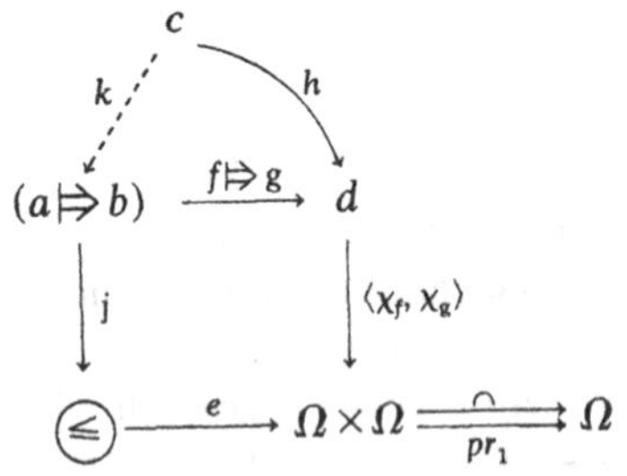

We have $h \subseteq f \mapsto g$ precisely when there is an arrow k as shown making the top triangle commute. Since the square is a pullback, such a k exists precisely when $\langle \chi_f, \chi_g \rangle \circ h$ factors through e. By the universal property of e as an equaliser, this happens precisely when $pr_1 \circ \langle \chi_f, \chi_g \rangle \circ h = \cap \circ \langle \chi_f, \chi_g \rangle \circ h$, i.e. $\chi_f \circ h = \chi_{f \cap g} \circ h$. But this last equality holds iff $f \cap h \subseteq g$, by the last Corollary.

(2) We use part (1). Suppose $f \subseteq g$. Then for any h in Sub(d), $f \cap h \subseteq f \subseteq g$, so by (1), $h \subseteq f \mapsto g$. This makes $f \mapsto g$ the unit 1_d of Sub(d). Conversely if $f \mapsto g \simeq 1_d$, then $f \subseteq f \mapsto g$, so $f \cap f \subseteq g$, i.e. $f \subseteq g$.

(3) From (2), and the definition of $\mapsto$, since $\chi_{1_d} = true_d$. □

EXERCISE. Give a categorial proof of part (2), by using the Corollary to Theorem 1 of §7.2 and the diagram

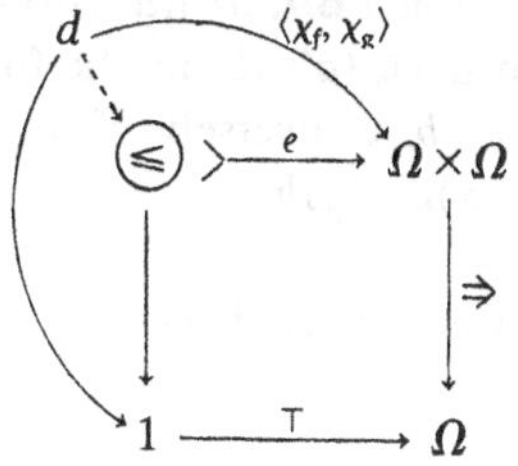

□

COROLLARY TO THEOREM 1. *In* Sub(d):

(1) $1_d \mapsto 1_d \simeq 0_d \mapsto 1_d \simeq 0_d \mapsto 0_d \simeq 1_d$.

(2) $1_d \mapsto 0_d \simeq 0_d$.

PROOF. (1) By part (2) of the Theorem, as $1_d \subseteq 1_d$, $0_d \subseteq 1_d$, $0_d \subseteq 0_d$.

(2) Since $1_d \mapsto 0_d \subseteq 1_d \mapsto 0_d$, part (1) gives

$$1_d \cap (1_d \mapsto 0_d) \subseteq 0_d,$$

i.e.

$$1_d \mapsto 0_d \subseteq 0_d, \qquad (1_d \text{ is maximum})$$

and hence

$$1_d \mapsto 0_d \simeq 0_d.$$ □

Now in $\mathcal{P}(D)$, $A \mapsto D$ is $-A \cup D$. (why?) The analogous situation does not obtain in all topoi. In $\mathbf{M_2}$, $\top \mapsto \top \simeq 1_\Omega$ in Sub(Ω) (by Theorem 1(2)), while $-\top \cup \top = \bot \cup \top = \top \cup \bot$, and we saw in §7.2 that $\top \cup \bot \neq 1_\Omega$ in $\mathbf{M_2}$.

To determine the conditions under which $\mapsto$ can be defined from $\cup$ and $-$ we need

LEMMA 2. (1) *In any lattice, if m and n satisfy*

(i) $x \sqsubseteq m$ *iff* $a \sqcap x \sqsubseteq b$, *all* x

(ii) $x \sqsubseteq n$ *iff* $a \sqcap x \sqsubseteq b$, *all* x

then $m = n$.

(2) *In a Boolean algebra,*

$$x \sqsubseteq (a' \sqcup b) \quad \textit{iff} \quad a \sqcap x \sqsubseteq b,$$

and so the only m that satisfies the condition of (1)(i) *is* $m = a' \sqcup b$.

PROOF. (1) Exercise – use $m \sqsubseteq m$ etc.

(2) First, by properties of l.u.b.'s and g.l.b.'s, note that if $x \sqsubseteq z$, then $y \sqcap x \sqsubseteq y \sqcap z$ (any x, y, z). Next note that in a **BA**, $a \sqcap (a' \sqcup b) = (a \sqcap a') \sqcup (a \sqcap b) = 0 \sqcup (a \sqcap b) = a \sqcap b \sqsubseteq b$ so that if $x \sqsubseteq (a' \sqcup b)$ by the foregoing we have $a \sqcap x \sqsubseteq a \sqcap (a' \sqcup b) \sqsubseteq b$, i.e. $a \sqcap x \sqsubseteq b$. Conversely, if $a \sqcap x \sqsubseteq b$ then $x = 1 \sqcap x = (a' \sqcup a) \sqcap x = (a' \sqcap x) \sqcup (a \sqcap x) \sqsubseteq a' \sqcup b$. □

THEOREM 2. *In any topos* $\mathscr{E}$, *the following are equivalent:*

(1) $\mathscr{E}$ *is Boolean*
(2) *In each* Sub(d), $f \mapsto g \simeq -f \cup g$
(3) *In* Sub(Ω), $f \mapsto g \simeq -f \cup g$
(4) $\top \mapsto \top = \top \cup \bot$.

PROOF. (1) implies (2): Theorem 1(1) states that in the lattice Sub(d), $h \subseteq f \mapsto g$ iff $f \cap h \subseteq g$. But if Sub(d) is a **BA**, Lemma 2(2) tells us that $h \subseteq -f \cup g$ iff $f \cap h \subseteq g$. Lemma 2(1) then implies that $f \mapsto g = -f \cup h$.

(2) implies (3): obvious.

(3) implies (4): $-\top \cup \top = \top \cup \bot$ as noted prior to Lemma 2.

(4) implies (1): We always have $\top \mapsto \top \simeq \mathbf{1}_\Omega$. Use part (5) of the Theorem in §7.3. □

So we see that in a non-Boolean topos, $\mapsto$ does not behave like a Boolean implication operator. What its behaviour *is* like in general will be revealed in the next chapter. Before proceeding to that however, we pause for the purpose of

7.6. Filling two gaps

1. Theorem 1 of §6.7 gave some tables for the behaviour of the truth-values $\top$ and $\bot$ under the arrows $\cap$, $\cup$, and $\Rightarrow$. We are now in a position to show why these tables are correct.

The key lies in the lattice structure of Sub(1), where the unit is 1_1 and the zero 0_1. Thus we have $1_1 \cap 1_1 \simeq 1_1$, while $1_1 \cap 0_1 \simeq 0_1 \cap 1_1 \simeq 0_1 \cap 0_1 \simeq$

0_1. But $\chi_{1_1} = \top$ and $\chi_{0_1} = \bot$, so we have

$$\top \cap \bot = \chi_{1_1} \cap \chi_{0_1} = \chi_{1_1 \cap 0_1} = \chi_{0_1} = \bot$$

$$\top \cap \top = \chi_{1_1} \cap \chi_{1_1} = \chi_{1_1 \cap 1_1} = \chi_{1_1} = \top,$$

and so on, yielding the table

$\cap$	$\top$	$\bot$
$\top$	$\top$	$\bot$
$\bot$	$\bot$	$\bot$

Now using the Corollary to Theorem 1 of §7.5 we find $\top \Rightarrow \bot = \chi_{1_1} \Rightarrow \chi_{0_1} = \chi_{1_1 \Rightarrow 0_1} = \chi_{0_1} = \bot$, $\bot \Rightarrow \top = \chi_{0_1 \Rightarrow 1_1} = \chi_{1_1} = \top$ etc. leading to

$\Rightarrow$	$\top$	$\bot$
$\top$	$\top$	$\bot$
$\bot$	$\top$	$\top$

EXERCISE. Derive the table

$\cup$	$\top$	$\bot$
$\top$	$\top$	$\top$
$\bot$	$\top$	$\bot$

□

2. Theorem 5 of §5.4 asserted without proof that a classical ($1+1 \cong \Omega$) topos in which every non-zero object is non-empty is in fact well-pointed. Now if $\mathscr{E}$ is classical, we now know it to be Boolean by §7.3. So let us take a pair of *distinct* parallel arrows $f, g : a \rightrightarrows b$ in $\mathscr{E}$ and look for an element $x : 1 \to a$ that distinguishes them, i.e. has $f \circ x \neq g \circ x$. We let $h : c \rightarrowtail a$ be the equaliser of f and g, and $-h : -c \rightarrowtail a$ the complement of h in Sub(a) (remember $\mathscr{E}$ is Boolean). Then $-c$ is non-zero (in **Set**, $-c \neq \emptyset$ as f and g differ at some point of a). For, if $-c \cong 0$, then $-h \simeq 0_a$, so $h \simeq h \cup 0_a \simeq h \cup -h \simeq 1_a$, whence h is iso and since $f \circ h = g \circ h$ we would get $f = g$.

Now if all non-zero $\mathscr{E}$-objects are non-empty there must then be an arrow $y : 1 \to -c$. Then let x be $-h \circ y : 1 \to a$. Then if $f \circ x = g \circ x$, as h

equalises f and g there would be some $z:1 \to c$ such that $h \circ z = x$. Hence the boundary of

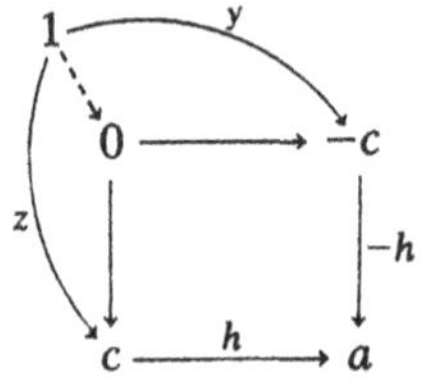

would commute, giving an arrow $1 \to 0$. But this would make $\mathscr{E}$ degenerate, contrary to the fact that $c \not\cong 0$. We conclude $f \circ x \neq g \circ x$.

7.7. Extensionality revisited

In Chapter 5 we considered well-pointedness as a categorial formulation of the extensionality principle for functions. For sets themselves, extensionality simply means that sets with the same elements are identical. It follows from this that identity of sets is characterised by the set inclusion relation: $A = B$ iff $A \subseteq B$ and $B \subseteq A$, since

$A \subseteq B$ iff every member of A is a member of B.

This definition of the subset relation is readily lifted to the general category. If $f: a \rightarrowtail d$ is a subobject of d, and $x: 1 \to d$ an element of d, then as in §4.8 we say that x *is an element of* f, $x \in f$, iff x factors through f.

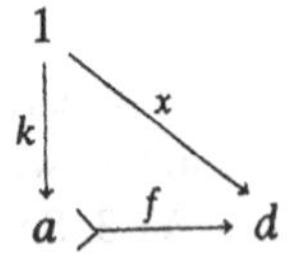

i.e. for some $k: 1 \to a$, $x = f \circ k$.

THEOREM 1. *In any topos* $\mathscr{E}$, *in* Sub(d) *we have*

$$x \in f \cap g \quad \textit{iff} \quad x \in f \quad \textit{and} \quad x \in g.$$

PROOF. If x factors through $f \cap g$, then since $f \cap g$ factors through both f and g, so too will x.

Conversely, suppose that $x \in f$ and $x \in g$, so that

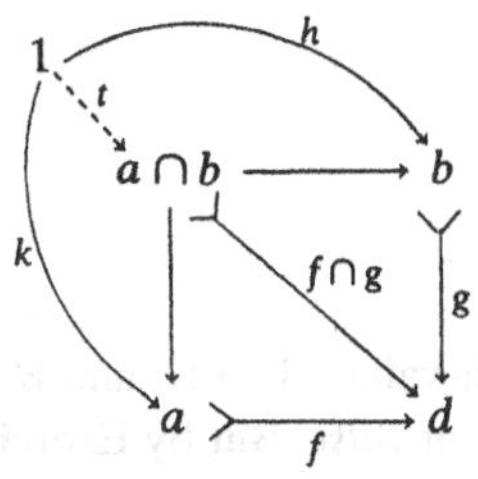

$x = f \circ k$ and $x = g \circ h$ for some elements $k: 1 \to a$ and $h: 1 \to b$. But the inner square of the diagram is a pullback (§7.1) so the arrow t exists as shown making $f \cap g \circ t = f \circ k = x$. This t factors x through $f \cap g$, giving $x \in f \cap g$. □

A topos in which subobjects are determined by their elements will be called *extensional*. That is, $\mathscr{E}$ is extensional iff for any $\mathscr{E}$-object d, the condition

$$f \subseteq g \quad \text{iff} \quad \text{for all} \quad x: 1 \to d,\ x \in f \quad \text{implies} \quad x \in g$$

holds in Sub(d).

THEOREM 2. *$\mathscr{E}$ is extensional iff well-pointed.*

PROOF. Let $f, g: a \rightrightarrows b$ be a pair of parallel $\mathscr{E}$-arrows, with $f \circ x = g \circ x$, all $x: 1 \to a$. Let $h: c \rightarrowtail a$ be the equaliser of f and g. Then if $x \in 1_a$,

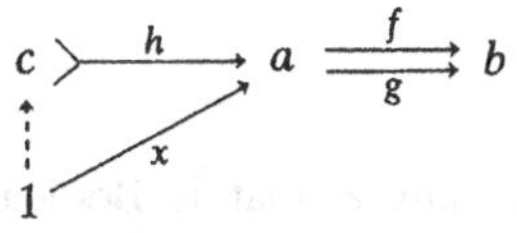

(which holds for any $x: 1 \to a$), we get $x \in h$ by the universal property of h as equaliser. Extensionality of $\mathscr{E}$ then gives $1_a \subseteq h$, and so $h \circ k = 1_a$, for some k. Since $f \circ h = g \circ h$, this yields $f = h$ upon composition with k.

Conversely, suppose that $\mathscr{E}$ is well-pointed. The "only if" part of the extensionality condition is straightforward and holds in any category. For the "if" part, suppose that every $x \in f$ has $x \in g$. In order to establish $f \subseteq g$, it suffices to show $f \cap g \simeq f$, i.e. $\chi_{f \cap g} = \chi_f$. Since in general $f \cap g \subseteq f$, Theorem 7.5.1 (3), gives

$$\Rightarrow \circ \langle \chi_{f \cap g}, \chi_f \rangle = true_d.$$

Then if $x:1 \to d$ is any element of d,

$$\Rightarrow \circ \langle \chi_{f \cap g}, \chi_f \rangle \circ x = true_d \circ x$$

i.e.

$$\chi_{f \cap g} \circ x \Rightarrow \chi_f \circ x = true$$

(Exercise 3.8.3 and 4.2.3).

Now $\chi_{f \cap g} \circ x$ and $\chi_f \circ x$ are both truth-values $1 \to \Omega$, and $\mathscr{E}$ is bivalent (being well-pointed), so that each is *true* or *false*. But by Exercise 4.8.2, if $\chi_f \circ x = true$, then $x \in f$, so by our hypothesis $x \in g$, and hence by Theorem 1, $x \in f \cap g$, yielding $\chi_{f \cap g} \circ x = true$. In view of the last equation derived above, and the table for $\Rightarrow$ established in §7.6, $\chi_{f \cap g} \circ x$ and $\chi_f \circ x$ must be either both *true*, or both *false*.

What we have shown then is that the parallel arrows $\chi_{f \cap g}, \chi_f : d \to \Omega$ are not distinguished by any element $x:1 \to d$ of their domain. Since $\mathscr{E}$ is well-pointed, this implies $\chi_{f \cap g} = \chi_g$ as required. □

Theorem 2 points up the advance of topos theory over Lawvere's earlier work [64] on a theory of the category of sets. That system included well-pointedness as an axiom, but the derivation of extensionality required an essential use of a version of the "axiom of choice" (cf. Chapter 12).

It is noteworthy that the analogues of Theorem 1 for the other set operations, viz

(a) $x \in -f$ iff not $x \in f$

and

(b) $x \in f \cup g$ iff $x \in f$ or $x \in g$

fail in some topoi. Take for instance any $\mathscr{E}$ that is Boolean but not bivalent – the simplest example would be the topos $\mathbf{Set}^2$ of pairs of sets. Then $\mathscr{E}$ has a truth value $x:1 \to \Omega$ distinct from $\top$ and $\bot$. Then neither of

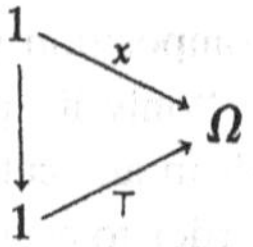

and

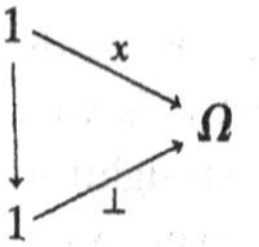

commute, so $x \notin \top$ and $x \notin -\top$ (since $\bot = -\top$ always). Moreover as $\mathscr{E}$ is Boolean, $\top \cup -\top \simeq 1_\Omega$, and so $x \in \top \cup -\top$. Hence both (a) and (b) fail.

THEOREM 3. *$\mathscr{E}$ is bivalent iff* (a) *holds in every* Sub(d).

PROOF. The argument just given to show that (a) fails at least in Sub(Ω) if $\mathscr{E}$ is not bivalent works in any $\mathscr{E}$. On the other hand if $\mathscr{E}$ is bivalent, then if $y:1 \to \Omega$ is a truth-value with $y \neq \top$, then $y = \perp$ and so $\neg \circ y = \top$. Using this, we find, for $f: a \rightarrowtail d$ and $x: 1 \to d$,

$$
\begin{aligned}
x \in -f \quad &\text{iff} \quad \chi_{-f} \circ x = \top \qquad \text{(Exercise 4.8.2)}\\
&\text{iff} \quad \neg \circ \chi_f \circ x = \top\\
&\text{iff} \quad \chi_f \circ x \neq \top\\
&\text{iff} \quad \text{not } x \in f.
\end{aligned}
$$

□

THEOREM 4. *$\mathscr{E}$ satisfies* (b) *for all $\mathscr{E}$-objects d iff $\mathscr{E}$ satisfies the condition* (c):

For any truth values $y: 1 \to \Omega$ and $z: 1 \to \Omega$, $y \cup z = true$ iff $y = true$ or $z = true$.

PROOF. If (b) holds in Sub(1), then let $f: a \rightarrowtail 1$ and $g: b \rightarrowtail 1$ be such that $\chi_f = y$, $\chi_g = z$. Then taking $x: 1 \to 1$, i.e. $x = 1_1$,

$$
\begin{aligned}
y \cup z = \top \quad &\text{iff} \quad (y \cup z) \circ x = \top\\
&\text{iff} \quad \chi_{f \cup g} \circ x = \top\\
&\text{iff} \quad x \in f \cup g\\
&\text{iff} \quad x \in f \quad \text{or} \quad x \in g\\
&\text{iff} \quad \chi_f \circ x = \top \quad \text{or} \quad \chi_g \circ x = \top\\
&\text{iff} \quad y = \top \quad \text{or} \quad z = \top.
\end{aligned}
$$

Conversely if (c) holds, then in any Sub(d) we find that

$$
\begin{aligned}
x \in f \cup g \quad &\text{iff} \quad \chi_{f \cup g} \circ x = \top\\
&\text{iff} \quad \cup \circ \langle \chi_f, \chi_g \rangle \circ x = \top\\
&\text{iff} \quad \cup \circ \langle \chi_f \circ x, \chi_g \circ x \rangle = \top\\
&\text{iff} \quad \chi_f \circ x = \top \quad \text{or} \quad \chi_g \circ x = \top\\
&\text{iff} \quad x \in f \quad \text{or} \quad x \in g.
\end{aligned}
$$

□

A topos satisfying (c), equivalently (b), will be called *disjunctive*. Obviously every bivalent topos is disjunctive. However, the converse is not true, and so (b) does not imply (a) in general. The category **Set$^{\to}$** of set functions has three truth values, and so violates (a). However, it does satisfy (c), since the disjunction arrow yields the table

$\cup$	$\top$	x	$\bot$
$\top$	$\top$	$\top$	$\top$
x	$\top$	x	x
$\bot$	$\top$	x	$\bot$

where x is the third element of Ω. This will perhaps be easier to see from the alternative description of $\mathbf{Set}^{\to}$ to emerge from Chapters 9 and 10. Indeed, Exercise 4 of §10.6 will provide a method of constructing an infinity of disjunctive, non-bivalent, and non-Boolean topoi.

THEOREM 5. *If $\mathscr{E}$ is Boolean and non-degenerate, then $\mathscr{E}$ is disjunctive iff $\mathscr{E}$ is bivalent.*

PROOF. Since $f \cup -f \simeq 1_d$ in a Boolean topos, for any $x : 1 \to d$ we have $x \in f \cup -f$. Thus if $\mathscr{E}$ is disjunctive, from (b) we get $x \in f$ or $x \in -f$. However, we cannot have $x \in f$ and $x \in -f$, for then $x \in f \cap -f \simeq 0_d$, and so $1 \cong 0$. Thus exactly one of "$x \in -f$" and "$x \in f$" obtains, making $\mathscr{E}$ bivalent. □

EXERCISE. Suppose that $\mathscr{E}$ is well-pointed, and $x \in f$ implies $x \in g$. Use Theorem 5.5.1 to show that the pullback h

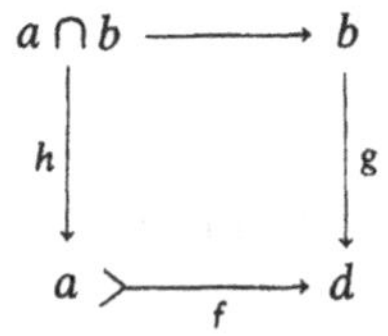

of g along f is iso, making $f \cap g \simeq f$. Hence give an alternative proof that any well-pointed topos is extensional.

CHAPTER 8

INTUITIONISM AND ITS LOGIC

"Let those who come after me wonder why I built up these mental constructions and how they can be interpreted in some philosophy; I am content to build them in the conviction that in some way they will contribute to the clarification of human thought."

L. E. J. Brouwer

8.1. Constructivist philosophy

For a considerable period after the Calculus was discovered by Newton and Leibnitz in the late 17th century, there was controversy and disagreement over its fundamental concepts. Notions of infinitely small quantities, and limits of infinite sequences remained shrouded in mystery, and some of the statements made them look rather strange today (e.g. "A quantity that is increased or decreased infinitely little is neither increased nor decreased" (J. Bernoulli)). The subject acquired a rigorous footing in the 19th century, initially through the development by Cauchy of precise definitions of the concepts of limit and convergence. Later came the "arithmetisation of analysis" by Weierstrass and others, that produced a purely algebraic treatment of the real number system. A significant consequence of this was that analysis began to be separated from its grounding in physical intuition (cf. Weierstrass' proof of the existence of a (counter intuitive?) continuous nowhere-differentiable function). This, along with other factors like the development of non-Euclidean geometry, contributed to the recognition that mathematical structures have an abstract conceptual reality quite independently of the physical world.

Also important during this time was the work of Dedekind and Peano on the number systems. The real numbers were constructed from the rationals, the rationals from the integers, and the integers in turn from the natural numbers. Then the Peano axioms gave an abstract account of the nature of the natural numbers themselves. This kind of reduction contributed to the development of the idea that the whole of mathematics could be presented in one grand axiom system that was itself founded on a few basic notions and principles. This conception has been central to foundational thinking ever since. It takes its extreme form in the "logicist" thesis of Frege and Russell, that mathematics is a part of logic and that mathematical truths are derivable from purely logical principles. It appears also in the work of Hilbert, who attempted to axiomatise mathematics, and prove the consistency of these axioms by finitary methods.

By the time Cantor appeared on the scene it was recognised that references to the infinite, as in "the sequence n^2 tends to infinity as n tends to infinity", could be taken as picturesque articulations of precise, albeit complex, statements about properties of real numbers ("for all ε there exists a δ..." etc.) Cantors set theory transcended this by treating the actual infinite as an object of mathematical investigation. An infinite collection became a "thing-in-itself" that could serve as an *element* of some other collection. The notion of number was extended from the finite to the infinite by the development of a theory of "transfinite" cardinal and ordinal numbers, whose arithmetic involved operations on infinite sets. Cantor's attitude was that as long as statements are grammatically correct and deductions logically sound, such statements have conceptual significance even if they go beyond our basic intuitions about finite numbers and collections.

The theory of sets has been enormously successful, but it has not been without its critics. Leopold Kronecker, well known for having said "God made the integers, all the rest is the work of man", rejected the notions of infinite set and irrational number as being mystical, not mathematical. He maintained that the logical correctness of a theory does not imply the existence of the entities it purported to describe. They remain devoid of any significance unless they can be actually *produced*. Numbers, and operations on them, must, said Kronecker, be "intuitively founded". Definitions and proofs must be "constructive" in a quite literal sense. The definition must show explicitly how to construct the object defined, using objects already known to exist. In classical mathematics an "existence proof" often proceeds by showing that the assumption of the non-existence of an entity of a certain kind leads to contradiction. From the constructivist stand-point this is not a proof of existence at all, since the latter, to be legitimate, must explicitly exhibit the particular object in question. Kronecker believed that the natural numbers could be given such a foundation, but not so for the reals. He actually attempted to rewrite parts of mathematics from this viewpoint.

The conception of things as being "built-up" from already given entities appears also in the reaction of Henri Poincaré to the paradoxes of set theory. He took the view that the source of contradiction lay in the use of *impredicative* definitions. These are circular, self-referential definitions that specify an object X by reference to sets whose own existence depends on that of X. Poincaré held that such definitions were inadmissible and that a set could not be specified until each of its elements had been specified. Thus one half of Russell's paradox (§1.1) consists in showing that $R \in R$. So, on this view, the definition of R is circular, since it can only be given if R has already been defined. Poincaré maintained that mathematics should be founded on the natural number system and developed without impredicative definitions. Thus the Russell class R would not even arise as an object of legitimate study. As it turns out a great deal more would disappear, as significant parts of the classical analysis of the real number system depend on impredicative definitions.

The constructivist attitude, reflected in the views of Kronecker and Poincaré, finds its most spirited expression in the philosophy of Intuitionism, pioneered by the Dutch mathematician L. E. J. Brouwer at the beginning of this century. Brouwer rejected non-constructive arguments, and the conception of infinite collections as things-in-themselves. But he went further than this, to deny traditional logic as a valid representation of mathematical reasoning. We have already noted that the so-called "argument by contradiction" (α is true, because otherwise a contradiction would follow) is constructively unacceptable in existence proofs. But to Brouwer it is not an acceptable principle of argument at all. The same goes for the law of excluded middle, $\alpha \vee {\sim}\alpha$.

Now the classical account of truth as examined in Chapter 6 regards a proposition as being always either true or false, whether we happen to know which is the case. Moreover ${\sim}\alpha$ is true provided only that α is false. Thus "$\alpha \vee {\sim}\alpha$" can be interpreted as saying "either α is true or false" and this last sentence is true on the classical theory. To the intuitionist however a statement is the record of a construction. Asserting the truth of α amounts to saying "I have made a (mental) construction of that which α describes". Likewise ${\sim}\alpha$ records a construction, one that demonstrates that α cannot be the case. From this view, the law of excluded middle has the reading:

> "either I have constructively demonstrated α, or I have constructively demonstrated that α is false."

Now if we take α to be some undecided statement, like Fermat's Last Theorem, then $\alpha \vee \sim\alpha$ is not true on this reading. The Theorem has not been shown to be either true, or false, at the present time.

Thus according to Brouwer we cannot assert "α is true" or "α is false" unless we constructively know which is the case. To say that α is not true means only that I have not at this time constructed α, which is not the same as saying α is false. I may well find a construction tomorrow.

The argument by contradiction mentioned earlier can be classically formalised by the tautology $\sim\sim\alpha \supset \alpha$. To prove α, show that it cannot be that α is false, i.e. show $\sim\sim\alpha$ is true, and then conclude that α holds. Now the intuitionist account of implication is that to assert the truth of $\alpha \supset \beta$ is to assert "I have developed a construction which when appended to a construction for α yields a construction for β". But then to show that it is contradictory to assume a certain thing does not exist ($\sim\sim\alpha$) does not itself amount to producing that thing (α). Hence $\sim\sim\alpha \supset \alpha$ is not valid under the constructive interpretation.

Brouwer's view of the history of logic is that the logical laws were obtained by abstraction of the structure of mathematical deductions at a time when the latter were concerned with the world of the finite. These principles of logic were then ascribed an a priori independent existence. Because of this they have been indiscriminately applied to all subsequent developments, including manipulation of infinite sets. Thus contemporary mathematics is based on and uses procedures that are only valid in a more restricted domain. To obtain genuine mathematical knowledge and determine what the correct modes of reasoning are we must go back to the original source of mathematical truth.

Brouwer maintained that this source is found in our primary intuitions about mathematical objects. For him mathematics is an *activity* – autonomous, self-sufficient, and not dependent on language. The essence of this activity lies in mental acts performed by the mathematician – mental constructions of intuitive systems of entities. Language is secondary, and serves only to communicate mathematical understanding. It arises by the formation of verbal parallels of mathematical thinking. This language is then analysed and from that develops formal languages and axiom systems.

Thus logic analyses the structure of the language that parallels mathematical thought. None of this linguistic activity is however to be regarded as part of mathematics itself. It has practical functions in describing and communicating, but is not prerequisite to the activity of performing mental constructions. The essential content of mathematics remains intuitive, not formal.

Having rejected classical mathematics and logic, Brouwer erected in its place a positive and vigorous philosophy of his own. He distinguished what he called the "two acts" of intuitionism. The first act, which demarcates mathematics as a languageless activity, is an intuitive construction in the mind of "two-ness" – the distinction of one thing and then another in time. Our direct awareness of two states of mind – one succeeding the other, lies at the heart of our intuition of objects. The second act recognises the prospect of repetition of a construction once completed. By such iteration we are lead to an infinitely proceeding sequence. Thus with the first act of distinguishing two states of awareness, and the second act of repeating this process, we obtain a linear series, and the sequence of natural numbers emerges as a product of our primary intuitive awareness. There is no such thing to the intuitionist as an actual completed infinite collection. However, by the generation of endlessly proceeding sequences we are lead to a mathematics of the *potentially* infinite, as embodied in the notion of constructions which, although finite at any given stage, can be continued in an unlimited fashion.

From these ideas Brouwer and his followers have built up an extensive treatment of constructive mathematics which is not merely a subsystem of the classical theory, but has a character and range of concepts all of its own, and is the subject of current research interest. The reader may find out more about it in Heyting [66] (cf. also Bishop [67] for a constructive approach even "stricter" than Brouwer's). Another introductory reference is Dummett [77].

8.2. Heyting's calculus

In 1930 an event occurred that greatly enhanced the general understanding of intuitionism. Arend Heyting produced an axiomatic system of propositional logic which was claimed to generate as theorems precisely those sentences that are valid according to the intuitionistic conception of truth. This system is based on the same language PL as used in Chapter 6. Its axioms are the forms I–XI of the CL axioms (i.e. it has all the CL axioms except $\alpha \vee {\sim}\alpha$). Its sole rule of inference is *Detachment*. We shall refer to this system as IL.

Of course the intuitionist only accepts formal systems as imperfect tools for description and communication. He leaves open the possibility that his intuitive deliberations will one day reveal as yet unheard of principles of reasoning. According to Heyting, "in principle it is impossible to set up a formal system which would be equivalent to intuitionist mathematics ... it can never be proved with mathematical rigour that the system of axioms really embraces every valid method of proof." Nonetheless the investigation of the system IL has proven invaluable in uncovering connections between intuitionistic principles and aspects of topology, recursive functions and computability, models of set theory (forcing), sheaves, and now category theory. Whatever status one attaches to the constructivist view of mathematical reality, there is no doubt that Brouwer's efforts have lead to the elucidation of a significant area of human thought.

Amongst the tautologies that are not IL-theorems are $\alpha\vee{\sim}\alpha$, ${\sim}{\sim}\alpha \supset \alpha$, ${\sim}\alpha \vee {\sim}{\sim}\alpha$. On the other hand $\alpha \supset {\sim}{\sim}\alpha$, ${\sim}{\sim}{\sim}\alpha \supset {\sim}\alpha$, and ${\sim}{\sim}(\alpha\vee{\sim}\alpha)$ *are* derivable. None of the connectives $\sim$, $\wedge$, $\vee$, $\supset$ are definable in terms of each other in IL.

The demonstration of such things is facilitated by the use of a semantical theory that links to IL-derivability. There are several of these available – topological, algebraic, and set-theoretic. The topological aspects of intuitionist logic were discovered independently by Alfred Tarski [38] and Marshall Stone [37]. There it is shown that the open sets of a topological space form an "algebra of sets" in which there are operations satisfying laws corresponding to the axioms of IL. This theme was taken up by J. C. C. McKinsey and Tarski in their study of the algebra of topology [44, 46]. This work involved *closure algebras*, which are BA's with an additional operator whose properties are abstracted from the operation of forming the closure of a set in a topological space. Within a closure algebra there is a special set of elements possessing operations $\sqcap$, $\sqcup$, $\Rightarrow$, $\neg$ obeying intuitionistic principles. McKinsey and Tarski singled these algebras out for special attention, gave an independent axiomatisation of them, and dubbed them *Brouwerian* algebras. Subsequently in [48] they showed that the class of Brouwerian algebras characterises IL in the same way that the class of Boolean algebras characterises CL.

The McKinsey–Tarski approach to algebraic semantics is dual to the one used in §6.5 (an IL-theorem is always assigned 0, rather than 1, etc.). To facilitate comparison with what we have already done we shall discuss, not Brouwerian algebras, but their duals, which are known as

8.3. Heyting algebras

To define these algebras we need to extend our concept of *least upper bound* to sets, rather than just pairs of elements.

If A is a subset of a lattice $\mathbf{L} = (L, \sqsubseteq)$, then $x \in L$ is an *upper bound* of A, denoted $A \sqsubseteq x$, if $y \sqsubseteq x$ whenever $y \in A$. If moreover $x \sqsubseteq z$ whenever $A \sqsubseteq z$, then x is a *least upper bound* (l.u.b.) of A.

EXERCISE 1. A has at most one l.u.b.

EXERCISE 2. Define the notion of g.l.b. of A. □

We say that x is the *greatest element* of A if x is a l.u.b. of A and also a member of A. Thus A *has* a greatest element precisely when one of its members is a l.u.b. of A.

EXERCISE 3. A g.l.b. of A is the greatest element of the set of lower bounds of A.

EXERCISE 4. Define the *least* element of A. □

Now in the powerset lattice $(\mathcal{P}(D), \subseteq)$, $-A$ is the greatest element disjoint from A. That is, $-A$ is disjoint from A, $A \cap -A = \emptyset$, and whenever $A \cap B = \emptyset$, then $B \subseteq -A$. This description of complements can be set out in any lattice and sometimes it leads to a non-Boolean operation. Hence it is given a different name, as follows:

If $\mathbf{L} = (L, \sqsubseteq)$ is a lattice with a zero 0, and $a \in L$, then $b \in L$ is the *pseudo-complement* of a iff b is the greatest element of L disjoint from a, i.e. b is the greatest element of the set $\{x \in L: a \sqcap x = 0\}$. If every member of L has a pseudo-complement, $\mathbf{L}$ is a *pseudo-complemented lattice*.

Using these definitions it is not hard to verify the

EXERCISE 5. b is the pseudo-complement of a precisely when it satisfies the condition:

$$\text{for all } x \in L, \quad x \sqsubseteq b \text{ iff } a \sqcap x = 0.$$ □

EXAMPLE 1. $(\mathcal{P}(D), \subseteq)$: $-A$ is the pseudo-complement of A.

EXAMPLE 2. $\mathbf{B} = (B, \sqsubseteq)$: in any BA,

$$x \sqsubseteq a' \text{ iff } a \sqcap x = 0 \qquad \text{(cf. Exercise 6.4.2)}$$

so the Boolean complement is always a pseudo-complement.

EXAMPLE 3. $(L_M, \subseteq)$: In the lattice of left ideals of monoid **M**, $\neg B = \{m : \omega_m(B) = \emptyset\}$ is the pseudo-complement of B. (why is $C \subseteq \neg B$ iff $B \cap C = \emptyset$?)

EXAMPLE 4. $(\Theta, \subseteq)$: In the lattice of open sets of a topological space, $U \in \Theta$ has a pseudo-complement, namely $(-U)^\circ$, the *interior* of $-U$ (i.e. the largest open subset of the complement of U). We have $V \subseteq (-U)^\circ$ iff $U \cap V = \emptyset$, for all open V.

EXAMPLE 5. **Sub**(d): In Sub(d), for any topos, $-f : -a \rightarrowtail d$ is the pseudo-complement of $f : a \rightarrowtail d$.

PROOF: We have to show that

$$g \subseteq -f \quad \text{iff} \quad f \cap g \simeq 0_d.$$

Now if $g \subseteq -f$, then by lattice properties, $f \cap g \subseteq f \cap -f \simeq 0_d$ (Theorem 7.2.3), and so $f \cap g \simeq 0_d$.

Conversely suppose $f \cap g \simeq 0_d$. Then the top square of

$$\begin{array}{ccc} 0 & \xrightarrow{0_b} & b \\ \downarrow & & \downarrow g \\ a & \xrightarrow{f} & d \\ \downarrow & & \downarrow \chi_f \\ 1 & \xrightarrow{\top} & \Omega \end{array}$$

is a pullback. But so is the bottom square, hence the PBL gives the outer rectangle as a pullback. By the Ω-axiom then,

$$\chi_f \circ g = \chi_{0_b} = \bot \circ !_b \qquad \text{(Exercise 5.4.3)}$$

Thus

$$\neg \circ \chi_f \circ g = \neg \circ \bot \circ !_b = \top \circ !_b.$$

But $\top \circ !_b = \chi_g \circ g$ (Ω-axiom) and $\neg \circ \chi_f = \chi_{-f}$, so altogether we have

$$\chi_{-f} \circ g = \chi_g \circ g.$$

But then Lemma 1 of §7.5 gives

$$-f \cap g \simeq g \cap g \simeq g.$$

Hence $g \simeq -f \cap g \subseteq -f$, as required. □

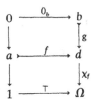

EXAMPLE 6. *Germs*. The collection $\Theta/\sim_i = \{[U]_i: U \text{ open in } I\}$ of germs of open sets at i (cf. the definition of Ω in **Top**(I)) is a pseudo-complemented lattice in which

$$0 = [\emptyset]_i, \qquad \text{the germ of } \emptyset$$
$$[U]_i \sqcap [V]_i = [U \cap V]_i$$
$$[U]_i \sqcup [V]_i = [U \cup V]_i$$

and the pseudo-complement of $[U]_i$ is $[(-U)^0]_i$ (i.e. we have the standard quotient lattice construction).

These operations yield the associated truth functions in **Top**(I). There, $\neg: \Omega \to \Omega$ is the function from $\hat{I}$ to $\hat{I}$ taking the germ of U at i to the germ of $(-U)^0$ at i. The conjunction and disjunction arrows from $\Omega \times \Omega$ to Ω are the above meet and join operations acting on each stalk. □

The notion of pseudo-complement can be generalised by replacing the zero 0 by some other element b of the lattice, to obtain the pseudo-complement of a *relative to* b. This, if it exists, is the greatest element of the set $\{x: a \sqcap x \sqsubseteq b\}$. In other words the pseudo-complement of a relative to b is the greatest element c such that $a \sqcap c \sqsubseteq b$. It is readily seen that

EXERCISE 6. c is the pseudo-complement of a relative to b precisely when it satisfies

$$\text{for all } x, \quad x \sqsubseteq c \quad \text{iff} \quad a \sqcap x \sqsubseteq b. \qquad \square$$

EXAMPLE 1. $(\mathcal{P}(D), \subseteq)$: $-A \cup B$ is the pseudo-complement of A relative to B.

EXAMPLE 2. $\mathbf{B} = (B, \sqsubseteq)$: In any **BA**, (Lemma 2(2), §7.5)

$$x \sqsubseteq a' \sqcup b \quad \text{iff} \quad a \sqcap x \sqsubseteq b.$$

EXAMPLE 3. $(L_M, \subseteq)$: $B \Rightarrow C = \{m: \omega_m(B) \subseteq \omega_m(C)\}$ has

$$X \subseteq B \Rightarrow C \quad \text{iff} \quad B \cap X \subseteq C, \qquad \text{all left ideals } X.$$

EXAMPLE 4. $(\Theta, \subseteq)$: The pseudo-complement of U relative to V is $(-U \cup V)^0$, the largest open subset of $-U \cup V$.

$$\text{Whenever } W \text{ is open,} \quad W \subseteq (-U \cup V)^0 \quad \text{iff} \quad U \cap W \subseteq V.$$

EXAMPLE 5. (**Sub**(d)): Theorem 1 of §7.5 states that

$$h \subseteq f \mapsto g \quad \text{iff} \quad f \cap h \subseteq g,$$

hence $\mapsto$ is an operation of relative pseudo-complementation.

EXAMPLE 6. *Germs.* In the lattice $\Theta/\sim_i$ of germs of open sets at i, $[(-U \cup V)^o]_i$ provides $[U]_i$ with a pseudo-complement relative to $[V]_i$. This operation, acting on each stalk, yields the truth-arrow $\Rightarrow : \Omega \times \Omega \to \Omega$ in the topos **Top**(I). □

In a general lattice **L**, the pseudo-complement of a relative to b, when it exists, will be denoted $a \Rightarrow b$. If $a \Rightarrow b$ exists for every a and b in **L**, we will say that **L** is a *relatively pseudo-complemented* (r.p.c.) lattice.

The theory of r.p.c. lattices is thoroughly discussed in Rasiowa–Sikorski [63] and Rasiowa [74]. We list here some basic facts which the reader may care to treat as

Exercises

If **L** is a r.p.c. lattice:

EXERCISE 7. **L** has a unit 1, and for each $a \in L$, $a \Rightarrow a = 1$.

EXERCISE 8. $a \sqsubseteq b$ iff $a \Rightarrow b = 1$.

EXERCISE 9. $b \sqsubseteq a \Rightarrow b$.

EXERCISE 10. $a \sqcap (a \Rightarrow b) = a \sqcap b \sqsubseteq b$.

EXERCISE 11. $(a \Rightarrow b) \sqcap b = b$.

EXERCISE 12. $(a \Rightarrow b) \sqcap (a \Rightarrow c) = a \Rightarrow (b \sqcap c)$.

EXERCISE 13. $(a \Rightarrow b) \sqsubseteq ((a \sqcap c) \Rightarrow (b \sqcap c))$.

EXERCISE 14. if $b \sqsubseteq c$ then $a \Rightarrow b \sqsubseteq a \Rightarrow c$.

EXERCISE 15. $(a \Rightarrow b) \sqcap (b \Rightarrow c) \sqsubseteq (a \Rightarrow c)$.

EXERCISE 16. $(a \Rightarrow b) \sqcap (b \Rightarrow c) \sqsubseteq (a \sqcup b) \Rightarrow c$.

EXERCISE 17. $a \Rightarrow (b \Rightarrow c) \sqsubseteq (a \Rightarrow b) \Rightarrow (a \Rightarrow c)$. □

The definition of r.p.c. lattice does not require the presence of a zero. A *Heyting algebra* (**HA**) is, by definition, an r.p.c. lattice that has a zero 0. If $\mathbf{H} = (H, \sqsubseteq)$ is a Heyting algebra, we define $\neg: H \to H$ by $\neg a = a \Rightarrow 0$. Then $\neg a$ is the l.u.b. of $\{x: a \sqcap x = 0\}$, i.e. $\neg a$ is the pseudo-complement of a.

Again the reader may consult Rasiowa and Sikorski [63] for details of the

Exercises

In any **HA** $\mathbf{H} = (H, \sqsubseteq)$:

Exercise 18. $\neg 1 = \neg(a \Rightarrow a) = 0$.

Exercise 19. $\neg 0 = 1$, and if $\neg a = 1$, then $a = 0$.

Exercise 20. $a \sqsubseteq \neg\neg a$.

Exercise 21. $(a \Rightarrow b) \sqsubseteq (\neg b \Rightarrow \neg a)$.

Exercise 22. $\neg a = \neg\neg\neg a$.

Exercise 23. $a \sqcap \neg a = 0$.

Exercise 24. $\neg(a \sqcup b) = \neg a \sqcap \neg b$.

Exercise 25. $\neg a \sqcup \neg b \sqsubseteq \neg(a \sqcap b)$.

Exercise 26. $\neg a \sqcup b \sqsubseteq a \Rightarrow b$.

Exercise 27. $\neg\neg(a \sqcup \neg a) = 1$.

Exercise 28. $\neg a \sqsubseteq (a \Rightarrow b)$.

Exercise 29. $(a \Rightarrow b) \sqcap (a \Rightarrow \neg b) = \neg a$. □

The six major examples of this section are all Heyting algebras. In the case of the topos **Top**(I) of sheaves over a topological space we can now describe Ω as a topological bundle of Heyting algebras, indexed by I, each of them a quotient of the **HA** of open sets in I.

Now that we know Sub(d) to be an **HA** we can return to the assertion of §7.2 that Sub(d) is a distributive lattice. The point is simply that every r.p.c. lattice is distributive. A proof may be found in Rasiowa and Sikorski, p. 59.

Now in a **BA**, the complement satisfies $x=(x')'$. The analogous property does not occur in all **HA**'s. In our example $\mathbf{M}_2$, in Sub(Ω) we have $\top \not\simeq -\bot$, since $\top$ corresponds to the subset $\{2\}$ of L_2, while $-\bot$ corresponds to $\{2,\{0\}\}$ (the character of $-\bot$ is $\neg \circ \chi_\bot = \neg\circ\neg$, which is the function f_Ω of §5.4). Since $\bot \simeq -\top$ in general, we get in $\mathbf{M}_2$ that $\top \not\simeq --\top$.

In the general **HA** we always have $x \sqsubseteq \neg\neg x$, but possibly not $\neg\neg x \sqsubseteq x$ (corresponding to $\sim\sim\alpha \supset \alpha$ not being an IL-theorem). Indeed the situation is as follows:

EXERCISE 30. If an **HA H** satisfies $\neg\neg x \sqsubseteq x$, all $x \in H$, then **H** is a Boolean algebra, i.e. $\neg x$ is an actual complement of x. (Hint: use Exercise 27.) □

In CL, α is logically equivalent to $\sim\sim\alpha$, as reflected in the fact that $x=\neg\neg x$ in 2. In the internal logic of **Set** this means that

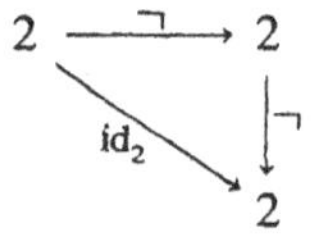

commutes, i.e. $\neg\circ\neg = \mathrm{id}_2$. The analogous diagram does not commute in all topoi, e.g. in $\mathbf{M}_2$, $\neg\circ\neg$ is the function f_Ω of §5.4 that has output 2 for input $\{0\}$, hence $\neg\circ\neg \neq 1_\Omega$. These deliberations are brought together in

THEOREM 1. *In any topos $\mathscr{E}$ the following are equivalent*

(1) *$\mathscr{E}$ is Boolean*

(2) *In* Sub(Ω), $\top \simeq -\bot$

(3) $\neg\circ\neg = 1_\Omega$.

PROOF. (1) implies (2): In general $\bot \cap \top \simeq 0_\Omega$ as shown by the pullback

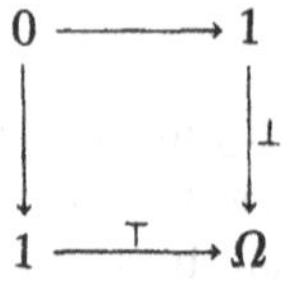

defining $\bot$. But if $\mathscr{E}$ is Boolean, $\bot \cup \top \simeq 1_\Omega$ (cf. §7.3), so that $\top$ is the unique complement of $\bot$ and hence is the pseudo-complement $-\bot$.

(2) implies (3): If $\top \simeq -\bot$, then $\chi_\top = \chi_{-\bot}$, i.e.

$$1_\Omega = \neg \circ \chi_\bot = \neg \circ \neg.$$

(3) implies (1): Let f be a subobject of d. Then

$$\begin{aligned}\chi_{--f} &= \neg \circ \neg \circ \chi_f \\ &= \chi_f, \quad \text{if} \quad \neg \circ \neg = 1_\Omega\end{aligned}$$

so $--f \simeq f$, making Sub(d) a **BA** by the last exercise. □

Algebraic semantics

If $\mathbf{H} = (H, \sqsubseteq)$ is a Heyting algebra (also known as a *pseudo-Boolean* algebra) then an **H**-*valuation* is a function $V: \Phi_0 \to H$. This may be extended to all sentences using joins $\sqcup$, meets $\sqcap$, relative pseudo-complements $\Rightarrow$, and pseudo-complements $\neg$, to "interpret" the connectives $\vee$, $\wedge$, $\supset$, $\sim$, exactly as for **BA**-valuations in §6.5. A sentence α is **H**-*valid* when $V(\alpha) = 1$ for every **H**-valuation V. α is **HA**-*valid* if valid in every Heyting algebra. We have the following characterisation result:

$$\alpha \textit{ is } \mathbf{HA}\textit{-valid iff } \vdash_{\mathrm{IL}} \alpha.$$

The "soundness" part of this consists in showing that the axioms I–XI are **HA**-valid and that *Detachment* preserves this property. For the latter observe by Exercise 8 above that if $V(\alpha) = V(\alpha \supset \beta) = 1$ then $V(\alpha) \sqsubseteq V(\beta)$ so $V(\beta) = 1$. The validity of I–XI is given by various other of the Exercises in combination with 8, e.g. 15. for Axiom IV, 16. for IX, 29. for XI etc.

The completeness of IL with respect to **HA**-validity can be shown by the Lindenbaum algebra method of the Exercise 2 in §6.5. The relation

$$\alpha \sim_{\mathrm{IL}} \beta \quad \text{iff} \quad \vdash_{\mathrm{IL}} \alpha \supset \beta \quad \text{and} \quad \vdash_{\mathrm{IL}} \beta \supset \alpha$$

is an equivalence on Φ. The Lindenbaum algebra for IL is $\mathbf{H}_{\mathrm{IL}} = (\Phi/{\sim_{\mathrm{IL}}}, \sqsubseteq)$ where

$$[\alpha] \sqsubseteq [\beta] \quad \text{iff} \quad \vdash_{\mathrm{IL}} \alpha \supset \beta$$

$\mathbf{H}_{\mathrm{IL}}$ is an **HA** with $\sqcap$, $\sqcup$ as in the Boolean case, and

$$\begin{aligned}&[\alpha] \Rightarrow [\beta] = [\alpha \supset \beta] \\ &\neg[\alpha] = [\sim\alpha]\end{aligned}$$

The valuation $V(\alpha) = [\alpha]$ can be used to show

$$\vdash_{\mathrm{IL}} \alpha \quad \text{iff} \quad \mathbf{H}_{\mathrm{IL}} \models \alpha,$$

hence any **HA**-valid sentence will be $\mathbf{H}_{\mathrm{IL}}$-valid and so an IL-theorem.

Now the Ω-axiom, through the assignment of χ_f to f establishes, (§4.2) a bijection

$$\mathrm{Sub}(d) \cong \mathscr{E}(d, \Omega)$$

which transfers the **HA** structure of Sub(d) to $\mathscr{E}(d, \Omega)$. Indeed the partial ordering on the latter was described in §7.2 (Theorem 1, Corollary): $\chi_f \sqsubseteq \chi_g$ precisely when $\langle \chi_f, \chi_g \rangle$ factors through $e : \circledS \rightarrowtail \Omega \times \Omega$. The Heyting operations on $\mathscr{E}(d, \Omega)$ are given by application of the truth-arrows. Thus the lattice meet operation in $\mathscr{E}(d, \Omega)$ assigns to two arrows $f, g : d \rightrightarrows \Omega$, the arrow $f \frown g = \cap \circ \langle f, g \rangle$, the join assigns to them $f \smile g = \cup \circ \langle f, g \rangle$ and so on. The *definition* of the operations $\cap$, $\cup$ etc. on Sub(d) shows that algebraically the two structures look the same, i.e. Sub(d) and $\mathscr{E}(d, \Omega)$ are isomorphic **HA**'s, from which one sees that they validate the same sentences.

The link between topos semantics and the present theory is that in any $\mathscr{E}$, we have

$$\mathscr{E} \models \alpha \quad \textit{iff} \quad \mathscr{E}(1, \Omega) \models \alpha \quad \textit{iff} \quad \mathrm{Sub}(1) \models \alpha$$

(which clarifies further the situation described in Theorem 2 of §7.4).

Thus topos validity in $\mathscr{E}$ amounts to **HA**-validity in the **HA**'s $\mathscr{E}(1, \Omega)$ and Sub(1). The point is that an $\mathscr{E}$-valuation is the same thing as an $\mathscr{E}(1, \Omega)$-valuation, and that $\mathscr{E}$-validity and $\mathscr{E}(1, \Omega)$-validity come to the same thing, since the unit of the **HA** $\mathscr{E}(1, \Omega)$ is $\top : 1 \to \Omega$. This provides the basis of Exercise 2 of §6.7, viz

$$\mathbf{Bn}(I) \models \alpha \quad \textit{iff} \quad (\mathscr{P}(I), \subseteq) \models \alpha,$$

since the truth-values in **Bn**(I) are "essentially" subsets of I. Recalling further that truth-values in **Top**(I) are essentially *open* subsets of I we find that

$$\mathbf{Top}(I) \models \alpha \quad \text{iff} \quad (\Theta, \subseteq) \models \alpha,$$

i.e. validity in the topos of sheaves over I is equivalent to **HA**-validity in the algebra of open subsets of I.

SOUNDNESS FOR $\mathscr{E}$-VALIDITY. *If* $\vdash_{\mathrm{IL}} \alpha$ *then* $\mathscr{E} \models \alpha$, *for all topoi* $\mathscr{E}$.

PROOF. If α is an IL-theorem then α is **HA**-valid. In particular then, $\mathscr{E}(1, \Omega) \models \alpha$, and so $\mathscr{E} \models \alpha$, by the above. □

EXERCISE 31. Give an algebraic reason why bivalent topoi always validate $\alpha \vee {\sim}\alpha$. □

Exponentials

The condition $x \sqsubseteq a \Rightarrow b$ iff $a \sqcap x \sqsubseteq b$ means that in an r.p.c. lattice, when considered as a poset category, there is a bijective correspondence between arrows $x \to (a \Rightarrow b)$ and arrows $a \sqcap x \to b$ (either one, or no, arrows in each case). This is reminiscent (§3.16) of the situation in a category $\mathscr{C}$ with exponentiation where there is a bijection $\mathscr{C}(x, b^a) \cong \mathscr{C}(x \times a, b)$. Now in a lattice $a \sqcap x = x \sqcap a$ is the product $x \times a$, and indeed in an r.p.c. lattice $a \Rightarrow b$ provides the exponential b^a. The evaluation arrow $ev: b^a \times a \to b$ is the unique arrow $(a \Rightarrow b) \sqcap a \to b$, which exists by Exercise 10 above. Conversely, exponentials provide relative pseudo-complements, and we find that categorially a *Heyting algebra is no more nor less than a Cartesian closed and finitely co-complete poset.*

The approach we have used in eliciting the **HA** structure of Sub(d) differs from the original method, as described in Freyd [72]. There, $|\!\Rightarrow$ is obtained via the Fundamental Theorem, and some complex machinery that we have not even begun to consider (limit preserving functors). The aim is to show that Sub(d) as a poset is Cartesian closed, since exponentials in posets provide r.p.c.'s. By using the truth-arrow $\Rightarrow$ to define $|\!\Rightarrow$ we have, apart from showing how the logic of $\mathscr{E}$ determines its subobject behaviour, come in an easier fashion to exactly the same point. For, as Lemma 2(1) of §7.5 indicates, a lattice can be relatively pseudo-complemented in one and only one way.

EXERCISE 32. Show that any chain (linearly ordered poset) with a maximum 1 is r.p.c., with

$$p \Rightarrow q = \begin{cases} 1 & \text{if } p \sqsubseteq q \\ q & \text{otherwise.} \end{cases}$$

(This is the origin of Example 2, §3.16).

EXERCISE 33. Distinguish between, say, $\top |\!\Rightarrow \top$ and $\top \Rightarrow \top$ in Sub(Ω) (this is why the special symbol "$|\!\Rightarrow$" is being used). □

8.4. Kripke semantics

In 1965 Saul Kripke published a new formal semantics for intuitionistic logic in which PL-sentences are interpreted as subsets of a poset. This theory arose as a sequel to a semantical analysis that Kripke had developed for modal logic. Briefly, modal logic is concerned with the concept of *necessity*, and on the propositional level uses the language PL enriched by a connective whose interpretation is "it is necessarily the case that". The appropriate algebraic "models" here are **BA**'s with an additional operation for this new connective. There is a particular modal axiom system, known as *S4*, that is characterised algebraically by the class of closure algebras. McKinsey and Tarski [48] used this fact to develop a translation of PL-sentences into modal sentences in such a way that IL-theorems correspond to *S4*-theorems. The mechanism of this translation when seen in the light of the Kripke models for *S4*, leads to a new way of giving formal "meaning" to IL sentences.

One attractive feature of the new theory is that its structures, apart from being generally more tractable than the algebraic ones, have an informal interpretation that accords well with the intuitionistic account of the nature of validity. In the latter, truth is temporally conditioned. A sentence is not true or false per se, as in classical logic, but is only so at certain times, i.e. those times at which it has been constructively determined. Now each moment of time is associated with a particular *stage*, or *state* of knowledge. This comprises all the facts that have been constructively established at that time. Sentences then true are so in view of the existing state of knowledge. We thus speak of sentences as being "true at a certain stage" or "true at a certain state of knowledge". The collection of all states of knowledge is ordered by its temporal properties. We speak of one state as coming after, or being later than, another state in time. A sentence true at a certain stage will be held to be true at all later (future) stages. This embodies the idea that constructive knowledge, once established, exists forever more. Having proven α, we cannot later show α to be false.

Now the temporal ordering of states is a partial ordering, not necessarily linear. The states we consider do not always follow one another in a linear sequence because they are *possible* states of knowledge, not just those that do actually occur. Thus at the present moment we may look to the future and contemplate two possible states of knowledge, one in which Fermat's Last Theorem is determined to be true, and one in which it is shown false. These states are incompatible with each other, so in view of the "persistence of truth in time" they cannot be connected by the ordering of states. We cannot proceed from the present to one, and then the other.

Altogether then, the collection of possible states of knowledge is a poset under the ordering of time. A sentence corresponds to a particular subset of this poset, consisting of the states at which the sentence is true.

In view of the persistence of truth in time, this set has a special property: given a particular state in the set, all states in the future of that state belong to the set as well. With these ideas in mind we move to the formal details of Kripke's semantics.

Let $\mathbf{P} = (P, \sqsubseteq)$ be a poset (also called a *frame* in this context). A set $A \subseteq P$ is *hereditary* in $\mathbf{P}$ if it is closed "upwards" under $\sqsubseteq$, i.e. if we have that

$$\text{whenever} \quad p \in A \quad \text{and} \quad p \sqsubseteq q, \quad \text{then} \quad q \in A.$$

The collection of hereditary subsets of $\mathbf{P}$ will be denoted $\mathbf{P}^+$. A $\mathbf{P}$-*valuation* is a function $V: \Phi_0 \to \mathbf{P}^+$, assigning to each π_i an *hereditary* subset $V(\pi_i) \subseteq P$. A *model based on* $\mathbf{P}$ is a pair $\mathcal{M} = (\mathbf{P}, V)$, where V is a $\mathbf{P}$-valuation. This notion formally renders the intuitive ideas sketched above. P is a collection of stages of knowledge temporally ordered by $\sqsubseteq$. $V(\pi_i)$ is the set of stages at which π_i is true. The requirement that $V(\pi_i)$ be hereditary formalises the "persistence in time of truth". We now extend the notion of truth at a particular stage to all sentences. The expression "$\mathcal{M} \models_p \alpha$" is to be read "$\alpha$ is true in $\mathcal{M}$ at p", and is defined inductively as follows:

(1) $\mathcal{M} \models_p \pi_i$ iff $p \in V(\pi_i)$

(2) $\mathcal{M} \models_p \alpha \wedge \beta$ iff $\mathcal{M} \models_p \alpha$ and $\mathcal{M} \models_p \beta$

(3) $\mathcal{M} \models_p \alpha \vee \beta$ iff either $\mathcal{M} \models_p \alpha$ or $\mathcal{M} \models_p \beta$

(4) $\mathcal{M} \models_p {\sim}\alpha$ iff for all q with $p \sqsubseteq q$, not $\mathcal{M} \models_q \alpha$

(5) $\mathcal{M} \models_p \alpha \supset \beta$ iff for all q with $p \sqsubseteq q$, if $\mathcal{M} \models_q \alpha$ then $\mathcal{M} \models_q \beta$.

Thus at stage p, ${\sim}\alpha$ is true if α is never established at any later stage, and $\alpha \supset \beta$ is true if β holds at all later stages that α is true at.

α is *true* (holds) in the model $\mathcal{M}$, denoted $\mathcal{M} \models \alpha$, if $\mathcal{M} \models_p \alpha$ for every $p \in P$. α is *valid* on the frame $\mathbf{P}$, $\mathbf{P} \models \alpha$, if α is true in every model $\mathcal{M} = (\mathbf{P}, V)$ based on $\mathbf{P}$.

"$\mathcal{M} \not\models_p \alpha$" will abbreviate "not $\mathcal{M} \models_p \alpha$". Similarly "$\mathbf{P} \not\models \alpha$".

EXAMPLE. Let $\mathbf{P}$ be $\mathbf{2} = (\{0, 1\}, \leqslant)$ $(0 \leqslant 1$ as usual). Take a V with $V(\pi) = \{1\}$ (which is hereditary). Then with $\mathcal{M} = (\mathbf{2}, V)$ we have by (1), $\mathcal{M} \not\models_0 \pi$. But $\mathcal{M} \models_1 \pi$ and $0 \leqslant 1$ so by (4), $\mathcal{M} \not\models_0 {\sim}\pi$. Thus by (3), $\mathcal{M} \not\models_0 \pi \vee {\sim}\pi$, so the law of excluded middle is not valid on this frame. Notice also $\mathcal{M} \not\models_1 {\sim}\pi$, hence $\mathcal{M} \models_0 {\sim}{\sim}\pi$. Since $0 \leqslant 0$, (5) then gives $\mathcal{M} \not\models_0 {\sim}{\sim}\pi \supset \pi$, hence $\mathbf{2} \not\models {\sim}{\sim}\pi \supset \pi$.

If we denote by $\mathcal{M}(\alpha)$ the set of points at which α is true in $\mathcal{M}$, i.e. $\mathcal{M}(\alpha) = \{p: \mathcal{M} \models_p \alpha\}$ then the semantic clauses (1), (2) and (3) can be expressed as

(1′) $\mathcal{M}(\pi_i) = V(\pi_i)$

(2′) $\mathcal{M}(\alpha \wedge \beta) = \mathcal{M}(\alpha) \cap \mathcal{M}(\beta)$

(3′) $\mathcal{M}(\alpha \vee \beta) = \mathcal{M}(\alpha) \cup \mathcal{M}(\beta)$.

To re-express (4) and (5) we define, for hereditary S, T,

$$\neg S = \{p\text{: for all } q \text{ such that } p \sqsubseteq q,\ q \notin S\}$$

and

$$S \Rightarrow T = \{p\text{: for all } q \text{ with } p \sqsubseteq q, \text{ if } q \in S \text{ then } q \in T\}.$$

We then have

(4′) $\mathcal{M}(\sim\alpha) = \neg\mathcal{M}(\alpha)$

(5′) $\mathcal{M}(\alpha \supset \beta) = \mathcal{M}(\alpha) \Rightarrow \mathcal{M}(\beta)$.

The notation is of course not accidental. The intersection and union of two hereditary sets are both hereditary, so the poset $\mathbf{P}^+ = (P^+, \subseteq)$ of hereditary sets under the inclusion ordering is a (bounded distributive) lattice with meets and joins given by $\cap$ and $\cup$. $\mathbf{P}^+$ is indeed a Heyting algebra, with $S \Rightarrow T$ being the pseudo-complement of S relative to T. We have

$$U \subseteq S \Rightarrow T \quad \text{iff} \quad S \cap U \subseteq T, \qquad \text{all hereditary } U,$$

and

$$\neg S = S \Rightarrow \emptyset,$$

the pseudo-complement of S (many exercises here for the reader).

Now a **P**-valuation $V : \Phi_0 \to \mathbf{P}^+$ for the frame **P** is also by definition a $\mathbf{P}^+$-valuation for the **HA** $\mathbf{P}^+$. This may be extended, using $\cap$, $\cup$, $\neg$, $\Rightarrow$ to obtain elements $V(\alpha)$ of the algebra $\mathbf{P}^+$ in the usual way. But V also yields a model $\mathcal{M} = (\mathbf{P}, V)$ and hence the set $\mathcal{M}(\alpha)$ for each α. By induction, using the two sets of semantic rules above, we find that for any α,

$$\mathcal{M}(\alpha) = V(\alpha),$$

and so

$$\mathcal{M} \models \alpha \quad \text{iff} \quad \mathcal{M}(\alpha) = P \quad \text{iff} \quad V(\alpha) = P.$$

But P is the unit of the lattice $\mathbf{P}^+$, and since this analysis holds for all V, we find for all α that

$$\mathbf{P} \models \alpha \quad \text{iff} \quad \mathbf{P}^+ \models \alpha,$$

i.e. Kripke-validity on the frame **P** is the same as **HA**-validity on the algebra $\mathbf{P}^+$. This contributes to the verification of the basic characterisation theorem for frame validity, which is that for any α.

$$\vdash_{\mathrm{IL}} \alpha \quad \textit{iff} \quad \alpha \textit{ is valid on every frame.}$$

For the soundness part, we note that if $\vdash_{\mathrm{IL}} \alpha$ then α is **HA**-valid, so for any $\mathbf{P}$, $\mathbf{P}^+ \models \alpha$, hence $\mathbf{P} \models \alpha$. One way of proving the completeness part would be to use the representation theory of Stone [37] to turn **HA**'s into frames. The original proof of Kripke used a "semantic tableaux" technique. An alternative approach, based on methods first used in classical logic by Leon Henkin [49], has subsequently been developed, and we now describe it briefly.

First, observe that if p is an element of model $\mathcal{M}$, then $\Gamma_p = \{\alpha : \mathcal{M} \models_p \alpha\}$, the set of sentences true in $\mathcal{M}$ at p, satisfies

(i) If $\vdash_{\mathrm{IL}} \alpha$ then $\alpha \in \Gamma_p$ (soundness)
(ii) If $\vdash_{\mathrm{IL}} \alpha \supset \beta$ and $\alpha \in \Gamma_p$, then $\beta \in \Gamma_p$ (closure under detachment)
(iii) there is at least one α such that $\alpha \notin \Gamma_p$ (consistency)
(iv) if $\alpha \vee \beta \in \Gamma_p$ then $\alpha \in \Gamma_p$ or $\beta \in \Gamma_p$ (Γ_p is "prime").

Γ_p could be called a "state-description". It describes the state p by specifying which sentences are true at p. A set $\Gamma \subseteq \Phi$ that satisfies these four conditions will be called *full*. In general a full set can be construed as a state-description, namely the description of that state in which all members of Γ are known to be true and all sentences not in Γ are not known to be true. This introduces us to the *canonical frame for* IL, which is the poset

$$\mathbf{P}_{\mathrm{IL}} = (P_{\mathrm{IL}}, \subseteq),$$

where P_{IL} is the collection of all full sets, and $\subseteq$ as usual is the subset relation. The *canonical model* for IL is $\mathcal{M}_{\mathrm{IL}} = (\mathbf{P}_{\mathrm{IL}}, V_{\mathrm{IL}})$, where

$$V_{\mathrm{IL}}(\pi_i) = \{\Gamma : \pi_i \in \Gamma\},$$

the set of full sets having π_i as a member.

An inductive proof, using facts about IL-derivability and properties of full sets, shows that for any α and Γ,

$$\mathcal{M}_{\mathrm{IL}} \models_{\Gamma} \alpha \quad \textit{iff} \quad \alpha \in \Gamma$$

To derive the completeness theorem we need the further result:

LINDENBAUM'S LEMMA. $\vdash_{\mathrm{IL}} \alpha$ *iff* α *is a member of every full set,* so that we can conclude

$$\vdash_{\mathrm{IL}} \alpha \quad \textit{iff} \quad \mathcal{M}_{\mathrm{IL}} \models \alpha.$$

From this we get

$$\vdash_{\mathrm{IL}} \alpha \quad \textit{iff} \quad \mathbf{P}_{\mathrm{IL}} \models \alpha$$

and this yields the completeness theorem. (It will also yield, in Chapter 10, a characterisation of the class of topos-valid sentences).

One of the great advantages of the Kripke semantics is that the validity of sentences can be determined by simple conditions on frames. For example, on the poset

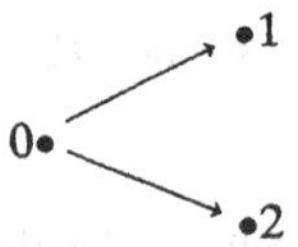

if $V(\pi_1)=\{1\}$ and $V(\pi_2)=\{2\}$, then the tautology $(\pi_1 \supset \pi_2) \vee (\pi_2 \supset \pi_1)$ is not true at 0. Notice that this frame is not linearly ordered. In fact it can be shown that:

$\mathbf{P} \models (\alpha \supset \beta) \vee (\beta \supset \alpha)$ iff $\mathbf{P}$ is *weakly linear*, i.e. whenever $p \sqsubseteq q$ and $p \sqsubseteq r$, then $q \sqsubseteq r$ or $r \sqsubseteq q$.

Adjunction of the axiom $(\alpha \supset \beta) \vee (\beta \supset \alpha)$ to IL yields a system, known as LC, first studied by Michael Dummett [59]. The canonical frame method can be adapted to show that the LC-theorems are precisely the sentences valid on all weakly linear frames.

Exercise 1. Show $\mathbf{P} \models \alpha \vee {\sim}\alpha$ iff $\mathbf{P}$ is discrete, i.e. has $p \sqsubseteq q$ iff $p = q$.

Exercise 2. $\mathbf{P} \models {\sim}\alpha \vee {\sim}{\sim}\alpha$ iff $\mathbf{P}$ is *directed*, i.e. if $p \sqsubseteq q$ and $p \sqsubseteq r$ then there is an s with $q \sqsubseteq s$ and $r \sqsubseteq s$.

Exercise 3. Construct models in which a sentence of the form $\alpha \supset \beta$ has a different truth value to ${\sim}\alpha \vee \beta$. Similarly for $\alpha \vee \beta$ and ${\sim}({\sim}\alpha \wedge {\sim}\beta)$.

Exercise 4. "$\mathbf{2} \models \alpha$" in Chapter 6 meant "$\alpha$ is valid on the **BA** $2 = \{0, 1\}$". Show this is the same as Kripke-validity on the *discrete* frame $2 = \{0, 1\}$, but different to validity on the non-discrete frame $(2, \leqslant)$ having $0 \leqslant 1$. □

The Kripke semantics is also closely related to the topological interpretation of intuitionism. On any frame **P**, the collection $\mathbf{P}^+$ of hereditary sets constitutes a topology (a rather special one, as the intersection of any family of open (hereditary) sets is open).

EXERCISE 5. Show that $\mathbf{P}^+$ is the Heyting algebra of open sets for the topology just described, i.e. $\neg S$ is the interior $(-S)^0$ of $-S$, the largest hereditary subset of $-S$, and $S \Rightarrow T$ is $(-S \cup T)^0$, the largest hereditary subset of $-S \cup T$. □

This last section has been a rather rapid survey of what is in fact quite an extensive theory. The full details are readily available in the literature, in the works e.g. of Segerberg [68], Fitting [69], and Thomason [68].

Beth models

Although the Kripke semantics has proven to be the most tractable for many investigations of intuitionistic logic, there is an alternative but related theory due to Evert Beth [56, 59] that is more useful for certain applications (cf. van Dalen [78]). The basic ideas of Beth models can be explained by modifying the semantic rules given in this section for Kripke models.

A *path through p* in a poset **P** is a subset A of P that contains p, that is linearly ordered (i.e. $q \sqsubseteq r$ or $r \sqsubseteq q$ for each $q, r \in A$), and that cannot be extended to a larger linearly ordered subset of P. A *bar for p* is a subset B of P with the property that every path through p intersects it. Intuitively, if P represents the possible states of knowledge that can be attained by a mathematician carrying out research, then a path represents a completed course of research. A bar for p is a set of possible states that is unavoidable for any course of research that yields p, i.e. any such course must lead to a state in B.

In a Beth model the connectives $\wedge$, $\sim$, $\supset$ are treated just as in the Kripke theory. The clauses for sentence letters and disjunction however are

$\mathcal{M} \models_p \pi_i$ iff there is a bar B for p with $B \subseteq V(\pi_i)$

$\mathcal{M} \models_p \alpha \vee \beta$ iff there is a bar B for p with $\mathcal{M} \models_q \alpha$ or $\mathcal{M} \models_q \beta$ for each $q \in B$.

For further discussion of Beth models in relation to Kripke semantics the reader should consult Kripke's paper and Dummett [77].

CHAPTER 9

FUNCTORS

> *"It should be observed first that the whole concept of a category is essentially an auxiliary one; our basic concepts are essentially those of a functor and of a natural transformation."*
>
> S. Eilenberg and S. MacLane

9.1. The concept of functor

A functor is a transformation from one category into another that "preserves" the categorial structure of its source. As the quotation from the founders of the subject indicates, the notion of functor is of the very essence of category theory. The original perspective has changed somewhat, and as far at least as this book is concerned functors are not more important than categories themselves. Indeed the viability of the topos concept as a foundation for mathematics pivots on the fact that it can be *defined* without reference to functors. However we have now reached the stage where we can ignore them no longer. They provide the necessary language for describing the relationship between topoi and Kripke models, and between topoi and models of set theory.

A *functor* F from category $\mathscr{C}$ to category $\mathscr{D}$ is a function that assigns

(i) to each $\mathscr{C}$-object a, a $\mathscr{D}$-object $F(a)$;

(ii) to each $\mathscr{C}$-arrow $f: a \to b$ a $\mathscr{D}$-arrow $F(f): F(a) \to F(b)$,

such that

(a) $F(1_a) = 1_{F(a)}$, all $\mathscr{C}$-objects a, i.e. the identity arrow on a is assigned the identity on $F(a)$,

(b) $F(g \circ f) = F(g) \circ F(f)$, whenever $g \circ f$ is defined.

This last condition states that the F-image of a composite of two arrows is the composite of their F-images, i.e. whenever

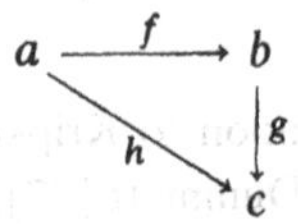

commutes in $\mathscr{C}$ ($h = g \circ f$), then

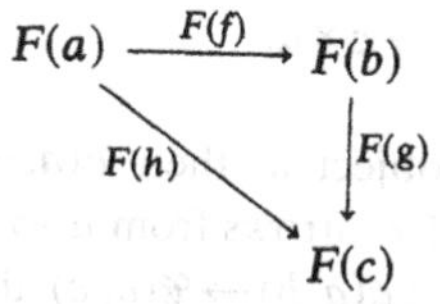

commutes in $\mathscr{D}$. We write $F:\mathscr{C}\to\mathscr{D}$ or $\mathscr{C}\xrightarrow{F}\mathscr{D}$ to indicate that F is a functor from $\mathscr{C}$ to $\mathscr{D}$. Briefly then a functor is a transformation that "preserves" dom's, cod's, identities and composites.

Example 1. The *identity functor* $1_{\mathscr{C}}:\mathscr{C}\to\mathscr{C}$ has $1_{\mathscr{C}}(a)=a$, $1_{\mathscr{C}}(f)=f$. The same rule provides an *inclusion* functor $\mathscr{C}\hookrightarrow\mathscr{D}$ when $\mathscr{C}$ is a subcategory of $\mathscr{D}$.

Example 2. *Forgetful functors:* Let $\mathscr{C}$ be any of the categories in the original list of §2.3, say $\mathscr{C}=\mathbf{Top}$. Then a $\mathscr{C}$-object is a set carrying some additional structure. The forgetful functor $U:\mathscr{C}\to\mathbf{Set}$ takes each $\mathscr{C}$-object to its underlying set, and each $\mathscr{C}$-arrow to itself. Thus U "forgets" the structure on $\mathscr{C}$-objects and remembers only that $\mathscr{C}$-arrows are set functions.

Example 3. *Power set Functor:* $\mathscr{P}:\mathbf{Set}\to\mathbf{Set}$ maps each set A to its powerset $\mathscr{P}(A)$, and each function $f:A\to B$ to the function $\mathscr{P}(f):\mathscr{P}(A)\to\mathscr{P}(B)$ from $\mathscr{P}(A)$ to $\mathscr{P}(B)$ that assigns to each $X\subseteq A$ its f-image $f(X)\subseteq B$.

Example 4. If $\mathbf{P}$ and $\mathbf{Q}$ are posets, then a functor $F:\mathbf{P}\to\mathbf{Q}$ is simply a function $F:P\to Q$ that is *monotonic*, i.e. whenever $p\sqsubseteq q$ in P then $F(p)\sqsubseteq F(q)$ in Q. As a special case of this consider the powerset as a poset $(\mathscr{P}(A),\subseteq)$. Given $f:A\to B$ and X, Y subsets of A, then $X\subseteq Y$ only if $f(X)\subseteq f(Y)$. Thus the function $\mathscr{P}(f):\mathscr{P}(A)\to\mathscr{P}(B)$ is itself a functor between (poset) categories.

Example 5. *Monoid homomorphisms:* A functor between monoids $(M, *, e)$ and $(N, \square, e')$, when these are construed as one-object categories, is essentially a monoid *homomorphism*, i.e. a function $F:M\to N$ that has

$$F(e)=e'$$

$$F(x * y)=F(x)\square F(y).$$

Example 6. If $\mathscr{C}$ has products, each $\mathscr{C}$-object a determines a functor $-\times a:\mathscr{C}\to\mathscr{C}$ which takes each object b to the object $b\times a$, and each arrow $f:b\to c$ to the arrow $f\times 1_a:b\times a\to c\times a$.

Example 7. *Hom-functors:* Given a $\mathscr{C}$-object a, then $\mathscr{C}(a,-):\mathscr{C}\to\mathbf{Set}$ takes each $\mathscr{C}$-object b to the set $\mathscr{C}(a, b)$ of $\mathscr{C}$-arrows from a to b and each $\mathscr{C}$-arrow $f:b\to c$ to the function $\mathscr{C}(a, f):\mathscr{C}(a, b)\to\mathscr{C}(a, c)$ that outputs $f\circ g$ for input g

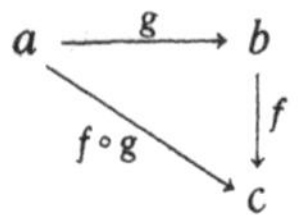

$\mathscr{C}(a,-)$ is called a *hom-functor* because of the use of the word "homomorphism" in some contexts for "arrow". $\mathscr{C}(a, b)=\hom_{\mathscr{C}}(a, b)$ is known as a *hom-set*. There is a restriction as to when this hom-functor is defined. The hom-sets of $\mathscr{C}$ have to be *small*, i.e. actual sets, and not proper classes. □

Contravariant functors

The above examples are all what are known as *covariant* functors. They preserve the "direction" of arrows, in that the domain of an arrow is assigned the domain of the image arrow, and similarly for codomains. A *contravariant* functor is one that *reverses* direction by mapping domains to codomains and vice versa.

Thus $F:\mathscr{C}\to\mathscr{D}$ is a contravariant functor if it assigns to $f:a\to b$ an arrow $F(f):F(b)\to F(a)$, so that $F(1_a)=1_{F(a)}$ as before, but now

$$F(g\circ f)=F(f)\circ F(g),$$

i.e. commuting

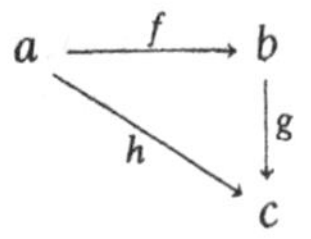

goes to commuting

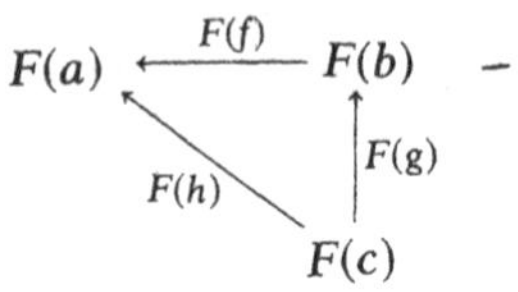

EXAMPLE 8. A contravariant functor between posets is a function $F: P \to Q$ that is *antitone*, i.e.

$$\text{if} \quad p \sqsubseteq q \quad \text{in } \mathbf{P}, \text{ then} \quad F(q) \sqsubseteq F(p) \quad \text{in } \mathbf{Q}.$$

EXAMPLE 9. *Contravariant powerset functor:*

$$\bar{\mathcal{P}}: \mathbf{Set} \to \mathbf{Set}$$

takes each set A to its powerset $\mathcal{P}(A)$, and each $f: A \to B$ to the function $\bar{\mathcal{P}}(f): \mathcal{P}(B) \to \mathcal{P}(A)$ that assigns to $X \subseteq B$ its *inverse image* $f^{-1}(X) \subseteq A$.

EXAMPLE 10. *Contravariant hom-functor:* $\mathscr{C}(-, a): \mathscr{C} \to \mathbf{Set}$, for fixed object a, takes object b to $\mathscr{C}(b, a)$, and $\mathscr{C}$-arrow $f: b \to c$ to function $\mathscr{C}(f, a): \mathscr{C}(c, a) \to \mathscr{C}(b, a)$ that outputs $g \circ f$ for input g

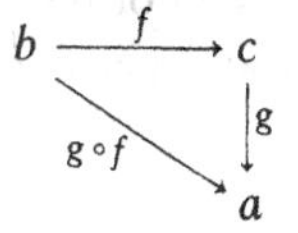

EXAMPLE 11. $\text{Sub}: \mathscr{C} \to \mathbf{Set}$ is the functor taking each $\mathscr{C}$-object a to its collection $\text{Sub}(a)$ of subobjects in $\mathscr{C}$, and each $\mathscr{C}$-arrow $f: a \to b$ to the function $\text{Sub}(f): \text{Sub}(b) \to \text{Sub}(a)$, assigning to $g: c \rightarrowtail b$ the pullback $h: d \rightarrowtail a$ of g along f. Of course this construction is only possible if $\mathscr{C}$ has

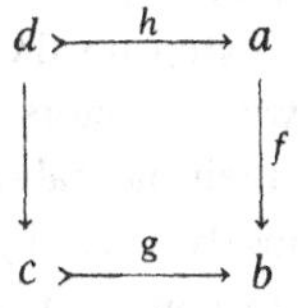

pullbacks. It generalises Example 9. □

EXERCISE Verify that (1)–(11) really are functors. □

The word "functor" used by itself will always mean "covariant functor". In principle contravariant $F: \mathscr{C} \to \mathscr{D}$ can be replaced by covariant $\bar{F}: \mathscr{C}^{\text{op}} \to \mathscr{D}$, where $\bar{F}(a) = F(a)$, and for $f^{\text{op}}: b \to a$ in $\mathscr{C}^{\text{op}}$ (where $f: a \to b$ in $\mathscr{C}$), $\bar{F}(f^{\text{op}}) = F(f): F(b) \to F(a)$. We will not consider contravariant functors again until Chapter 14.

Now given functors $F:\mathscr{C}\to\mathscr{D}$, $G:\mathscr{D}\to\mathscr{F}$, *functional* composition of F and G yields a functor $G\circ F:\mathscr{C}\to\mathscr{F}$, and this operation is associative,

$$H\circ(G\circ F)=(H\circ G)\circ F.$$

We can thus consider functors as arrows between categories. We intuitively envisage a category **Cat**, the category of categories, whose objects are the categories, and arrows the functors. The identity arrows are the identity functors $1_{\mathscr{C}}$ of Example 1.

The notion of **Cat** leads us to some foundational problems. **Set** could not be an element of the class of **Cat**-objects (if we regard these as forming a class), since **Set** as a collection of things is a proper class, and not a member of any collection. Moreover contemplation of the question "is **Cat** a **Cat**-object?" leads us to the brink of Russell's paradox. Generally **Cat** is understood to be the category of *small* categories, i.e. ones whose collection of arrows is a set. Further discussion of these questions may be found in Hatcher [68] Chapter 8, (cf. also a paper by Lawvere [66] on **Cat** as a foundation for mathematics).

9.2. Natural transformations

Having originally defined categories as collections of objects with arrows between them, by introducing functors we took a step up the ladder of abstraction to consider categories as objects, with functors as arrows between them. Readers are now invited to fasten their mental safety-belts as we climb even higher, to regard functors themselves as objects!

Given two categories $\mathscr{C}$ and $\mathscr{D}$ we are going to construct a category, denoted Funct$(\mathscr{C},\mathscr{D})$, or $\mathscr{D}^{\mathscr{C}}$, whose objects are the functors from $\mathscr{C}$ to $\mathscr{D}$. We need a definition of arrow from one functor to another. Taking $F:\mathscr{C}\to\mathscr{D}$ and $G:\mathscr{C}\to\mathscr{D}$, we think of the functors F and G as providing different "pictures" of $\mathscr{C}$ inside $\mathscr{D}$. A reasonably intuitive idea of "transformation" from F to G comes if we image ourselves trying to superimpose or "slide" the F-picture onto the G-picture, i.e. we use the structure of $\mathscr{D}$ to translate the former into the latter. This could be done by assigning to each $\mathscr{C}$-object a an arrow in $\mathscr{D}$ from the F-image of a to the G-image of a. Denoting this arrow by τ_a, we have $\tau_a:F(a)\to G(a)$. In order for this process to be "structure-preserving" we require that each $\mathscr{C}$-*arrow* $f : a \to b$ gives rise to a diagram

$$\begin{array}{ccccc} a & \qquad & F(a) & \xrightarrow{\tau_a} & G(a) \\ \Big\downarrow f & & \Big\downarrow F(f) & & \Big\downarrow G(f) \\ b & & F(b) & \xrightarrow{\tau_b} & G(b) \end{array}$$

that commutes. Thus τ_a and τ_b provide a categorial way of turning the F-picture of $f : a \to b$ into its G-picture.

In summary then, a *natural transformation* from functor $F : \mathscr{C} \to \mathscr{D}$ to functor $G : \mathscr{C} \to \mathscr{D}$ is an assignment τ that provides, for each $\mathscr{C}$-object a, a $\mathscr{D}$-arrow $\tau_a : F(a) \to G(a)$, such that for any $\mathscr{C}$-arrow $f : a \to b$, the above diagram commutes in $\mathscr{D}$, i.e. $\tau_b \circ F(f) = G(f) \circ \tau_a$. We use the symbolism $\tau : F \dot{\to} G$, or $F \overset{\tau}{\dot{\to}} G$, to denote that τ is a natural transformation from F to G. The arrows τ_a are called the *components* of τ.

Now if each component τ_a of τ is an iso arrow in $\mathscr{D}$ then we can interpret this as meaning that the F-picture and the G-picture of $\mathscr{C}$ look the same in $\mathscr{D}$, and in this case we call τ a *natural isomorphism*. Each $\tau_a : F(a) \to G(a)$ then has an inverse $\tau_a^{-1} : G(a) \to F(a)$, and these τ_a^{-1}'s form the components of a natural isomorphism $\tau^{-1} : G \dot{\to} F$. We denote natural isomorphism by $\tau : F \cong G$.

Example 1. The identity natural transformation $1_F : F \dot{\to} F$ assigns to each object a, the identity arrow $1_{F(a)} : F(a) \to F(a)$. This is clearly a natural isomorphism.

Example 2. In **Set**, as noted in §3.4, we have $A \cong A \times 1$, for each set A. This isomorphism is a natural one, as we can see by using the functor $- \times 1 : \mathbf{Set} \to \mathbf{Set}$, as described in Example 6 of the last section. Given $f : A \to B$ then the diagram

$$\begin{array}{ccccc} A & \qquad & A & \xrightarrow{\tau_A} & A \times 1 \\ \Big\downarrow f & & \Big\downarrow f & & \Big\downarrow f \times \mathrm{id}_1 \\ B & & B & \xrightarrow{\tau_B} & B \times 1 \end{array}$$

commutes, where $\tau_A(x) = \langle x, 0 \rangle$, and similarly for τ_B. (i.e. $\tau_A = \langle \mathrm{id}_A, !_A \rangle$). The left side of the square is the image of f under the identity functor. Thus the bijections τ_A are the components of a natural isomorphism τ from $1_{\mathbf{Set}}$ to $- \times 1$.

EXAMPLE 3. Again in **Set**, we have $A \times B \cong B \times A$ by the "twist" map $tw_B : A \times B \to B \times A$ given by the rule $tw_B(\langle x, y\rangle) = \langle y, x\rangle$. Now for given object A, as well as the "right product" functor $- \times A : \mathbf{Set} \to \mathbf{Set}$ we have a left-product functor $A \times - : \mathbf{Set} \to \mathbf{Set}$, taking B to $A \times B$, and $f : B \to C$ to $1_A \times f : A \times B \to A \times C$. Now for any $f : B \to C$, the diagram

$$
\begin{array}{ccccc}
B & \qquad & A \times B & \xrightarrow{tw_B} & B \times A \\
\Big\downarrow f & & \Big\downarrow 1_A \times f & & \Big\downarrow f \times 1_A \\
C & & A \times C & \xrightarrow{tw_C} & C \times A
\end{array}
$$

commutes, showing that the bijections tw_B are the components of a natural isomorphism from $A \times -$ to $- \times A$. □

Equivalence of categories

When do two categories look the same? One possible answer is when they are isomorphic as objects in **Cat**. We say that functor $F : \mathscr{C} \to \mathscr{D}$ is *iso* if it has an inverse, i.e. a functor $G : \mathscr{D} \to \mathscr{C}$ such that $G \circ F = 1_{\mathscr{C}}$ and $F \circ G = 1_{\mathscr{D}}$. We then say that $\mathscr{C}$ and $\mathscr{D}$ are isomorphic, $\mathscr{C} \cong \mathscr{D}$, if there is an iso functor $F : \mathscr{C} \to \mathscr{D}$.

This notion of "sameness" is stricter than it need be. If F has inverse G then for given $\mathscr{C}$-object a we have $a = G(F(a))$, and for $\mathscr{D}$-object b, $b = F(G(b))$. In view of the basic categorial principle of indistinguishability of isomorphic entities we might still regard $\mathscr{C}$ and $\mathscr{D}$ as "essentially the same" if we just had $a \cong G(F(a))$ in $\mathscr{C}$ and $b \cong F(G(b))$ in $\mathscr{D}$. In other words $\mathscr{C}$ and $\mathscr{D}$ are to be categorially equivalent if they are "isomorphic up to isomorphism". This will occur when the isomorphisms $a \to G(F(a))$ and $b \to F(G(b))$ are natural.

Thus a functor $F : \mathscr{C} \to \mathscr{D}$ is called an *equivalence of categories* if there is a functor $G : \mathscr{D} \to \mathscr{C}$ such that there are natural isomorphisms $\tau : 1_{\mathscr{C}} \cong G \circ F$, and $\sigma : 1_{\mathscr{D}} \cong F \circ G$, from the identity functor on $\mathscr{C}$ to $G \circ F$, and from the identity functor on $\mathscr{D}$ to $F \circ G$.

Categories $\mathscr{C}$ and $\mathscr{D}$ are *equivalent*, $\mathscr{C} \simeq \mathscr{D}$, when there exists an equivalence $F : \mathscr{C} \to \mathscr{D}$.

EXAMPLE. **Finord** $\simeq$ **Finset**. Let $F : \mathbf{Finord} \hookrightarrow \mathbf{Finset}$ be the inclusion functor. For each finite set X, let $G(X) = n$, where n is the number of elements in X. For each X, let τ_X be a bijection from X to $G(X)$, with τ_X

being the identity when X is an ordinal. Given $f:X\to Y$, put $G(f)=\tau_Y\circ f\circ\tau_X^{-1}$. Then G is a functor from **Finset** to **Finord**.

Since

$$\begin{array}{ccc} X & \xrightarrow{\tau_X} & F(G(X)) \\ {\scriptstyle f}\downarrow & & \downarrow{\scriptstyle F(G(f))} \\ Y & \xrightarrow{\tau_Y} & F(G(Y)) \end{array}$$

commutes, by definition of $G(f)=F(G(f))$, the τ_X's are the components of a natural isomorphism $\tau:1\to F\circ G$. But also $G\circ F$ is the identity functor on **Finord**. □

The notion of equivalence of categories can be clarified by considering *skeletal* categories. Recall from §3.4 that these are categories in which isomorphic objects are identical, $a\cong b$ only if $a=b$. **Finord** is skeletal, since isomorphic *finite* sets have the same number of elements. A *skeleton* of a category $\mathscr{C}$ is a full subcategory $\mathscr{C}_0$ of $\mathscr{C}$ that is skeletal, and such that each $\mathscr{C}$-object is isomorphic to one (and only one) $\mathscr{C}_0$-object. **Finord** is a skeleton of **Finset**. In general a skeleton $\mathscr{C}_0$ of $\mathscr{C}$ exhibits the essential categorial structure of $\mathscr{C}$. $\mathscr{C}_0$ is equivalent to $\mathscr{C}$, and the equivalence is provided by the inclusion functor $\mathscr{C}_0\hookrightarrow\mathscr{C}$, as may be shown by the method of the last Example.

Any category $\mathscr{C}$ has a skeleton. The relation of isomorphism partitions the collection of $\mathscr{C}$-objects into equivalence classes. Choose one object from each equivalence class and let $\mathscr{C}_0$ be the full subcategory of $\mathscr{C}$ based on this collection of choices. $\mathscr{C}_0$ is a skeleton of $\mathscr{C}$ (cf. Chapter 12 for a discussion of the legitimacy of such a selection process in set-theory). Equivalence of categories is described in these terms by:

> *categories $\mathscr{C}$ and $\mathscr{D}$ are equivalent iff they have isomorphic skeletons ($\mathscr{C}\simeq\mathscr{D}$ iff $\mathscr{C}_0\cong\mathscr{D}_0$),*

and in this sense equivalent categories are categorially "essentially the same". Note however that they need not be in bijective correspondence, indeed need not be comparable in size at all. The collection of finite ordinals is small, i.e. a set, identifiable with the set of natural numbers, whereas the objects of **Finset** form a proper class (e.g. it includes $\{x\}$, for each set x).

EXERCISE 1. Any two skeletons of a given category are isomorphic.

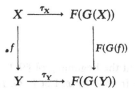

EXERCISE 2. In a topos $\mathscr{E}$, for each object d there is a bijection $\mathrm{Sub}(d) \cong \mathscr{E}(d, \Omega)$ (§4.2). Show that these bijections form a natural isomorphism between the functors $\mathrm{Sub}: \mathscr{E} \to \mathbf{Set}$ and $\mathscr{E}(-, \Omega): \mathscr{E} \to \mathbf{Set}$ (this is a functorial statement of the Ω-axiom).

9.3. Functor categories

We return now to the intention stated at the beginning of §9.2 – to define the functor category $\mathscr{D}^{\mathscr{C}}$ of all functors from $\mathscr{C}$ to $\mathscr{D}$. Let F, G, H be such functors, with natural transformations $\tau: F \dot{\to} G$, $\sigma: G \dot{\to} H$. Then for any $\mathscr{C}$-arrow $f: a \to b$ we get a diagram

$$\begin{array}{ccccccc}
a & & F(a) & \xrightarrow{\tau_a} & G(a) & \xrightarrow{\sigma_a} & H(a) \\
{\scriptstyle f}\downarrow & & {\scriptstyle F(f)}\downarrow & & {\scriptstyle G(f)}\downarrow & & {\scriptstyle H(f)}\downarrow \\
b & & F(b) & \xrightarrow{\tau_b} & G(b) & \xrightarrow{\sigma_b} & H(b)
\end{array}$$

We wish to define the composite $\sigma \circ \tau$ of τ and σ, and have it as a natural transformation. The diagram indicates what to do. For each a, put $(\sigma \circ \tau)_a = \sigma_a \circ \tau_a$. Now each of the two squares in the diagram commutes, so the outer rectangle commutes, giving $(\sigma \circ \tau)_b \circ F(f) = H(f) \circ (\sigma \circ \tau)_a$, and thus the $(\sigma \circ \tau)_a$'s are the components of a natural transformation $\sigma \circ \tau: F \dot{\to} H$. This then provides the operation of composition in the functor category $\mathscr{D}^{\mathscr{C}}$. For each functor $F: \mathscr{C} \to \mathscr{D}$ the identity transformation $1_F: F \dot{\to} F$ (Example 1, §9.2) is the identity arrow on the $\mathscr{D}^{\mathscr{C}}$-object F.

EXERCISE 1. The natural isomorphisms are precisely the iso arrows in $\mathscr{D}^{\mathscr{C}}$.

EXERCISE 2. Let C and D be sets, construed as discrete categories with only identity arrows. Show that for $F, G: C \to D$ there is a transformation $F \dot{\to} G$ iff $F = G$, and that the functor category D^C is the set of functions $C \to D$.

EXERCISE 3. $\tau: F \dot{\to} G$ is monic in $\mathscr{D}^{\mathscr{C}}$ if τ_a is monic in $\mathscr{D}$ for all a. □

A number of the topoi described in Chapter 4 can be construed as "set-valued functor" categories, as follows.

(1) $\mathbf{Set}^2$. The set $2 = \{0, 1\}$ is a discrete category. A functor $F: 2 \to \mathbf{Set}$ assigns a set F_0 to 0 and a set F_1 to 1. Since F as a functor is required to preserve identity arrows, and 2 only has identities, we can suppress all mention of arrows, and identify F with the pair $\langle F_0, F_1 \rangle$. Thus functors $2 \to \mathbf{Set}$ are essentially objects in the category $\mathbf{Set}^2$ of pairs of sets. Now given two such functors F and G, identified with $\langle F_0, F_1 \rangle$ and $\langle G_0, G_1 \rangle$, a natural transformation $\tau: F \dot{\to} G$ has components $\tau_0: F_0 \to G_0$, $\tau_1: F_1 \to G_1$. We may thus identify τ with the pair $\langle \tau_0, \tau_1 \rangle$, which is none other than a $\mathbf{Set}^2$-arrow from $\langle F_0, F_1 \rangle$ to $\langle G_0, G_1 \rangle$.

(2) $\mathbf{Set}^{\to}$. Consider the poset category $\mathbf{2} = \{0, 1\}$ with non-identity arrow $0 \to 1$. A functor $F: \mathbf{2} \to \mathbf{Set}$ comprises two sets F_0, F_1, and a function $f: F_0 \to F_1$. Thus F is "essentially" an arrow f in **Set**, i.e. an object in $\mathbf{Set}^{\to}$. Now given another such functor G, construed as $g: G_0 \to G_1$, then a $\tau: F \dot{\to} G$ has components τ_0, τ_1 that make

$$\begin{array}{ccc} 0 & \quad & F_0 \xrightarrow{\ \tau_0\ } G_0 \\ \downarrow & & f \downarrow \qquad \downarrow g \\ 1 & & F_1 \xrightarrow{\ \tau_1\ } G_1 \end{array}$$

commute. We see then that τ, identified with $\langle \tau_0, \tau_1 \rangle$ becomes an arrow from f to g in $\mathbf{Set}^{\to}$, and so the latter "is" the category $\mathbf{Set}^{\mathbf{2}}$ of functors from **2** to **Set**.

(3) **M-Set**. Let $\mathbf{M} = (M, *, e)$ be a monoid. An **M**-set is a pair (X, λ) where X is a set and λ assigns to each $m \in M$ a function $\lambda_m: X \to X$, so that

(i) $\lambda_e = \mathrm{id}_X$, and

(ii) $\lambda_m \circ \lambda_p = \lambda_{m*p}$.

Now **M** is a category with one object, say M, arrows the members m of M, $*$ as a composition, and $e = \mathrm{id}_M$. Then λ becomes a functor $\lambda: \mathbf{M} \to \mathbf{Set}$, with $\lambda(M) = X$ for the one object, and $\lambda(m) = \lambda_m$, each arrow m. Indeed (i), (ii) are precisely the conditions for λ to be a functor. Now given any other functor $\mu: \mathbf{M} \to \mathbf{Set}$, with $\mu(M) = Y$, then a $\tau: \lambda \dot{\to} \mu$ assigns to M a function $f: X \to Y$ so that

$$\begin{array}{ccc} M & \quad & X \xrightarrow{\ f\ } Y \\ m \downarrow & & \lambda_m \downarrow \qquad \downarrow \mu_m \\ M & & X \xrightarrow{\ f\ } Y \end{array}$$

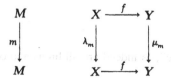

commutes for each $m \in M$. But this says precisely that f is an equivariant map from (X, λ) to (Y, μ). Thus **M-Set** is the category $\mathbf{Set}^{\mathbf{M}}$ of functors from **M** to **Set**.

(4) **Bn**(I). Taking the set I as a discrete category, a functor $F: I \rightarrow \mathbf{Set}$ assigns to each $i \in I$ a set F_i. So we can identify such functors with collections $\{F_i : i \in I\}$ of sets indexed by I.

An object (X, f) in **Bn**(I) (i.e. a function $f: X \rightarrow I$) gives a functor $\bar{f}: I \rightarrow \mathbf{Set}$, with $\bar{f}(i) = f^{-1}(\{i\})$, the stalk of f over i.

An arrow $h: (X, f) \rightarrow (Y, g)$ is a function that maps the f-stalk over i to the g-stalk over i, hence determines a function $h_i : \bar{f}(i) \rightarrow \bar{g}(i)$. These h_i's are the components for $\bar{h}: \bar{f} \dot{\rightarrow} \bar{g}$. Thus each bundle can be turned into a functor from I to **Set**. The converse will only work if the F_i's are pairwise disjoint. So given $F: I \rightarrow \mathbf{Set}$ we define a new functor $\bar{F}: I \rightarrow \mathbf{Set}$ by putting $\bar{F}(i) = F(i) \times \{i\}$ and then turn $\{\bar{F}(i) : i \in I\}$ into a bundle over I. Since $F(i) \cong F(i) \times \{i\}$, the functors F and $\bar{F}$ are naturally isomorphic. What this all boils down to is that the passage from (X, f) to $\bar{f}$ is an equivalence of categories. The category **Bn**(I) of bundles over I is equivalent to the category $\mathbf{Set}^I$ of set-valued functors defined on I. □

These last four examples illustrate a construction that provides us with many topoi. We have:

> *for any "small" category $\mathscr{C}$, the functor category* $\mathbf{Set}^{\mathscr{C}}$ *is a topos!*

We devote the rest of this chapter to describing the topos structure of $\mathbf{Set}^{\mathscr{C}}$.

Terminal object

In $\mathbf{Set}^{\mathscr{C}}$ this is the constant functor $1: \mathscr{C} \rightarrow \mathbf{Set}$ that takes every $\mathscr{C}$-object to the one-element set $\{0\}$, and every $\mathscr{C}$-arrow to the identity on $\{0\}$. For any $F: \mathscr{C} \rightarrow \mathbf{Set}$ the unique arrow $F \dot{\rightarrow} 1$ *in* $\mathbf{Set}^{\mathscr{C}}$ is the natural transformation whose components are the unique functions $!: F(a) \rightarrow \{0\}$ for each $\mathscr{C}$-object a.

Pullback

This is defined "componentwise", as indeed are all limits and colimits in $\mathbf{Set}^{\mathscr{C}}$.

Given $\tau : F \dot{\to} H$ and $\sigma : G \dot{\to} H$, then for each $\mathscr{C}$-object a, form the pullback

$$\begin{array}{ccc} K(a) & \xrightarrow{\mu_a} & G(a) \\ {\scriptstyle\lambda_a}\big\downarrow & & \big\downarrow{\scriptstyle\sigma_a} \\ F(a) & \xrightarrow{\tau_a} & H(a) \end{array}$$

in **Set** of the components τ_a and σ_a. The assignment of $K(a)$ to a establishes a functor $K : \mathscr{C} \to \mathbf{Set}$. Given $\mathscr{C}$-arrow $f : a \to b$, $K(f)$ is the unique arrow $K(a) \to K(b)$ in the "cube"

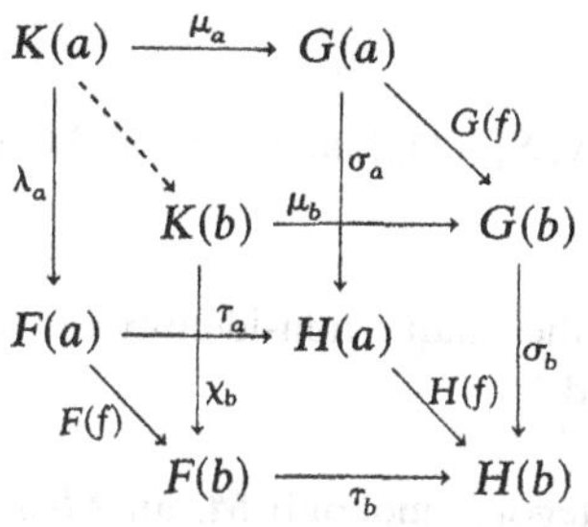

given by the universal property of the front face as pullback. The λ_a's and μ_a's are components for $\lambda : K \dot{\to} F$ and $\mu : K \dot{\to} G$ that make

$$\begin{array}{ccc} K & \xrightarrow{\mu} & G \\ {\scriptstyle\lambda}\big\downarrow & & \big\downarrow{\scriptstyle\sigma} \\ F & \xrightarrow{\tau} & H \end{array}$$

a pullback in $\mathbf{Set}^{\mathscr{C}}$.

EXERCISE 4. Define the product $F \times G : \mathscr{C} \to \mathbf{Set}$ of two objects in $\mathbf{Set}^{\mathscr{C}}$. □

Subobject classifier

To define this we introduce a new notion. For a given $\mathscr{C}$-object a, let S_a be the collection of all $\mathscr{C}$-arrows with domain a,

$$S_a = \left\{ f : \text{for some } b,\ a \xrightarrow{f} b \text{ in } \mathscr{C} \right\}$$

(S_a is the class of objects for the category $\mathscr{C} \uparrow a$ of "objects under a" described in Chapter 3).

We note that S_a is "closed under left composition", i.e. if $f \in S_a$, then for any $\mathscr{C}$-arrow $g: b \to c$, $g \circ f \in S_a$

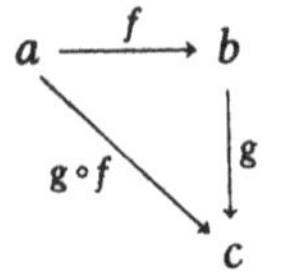

We define a *sieve on a*, or an *a-sieve* to be a subset S of S_a that is itself closed under left composition, i.e. has $g \circ f \in S$ whenever $f \in S$. For any object a there are always at least two a-sieves S_a and $\emptyset$ (the empty sieve).

EXAMPLE 1. In a discrete category, $S_a = \{1_a\}$, and so S_a and $\emptyset$ are the only a-sieves.

EXAMPLE 2. In **2**, with $f: 0 \to 1$ the unique non-identity arrow there are three 0-sieves, $\emptyset$, $S_0 = \{1_0, f\}$, and $\{f\}$.

EXAMPLE 3. In a one-object category (monoid) **M**, an M-sieve is a set $S \subseteq M$ of arrows closed under left composition = left multiplication. The sieves are just the left-ideals of **M**. □

Now we define $\Omega: \mathscr{C} \to \mathbf{Set}$ by

$$\Omega(a) = \{S : S \text{ is an } a\text{-sieve}\}$$

and for $\mathscr{C}$-arrow $f: a \to b$, let $\Omega(f): \Omega(a) \to \Omega(b)$ be the function that takes the a-sieve S to the b-sieve $\{b \xrightarrow{g} c : g \circ f \in S\}$ (why *is* this a sieve?)

Thus in $\mathbf{Set}^{\mathbf{M}}$, we find that $\Omega(M) = L_M$, the set of left ideals in **M**, and for arrow $m: M \to M$, $\Omega(m): L_M \to L_M$ takes S to $\{n : n * m \in S\} = \omega(m, S)$. So Ω becomes the action (L_M, ω) that is the codomain of the subobject classifier.

In $\mathbf{Set}^{\mathscr{C}}$ we define $\top: 1 \dot{\to} \Omega$ to be the natural transformation that has components $\top_a: \{0\} \to \Omega(a)$ given by $\top_a(0) = S_a$, the "largest" a-sieve. This arrow is the classifier for $\mathbf{Set}^{\mathscr{C}}$. To see how $\top$ works, suppose that $\tau: F \dot{\to} G$ is a monic arrow in $\mathbf{Set}^{\mathscr{C}}$. Then for each $\mathscr{C}$-object a, the component $\tau_a: F(a) \to G(a)$ is monic in **Set** (Exercise 3) and we will suppose it to be the inclusion $F(a) \hookrightarrow G(a)$. Now the character $\chi_\tau: G \dot{\to} \Omega$

of τ is to be a natural transformation with the component $(\chi_\tau)_a$ a set function from $G(a)$ to $\Omega(a)$. Thus $(\chi_\tau)_a$ assigns to each $x \in G(a)$, an a-sieve $(\chi_\tau)_a(x)$. The question then is to decide when an arrow $f: a \to b$ with domain a is in $(\chi_\tau)_a(x)$. For such an f, we have a commutative diagram

$$\begin{array}{ccc} F(a) & \xrightarrow{\tau_a} & G(a) \\ {\scriptstyle F(f)}\downarrow & & \downarrow{\scriptstyle G(f)} \\ F(b) & \xrightarrow{\tau_b} & G(b) \end{array}$$

so that $F(f)$ is the restriction of $G(f)$ to $F(a)$. We put f in $(\chi_\tau)_a(x)$ if and only if $G(f)$ maps x into $F(b)$. (Compare this with the picture for **Set**$^{\to}$ in §4.4). Thus $(\chi_\tau)_a(x) = \{f: a \to b: G(f)(x) \in F(b)\}$.

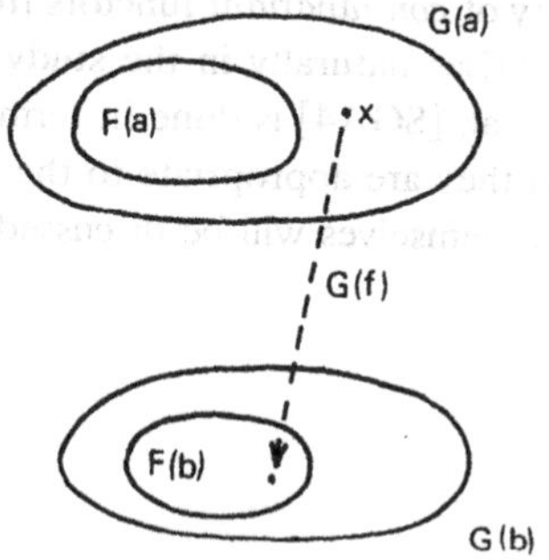

Fig. 9.1.

More generally, assuming only that τ_a is a function, perhaps not an inclusion, we put

$$(\chi_\tau)_a(x) = \left\{a \xrightarrow{f} b: G(f)(x) \in \tau_b(F(b))\right\}$$

$$= \left\{a \xrightarrow{f} b: \text{for some } y \in F(b),\ G(f)(x) = \tau_b(y)\right\}$$

EXERCISE 5. Verify that $(\chi_\tau)_a(x)$ is an a-sieve, and that this construction satisfies the Ω-axiom. (see §10.3)

EXERCISE 6. Show that it produces the classifiers for **Set**^{2}, **Set**$^{\to}$ and **Bn**(I).

EXERCISE 7. Let S be an a-sieve. Define $\bar{S}: \mathscr{C} \to$ **Set** by $\bar{S}(b) = S \cap \mathscr{C}(a, b)$. Show that the inclusions $\bar{S}(b) \hookrightarrow \mathscr{C}(a, b)$ are the components of a monic **Set**$^{\mathscr{C}}$-arrow $\bar{S} \rightarrowtail \mathscr{C}(a, -)$. Show that in fact the a-sieves are in bijective correspondence with the subobjects of the homfunctor $\mathscr{C}(a, -)$ in $\mathbf{Set}^{\mathscr{C}}$.

EXERCISE 8. Show that for each $\mathscr{C}$-object a, $(\Omega(a), \subseteq)$ is a Heyting algebra of subsets of S_a, with

$$\neg S = \left\{a \xrightarrow{f} b\text{: for any } b \xrightarrow{g} c,\ g \circ f \notin S\right\}$$

$$S \Rightarrow T = \{f\text{: whenever } g \circ f \in S, \text{ then } g \circ f \in T\}$$

Show that $\neg S$ is the largest (union) of all the a-sieves contained in $-S$, and $S \Rightarrow T$ is the largest a-sieve contained in $-S \cup T$. □

The dual to the notion of sieve is called an *a-crible*. This is a collection of arrows with codomain a that is closed under right-composition. Cribles are used to show that the category of *contravariant* functors from $\mathscr{C}$ to **Set** is a topos. This type of functor arises naturally in the study of sheaves, and the work of Grothendieck et al. [SGA4] is done in terms of cribles. We have used co-cribles because they are appropriate to the conventions of the Kripke semantics. Cribles themselves will be discussed in Chapter 14.

Exponentiation in $\mathbf{Set}^{\mathscr{C}}$

Let $F: \mathscr{C} \to \mathbf{Set}$. For each $\mathscr{C}$-object a, define a "forgetful" functor $F_a: \mathscr{C}\uparrow a \to \mathbf{Set}$ that takes $f: a \to b$ to $F(b)$, and $h: f \to g$ where

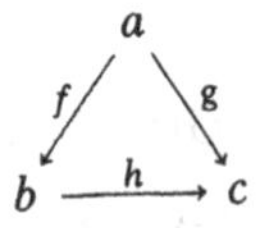

commutes, to $F(h)$.

Now given $F,G: \mathscr{C} \to \mathbf{Set}$, define $G^F: \mathscr{C} \to \mathbf{Set}$ by

$$G^F(a) = \mathrm{Nat}[F_a, G_a],$$

the collection of natural transformations from F_a to G_a.

Acting on arrows, G^F takes $k: a \to d$ to a function $G^F(k)$ from $\mathrm{Nat}[F_a, G_a]$ to $\mathrm{Nat}[F_d, G_d]$. This takes $\tau: F_a \dot{\to} G_a$ to $\tau': F_d \dot{\to} G_d$ that has components $\tau'_f = \tau_{f \circ k}$, for

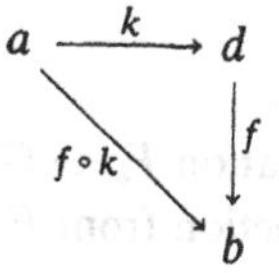

f an object in $\mathscr{C} \uparrow d$.

EXAMPLE. Let F and G be functors $\mathbf{2} \to \mathbf{Set}$, thought of as functions $f: A \to B$ and $g: C \to D$ (i.e. $\mathbf{Set}^{\to}$-objects). Now $\mathbf{2} \uparrow 1$ is the discrete one-object category. So F_1 is identifiable with $F(1) = B$, likewise G_1 "is" D, and

$$G^F(1) = D^B, \quad \text{the set of functions } B \to D.$$

Now $\mathbf{2} \uparrow 0$ is isomorphic to $\mathbf{2}$ itself, so F_0 and G_0 can be taken as just F and G. Then

$$G^F(0) = \mathrm{Nat}[F, G] \text{ "=" } E,$$

where E is the set of $\mathbf{Set}^{\to}$-arrows from f to g. Finally G^F takes $!: 0 \to 1$ to

$$E \xrightarrow{g^f} D^B, \text{ as follows:}$$

Given $\tau: F \dot{\to} G$, corresponding to the $\mathbf{Set}^{\to}$-arrow $\langle \tau_0, \tau_1 \rangle$ from f to g, $G^F(\tau)$ is the transformation $F_1 \dot{\to} G_1$ whose sole component is τ_1, since 1 corresponds to the unique member 1_1 of $\mathbf{2} \uparrow 1$.

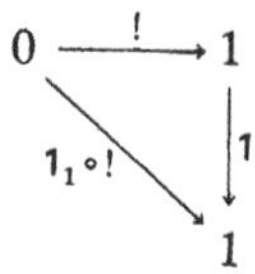

Thus $g^f(\langle \tau_0, \tau_1 \rangle) = \tau_1$, and this very complex construction has yielded the exponential object in $\mathbf{Set}^{\to}$. □

We have yet to define the evaluation arrow $ev: G^F \times F \dot{\to} G$ in $\mathbf{Set}^{\mathscr{C}}$. This has components $ev_a: G^F(a) \times F(a) \to G(a)$, where $ev_a(\langle \tau, x \rangle) = \tau_{1_a}(x)$ whenever $x \in F(a)$ and $\tau \in G^F(a)$, i.e. $\tau: F_a \dot{\to} G_a$ (note that the component τ_{1_a} of the $\mathscr{C} \uparrow a$-object 1_a is indeed a function from $F(a)$ to

$G(a)$. Now for a $\mathbf{Set}^{\mathscr{C}}$ arrow $\tau : H \times F \dot{\to} G$, the exponential adjoint $\hat{\tau} : H \dot{\to} G^F$ has components that are functions of the form

$$\hat{\tau}_a : H(a) \to G^F(a).$$

For each y in $H(a)$, $\hat{\tau}_a(y)$ is a natural transformation $F_a \dot{\to} G_a$. For each $\mathscr{C}\uparrow a$-object $f : a \to b$, $\hat{\tau}_a(y)$ assigns to f that function from $F(b)$ to $G(b)$ that for input $x \in F(b)$ gives output

$$\tau_b(\langle H(f)(y), x\rangle)$$

(note that τ_b: $H(b) \times F(b) \to G(b)$ and $H(f) : H(a) \to H(b)$).

The reader who has the head for such things may check out the details of this construction and relate it to exponentials in **M-Set**, **Bn**(I) etc. We shall need it only for the description of power objects in a special topos of Kripke models in Chapter 11. Our major concern will be with the subobject classifier of "set-valued" functor categories.

CHAPTER 10

SET CONCEPTS AND VALIDITY

> "... *a natural and useful generalisation of set theory to the consideration of 'sets which internally develop'* "
>
> F. W. Lawvere

10.1. Set concepts

We saw in Chapter 1 that a statement $\varphi(x)$, pertaining to individuals x, determines a set, viz the set $\{x: \varphi(x)\}$ of all things of which the statement is true. But according to the constructivist attitude outlined in Chapter 8, truth is not something ascribed to a statement *absolutely*, but rather is a "context-dependent" attribute. The truth-value of a sentence varies according to the state of knowledge existing at the time of assertion of the sentence. In these terms we might regard φ not as determining a set per se, but rather as determining, for each state p, the collection

$$\varphi_p = \{x: \varphi(x) \text{ is known at } p \text{ to be true}\}.$$

φ_p will be called the *extension* of φ at p.

Thus, given a frame **P** of states of knowledge, the assignment of φ_p to p determines a function $P \rightarrow \mathbf{Set}$. Moreover, if truth is taken to "persist in time", then if $x_0 \in \varphi_p$ and $p \sqsubseteq q$, we have $\varphi(x_0)$ true also at q, so $x_0 \in \varphi_q$. Thus

$$(*) \qquad p \sqsubseteq q \quad \text{implies} \quad \varphi_p \subseteq \varphi_q$$

This means that φ determines a functor $\mathbf{P} \rightarrow \mathbf{Set}$, which assigns the inclusion arrow $\varphi_p \hookrightarrow \varphi_q$ to each $p \rightarrow q$ in **P**.

EXAMPLE. Let $\varphi(x)$ be the statement "x is an integer greater than 2, and there are no non-zero integers a, b, c with $a^x + b^x = c^x$". Fermat's celebrated "last theorem" asserts that $\varphi(x)$ holds for every integer $x \geqslant 2$. At the present moment it is not known if this is correct, although it is known that φ is true for all $x \leqslant 25{,}000$. Until Fermat's "theorem" is decided either way we may expect the extension of φ to increase with time.

So, corresponding to an expression φ we have an object in the functor category $\mathbf{Set}^{\mathbf{P}}$. Such an object might be though of as a "variable set", as in Lawvere [75, 76]. We might also call it an *intensional set*, or a *set concept*. This terminology derives from semantic theories of the type set out by Rudolf Carnap [47]. In such theories the *extension* of an individual expression is taken to be the actual thing, or collection of things, to which it refers. The *intension* on the other hand is a somewhat more elusive entity, which is sometimes described as being the *meaning* of the expression. Carnap ([47], p. 41) defines the intension of an individual expression to be the "individual concept expressed by it". Thus if $\varphi(x)$ is the statement "x is a finite ordinal" then the intension of φ is the *concept* of a finite ordinal. This is represented by the functor that assigns to each p the set of things known at stage p to be finite ordinals. This functor can also be said to represent the concept of the set of finite ordinals. In this way we construe $\mathbf{Set}^{\mathbf{P}}$ as being a category of set concepts.

There are some difficulties with the theme just developed. Consider the expression "the smallest non-finite ordinal". This expresses quite a different concept to "the set of finite ordinals", and yet the two have the same extension, i.e. the set of finite ordinals *is* the smallest non-finite ordinal. Thus two different concepts might well be represented in $\mathbf{Set}^{\mathbf{P}}$ by the same object, i.e. $\mathbf{Set}^{\mathbf{P}}$ does not *faithfully* represent all concepts (for a more basic example consider the expressions "2 plus 2" and "2 times 2").

Another difficulty relates to the derivation of the principle $(*)$ above. The argument would seem to be simply fallacious in the event that x_0 is itself the extension of some set concept, i.e. $x_0 = \psi_p$ for some expression $\psi(x)$. Suppose for example that $\varphi(x)$ is the statement "$x = \{y : \psi(y)\}$". Then $\varphi_p = \{\psi_p\}$, the set whose only member is $\psi_p = x_0$, while $\varphi_q = \{\psi_q\}$. If $\psi_p \neq \psi_q$, then $x_0 \notin \varphi_q$. We do salvage from this however the fact that if $\psi_p \in \varphi_p$, then $\psi_q \in \varphi_q$. Perhaps we should then replace the inclusion function of $(*)$ by the map taking each element of φ_p to its counter-part in φ_q. In this way φ would still determine a functor. Unfortunately the notion of counterpart is ambiguous here – x_0 may also be the extension of some other expression $\theta(x)$ $(x_0 = \psi_p = \theta_p)$ whose extension at q differs from ψ $(\psi_q \neq \theta_q)$.

In spite of these problems, the notion of set concept would still seem appropriate to an understanding of the objects in $\mathbf{Set}^{\mathbf{P}}$, and to the viewpoint that $\mathbf{Set}^{\mathbf{P}}$ is the universe for a generalised "non-extensional" set theory. Indeed the study of $\mathbf{Set}^{\mathbf{P}}$ may help to clarify the philosophically difficult notions of "individual concept" and "intensional object" (for an indication of how intractable these ideas are, read Scott [70i]).

Certainly the notion of "variable structure" is a mathematically significant one. One thinks of the concept of "neighbourhood system" as represented by the assignment to each point in a topological space of its set of neighbourhoods – or the concept of "tangent space" as represented by the assignment to each point in a manifold of the space of vectors tangent to the manifold at that point.

In this chapter we propose to look in depth at the topos structure of $\mathbf{Set}^{\mathbf{P}}$, and in particular the nature of its truth arrows. The conclusion we will reach is that "the logic of variable sets is intuitionistic".

10.2. Heyting algebras in P

Let $\mathbf{P}=(P, \sqsubseteq)$ be a poset. For each $p \in P$, let

$$[p)=\{q: p \sqsubseteq q\}$$

be the set of P-elements "above" p in the ordering $\sqsubseteq$. If $q \in [p)$ and $q \sqsubseteq r$, then, by the transitivity of $\sqsubseteq$, $r \in [p)$. Thus $[p)$ is hereditary in $\mathbf{P}$ ($[p) \in \mathbf{P}^+$), and will be called the *principal* $\mathbf{P}$*-hereditary set generated by* p. Principal sets are very useful in describing the structure of the **HA** $\mathbf{P}^+$, as seen in the following

Exercises

Cf. §8.4 for notation.

EXERCISE 1. For any $S \subseteq P$, if $[p) \subseteq S$ then $p \in S$.

EXERCISE 2. $p \sqsubseteq q$ iff $[q) \subseteq [p)$.

EXERCISE 3. The following are equivalent, for any $S \subseteq P$:

(i) S is $\mathbf{P}$-hereditary;
(ii) for all $p \in P$, $p \in S$ iff $[p) \subseteq S$;
(iii) for all $p \in P$, $p \in S$ implies $[p) \subseteq S$.

EXERCISE 4. For any $S, T \in \mathbf{P}^+$,

$$S \Rightarrow T = \{p: S \cap [p) \subseteq T\}$$

$$\neg S = \{p: [p) \subseteq -S\} = \{p: [p) \cap S = \emptyset\}. \qquad \square$$

Now the relation $\sqsubseteq$ when restricted to the members of $[p)$ is still a partial ordering, and so we have a poset $([p), \sqsubseteq)$, and a collection $[p)^+$

consisting of all the sets that are hereditary in $[p)$. Now if $q \in [p)$, then the principal set generated in $[p)$ by q is

$$\begin{aligned}[q)_p &= \{r: r \in [p) \text{ and } q \sqsubseteq r\} \\ &= [p) \cap [q).\end{aligned}$$

But by Exercise 2, this is just $[q)$. In other words, the principal set of q in $\mathbf{P}$ is the same as the principal set of q with respect to $[p)$, $[q) = [q)_p$. From this we obtain a detailed account of the relationship between $\mathbf{P}^+$ and $[p)^+$.

If S is *any* subset of P, put

$$\begin{aligned}S_p &= S \cap [p) \\ &= \{q: q \in S \text{ and } p \sqsubseteq q\}.\end{aligned}$$

THEOREM 1. (1) *If* $S \subseteq [p)$, *then* $S = S_p$, *and* $S \in [p)^+$ *iff* $S \in \mathbf{P}^+$;
(2) *If* $S \in \mathbf{P}^+$ *then* $S_p \in [p)^+$;
(3) $T \in [p)^+$ *iff for some* $S \in \mathbf{P}^+$, $T = S_p$;
(4) *If* $S \in \mathbf{P}^+$, *then* $S = \bigcup\{S_p : p \in P\}$.

PROOF. (1) Clearly if $S \subseteq [p)$, then $S = S \cap [p)$. Moreover, by Exercise 3 (iii),

$$S \in [p)^+ \quad \text{iff} \quad q \in S \quad \text{implies} \quad [q)_p \subseteq S$$

while

$$S \in \mathbf{P}^+ \quad \text{iff} \quad q \in S \quad \text{implies} \quad [q) \subseteq S.$$

But since $S \subseteq [p)$, $q \in S$ implies $[q)_p = [q)$.

(2) Since $[p) \in \mathbf{P}^+$, $S \in \mathbf{P}^+$ implies $S \cap [p) \in \mathbf{P}^+$, i.e. $S_p \in \mathbf{P}^+$. Since $S_p \subseteq [p)$, the result follows by part (1).

(3) Exercise.

(4) We have to show that

$$q \in S \quad \text{iff for some } p, \quad q \in S_p = S \cap [p).$$

Since in general, $S_p \subseteq S$, the implication from right to left is immediate. Conversely, if $q \in S$ then if S is hereditary we have $q \in [q) \subseteq S$, and so $q \in S \cap [q)$, i.e. the proof is completed by taking $p = q$. □

Now we know from §8.4 that the poset $([p)^+, \subseteq)$ of hereditary subsets of $[p)$ under the *subset* ordering is a Heyting algebra (in fact – for the interest of the reader familiar with such things – $[p)^+$ is a subdirectly irreducible **HA**). The lattice meet $\cap_p$ and join $\cup_p$ are simply the operations $\cap$ and $\cup$ of set intersection and union. The pseudo-complement

$\neg_p : [p)^+ \to [p)^+$ is defined for $S \subseteq [p)$ by

$$\neg_p S = \{q : q \in [p) \text{ and } [q)_p \subseteq -S\}$$

while the relative pseudo-complement $\Rightarrow_p : [p)^+ \times [p)^+ \to [p)^+$ has

$$S \Rightarrow_p T = \{q : q \in [p) \text{ and } S \cap [q)_p \subseteq T\}.$$

Now given any $S \subseteq P$, we may first relativise S to $[p)$, i.e. form S_p, and then apply $\neg_p$, or we may apply $\neg$ to S first, and then relativise. The two procedures prove to be commutative, for **P**-hereditary S, and more generally we have

THEOREM 2. *For any* $S, T \in \mathbf{P}^+$

(1) $(S_p) \cap_p (T_p) = (S \cap T)_p$;
(2) $(S_p) \cup_p (T_p) = (S \cup T)_p$;
(3) $\neg_p (S_p) = (\neg S)_p$;
(4) $(S_p) \Rightarrow_p (T_p) = (S \Rightarrow T)_p$.

PROOF. (1) Exercise.

(2)
$$\begin{aligned} S_p \cup_p T_p &= S_p \cup T_p \\ &= (S \cap [p)) \cup (T \cap [p)) \\ &= (S \cup T) \cap [p) \quad \text{(distributive law)} \\ &= (S \cup T)_p. \end{aligned}$$

(3) Since $[q) = [q)_p$ for $p \sqsubseteq q$, we have

$$\begin{aligned} \neg_p (S_p) &= \{q : q \in [p) \text{ and } [q) \subseteq -S\} \\ &= [p) \cap \neg S \\ &= (\neg S)_p. \end{aligned}$$

(4) Exercise. □

The algebraically minded reader will note that Theorem 2 states that the assignment of S_p to S is an **HA** homomorphism from $\mathbf{P}^+$ to $[p)^+$, which is surjective by Theorem 1 (3).

10.3. The subobject classifier in $\mathbf{Set}^{\mathbf{P}}$

That $\mathbf{Set}^{\mathbf{P}}$ is a topos is a special case of the fact that $\mathbf{Set}^{\mathscr{C}}$ is a topos for any small category $\mathscr{C}$. The definition of the subobject classifier for $\mathbf{Set}^{\mathscr{C}}$ given in §9.3 proves in the case $\mathscr{C} = \mathbf{P}$ to be expressible in terms of the **HA**'s of the form $[p)^+$. According to §9.3, $\Omega : \mathbf{P} \to \mathbf{Set}$ has

$$\Omega(p) = \text{the set of } p\text{-sieves.}$$

Now a p-sieve is a subset S of

$$\mathbf{P}_p = \left\{ f: \text{for some } q, p \xrightarrow{f} q \text{ in } \mathbf{P} \right\}$$

that is closed under left multiplication, i.e. has $g \circ f \in S$ whenever $f \in S$ and $g: q \rightarrow r$ is a $\mathbf{P}$-arrow. But as $\mathbf{P}$ is a preorder category, there is at most one arrow from p to q, and this exists precisely when $p \sqsubseteq q$. So for a fixed p, we may identify the arrow $f: p \rightarrow q$ with its codomain q. Hence $\mathbf{P}_p$ becomes

$$\{q: p \sqsubseteq q\} = [p)!,$$

and the description of S as a p-sieve becomes

$$r \in S \quad \text{whenever} \quad q \in S \quad \text{and} \quad q \sqsubseteq r$$

i.e. S is $[p)$-hereditary!

Thus $\Omega(p) = [p)^+$, the collection of hereditary subsets of $[p)$.

In general for a functor $F: \mathbf{P} \rightarrow \mathbf{Set}$ we will write F_p for the image $F(p)$ of p in **Set**. Whenever $p \sqsubseteq q$, F yields a function from F_p to F_q, which will be denoted F_{pq}. We may thus view F as a collection $\{F_p : p \in P\}$ of sets indexed by P and provided with "transition maps" $F_{pq}: F_p \rightarrow F_q$ whenever $p \sqsubseteq q$. In particular F_{pp} is the identity function on F_p.

In the case of Ω, the modification as above of the definition of §9.3 shows that when $p \sqsubseteq q$, $\Omega_{pq}: \Omega_p \rightarrow \Omega_q$ takes $S \in [p)^+$ to $S \cap [q) \in [q)^+$, i.e.

$$\Omega_{pq}(S) = S_q.$$

The terminal object for $\mathbf{Set}^{\mathbf{P}}$ is the "constant" functor $1: \mathbf{P} \rightarrow \mathbf{Set}$ having $1_p = \{0\}$, all $p \in P$, and $1_{pq} = \mathrm{id}_{\{0\}}$ for $p \sqsubseteq q$. The subobject classifier $true: 1 \rightarrow \Omega$ is the natural transformation whose "p-th" component $true_p: \{0\} \rightarrow \Omega_p$ is given by

$$true_p(0) = [p).$$

Thus *true* picks out the unit element from each **HA** $[p)^+$.

Now if $\tau: F \rightarrowtail G$ is a subobject of G in $\mathbf{Set}^{\mathbf{P}}$ then each component τ_p will be injective, and will whenever convenient be assumed to be the inclusion function $F_p \hookrightarrow G_p$. Again by modifying the §9.3 definition we find that the character $\chi_\tau: G \rightarrow \Omega$ has p-th component $(\chi_\tau)_p: G_p \rightarrow [p)^+$ given by

$$\text{for each} \quad x \in G_p, \qquad (\chi_\tau)_p(x) = \{q: p \sqsubseteq q \text{ and } G_{pq}(x) \in F_q\}$$

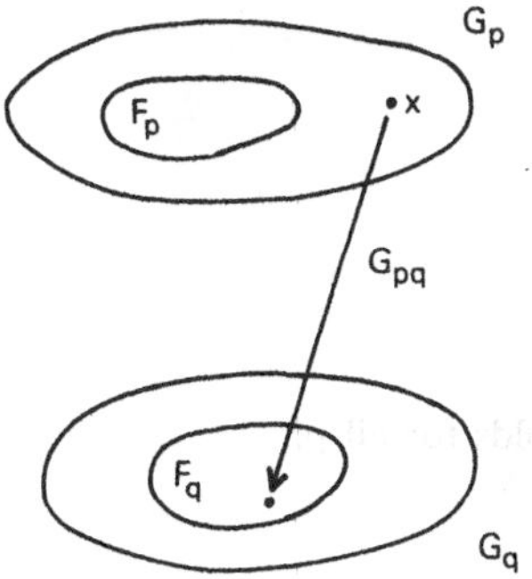

Fig. 10.1.

EXERCISE 1. Show that $(\chi_\tau)_p(x)$ is hereditary in $[p)$.

EXERCISE 2. Show that χ_τ *is* a natural transformation from G to Ω, i.e. that

$$\begin{array}{ccc} G_p & \xrightarrow{(\chi_\tau)_p} & \Omega_p \\ {\scriptstyle G_{pq}}\downarrow & & \downarrow{\scriptstyle \Omega_{pq}} \\ G_q & \xrightarrow{(\chi_\tau)_q} & \Omega_q \end{array}$$

commutes whenever $p \sqsubseteq q$. □

Notice that if $x \in F_p$, then for any $q \in [p)$, since

$$\begin{array}{ccc} F_p & \overset{\tau_p}{\hookrightarrow} & G_p \\ {\scriptstyle F_{pq}}\downarrow & & \downarrow{\scriptstyle G_{pq}} \\ F_q & \overset{\tau_q}{\hookrightarrow} & G_q \end{array}$$

commutes we must have $G_{pq}(x) = F_{pq}(x) \in F_q$, and so $q \in (\chi_\tau)_p(x)$. On the other hand if $x \notin F_p$, then $G_{pp}(x) = x \notin F_p$, and so $p \notin (\chi_\tau)_p(x)$, i.e. $(\chi_\tau)_p(x) \neq [p)$. Altogether then we have that

$$F_p = \{x\colon (\chi_\tau)_p(x) = [p)\} \cong \{\langle 0, x\rangle\colon (\chi_\tau)_p(x) = true_p(0)\}$$

and hence

$$\begin{array}{ccc} F_p & \xhookrightarrow{\tau_p} & G_p \\ \big\downarrow & & \big\downarrow{\scriptstyle (\chi_\tau)_p} \\ \{0\} & \xhookrightarrow{true_p} & [p)^+ \end{array}$$

is a pullback in **Set**. Since this holds for all p,

$$\begin{array}{ccc} F & \overset{\tau}{\rightarrowtail} & G \\ \big\downarrow & & \big\downarrow{\scriptstyle \chi_\tau} \\ 1 & \xrightarrow{true} & \Omega \end{array}$$

is a pullback in $\mathbf{Set}^{\mathbf{P}}$. The verification of the rest of the Ω-axiom is rather delicate. Suppose $\sigma : G \to \Omega$ makes

$$\begin{array}{ccc} F & \xrightarrow{\tau} & G \\ \big\downarrow & & \big\downarrow{\scriptstyle \sigma} \\ 1 & \xrightarrow{true} & \Omega \end{array}$$

a pullback. Then for each q,

$$\begin{array}{ccc} F_q & \hookrightarrow & G_q \\ \big\downarrow & & \big\downarrow{\scriptstyle \sigma_q} \\ \{0\} & \xrightarrow{true_q} & \Omega_q \end{array}$$

will be a pullback, and so by the nature of pullbacks in **Set** we may assume

(∗) $\qquad F_q = \{x : \sigma_q(x) = [q)\}$

Now let us take a particular p. Then whenever $p \sqsubseteq q$,

$$\begin{array}{ccc} G_p & \xrightarrow{\sigma_p} & \Omega_p \\ {\scriptstyle G_{pq}}\big\downarrow & & \big\downarrow{\scriptstyle \Omega_{pq}} \\ G_q & \xrightarrow{\sigma_q} & \Omega_q \end{array}$$

commutes, and hence

$$\begin{aligned} q \in (\chi_\tau)_p(x) \quad &\text{iff} \quad G_{pq}(x) \in F_q \\ &\text{iff} \quad \sigma_q(G_{pq}(x)) = [q) && \text{(by (*))} \\ &\text{iff} \quad \Omega_{pq}(\sigma_p(x)) = [q) && \text{(last diagram)} \\ &\text{iff} \quad \sigma_p(x) \cap [q) = [q) && \text{(definition } \Omega_{pq}) \\ &\text{iff} \quad [q) \subseteq \sigma_p(x) \\ &\text{iff} \quad q \in \sigma_p(x) && \text{(Exercise 10.2.3)} \end{aligned}$$

Thus $(\chi_\tau)_p(x) = \sigma_p(x)$. Since this holds of all $p \in P$ and all $x \in G_p$, it follows that $\sigma = \chi_\tau$.

EXAMPLE 1. We saw in §9.3 that the topos **Set**$^{\rightarrow}$ of set functions is essentially the same as **Set**$^{\mathbf{2}}$ where **2** is the poset category {0, 1} with $0 \sqsubseteq 1$. In **2** we have

$$\Omega_0 = \{\{0, 1\}, \{1\}, \emptyset\}$$
$$\Omega_1 = \{\{1\}, \emptyset\}$$

and Ω_{01} maps {0, 1} and {1} to {1}, and $\emptyset$ to $\emptyset$. If we denote {0, 1}, {1} and $\emptyset$ by 1, $\frac{1}{2}$, and 0 respectively in Ω_0, and {1} and $\emptyset$ in Ω_1 by 1 and 0, Ω_{01} becomes the function t providing the **Set**$^{\rightarrow}$-classifier defined in §4.4.

EXAMPLE 2. Let $\boldsymbol{\omega} = (\omega, \leqslant)$ be the poset of all finite ordinals $0, 1, 2, \ldots, m, \ldots$, under their natural ordering. **Set**$^{\boldsymbol{\omega}}$ is described by Maclane [75] as the category of "sets through time", an object being thought of as a string

$$F_0 \xrightarrow{F_{01}} F_1 \xrightarrow{F_{12}} F_2 \longrightarrow \ldots \longrightarrow F_m \xrightarrow{F_{mm+1}} F_{m+1} \longrightarrow \ldots$$

Now in $\boldsymbol{\omega}$, $[m) = \{m, m+1, m+2, \ldots\}$. Moreover if $S \subseteq \omega$ is non-empty, S has a first member m_S, so that if S is hereditary, $S = [m_S)$. Thus all non-empty hereditary sets are principal and can be identified with their first elements. Introducing a symbol ∞ to stand for the empty set we may then simplify Ω by identifying $\boldsymbol{\omega}^+$ with

$$\{0, 1, 2, \ldots, m, \ldots, \infty\}$$

and for $m \in \omega$, putting

$$\Omega_m = \{m, m+1, \ldots, \infty\}.$$

Whenever $m \leqslant n$, Ω_{mn} becomes

$$\Omega_{mn}(p) = \begin{cases} n & \text{if} \quad m \leqslant p \leqslant n \\ p & \text{if} \quad n \leqslant p \\ \infty & \text{if} \quad p = \infty, \end{cases}$$

while $true_m(0) = m$, for each $m \in \omega$.

Given $\tau : F \rightarrowtail G$, the character χ_τ has $(\chi_\tau)_m : F_m \to G_m$, given by

$$(\chi_\tau)_m(x) = \text{the first } n \text{ after } m \text{ that has } G_{mn}(x) \in F_n, \text{ if such exists,}$$

while

$$(\chi_\tau)_m(x) = \infty \text{ if } G_{mn}(x) \notin F_n \text{ whenever } m \leqslant n.$$

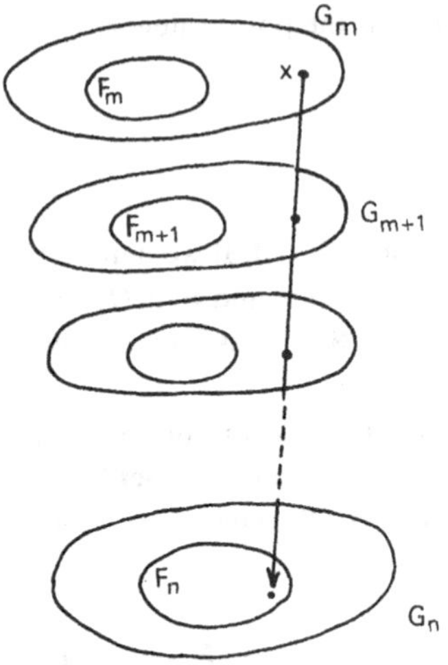

Fig. 10.2.

Thus $(\chi_\tau)_m(x)$ denotes the first time that x lands in the subobject F, the "time till truth" as Maclane puts it. Maclane's description of the subobject classifier for $\mathbf{Set}^\omega$ is even simpler than the one just given. The effect of the map Ω_{mm+1} can be displayed as

$$\begin{array}{ccccccc} \Omega_m & m & m+1 & m+2 & m+3 & \cdots & \infty \\ & \downarrow & \swarrow & \swarrow & \swarrow & & \downarrow \\ \Omega_{m+1} & m+1 & m+2 & m+3 & \cdots & & \infty \end{array}$$

The picture looks the same for each m, and indeed it is the structure of the map that is significant, not the labelling of the entries in the "order-isomorphic" sequences Ω_m and Ω_{m+1}. We may replace each Ω_m by the single set

$$\Omega = \{0, 1, 2, \dots, \infty\}$$

and each Ω_{mm+1} by the single map $t:\Omega \to \Omega$, displayed as

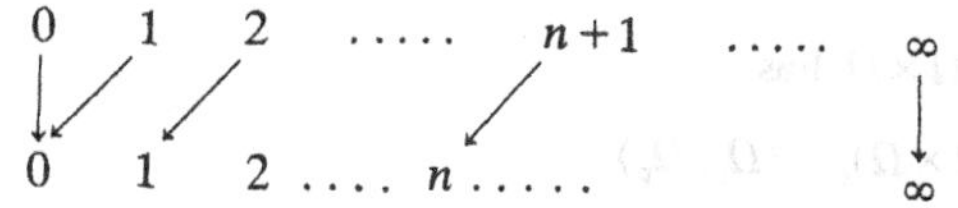

Then the object of truth values becomes, as in Maclane, the constant functor $\Omega \xrightarrow{t} \Omega \xrightarrow{t} \Omega \to \dots$ and the arrow *true* has the inclusion $\{0\} \hookrightarrow \Omega$ for each component.

So now we have seen three set-theoretically distinct objects in $\mathbf{Set}^{\omega}$ that serve as objects of truth-values, underlining again the point that the Ω-axiom characterises $\top : 1 \to \Omega$ uniquely up to isomorphism only.

10.4. The truth arrows

I. False

The initial object $0 : \mathbf{P} \to \mathbf{Set}$ in $\mathbf{Set}^{\mathbf{P}}$ is the constant functor having $0_p = \emptyset$ and $0_{pq} = \mathrm{id}_{\emptyset}$ for $p \sqsubseteq q$. The unique transformation $0 \to 1$ has components $\emptyset \hookrightarrow \{0\}$ (i.e. the same component for each p). The character of $! : 0 \to 1$ is $false : 1 \to \Omega$, with $false_p : \{0\} \to \Omega_p$ having

$$\begin{aligned} false_p(0) &= \{q : p \sqsubseteq q \text{ and } 1_{pq}(0) \in 0_q\} \\ &= \{q : p \sqsubseteq q \text{ and } 0 \in \emptyset\} \\ &= \emptyset. \end{aligned}$$

Thus false picks out the zero element from each **HA** $[p)^+$.

II. Negation

$\neg : \Omega \to \Omega$ is the character of *false*. Identifying $false_p$ with $\{\emptyset\} \subseteq \Omega_p$ we find then the p-th component $\neg_p : \Omega_p \to \Omega_p$ of $\neg$ has

$$\begin{aligned} \neg_p(S) &= \{q : p \sqsubseteq q \text{ and } \Omega_{pq}(S) \in \{\emptyset\}\} \\ &= \{q : p \sqsubseteq q \text{ and } S \cap [q) = \emptyset\} \\ &= [p) \cap \neg S \\ &= (\neg S)_p. \end{aligned}$$

We had already used the symbol $\neg_p$ in §10.2 to denote the pseudo-complementation operation in $[p)^+$. The equation just derived shows that *the latter operation is precisely the same as the p-th component of the negation truth arrow in* **Set**$^{\mathbf{P}}$, and so the notation remains consistent.

III. Conjunction

The functor $\Omega \times \Omega$ has

$$(\Omega \times \Omega)_p = \langle \Omega_p, \Omega_p \rangle$$

and for $p \sqsubseteq q$, $(\Omega \times \Omega)_{pq}$ is the product map $\Omega_{pq} \times \Omega_{pq}$ (cf. §3.8).

The arrow $\langle \top, \top \rangle : 1 \to \Omega \times \Omega$ in **Set**$^{\mathbf{P}}$ has components

$$\langle \top, \top \rangle_p : \{0\} \to \Omega_p \times \Omega_p$$

given by $\langle \top, \top \rangle_p(0) = \langle [p), [p) \rangle$.

Its character is the conjunction arrow

$$\Omega \times \Omega \xrightarrow{\cap} \Omega$$

with components $\cap_p : \Omega_p \times \Omega_p \to \Omega_p$ having

$$\begin{aligned}
\cap_p(\langle S, T \rangle) &= \{q : p \sqsubseteq q \text{ and } \langle \Omega_{pq}(S), \Omega_{pq}(T) \rangle = \langle [q), [q) \rangle\} \\
&= \{q : p \sqsubseteq q \text{ and } S \cap [q) = [q) = T \cap [q)\} \\
&= \{q : p \sqsubseteq q \text{ and } [q) \subseteq S \text{ and } [q) \subseteq T\} \\
&= \{q : p \sqsubseteq q \text{ and } q \in S \text{ and } q \in T\} \\
&= S \cap T \cap [p) \\
&= (S \cap T)_p \\
&= S \cap T. \qquad \text{(Theorem 10.2.1)}
\end{aligned}$$

IV. Implication

The equaliser $e : \circledleq \rightarrowtail \Omega \times \Omega$ of $\cap : \Omega \times \Omega \to \Omega$ and $pr_1 : \Omega \times \Omega \to \Omega$, has as domain the functor $\circledleq : \mathbf{P} \to \mathbf{Set}$, with

$$\begin{aligned}
\circledleq_p &= \{\langle S, T \rangle : \cap_p(\langle S, T \rangle) = S\} \\
&= \{\langle S, T \rangle : S \subseteq T\} \subseteq \Omega_p \times \Omega_p,
\end{aligned}$$

and $\circledleq_{pq}$, for $p \sqsubseteq q$, giving output $\langle S_q, T_q \rangle$ for input $\langle S, T \rangle$.

The components of e are the inclusions $e_p : \circledleq_p \hookrightarrow \Omega_p \times \Omega_p$.

The implication arrow $\Rightarrow : \Omega \times \Omega \to \Omega$, being the character of e, has component $\Rightarrow_p$ given by

$$\begin{aligned}\Rightarrow_p(\langle S, T\rangle) &= \{q : p \sqsubseteq q \text{ and } \langle \Omega_{pq}(S), \Omega_{pq}(T)\rangle \in \circledS_p\} \\ &= \{q : p \sqsubseteq q \text{ and } S \cap [q) \subseteq T \cap [q)\} \\ &= \{q : p \sqsubseteq q \text{ and } S \cap [q) \subseteq T\} \\ &= (S \Rightarrow T) \cap [p) \\ &= (S \Rightarrow T)_p.\end{aligned}$$

Thus the p-th component of the implication arrow is the relative pseudo-complementation for the **HA** $[p)^+$.

V. Disjunction

EXERCISE 1. Show that the p-th component of the transformation

$$[\langle \top_\Omega, 1_\Omega\rangle, \langle 1_\Omega, \top_\Omega\rangle]$$

is "essentially" the set

$$\{\langle [p), S\rangle : S \in \Omega_p\} \cup \{\langle S, [p)\rangle : S \in \Omega_p\}$$

and hence that *the disjunction arrow* $\cup : \Omega \times \Omega \to \Omega$ *has components* $\cup_p(\langle S, T\rangle) = S \cup T$.

It is worth pausing here to reflect on what has been accomplished. We now know that the truth arrows in $\mathbf{Set}^\mathbf{P}$ are precisely those natural transformations whose components interpret the corresponding connectives on the Heyting algebras in **P**. But remember that the truth arrows were defined long before intuitionistic logic and **HA**'s were mentioned. They arose from a categorial description of the classical truth functions in **Set**. *Subsequently*, when interpreted in the particular topos $\mathbf{Set}^\mathbf{P}$, they yield the intuitionistic truth functions. Thus the theory of "topos logic" abstracts a structure common to classical and intuitionistic logic. What better example could there be of the advancement of understanding through the interplay of generalisation and specialisation (§2.4)?

10.5. Validity

In view of the results of the last section one would anticipate an intimate relationship between validity in $\mathbf{Set}^\mathbf{P}$ and algebraic semantics on the **HA**'s

$[p)^+$. In fact the main result of this section, indeed of this chapter, is the

VALIDITY THEOREM. *For any poset* $\mathbf{P}$, *and propositional sentence* $\alpha \in \Phi$,

$$\mathbf{Set}^{\mathbf{P}} \models \alpha \quad \text{iff} \quad \mathbf{P} \models \alpha.$$

In the left-hand expression we mean topos-validity as defined in §6.7. The right-hand expression refers to Kripke-style validity as in §8.4.

There is some choice as to how we go about proving the Validity Theorem. We know from §8.4 that

$$\mathbf{P} \models \alpha \quad \text{iff} \quad \mathbf{P}^+ \models \alpha,$$

and from §8.3 that

$$\mathbf{Set}^{\mathbf{P}} \models \alpha \quad \text{iff} \quad \mathbf{Set}^{\mathbf{P}}(1, \Omega) \models \alpha \quad \text{iff} \quad \mathrm{Sub}(1) \models \alpha,$$

so we could proceed to establish relationships between the **HA**'s $\mathbf{P}^+$, $\mathbf{Set}^{\mathbf{P}}(1, \Omega)$, and Sub(1). Ultimately these are all variations on the same underlying theme. We choose to approach the Validity Theorem directly in terms of the definitions of validity concerned.

Let $\mathcal{M} = (\mathbf{P}, V)$ be a model based on $\mathbf{P}$, where $V: \Phi_0 \to \mathbf{P}^+$ is a $\mathbf{P}$-valuation. We use V to define a $\mathbf{Set}^{\mathbf{P}}$-valuation $V': \Phi_0 \to \mathbf{Set}^{\mathbf{P}}(1, \Omega)$ à la §6.7. V' assigns to each sentence letter π a truth value $V'(\pi): 1 \to \Omega$ in $\mathbf{Set}^{\mathbf{P}}$. The p-th component $V'(\pi)_p: \{0\} \to \Omega_p$ is defined by

$$(*) \qquad \begin{aligned} V'(\pi)_p(0) &= V(\pi) \cap [p) \\ &= V(\pi)_p \end{aligned}$$

Thus $V'(\pi)_p$ picks out the set of points in $[p)$ at which π is true in $\mathcal{M}$.

Now if $p \sqsubseteq q$ then $V(\pi) \cap [p) \cap [q) = V(\pi) \cap [q)$ (Exercise 10.2.2) and so

$$\begin{array}{ccc} \{0\} & \xrightarrow{V'(\pi)_p} & [p)^+ \\ \big\downarrow & & \big\downarrow \Omega_{pq} \\ \{0\} & \xrightarrow{V'(\pi)_q} & [q)^+ \end{array}$$

commutes. Hence $V'(\pi)$ *is* a natural transformation.

By the rules of §8.4 the model $\mathcal{M}$ produces for each sentence $\alpha \in \Phi$ a subset $\mathcal{M}(\alpha) = \{q: \mathcal{M} \models_q \alpha\}$ of P, and hence, for each $p \in P$, a subset $\mathcal{M}(\alpha)_p = \mathcal{M}(\alpha) \cap [p)$ of $[p)$. On the other hand by the rules of §6.7, V'

provides each α with a $\mathbf{Set^P}$-arrow $V'(\alpha): 1 \to \Omega$ and hence, for each $p \in P$, a function $V'(\alpha)_p : \{0\} \to \Omega_p$. We have

LEMMA 1. *For any α, the p-th component*

$$V'(\alpha)_p : \{0\} \to [p)^+$$

of $V'(\alpha)$ has $V'(\alpha)_p(0) = \mathcal{M}(\alpha)_p$.

PROOF. By induction on the formation of α. Since $\mathcal{M}(\pi) = V(\pi)$, for $\alpha = \pi$ the result is immediate from $(*)$. If $\alpha = {\sim}\beta$, and the result holds for β, then

$$\begin{aligned} V'({\sim}\beta)_p &= (\neg \circ V'(\beta))_p \\ &= \neg_p \circ V'(\beta)_p \end{aligned}$$

and so

$$\begin{aligned} V'(\alpha)_p(0) &= \neg_p(V'(\beta)_p(0)) \\ &= \neg_p(\mathcal{M}(\beta)_p) && \text{(induction hypothesis)} \\ &= (\neg \mathcal{M}(\beta))_p && \text{(Part II of §10.4, and Theorem 10.2.2(3))} \\ &= \mathcal{M}({\sim}\beta)_p && \text{((4'), §8.4)} \\ &= \mathcal{M}(\alpha)_p, \end{aligned}$$

hence the result holds for α. □

EXERCISE 1. Complete the proof of Lemma 1 for the cases of the connectives $\vee$, $\wedge$, $\supset$, using the other parts of §10.4, the rest of Theorem 2 of §10.2, and clauses (2'), (3'), and (5') from §8.4. □

COROLLARY 2. $\mathbf{Set^P} \models \alpha$ *only if* $\mathbf{P} \models \alpha$.

PROOF. Let $\mathcal{M} = (\mathbf{P}, V)$ be any $\mathbf{P}$-based model, and V' the $\mathbf{Set^P}$-valuation corresponding to V as in $(*)$. Since $\mathbf{Set^P} \models \alpha$, $V'(\alpha) = true$, and so for each p, $V'(\alpha)_p(0) = true_p(0) = [p)$. Since $p \in [p)$, Lemma 1 gives $p \in \mathcal{M}(\alpha)_p \subseteq \mathcal{M}(\alpha)$. Thus $\mathcal{M}(\alpha) = P$. As this holds for any model on $\mathbf{P}$, α is valid on $\mathbf{P}$.

To prove the converse of Corollary 2, we begin with a $\mathbf{Set^P}$-valuation $V' : \Phi_0 \to \mathbf{Set^P}(1, \Omega)$ and construct from it a $\mathbf{P}$-valuation $V : \Phi_0 \to \mathbf{P}^+$. The arrow $V'(\pi): 1 \to \Omega$ picks out, for each $q \in P$, an hereditary subset $V'(\pi)_q(0)$ of $[q)$. We form the union of all of these sets to get $V(\pi)$. Thus

$$V(\pi) = \bigcup\{V'(\pi)_q(0) : q \in P\}$$

i.e.

(**) $\quad r \in V(\pi)$ iff for some q, $\quad r \in V'(\pi)_q(0)$.

Having now obtained a **P**-valuation V we could apply $(*)$ to get another $\mathbf{Set}^{\mathbf{P}}$-valuation V'', with $V''(\pi)_p(0) = V(\pi) \cap [p)$. However this just gives us back the original V', as we see from

Lemma 3. *For any* $p \in P$,

$$V(\pi) \cap [p) = V'(\pi)_p(0),$$

where $V(\pi)$ *is defined by* (**).

Proof. It is clear from (**) that $V'(\pi)_p(0) \subseteq V(\pi)$. But since $V'(\pi): 1 \rightarrow \Omega$, $V'(\pi)_p : \{0\} \rightarrow \Omega_p$, and so $V'(\pi)_p(0) \subseteq [p)$. Hence $V'(\pi)_p(0) \subseteq V(\pi) \cap [p)$. Conversely, suppose $r \in V(\pi) \cap [p)$. Then $p \sqsubseteq r$, and for some q, $r \in V'(\pi)_q(0)$. Since $V'(\pi)_q(0) \subseteq [q)$, it follows that $q \sqsubseteq r$, and hence

$$\begin{array}{ccc} \{0\} & \xrightarrow{V'(\pi)_q} & \Omega_q \\ \Big\downarrow & & \Big\downarrow \Omega_{qr} \\ \{0\} & \xrightarrow{V'(\pi)_r} & \Omega_r \end{array}$$

commutes, because $V'(\pi)$ is a natural transformation. Thus $V'(\pi)_q(0) \cap [r) = V'(\pi)_r(0)$.

Analogously, since $p \sqsubseteq r$,

$$V'(\pi)_p(0) \cap [r) = V'(\pi)_r(0).$$

Then, knowing that $r \in V'(\pi)_q(0)$ and $r \in [r)$, we may apply these last two equations to conclude that $r \in V'(\pi)_p(0)$. Hence $V(\pi) \cap [p) \subseteq V'(\pi)_p(0)$. □

Now if V is a **P**-valuation, and V' is defined by $(*)$, i.e. $V'(\pi)_p(0) = V(\pi)_p$, then by Theorem 1(4) of §10.2,

$$\begin{aligned} \bigcup\{V'(\pi)_p(0): p \in P\} &= \bigcup\{V(\pi)_p : p \in P\} \\ &= V(\pi), \end{aligned}$$

so the application of (**) just gives us V back again. The upshot of this, and Lemma 3, is that the definitions $(*)$ and (**) are inverse to each other and establish a bijection between **P**-valuations and $\mathbf{Set}^{\mathbf{P}}$-valuation. Thus in Lemma 1 we may alternatively regard V as having been defined from V' by (**).

COROLLARY 4. $\mathbf{P} \models \alpha$ *only if* $\mathbf{Set}^{\mathbf{P}} \models \alpha$.

PROOF. Let V' be any $\mathbf{Set}^{\mathbf{P}}$-valuation, and $\mathcal{M} = (\mathbf{P}, V)$ the corresponding model defined by (**). Since $\mathbf{P} \models \alpha$, $\mathcal{M}(\alpha) = P$, and so for any p, $\mathcal{M}(\alpha)_p = \mathcal{M}(\alpha) \cap [p) = [p) = true_p(0)$. Thus by Lemma 1, $V'(\alpha)_p(0) = true_p(0)$. Hence $V'(\alpha) = true$. □

Corollaries 2 and 4 together give the Validity Theorem.

10.6. Applications

(1) The most important immediate consequence of the Validity Theorem is the characterisation of the class of topos-valid sentences. If $\mathbf{P}_{\mathrm{IL}}$ is the canonical frame for IL described in §8.4 then, for any $\alpha \in \Phi$

$$\vdash_{\mathrm{IL}} \alpha \quad \textit{iff} \quad \mathbf{P}_{\mathrm{IL}} \models \alpha,$$

and hence by the Validity Theorem

$$\vdash_{\mathrm{IL}} \alpha \quad \textit{iff} \quad \mathbf{Set}^{\mathbf{P}_{\mathrm{IL}}} \models \alpha.$$

From this we get the:

COMPLETENESS THEOREM FOR $\mathscr{E}$-VALIDITY. *If* α *is valid on every topos, then*

$$\vdash_{\mathrm{IL}} \alpha.$$

Together with the Soundness Theorem given in §8.3 this yields the result that the sentences valid on all topoi are precisely the IL-theorems.

(2) It was stated in §6.7 that the category $\mathbf{Set}^{\rightarrow}$ does not validate $\alpha \vee \sim\alpha$. To see this, recall that $\mathbf{Set}^{\rightarrow}$ is essentially the same as $\mathbf{Set}^{\mathbf{2}}$. But in the Example of §8.4 it was shown that $\mathbf{2} \not\models \alpha \vee \sim\alpha$. The Validity Theorem then gives $\mathbf{Set}^{\mathbf{2}} \not\models \alpha \vee \sim\alpha$.

(3) The logic LC, mentioned in §8.4, is generated by adjoining to the IL-axioms the classical tautology

$$(\alpha \supset \beta) \vee (\beta \supset \alpha)$$

LC is what is known as an *intermediate logic*, i.e. its theorems include all IL-theorems and are included in the CL-theorems.

Now it is known (cf. Dummett [59] or Segerberg [68]) that

$$\boldsymbol{\omega} \models \alpha \quad \textit{iff} \quad \vdash_{\text{LC}} \alpha,$$

and so we have

$$\vdash_{\text{LC}} \alpha \quad \textit{iff} \quad \mathbf{Set}^{\boldsymbol{\omega}} \models \alpha,$$

i.e. LC is the logic of the topos of "sets through time" described in §10.3. This is the appropriate context if time is considered to be made up of discrete moments. However the logic is not altered by the assumption that time is dense, or even continuous. If $\mathbb{Q}$ and $\mathbb{R}$ denote respectively the posets of rational, and of real, numbers under their natural (arithmetic) ordering, then from Section 5 of Segerberg we conclude that

$$\boldsymbol{\omega} \models \alpha \quad \textit{iff} \quad \mathbb{Q} \models \alpha \quad \textit{iff} \quad \mathbb{R} \models \alpha.$$

and so the topoi $\mathbf{Set}^{\boldsymbol{\omega}}$, $\mathbf{Set}^{\mathbb{Q}}$, and $\mathbf{Set}^{\mathbb{R}}$ all have the same logic.

In fact the most general conclusion we can make is that if $\mathbf{P}$ is *any* infinite linearly order poset (i.e. $p \sqsubseteq q$ or $q \sqsubseteq p$, for all $p, q \in P$), then

$$\mathbf{Set}^{\mathbf{P}} \models \alpha \quad \textit{iff} \quad \vdash_{\text{LC}} \alpha.$$

EXERCISE 1. Let $\{0, 1, 2, \ldots, \infty\}$ be the modified version of $\boldsymbol{\omega}^+$ described in §10.3. Define **HA** operations on this set by modifying the operations on $\boldsymbol{\omega}^+$. Relate these operations to the definition of the "LC-matrix" given in Dummett [59]. □

PROBLEM. Let $\mathscr{E}$ be any topos, and put

$$L_{\mathscr{E}} = \{\alpha : \mathscr{E} \models \alpha\}$$

then $L_{\mathscr{E}}$ is closed under *Detachment*, and is an intermediate logic. A canonical frame $\mathbf{P}_{L_{\mathscr{E}}}$ may be defined for $L_{\mathscr{E}}$ by replacing IL by $L_{\mathscr{E}}$ everywhere in the definition of $\mathbf{P}_{\text{IL}}$.

Is there a general categorial relationship between the topoi $\mathscr{E}$ and $\mathbf{Set}^{\mathbf{P}_{L_{\mathscr{E}}}}$?

□

Exercises (for Heyting-algebraists)

Exercise 2. Given a truth value $\tau : 1 \to \Omega$ in $\mathbf{Set}^{\mathbf{P}}$, define $S_\tau \in \mathbf{P}^+$ by

$$S_\tau = \bigcup\{\tau_p(0) : p \in P\}.$$

Show that the assignment of S_τ to τ gives a Heyting algebra isomorphism

$$\mathbf{Set}^{\mathbf{P}}(1, \Omega) \cong \mathbf{P}^+.$$

Exercise 3. Let $\sigma : F \rightarrowtail 1$ be a subobject of 1 in $\mathbf{Set}^{\mathbf{P}}$. Then for each p, σ_p can be taken as the inclusion $F_p \hookrightarrow \{0\}$, and so we have either $F_p = \emptyset$, or $F_p = \{0\} = 1$. Define

$$S_\sigma = \{p : F_p = 1\}.$$

Show that S_σ is hereditary and that the assignment of S_σ to σ yields an **HA** isomorphism

$$\mathrm{Sub}(1) \cong \mathbf{P}^+.$$

What is the inverse of this isomorphism?

Exercise 4. Suppose that the poset $\mathbf{P}$ has a least (initial) element. Show then that if $S, T \in \mathbf{P}^+$, $S \cup T = P$ iff $S = P$ or $T = P$.

Derive from this that the topos $\mathbf{Set}^{\mathbf{P}}$ is disjunctive, in the sense of §7.7. □

CHAPTER 11

ELEMENTARY TRUTH

> *"... a new theory, however special its range of application, is seldom or never just an increment to what is already known. Its assimilation requires the reconstruction of prior theory and the re-evaluation of prior fact, an intrinsically revolutionary process that is seldom completed by a single man and never overnight."*
>
> Thomas Kuhn.

This chapter marks a change in emphasis towards an approach that will be more descriptive than rigorous. Our major concern will as usual be to analyse classical notions and define their categorial counterparts, but the detailed attention to verification of previous chapters will often be foregone. The proof that these generalisations work "as they should" will thus at times be left to the reader.

11.1. The idea of a first-order language

The propositional language PL of §6.3 is quite inadequate to the task of expressing the most basic discourse about mathematical structures. Take for example a structure $\langle A, R\rangle$ consisting of a binary relation R on a set A (i.e. $R \subseteq A \times A$). Let c be a particular element of A and consider the sentence "if every x is related by R to c, then there is some x to which c is related by R". If the "range" of the variable x is A, then this sentence is certainly true. For, if everything is related to c, then in particular c is related to c, so c is related to something. To see the structure of the sentence a little more clearly let

α abbreviate "for all x, xRc"

and

β abbreviate "for some x, cRx".

Then the sentence is schematised as

$$\alpha \supset \beta.$$

Now the semantical theory developed for PL in Chapter 6 cannot analyse the above argument, i.e. it cannot tell us why $\alpha \supset \beta$ is true. To know the truth value of the whole sentence we must know the values of α and β. However these function as "atomic" sentences (like the letters π_i). Their structure cannot be expressed in the language PL, and the PL-semantics does not itself explain why β must have the value "true" if α does. In order then to formalise α and β we introduce the following symbols:

(i) a symbol $\forall$, known as the *universal quantifier*, and read "for all";

(ii) a symbol $\exists$, known as the *existential quantifier*, and read "for some" or "there exists";

(iii) a symbol $\mathbf{c}$, called an *individual constant*, which is a "name" for the element c;

(iv) a symbol $\mathbf{R}$, a (two placed) *relation symbol*, or *predicate letter*, which names the relation R;

(v) a symbol v, called an *individual variable* whose interpretation is, literally, variable. It may be taken to refer to any member of A. (We shall help ourselves to an infinite number of these variables shortly, but for now one will do).

We can now symbolise α as $(\forall v)v\mathbf{Rc}$, and β as $(\exists v)\mathbf{cR}v$.

A language of the type we are now developing is called a *first-order* or *elementary* language. The word "elementary" here means "of elements". The variables of a first-order language range over elements of a structure. In a *higher-order* language, quantifiers would be applied to variables ranging over, not just elements, but also sets of elements, sets of sets of elements, etc. However in saying that the sentence

$$(\forall v)v\mathbf{Rc} \supset (\exists v)\mathbf{cR}v$$

is true of the structure or "interpretation" $\langle A, R, c\rangle$ it is thereby understood that the variable v ranges over the elements of A. Thus we need not include in our first order language any symbolisations of locutions like "for all x *belonging* to A". That is, the use of an elementary language does not depend on a formalisation of set theory.

The language we have just sketched is but one among many first order languages. The one we use will depend on the nature of the mathematical

structure we wish to discuss. If we wanted to analyse **BA**'s we would need
- constants **0** and **1** to name zero and unit elements;
- *functions letters* for the Boolean operations. These would comprise a one-place letter **f** for complementation, with $\mathbf{f}(v)$ read "the complement of v", and a pair of two-placed function letters, **g** and **h**, for meets and joins, with $\mathbf{g}(v_1, v_2)$ read "the meet of v_1 and v_2", and $\mathbf{h}(v_1, v_2)$ read "the join of v_1 and v_2";
- the *identity symbol* $\approx$, with $v_1 \approx v_2$ read "v_1 is identical to v_2".

Then, for example, the sentences

$$(\forall v)(\mathbf{g}(v, \mathbf{f}(v)) \approx \mathbf{0})$$

and

$$(\forall v)(\mathbf{h}(v, \mathbf{f}(v)) \approx \mathbf{1})$$

would be true of any Boolean algebra – they simply express the defining property of the complement of an element.

In principle, functions can always be replaced by relations (their graphs). Correspondingly, instead of introducing a function letter, say **h** above, we could use a three place relation symbol **S**, with $\mathbf{S}(v_1, v_2, v_3)$ being read "v_1 is the join of v_2 and v_2". The last sentence would then be replaced by

$$(\forall v)\mathbf{S}(\mathbf{1}, v, \mathbf{f}(v))$$

The most important mathematical structure as far as this book is concerned is the notion of category. This too is a "first-order concept" and there is some choice in how we formalise it. We could introduce two different sorts of variables, one sort to range over objects and the other over arrows, and hence have what is called a "two-sorted language". Alternatively we could use one sort of variable and the following list of predicate letters:

$\mathbf{Ob}(v)$	"v is an object"
$\mathbf{Ar}(v)$	"v is an arrow"
$\mathbf{dom}(v_1, v_2)$	"$v_1 = \text{dom } v_2$"
$\mathbf{cod}(v_1, v_2)$	"$v_1 = \text{cod } v_2$"
$\mathbf{id}(v_1, v_2)$	"$v_1 = 1_{v_2}$"
$\mathbf{com}(v_1, v_2, v_3)$	"$v_1 = v_2 \circ v_3$"

Amongst the sentences we would need to formally axiomatise the concept of a category are

$$\forall v((\mathbf{Ob}(v) \vee \mathbf{Ar}(v)) \wedge \sim(\mathbf{Ob}(v) \wedge \mathbf{Ar}(v)))$$

$$(\forall v_2)(\mathbf{Ob}(v_2) \supset (\exists v_1)\mathbf{id}(v_1, v_2))$$

$$(\forall v_1)(\forall v_2)(\mathbf{dom}(v_1, v_2) \supset \mathbf{Ob}(v_1) \wedge \mathbf{Ar}(v_2))$$

$$(\forall v_1) \dots (\forall v_6)(\mathbf{com}(v_4, v_1, v_2) \wedge \mathbf{com}(v_5, v_4, v_3) \wedge \mathbf{com}(v_6, v_2, v_3) \supset \mathbf{com}(v_5, v_1, v_6))$$

The last sentence expresses the associative law – $(v_1 \circ v_2) \circ v_3 = (v_1 \circ (v_2 \circ v_3)$. The interpretation of the others is left to the reader.

Notice that with the aid of the identity symbol we can express the statement $\psi(v_1)$ that an individual v_1 is the only one having a certain property φ (this of course is vital to the description of universal properties). We put $\psi(v_1) = (\varphi(v_1) \wedge (\forall v_2)(\varphi(v_2) \supset v_1 \approx v_2))$, i.e. "$v_1$ *has* the property, and anything having it is equal to v_1". The formula $\exists v_1 \psi(v_1)$ is sometimes written $(\exists! v_1)\varphi(v_1)$ which is read, "there is exactly one v_1 such that $\varphi(v_1)$".

The language just outlined is rather cumbersome in distinguishing arrows from objects. A simpler approach, mentioned earlier, is to eliminate objects in favour of their identity arrows, and so assume all individuals are arrows. We would then use the predicate **com** as before, as well as the function letters $\mathbf{D}(v)$ – "dom v", and $\mathbf{C}(v)$ – "cod v". Thus dom v is now an arrow, namely an identity arrow. But the dom and cod of an identity arrow ought to be itself, so we can *define* $\mathbf{Ob}(v)$ to be an abbreviation of the expression

$$(\mathbf{D}(v) \approx v) \wedge (\mathbf{C}(v) \approx v).$$

An extensive development of this type of first-order language for categories is presented by W. S. Hatcher [68], who uses it to discuss Lawvere's earlier work [64] on an elementary theory of the category of sets. Hatcher also gives a rigorous proof of the Duality Principle, which after all is a principle of logic (caveat – composites in Hatcher are written the other way around, i.e. what we have been calling "$g \circ f$" is written "fg").

EXERCISE 1. Express the Identity Law in the above languages.

EXERCISE 2. Write down a first order sentence expressing each of the axioms for the notion of an elementary topos.

11.2. Formal language and semantics

All of the examples just given have a common core, one shared by all such languages.

Basic alphabet for elementary languages

(i) an infinite list $v_1, v_2, v_3, \ldots$ of individual variables;
(ii) propositional connectives $\wedge, \vee, \sim, \supset$;
(iii) quantifier symbols $\forall, \exists$;
(iv) identity symbol $\approx$;
(v) brackets), (.

Given this stock of symbols we can specify a particular language, intended to describe a particular kind of structure, by listing its relation symbols, function letters, and individual constants. Hence a first-order language is, by definition, a set of symbols of these three kinds. For **BA**'s we employ the language $\{\mathbf{0}, \mathbf{1}, \mathbf{f}, \mathbf{g}, \mathbf{h}\}$, while for categories we could use $\{\mathbf{com}, \mathbf{C}, \mathbf{D}\}$. In order to discuss semantic theories for elementary logic we will work throughout with a particularly simple language, namely

$$\mathscr{L} = \{\mathbf{R}, \mathbf{c}\}$$

having just one (two-place) relation symbol, and one individual constant. This will suffice to illustrate the main points while avoiding complexities that are technical rather than conceptual.

TERMS: These are expressions denoting individuals. For $\mathscr{L}$ the terms are the variables $v_1, v_2, \ldots$ and the constant **c**.

ATOMIC FORMULAE: These are the basic building blocks for sentences. For $\mathscr{L}$ they comprise all (and only) those expressions of the form $t \approx u$, and $t\mathbf{R}u$, where t and u are terms.

FORMULAE: These are built up inductively by the rules

(i) each atomic formula is a formula;

(ii) if φ and ψ are formulae, then so are $(\varphi \wedge \psi)$, $(\varphi \vee \psi)$, $(\varphi \supset \psi)$, $(\sim\varphi)$;

(iii) if φ is a formula and v an individual variable, then $(\forall v)\varphi$ and $(\exists v)\varphi$ are formulae.

SENTENCES: If a particular occurrence of a variable in a formula is within the scope of a quantifier, that is said to be a *bound* occurrence of the variable. Otherwise the occurrence is *free*. Thus the first occurrence of v_1 in $(v_1 \approx v_1) \vee \sim(\exists v_1) v_1 \mathbf{R} v_1$ is free, while its third occurrence is bound. A *sentence* is a formula in which every occurrence of a variable is bound. A formula that is not a sentence, i.e. has at least one free occurrence of a variable, is called an *open formula*.

We will write $\varphi(v)$ to indicate that the variable v has a free occurrence in φ – thereby formalising a notation we have used all along. This may be extended to $\varphi(v_{i_1}, \ldots, v_{i_n})$ to indicate several (or perhaps all) of the free variables of φ.

INTERPRETATIONS OF $\mathscr{L}$: To ascribe meanings to $\mathscr{L}$-sentences we need to give an interpretation of the symbols $\mathbf{R}$ and $\mathbf{c}$, and then use these to define interpretations of formulae by induction over their rules of formation.

A *model* for $\mathscr{L}$, or a *realisation* of $\mathscr{L}$, is a structure $\mathfrak{A} = \langle A, R, c \rangle$ comprising

(i) a non-empty set A;

(ii) a relation $R \subseteq A \times A$;

(iii) a particular individual $c \in A$.

Now if φ is the sentence $(\forall v_1) v_1 \mathbf{Rc}$, then we may ask whether φ is true or false with respect to $\mathfrak{A}$. The answer is – yes, if every element of A is R-related to c, and no otherwise. On the other hand if $\varphi(v_1)$ is the open formula $v_1 \mathbf{Rc}$ it makes no sense to ask whether φ is true or false simpliciter. We would have to give some interpretation to the free variable v_1. We could thus ask whether φ is true when v_1 is interpreted as referring to the individual c. The answer then is – yes, if cRc, and no otherwise. The general point then is that to give an open formula a truth value relative to a model we have first to assign to its free variables specific "values" in that model.

We now introduce a method of interpreting the variables "all at once" in $\mathfrak{A}$. Let x be a function that assigns to each positive integer n an element $x(n)$, or simply x_n, of A. Such a function is called an $\mathfrak{A}$-valuation, and is represented as an infinite sequence $x = \langle x_1, x_2, \ldots, x_i, \ldots \rangle$. The i-th member x_i of this sequence is the interpretation of the variable v_i provided by the valuation x. In what follows we will have occasion to alter valuations like x in one place only. We denote by $x(i/a)$ the valuation obtained by replacing x_i by the element $a \in A$. Thus

$$x(i/a) = \langle x_1, x_2, \ldots, x_{i-1}, a, x_{i+1}, \ldots \rangle.$$

Once variables have been interpreted, we can discuss matters of truth. We are going to give a rigorous definition of the statement "*the formula φ is satisfied in $\mathfrak{A}$ by the valuation x*", which is symbolised

$$\mathfrak{A} \models \varphi[x].$$

The definition of satisfaction is intuitively almost obvious, but to set it out precisely is rather laborious. That such a rigorous definition really is needed was first realised by Alfred Tarski, who gave one in [36], thereby opening up what has become a substantial branch of mathematical logic, known as model theory.

Atomic Formulae: Given a valuation x, each term t determines an element x_t of A, defined by

$$x_t = \begin{cases} x_i & \text{if} \quad t \text{ is the variable } v_i \\ c & \text{if} \quad t \text{ is the constant } \mathbf{c}. \end{cases}$$

Then

(1) $\mathfrak{A} \models t \approx u[x]$ iff x_t is the same element as x_u

(2) $\mathfrak{A} \models t\mathbf{R}u[x]$ iff $x_t R x_u$.

Thus the symbol $\approx$ has a fixed interpretation on any model. It denotes the identity relation $\Delta = \{\langle x, y \rangle : x = y\}$.

Formulae:

(3) $\mathfrak{A} \models \varphi \wedge \psi[x]$ iff $\mathfrak{A} \models \varphi[x]$ and $\mathfrak{A} \models \psi[x]$

(4) $\mathfrak{A} \models \varphi \vee \psi[x]$ iff $\mathfrak{A} \models \varphi[x]$ or $\mathfrak{A} \models \psi[x]$

(5) $\mathfrak{A} \models \sim\varphi[x]$ iff not $\mathfrak{A} \models \varphi[x]$

(6) $\mathfrak{A} \models \varphi \supset \psi[x]$ iff either not $\mathfrak{A} \models \varphi[x]$ or $\mathfrak{A} \models \psi[x]$

(7) $\mathfrak{A} \models (\forall v_i)\varphi[x]$ iff for every $a \in A$, $\mathfrak{A} \models \varphi[x(i/a)]$

(8) $\mathfrak{A} \models (\exists v_i)\varphi[x]$ iff for some $a \in A$, $\mathfrak{A} \models \varphi[x(i/a)]$.

In fact the satisfaction of a formula depends only on the interpretation of free variables in that formula, as shown by the

Exercise 1. If x and y are valuations with $x_i = y_i$ whenever v_i occurs free in φ, then

$$\mathfrak{A} \models \varphi[x] \quad \text{iff} \quad \mathfrak{A} \models \varphi[y].$$

□

In view of this fact, if φ is a sentence (no free variables) then one of two things can happen: either

(i) φ is satisfied by every valuation in $\mathfrak{A}$, or

(ii) φ is satisfied by no valuation in $\mathfrak{A}$.

In case (i), we simply write $\mathfrak{A} \models \varphi$, read "*$\varphi$ is true in $\mathfrak{A}$*", or "*$\mathfrak{A}$ is a model of φ*". In case (ii) we say that *φ is false in $\mathfrak{A}$*, or that *φ fails in $\mathfrak{A}$*.

Now there are some open formulae that we might want to say *are* simply true in $\mathfrak{A}$. One such example is $v_1 \approx v_1$ – it comes out true no matter how it is interpreted, i.e. it is satisfied by every valuation. To make this usage precise, and to reflect the fact that only interpretations of *free* variables are required we consider satisfaction of formulae by finite sequences. The *index* of a formula is defined to be the number of free variables that it has. If $\varphi(v_{i_1}, \ldots, v_{i_n})$ has index n, with $v_{i_1}, \ldots, v_{i_n}$ constituting all of its variables, we write $\mathfrak{A} \models \varphi[x_1, \ldots, x_n]$ if $\mathfrak{A} \models \varphi[y]$ for some (equivalently any) valuation y that has $y_{i_1} = x_1$, $y_{i_2} = x_2, \ldots, y_{i_n} = x_n$. This means that φ is satisfied when v_{i_1} is interpreted as x_1, v_{i_2} as x_2, etc. Then φ is said to be true in $\mathfrak{A}$, $\mathfrak{A} \models \varphi$, iff for any $x_1, \ldots, x_n \in A$, $\mathfrak{A} \models \varphi[x_1, \ldots, x_n]$.

EXERCISE 2. $\mathfrak{A} \models \varphi(v_{i_1}, \ldots, v_{i_n})$ iff $\mathfrak{A} \models (\forall v_{i_1})(\forall v_{i_2}) \ldots (\forall v_{i_n})\varphi$.

EXERCISE 3. $\mathfrak{A} \models (\forall v)\varphi[x]$ iff $\mathfrak{A} \models \sim(\exists v) \sim \varphi[x]$.

11.3. Axiomatics

An $\mathcal{L}$-formula φ is *valid* if it is true in all $\mathcal{L}$-models. To axiomatise the valid formulae we need to consider substitutions of a term t for a variable v in a formula φ. We write $\varphi(v/t)$ to denote the result of replacing every free occurrence of v in φ by t. This operation will "preserve truth" in general only if v is *free for t in φ*. This means either that t is the constant **c**, or that t is a variable and no free occurrence of v is within the scope of a t-quantifier. This means then that t does not become bound when substituted for a free occurrence of v.

The classical axioms for $\mathcal{L}$ are of three kinds.

PROPOSITIONAL AXIOMS: *All formulae that are instances of the schemata* I–XII *of* §6.3 *are axioms.*

QUANTIFIER AXIOMS: *For each formula* $\varphi(v)$, *and term* t *for which* v *is free in* φ,

(UI) $\forall v\varphi \supset \varphi(v/t)$,

(EG) $\varphi(v/t) \supset \exists v\varphi$

are axioms.

(The names stand for "universal instantiation" and "existential generalisation".)

IDENTITY AXIOMS: For any term t,

($I1$) $t \approx t$ is an axiom.

For any $\varphi(v)$, and terms t and u, for which v is free in φ,

($I2$) $(t \approx u) \wedge \varphi(v/t) \supset \varphi(v/u)$, is an axiom.

The rules of inference are,

DETACHMENT: *From* φ *and* $\varphi \supset \psi$ *infer* ψ,

and two quantifier rules:

($\forall$) *From* $\varphi \supset \psi$ *infer* $\varphi \supset (\forall v)\psi$, *provided* v *is not free in* φ

($\exists$) *From* $\varphi \supset \psi$ *infer* $(\exists v)\varphi \supset \psi$, *provided* v *is not free in* ψ.

Writing $\vdash_{CL} \varphi$ to mean that φ is derivable from the above axioms by the above rules, we have

$$\vdash_{CL} \varphi \quad \textit{iff} \quad \textit{for all } \mathscr{L}\textit{-models } \mathfrak{A},\ \mathfrak{A} \models \varphi.$$

This fact, that the class of valid $\mathscr{L}$-formulae is axiomatisable, is known as Gödel's Completeness Theorem, and was first proven for elementary logic by Gödel [30]. There are now several ways of proving it, and information about these may be found for example in Chang and Keisler [73] and Rasiowa and Sikorski [63].

EXERCISE. Show that the following are CL-theorems:

$$t \approx u \supset u \approx t, \qquad (t \approx u) \wedge (u \approx u') \supset (t \approx u'),$$

$$\sim(\exists v) \sim \varphi \supset (\forall v)\varphi, \qquad (\forall v)\varphi \supset \sim(\exists v) \sim \varphi.$$

11.4. Models in a topos

The interpretation of $\mathscr{L}$ in a topos is, like its classical counterpart, both natural in its conception, and arduous in its detail. It is based on a reformulation in arrow-language of the satisfaction relation

$$\mathfrak{A} \models \varphi[x_1, \ldots, x_n].$$

In fact it is convenient to deal first with a more general notion. An integer $m \geqslant 1$ will be called *appropriate* to φ if all of the variables of φ, free and bound, appear in the list $v_1, v_2, \ldots, v_m$. Notice that it is permitted that the list include other variables than those occurring in φ, so that if $m \leqslant l$, then l is also appropriate to φ. Now given an appropriate m, we can discuss satisfaction of φ by m-length sequences. We put $\mathfrak{A} \models \varphi[x_1, \ldots, x_m]$ iff $\mathfrak{A} \models \varphi[y]$ for some (equivalently any) valuation y that has $y_i = x_i$ whenever v_i is free in φ (such a v_i will then occur in the list $v_1, \ldots, v_m$).

Now given a model $\mathfrak{A} = \langle A, R, c\rangle$ and a particular m, each φ to which m is appropriate determines a subset, φ^m, of the m-fold product A^m. Namely,

$$\varphi^m = \{\langle x_1, \ldots, x_m\rangle : \mathfrak{A} \models \varphi[x_1, \ldots, x_m]\}$$

is the set of all m-length sequences satisfying φ in $\mathfrak{A}$.

To know all the φ^m's, for appropriate m's, is to know all about satisfaction of φ in $\mathfrak{A}$. Moreover the rules for satisfaction for the propositional connectives correspond to the Boolean set operations on subsets of A^m. Thus the complement of φ^m (i.e. the sequences not satisfying φ) is the set of sequences satisfying $\sim\varphi$, the intersecting of φ^m and ψ^m consists of the sequences satisfying $\varphi \wedge \psi$, and we get

$$(\sim\varphi)^m = -\varphi^m$$

$$(\varphi \wedge \psi)^m = \varphi^m \cap \psi^m$$

$$(\varphi \vee \psi)^m = \varphi^m \cup \psi^m \qquad \text{etc.}$$

(We see now the point of dealing with appropriate m's. If m is appropriate to φ and ψ it will be to $\varphi \wedge \psi$ also, although the three formulae might all have different indices.)

It would seem then that we could interpret φ in a topos as a subobject of a^m, for some object a, and then use the Heyting algebra structure of $\mathrm{Sub}(a^m)$ to interpret connectives, and hopefully quantifiers as well. This approach to categorial semantics has been set out in dissertations by students of Gonzalo Reyes and André Joyal at Montréal. The theory for elementary logic is presented by Monique Robitaille-Giguère [75].

The alternative approach is to switch from subobjects to their characteristic arrows. This accords with the propositional semantics of Chapter 6, and has the advantage for us that the interpretation of quantifiers is more accessible to a "first principles" treatment. This latter theory has been developed by Michael Brockway [76].

Returning to our $\mathscr{L}$-model $\mathfrak{A}$, we replace φ^m by its characteristic function $[\![\varphi]\!]^m : A^m \to 2$, where

$$[\![\varphi]\!]^m(\langle x_1, \ldots, x_m\rangle) = \begin{cases} 1 & \text{if } \mathfrak{A} \models \varphi[x_1, \ldots, x_m] \\ 0 & \text{otherwise} \end{cases}$$

Using the correspondence described in Theorem 1 of §7.1, we find that

$$\begin{aligned} [\![\sim\varphi]\!]^m &= \neg \circ [\![\varphi]\!]^m \\ [\![\varphi \wedge \psi]\!]^m &= [\![\varphi]\!]^m \cap [\![\psi]\!]^m \qquad (= \cap \circ \langle [\![\varphi]\!]^m, [\![\psi]\!]^m \rangle) \\ [\![\varphi \vee \psi]\!]^m &= [\![\varphi]\!]^m \cup [\![\psi]\!]^m \end{aligned}$$

where $\neg$, $\cap$, $\cup$ are the classical truth functions on 2.

To treat quantifiers in this manner we consider an example. Suppose that φ has just the variables v_1, v_2, and v_3 and (with $m = 3$), $[\![\varphi]\!]^3 : A^3 \to 2$ has been defined. We wish to define $[\![\forall v_2\varphi]\!]^3 : A^3 \to 2$. So, take a triple $\langle x_1, x_2, x_3\rangle \in A^3$ and let

$$\begin{aligned} B_2 &= \{x \in A : \mathfrak{A} \models \varphi[x_1, x, x_3]\} \\ &= \{x \in A : [\![\varphi]\!]^3(\langle x_1, x, x_3\rangle) = 1\}. \end{aligned}$$

The satisfaction definition tells us that

$$\mathfrak{A} \models \forall v_2\varphi[x_1, x_2, x_3] \quad \text{iff} \quad B_2 = A,$$

so we want

$$[\![\forall v_2\varphi]\!]^3(\langle x_1, x_2, x_3\rangle) = \begin{cases} 1 & \text{if } B_2 = A \\ 0 & \text{otherwise.} \end{cases}$$

Now the assignment of the subset B_2 of A to the triple $\langle x_1, x_2, x_3\rangle$ establishes a function $|\varphi|_2^3$ from A^3 to $\mathscr{P}(A)$. We now define a new function $\forall_A : \mathscr{P}(A) \to 2$ by putting

$$\forall_A(B) = \begin{cases} 1 & \text{if } B = A \\ 0 & \text{if } B \neq A \quad (\text{i.e. } B \subset A) \end{cases}$$

Then the definition of $[\![\forall v_2\varphi]\!]^3$ becomes

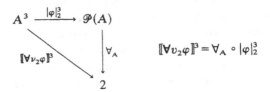

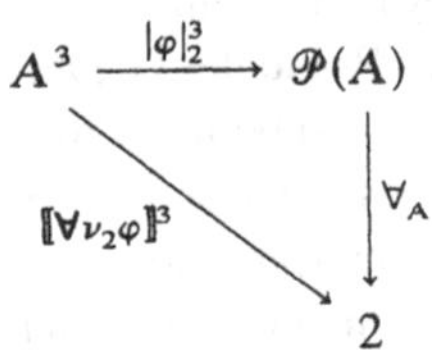

$$[\![\forall v_2\varphi]\!]^3 = \forall_A \circ |\varphi|_2^3$$

Under the isomorphism $\mathcal{P}(A)\cong 2^A$ we may construe $|\varphi|_2^3$ as a function $A^3\to 2^A$, and hence it becomes the exponential adjoint (cf. §3.16) of a function $f: A^3\times A\to 2$, i.e. $f: A^4\to 2$. Then f assigns a 1 or a 0 to a 4-tuple $\langle x_1, x_2, x_3, x_4\rangle\in A^4$ according as the function $|\varphi|_2^3(\langle x_1, x_2, x_3\rangle)=\chi_{B_2}$ assigns a 1 or a 0 to x_4, i.e. according as $[\![\varphi]\!]^3(\langle x_1, x_4, x_3\rangle)$ equals 1 or 0. Thus if we define $T_2^4: A^4\to A^3$ by $T_2^4(\langle x_1, x_2, x_3, x_4\rangle)=\langle x_1, x_4, x_3\rangle$, we have that

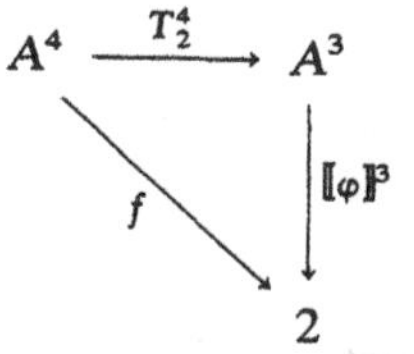

commutes. But T_2^4 can be given a categorial description. Recall from §3.8 that whenever $j\leqslant m$, we have a "j-th projection map" $pr_j^m: A^m\to A$ taking each m-sequence to its j-th member. In the present case, the effect of T_2^4 is to place the result of the 4-th projection of a 4-sequence in its 2nd position. But (§3.8) this process can be described as a product map – T_2^4 is the map

$$A^4 \xrightarrow{\langle pr_1^4, pr_4^4, pr_3^4\rangle} A^3.$$

Consequently we get a categorial definition of f, and hence of $|\varphi|_2^3$. To complete the picture we need such a definition for $\forall_A$. This was given by Lawvere in [72], where he described $\forall_A$ as "the characteristic map of the name of $true_A$". In §4.2 we described $\ulcorner true_A\urcorner: 1\to 2^A$, the name of $true_A$, as the arrow that picks $true_A$ out of 2^A. Since $true_A=\chi_A: A\to 2$, we identify $true_A$ with $\{A\}\subseteq\mathcal{P}(A)$. But the character of this last subobject is, by definition, $\forall_A$. $\ulcorner true_A\urcorner$ itself is the exponential adjoint of the composite

$$1\times A\xrightarrow{pr_A} A\xrightarrow{true_A} 2,\quad \text{where}\quad pr_A(\langle 0, x\rangle)=x.$$

In summary then, $[\![\forall v_2\varphi]\!]^3=\forall_A\circ|\varphi|_2^3$, where $\forall_A$ is the character of the exponential adjoint of $true_A\circ pr_A$, while $|\varphi|_2^3$ is the exponential adjoint of $[\![\varphi]\!]^3\circ\langle pr_1^4, pr_4^4, pr_3^4\rangle$.

For existential quantifiers, by analogy we have

$$\mathfrak{A}\models\exists v_2\varphi[x_1, x_2, x_3]\quad\text{iff}\quad B_2\neq\emptyset$$

and so we put

$$[\![\exists v_2\varphi]\!]^3(\langle x_1, x_2, x_3\rangle)=\begin{cases}1 & \text{if } B_2\neq\emptyset\\ 0 & \text{otherwise}\end{cases}$$

and hence

$$\begin{array}{ccc} A^3 & \xrightarrow{|\varphi|_2^3} & \mathcal{P}(A) \\ & \searrow^{[\![\exists v_2\varphi]\!]^3} & \downarrow \exists_A \\ & & 2 \end{array}$$

commutes where

$$\exists_A(B)=\begin{cases}1 & \text{if} \quad B\neq\emptyset\\ 0 & \text{if} \quad B=\emptyset.\end{cases}$$

It follows that $\exists_A$ is the character of the set

$$\begin{aligned} C&=\{B: B\neq\emptyset\}\\ &=\{B: \text{for some } x\in A, x\in B\}.\end{aligned}$$

But then if $\in_A \hookrightarrow \mathcal{P}(A)\times A$ is the membership relation on A (§4.7), i.e.

$$\in_A=\{\langle B, x\rangle : B\subseteq A, \text{ and } x\in B\},$$

we see that applying the first projection $p_A(\langle B, x\rangle)=B$ from $\mathcal{P}(A)\times A$ to $\mathcal{P}(A)$ yields $p_A(\in_A)=C$.

Thus $\exists_A$ is the character of the image of the composite

$$\in_A \hookrightarrow \mathcal{P}(A)\times A \xrightarrow{p_A} \mathcal{P}(A).$$

This places our account of quantifiers on an "arrows only" basis. The general definition of $[\![\forall v_i\varphi]\!]^m$, and $[\![\exists v_i\varphi]\!]^m$ comes from the above by putting m in place of 4, and i in place of 2.

The function $[\![t\approx u]\!]^m : A^m\to 2$ has

$$[\![t\approx u]\!]^m(\langle x_1,\ldots,x_m\rangle)=\begin{cases}1 & \text{if} \quad x_t=x_u\\ 0 & \text{otherwise}\end{cases}$$

so

$$\begin{array}{ccc} A^m & \xrightarrow{\langle p_t^m, p_u^m\rangle} & A^2 \\ & \searrow^{[\![t\approx u]\!]^m} & \downarrow \delta_A \\ & & 2 \end{array}$$

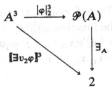

commutes where $\rho_t^m: A^m \to A$, $\rho_u^m: A^m \to A$, and δ_A have

$$\rho_t^m(\langle x_1, \ldots, x_m \rangle) = x_t$$
$$\rho_u^m(\langle x_1, \ldots, x_m \rangle) = x_u$$

and

$$\delta_A(\langle x, y \rangle) = \begin{cases} 1 & \text{if } x = y \\ 0 & \text{if } x \neq y \end{cases}, \qquad x, y \in A.$$

δ_A (the "Kronecker delta") is the character of the identity relation (diagonal) $\Delta = \{\langle x, y \rangle : x = y\} \subseteq A^2$. Notice that Δ can be identified with the monic $\langle 1_A, 1_A \rangle : A \to A^2$, that takes x to $\langle x, x \rangle$.

To define ρ_t^m, let $f_c : \{0\} \to A$ have $f_c(0) = c$.

Then

$$\rho_t^m = \begin{cases} pr_i^m : A^m \to A & \text{if } t = v_i \\ f_c \circ ! : A^m \xrightarrow{!} 1 \xrightarrow{f_c} A & \text{if } t = \mathbf{c}. \end{cases}$$

(Similarly for ρ_u^m).

To deal with the predicate letter $\mathbf{R}$, let $r : A^2 \to 2$ be the characteristic function of $R \subseteq A \times A$. Then

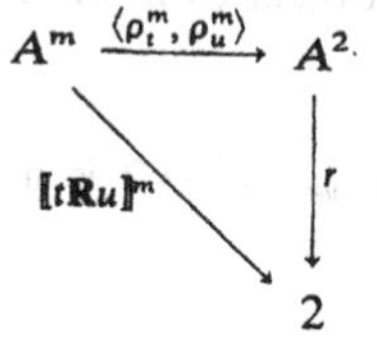

commutes. The final notion to be re-examined is truth in a model. If $\varphi(v_{i_1}, \ldots, v_{i_n})$ has index n, then defining $[\![\varphi]\!]_{\mathfrak{A}} : A^n \to 2$ by

$$[\![\varphi]\!]_{\mathfrak{A}}(\langle x_1, \ldots, x_n \rangle) = \begin{cases} 1 & \text{if } \mathfrak{A} \models \varphi[x_1, \ldots, x_n] \\ 0 & \text{otherwise} \end{cases}$$

we have

$$\begin{aligned} \mathfrak{A} \models \varphi \quad & \text{iff} \quad \text{for all } x_1, \ldots, x_n \in A, \quad [\![\varphi]\!]_{\mathfrak{A}}(\langle x_1, \ldots, x_n \rangle) = 1 \\ & \text{iff} \quad [\![\varphi]\!]_{\mathfrak{A}} = \chi_{A^n} \\ & \text{iff} \quad [\![\varphi]\!]_{\mathfrak{A}} = true_{A^n}. \end{aligned}$$

To describe $[\![\varphi]\!]_{\mathfrak{A}}$ by arrows, we observe that if m is appropriate to φ, $\mathfrak{A} \models \varphi[x_1, \ldots, x_n]$ iff for any $y_1, \ldots, y_m$ having

$$y_{i_1} = x_1, \ldots, y_{i_n} = x_n,$$
$$\mathfrak{A} \models \varphi[y_1, \ldots, y_m].$$

Thus

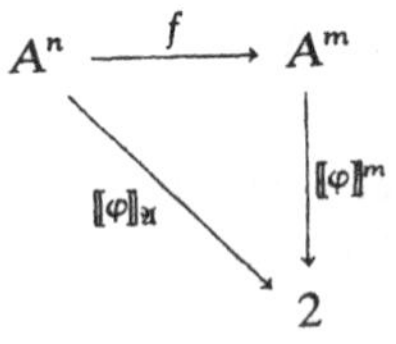

commutes for any f, provided only that

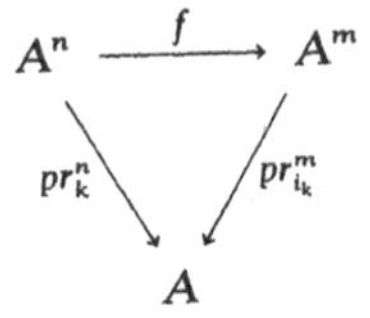

$$pr^m_{i_k} \circ f = pr^n_k \quad \text{for} \quad 1 \leqslant k \leqslant n.$$

This description fits in with the definition of truth of *sentences*. A member of A^n, i.e. an n-length sequence, can be thought of as a function from the ordinal $n = \{0, 1, \ldots, n-1\}$ to A. Thus, with $n = 0$, A^0 is the set of functions from the ordinal 0 (the initial object $\emptyset$) to A. Thus

$$A^0 = A^{\emptyset} = \{\emptyset\} = 1.$$

So if φ is a sentence, with index $n = 0$, $[\![\varphi]\!]_{\mathfrak{A}} : A^0 \to 2$ is a truth value $1 \to 2$. We have

$$[\![\varphi]\!]_{\mathfrak{A}} = \begin{cases} \mathit{true} & \text{if } \mathfrak{A} \models \varphi \\ \mathit{false} & \text{if not } \mathfrak{A} \models \varphi. \end{cases}$$

But then for any $m \geqslant 1$, *any* $f : 1 \to A^m$ makes

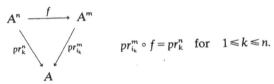

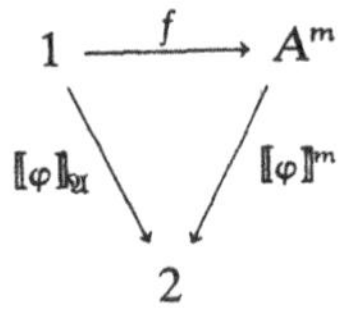

commute, for if $\mathfrak{A} \models \varphi$ then $[\![\varphi]\!]^m$ is the "constant" function that outputs only 1's, while if not $\mathfrak{A} \models \alpha$, then $[\![\varphi]\!]^m$ outputs only 0's.

EXERCISE 1. Suppose that $\varphi(v_{i_1}, \ldots, v_{i_n})$ has index n, and m is appropriate to φ. Explain why

$$\begin{array}{ccc} A^m & \xrightarrow{f} & A^n \\ & {\scriptstyle [\![\varphi]\!]^m}\searrow \quad \swarrow{\scriptstyle [\![\varphi]\!]_{\mathfrak{A}}} & \\ & 2 & \end{array}$$

commutes, where $f(\langle y_1, \ldots, y_m\rangle) = \langle y_{i_1}, \ldots, y_{i_n}\rangle$. □

The general definition

Let $\mathscr{E}$ be a topos, and a an $\mathscr{E}$-object. We define several arrows related to a.

DEFINITION 1. $\Delta_a : a \rightarrowtail a \times a$ is the product arrow $\langle 1_a, 1_a\rangle$
$\delta_a : a \times a \to \Omega$ is the character of Δ_a.

DEFINITION 2. $\forall_a : \Omega^a \to \Omega$ is the unique arrow making

$$\begin{array}{ccc} 1 & \xrightarrow{\ulcorner true_a \urcorner} & \Omega^a \\ \downarrow & & \downarrow {\scriptstyle \forall_a} \\ 1 & \xrightarrow{true} & \Omega \end{array}$$

a pullback, where $\ulcorner true_a \urcorner$ is the exponential adjoint of the composite $true_a \circ pr_a : 1 \times a \to a \to \Omega$.

DEFINITION 3. $\exists_a : \Omega^a \to \Omega$ is the character of the image arrow of the composite $p_a \circ \in_a : \in \rightarrowtail \Omega^a \times a \to \Omega^a$, where p_a is the first projection arrow, and $\in_a$ (§4.7) is the subobject of $\Omega^a \times a$ whose character is the evaluation arrow $ev_a : \Omega^a \times a \to \Omega$. Thus we have a diagram

$$\begin{array}{ccc} \in & \overset{\in_a}{\rightarrowtail} & \Omega^a \times a \\ \downarrow & & \downarrow {\scriptstyle p_a} \\ p_a \circ \in_a(\in) & \overset{\mathrm{im}(p_a \circ \in_a)}{\rightarrowtail} & \Omega^a \\ \downarrow & & \downarrow {\scriptstyle \exists_a} \\ 1 & \xrightarrow{true} & \Omega \end{array}$$

where the bottom square is a pullback, and the top an epi-monic factorisation.

DEFINITION 4. For each m and i, with $1 \leqslant i \leqslant m$, $T_i^{m+1}: a^{m+1} \to a^m$ is the product arrow

$$\langle pr_1^{m+1}, \ldots, pr_{i-1}^{m+1}, pr_{m+1}^{m+1}, pr_{i+1}^{m+1}, \ldots, pr_m^{m+1} \rangle.$$

An $\mathscr{E}$-model for $\mathscr{L}$ is a structure

$$\mathfrak{A} = \langle a, r, f_c \rangle, \quad \text{where}$$

(i) a is an $\mathscr{E}$-object that is non-empty, i.e. $\mathscr{E}(1, \mathrm{a}) \neq \emptyset$;
(ii) $r: a \times a \to \Omega$ is an $\mathscr{E}$-arrow;
(iii) $f_c: 1 \to a$ is an "$\mathscr{E}$-element" of a.

Then given a term t we associate with each appropriate m an arrow ρ_t^m, where,

$$\rho_t^m = \begin{cases} pr_i^m : a^m \to a & \text{if} \quad t = v_i \\ f_c \circ ! : a^m \to a & \text{if} \quad t = \mathbf{c}. \end{cases}$$

Then for each $\mathscr{L}$-formula φ and appropriate m we define an $\mathscr{E}$-arrow $[\![\varphi]\!]^m : a^m \to \Omega$ inductively as follows:

(1) $[\![t \approx u]\!]^m = \delta_a \circ \langle \rho_t^m, \rho_u^m \rangle$

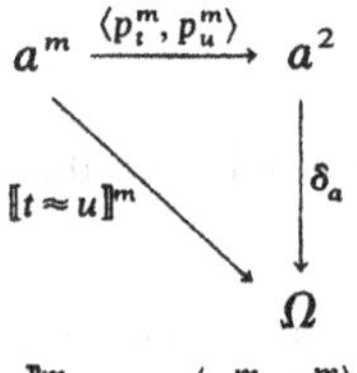

(2) $[\![t\mathbf{R}u]\!]^m = r \circ \langle \rho_t^m, \rho_u^m \rangle$

(3) $[\![\varphi \wedge \psi]\!]^m = [\![\varphi]\!]^m \cap [\![\psi]\!]^m = \cap \circ \langle [\![\varphi]\!]^m, [\![\psi]\!]^m \rangle$

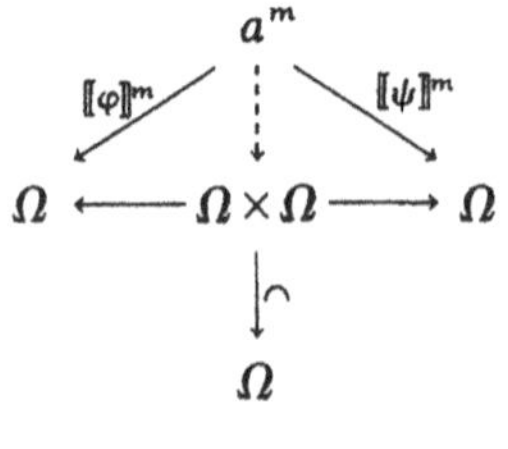

(4) $[\![\varphi \vee \psi]\!]^m = [\![\varphi]\!]^m \cup [\![\psi]\!]^m$
(5) $[\![\sim\varphi]\!]^m = \neg \circ [\![\varphi]\!]^m$

(6) $[\![\varphi \supset \psi]\!]^m = [\![\varphi]\!]^m \Rightarrow [\![\psi]\!]^m$

(7) $[\![\forall v_i\varphi]\!]^m = \forall_a \circ |\varphi|_i^m$

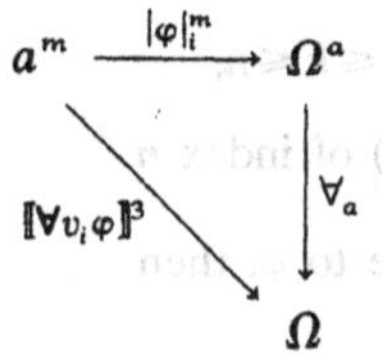

where $|\varphi|_i^m$ is the exponential adjoint of the composite of

$$a^{m+1} \xrightarrow{T_i^{m+1}} a^m \xrightarrow{[\![\varphi]\!]^m} \Omega$$

(8) $[\![\exists v_i\varphi]\!]^m = \exists_a \circ |\varphi|_i^m$.

Now let $\varphi(v_{i_1}, \ldots, v_{i_n})$ have index n. Then let g be any arrow from a^n to a. Choose a φ-appropriate m, and let $f: a^n \to a^m$ be the product arrow $\langle p_1, \ldots, p_m\rangle$, where

$$p_j = \begin{cases} pr_k^n : a^n \to a, & \text{if} \quad j = i_k, \quad \text{some} \quad 1 \leqslant k \leqslant n \\ g & \text{otherwise.} \end{cases}$$

Then define $[\![\varphi]\!]_{\mathfrak{A}} : a^n \to \Omega$ by

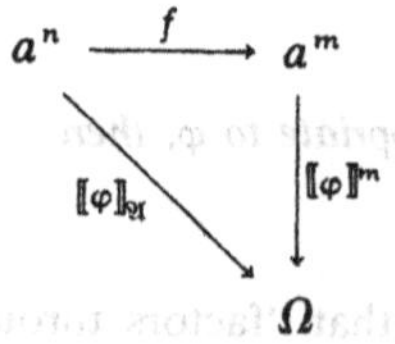

i.e. $[\![\varphi]\!]_{\mathfrak{A}} = [\![\varphi]\!]^m \circ f$. Then we define "$\mathfrak{A}$ is an $\mathscr{E}$-model of φ" by

$$\mathfrak{A} \models^{\mathscr{E}} \varphi \quad \text{iff} \quad [\![\varphi]\!]_{\mathfrak{A}} = true_{a^n}.$$

Notice that if $n \geqslant 1$, we could take g as any of the projection arrows $a^n \to a$, while if $n = 0$, we need the assumption that a is non-empty for there to be a $g: 1 \to a$ at all.

The demonstration that the definition of $[\![\varphi]\!]_{\mathfrak{A}}$ does not depend on which g is chosen, or which appropriate m, depends on some lengthy but straightforward exercises:

EXERCISE 2. If $f,h : a^n \rightrightarrows a^m$ have

$$pr^m_{i_k} \circ f = pr^m_{i_k} \circ h = pr^n_k, \quad \text{for all} \quad 1 \leq k \leq n,$$

then $[\![\varphi]\!]^m \circ f = [\![\varphi]\!]^m \circ h$, for $\varphi(v_{i_1}, \ldots, v_{i_n})$ of index n.

EXERCISE 3. If m and l are both appropriate to φ, then

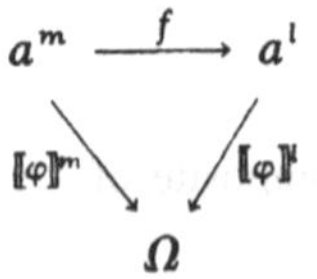

commutes provided that $pr^l_i \circ f = pr^m_i$, whenever v_i is free in φ. Show that such an f exists.

EXERCISE 4. If $\varphi(v_{i_1}, \ldots, v_{i_n})$ has index n, and m is appropriate to φ, then

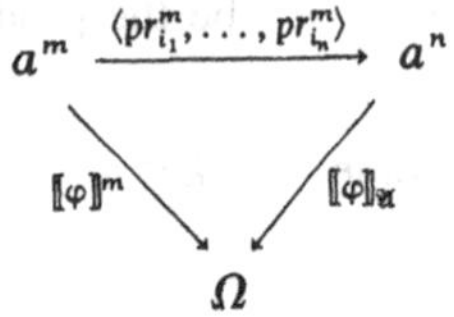

commutes (cf. Exercise 1). □

From these results we obtain:

THEOREM. *If φ has index n, and m is appropriate to φ, then*

$$\mathfrak{A} \models^{g} \varphi \quad \textit{iff} \quad [\![\varphi]\!]^m = true_{a^m}.$$

PROOF. By Exercise 3 of §4.2, any arrow that "factors through *true* is *true*", i.e. if

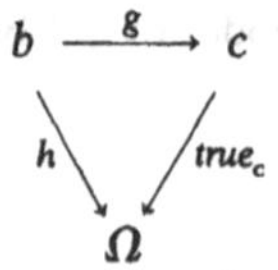

commutes, then $h = true_b$. But by the definition of $[\![\varphi]\!]_{\mathfrak{A}}$, and Exercise 4, each of $[\![\varphi]\!]_{\mathfrak{A}}$ and $[\![\varphi]\!]^m$ factor through each other, hence

$$[\![\varphi]\!]_{\mathfrak{A}} = true_{a^n} \quad \text{iff} \quad [\![\varphi]\!]^m = true_{a^m}.$$ □

11.5. Substitution and soundness

An $\mathscr{L}$-formula φ is called *$\mathscr{E}$-valid*, $\mathscr{E} \models \varphi$, if $\mathfrak{A} \models^{\mathscr{E}} \varphi$ holds for every $\mathscr{E}$-model $\mathfrak{A}$.

THEOREM 1. *If $\mathscr{E} \models \varphi$ and $\mathscr{E} \models \varphi \supset \psi$, then $\mathscr{E} \models \psi$.*

PROOF. Let $\mathfrak{A}$ be any $\mathscr{E}$-model. Then $\mathfrak{A} \models \varphi$ and $\mathfrak{A} \models \varphi \supset \psi$, and so taking an m appropriate to $(\varphi \supset \psi)$, we have $[\![\varphi]\!]^m \Rightarrow [\![\psi]\!]^m = [\![\varphi \supset \psi]\!]^m = true_{a^m}$ (by the Theorem at the end of the last section). But $true_{a^m}$ is the unit of the **HA** $\mathscr{E}(a^m, \Omega)$, so (Exercise 8.3.8) in that **HA**, $[\![\varphi]\!]^m \sqsubseteq [\![\psi]\!]^m$. But since m is also appropriate to φ, and $\mathfrak{A} \models \varphi$, we also have $[\![\varphi]\!]^m = true_{a^m}$. Thus in $\mathscr{E}(a^m, \Omega)$, $[\![\psi]\!]^m = true_{a^m}$ and so as m is appropriate to ψ, $\mathfrak{A} \models \psi$. □

So the rule of *Detachment* preserves $\mathscr{E}$-validity. Since the propositional connectives are interpreted as the truth arrows in a topos it should come as no surprise that any instance of the schemata I–XI is valid in any $\mathscr{E}$, while there are topos models in which XII fails (an example will be given later). We shall write $\vdash_{\mathrm{IL}} \varphi$ to mean that φ is derivable in the system that has all the rules and axioms of §11.3 except for XII. Without *I*1 and *I*2, this is the system of intuitionistic predicate logic of Heyting [66]. Axioms for identity equivalent to the ones given here are discussed by Rasiowa and Sikorski [63].

SOUNDNESS THEOREM. *If $\vdash_{\mathrm{IL}} \varphi$, then for any $\mathscr{E}$, $\mathscr{E} \models \varphi$.*

We will not prove all the Soundness Theorem, but will concentrate on setting up the machinery that lies behind it. The method as always is to show that the axioms are $\mathscr{E}$-valid and the rules of inference preserve this property. The strategy for the first part is to show that if φ is an axiom then relative to $\mathfrak{A}$, $[\![\varphi]\!]^m = true_{a^m}$, for some (or any) appropriate m. The Theorem of the last section then gives $\mathfrak{A} \models \varphi$.

To establish validity of the quantifier and identity axioms we must look at the categorial content of the substitution process. If $\psi = \varphi(v_i/t)$, then in **Set**, interpreting t in ψ as x_t is the same as interpreting v_i in φ as x_t, i.e. $\mathfrak{A} \models \psi[x_1, \ldots, x_m]$ iff $\mathfrak{A} \models \varphi[x_1, \ldots, x_{i-1}, x_t, x_{i+1}, \ldots, x_m]$, and so

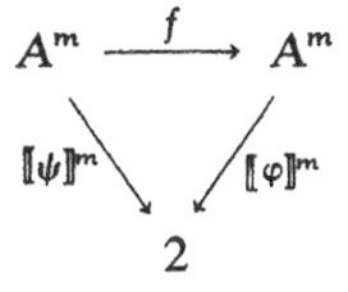

commutes, where $f(\langle x_1, \ldots, x_m\rangle) = \langle x_1, \ldots, x_{i-1}, x_t, x_{i+1}, \ldots, x_m\rangle$.

Correspondingly, in a general topos $\mathscr{E}$, if $i \leqslant m$ and t is a term to which m is appropriate (i.e. if $t = v_j$ then $j \leqslant m$), the arrow $\delta^m[i/t]: a^m \to a^m$ is defined to be the product arrow

$$\langle pr_1^m, \ldots, pr_{i-1}^m, \rho_t^m, pr_{i+1}^m, \ldots, pr_m^m\rangle.$$

Substitution Lemma. *In any topos, the diagram*

$$\begin{array}{ccc} a^m & \xrightarrow{\delta^m[i/t]} & a^m \\ & {}_{[\![\varphi(v_i/t)]\!]^m}\searrow \quad \swarrow_{[\![\varphi]\!]^m} & \\ & \Omega & \end{array}$$

commutes whenever v_i is free for t in φ. □

Exercise 1. $pr_i^m \circ \delta[i/v_j] = pr_j^m \circ \delta^m[i/v_j] = pr_j^m$.

Exercise 2. If $f: b \to a^m$ has $pr_i^m \circ f = pr_j^m \circ f$, then

$$\begin{array}{ccc} & b & \\ {}^{f}\swarrow & & \searrow^{f} \\ a^m & \xrightarrow{\delta^m[i/v_j]} & a^m \end{array}$$

commutes. (Interpret this in **Set**.)

Exercise 3. For $i, j \leqslant m$,

$$\begin{array}{ccc} a^{m+1} & \xrightarrow{T_j^{m+1}} & a^m \\ {\scriptstyle \langle T_j^{m+1}, pr_{m+1}^{m+1}\rangle}\downarrow & & \downarrow{\scriptstyle \delta[i/v_j]} \\ a^{m+1} & \xrightarrow{T_i^{m+1}} & a^m \end{array}$$

commutes.

Exercise 4. If v_j does not occur in φ, then

$$[\![\varphi(v_i/v_j)]\!]^m \circ T_j^{m+1} = [\![\varphi]\!]^m \circ T_i^{m+1},$$

and hence

$$[\![\exists v_j\varphi(v_i/v_j)]\!]^m = [\![\exists v_i\varphi]\!]^m$$
$$[\![\forall v_j\varphi(v_i/v_j)]\!]^m = [\![\forall v_i\varphi]\!]^m.$$

Consequently $[\![\varphi]\!]^m = [\![\psi]\!]^m$ if φ and ψ are "bound alphabetical variants" of each other. □

To use the Substitution Lemma to show validity of the identity axioms we examine the properties of the Kronecker delta.

THEOREM 2. *For any pair $f, g: b \to a$, $\delta_a \circ \langle f, g\rangle$ is the character of the equaliser of f and g.*

PROOF. Consider

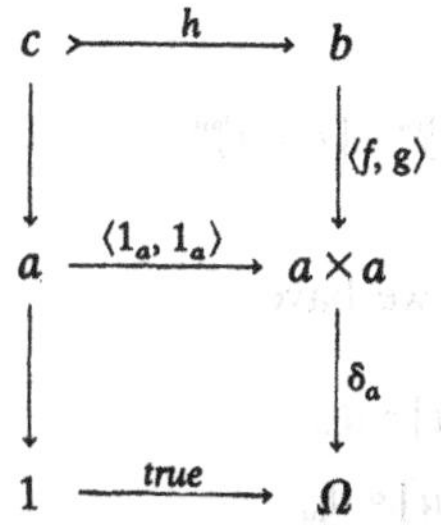

The top square is obtained by pulling $\langle 1_a, 1_a\rangle = \Delta_a$ back along $\langle f, g\rangle$. By the universal property of that square qua pullback, it is an easy exercise to show that h equalises f and g. But the bottom square is the pullback defining δ_a, so by the PBL and the Ω-axiom, $\delta_a \circ \langle f, g\rangle = \chi_h$. □

COROLLARY. *$\delta_a \circ \langle f, f\rangle = true_b$, for $f: b \to a$.*

PROOF. $true_b = \chi_{1_b}$ and 1_b equalises the pair $\langle f, f\rangle$. □

From this Corollary we obtain immediately the validity of $I1$, i.e. $\mathscr{E} \models t \approx t$. For, $[\![t \approx t]\!]^m = \delta_a \circ \langle \rho_t^m, \rho_t^m\rangle$, where $\rho_t^m : a^m \to a$.

Now in **Set**, the formula $(t \approx u)$ determines the set

$$D_{tu} = \{\langle x_1, \ldots, x_m\rangle : \mathfrak{A} \models (t \approx u)[x_1, \ldots, x_m]\}$$
$$= \{\langle x_1, \ldots, x_m\rangle : x_t = x_u\}.$$

Correspondingly in $\mathscr{E}$ we define $d_{tu} : d \rightarrowtail a^m$ to be the subobject whose character is $[\![t \approx u]\!]^m$.

THEOREM 3. *For appropriate m,*

$$\delta^m[i/t] \circ d_{tu} = \delta^m[i/u] \circ d_{tu}.$$

PROOF. Since $[\![t \approx u]\!]^m = \delta_a \circ \langle \rho_t, \rho_u \rangle : a^m \to \Omega$, Theorem 2 tells us that d_{tu} equalises ρ_t and ρ_u, hence $\rho_t \circ d_{tu} = \rho_u \circ d_{tu}$. Then

$$\begin{aligned}
&\langle pr_1, \dots, \rho_t, \dots, pr_m \rangle \circ d_{tu} \\
&\quad = \langle pr_1 \circ d_{tu}, \dots, \rho_t \circ d_{tu}, \dots, pr_m \circ d_{tu} \rangle \\
&\quad = \langle pr_1 \circ d_{tu}, \dots, \rho_u \circ d_{tu}, \dots, pr_m \circ d_{tu} \rangle \\
&\quad = \langle pr_1, \dots, \rho_u, \dots, pr_m \rangle \circ d_{tu}.
\end{aligned}$$

□

COROLLARY. *If m is appropriate to t, u, and* $\varphi(v_i)$, *with* v_i *free for t and u in* φ, *then*

$$[\![\varphi(v_i/t)]\!]^m \cap [\![t \approx u]\!]^m = [\![\varphi(v_i/u)]\!]^m \cap [\![t \approx u]\!]^m$$

PROOF. Using the Substitution Lemma, we have

$$\begin{aligned}
[\![\varphi(v_i/t)]\!]^m \circ d_{tu} &= [\![\varphi]\!]^m \circ \delta^m[i/t] \circ d_{tu} \\
&= [\![\varphi]\!]^m \circ \delta^m[i/u] \circ d_{tu} \\
&= [\![\varphi(v_i/u)]\!]^m \circ d_{tu}.
\end{aligned}$$

Since $\chi_{d_{tu}} = [\![t \approx u]\!]^m$, Lemma 1(2) of §7.5 yields the desired result. □

Now in order to have $\mathfrak{A} \models [(t \approx u) \wedge \varphi(v_i/t)] \supset \varphi(v_i/u)$ we require that for some appropriate m,

$$[\![t \approx u]\!]^m \cap [\![\varphi(v_i/t)]\!]^m \sqsubseteq [\![\varphi(v_i/u)]\!]^m$$

in the **HA** $\mathscr{E}(a^m, \Omega)$. But this follows from the Corollary, by lattice properties, and so the schema $I2$ is $\mathscr{E}$-valid.

We turn now to the validity of the quantifier axioms. For this we elicit the basic properties of the quantifier arrows.

THEOREM 4. (1) $(\forall_a \circ p_a) \Rightarrow ev_a = true_{\Omega^a \times a}$

(2) $ev_a \Rightarrow (\exists_a \circ p_a) = true_{\Omega^a \times a}$

PROOF. (1) Consider

$$
\begin{array}{ccc}
d & \xrightarrow{\ f\ } & \Omega^a \times a \\
\downarrow & & \downarrow{\scriptstyle p_a} \\
1 & \xrightarrow{\ \ulcorner true_a \urcorner\ } & \Omega^a \\
\downarrow & & \downarrow{\scriptstyle \forall_a} \\
1 & \xrightarrow{\ true\ } & \Omega
\end{array}
$$

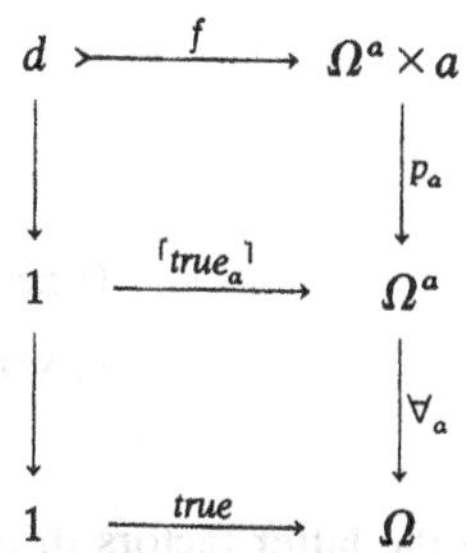

The top square is obtained by pulling $\ulcorner true_a \urcorner$ back along p_a. A now familiar argument tells then that $\chi_f = \forall_a \circ p_a$. But by definition of $\ulcorner true_a \urcorner$ as the exponential adjoint of $true_a \circ pr_a$, the diagram

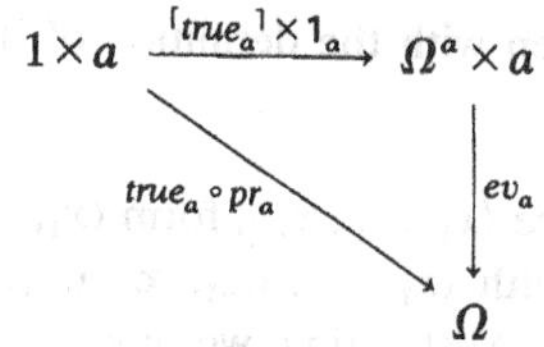

commutes, which says precisely that the perimeter of

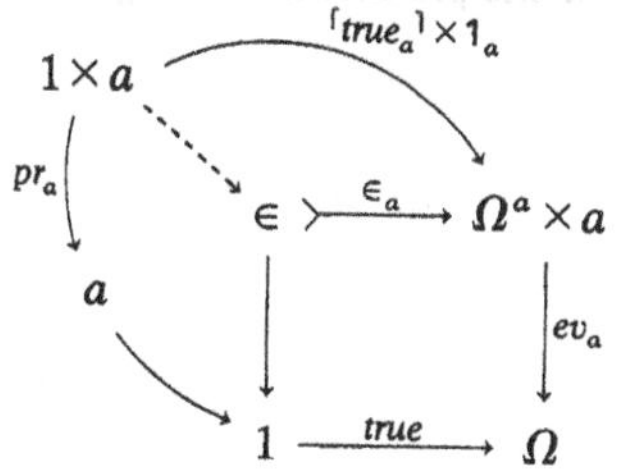

commutes, yielding an arrow $1 \times a \to \,\in$ that makes $\ulcorner true_a \urcorner \times 1_a$ factor through $\in_a$.

But consider the diagram

$$
\begin{array}{ccc}
d & \xrightarrow{\ f\ } & \\
{\scriptstyle \langle 1_d, p_2 \circ f \rangle}\downarrow & \searrow & \Omega^a \times a \\
& 1 \times a & \\
& \nearrow_{\ulcorner true_a \urcorner \times 1_a} &
\end{array}
$$

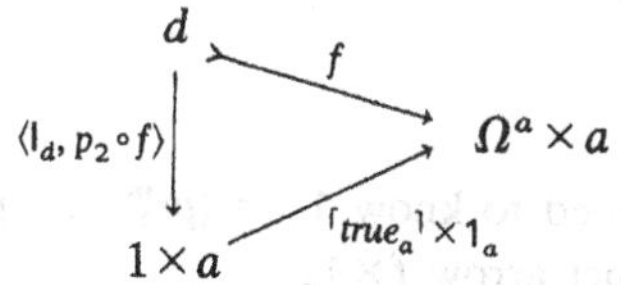

where $p_2 : \Omega^a \times a \to a$ is the 2nd projection.

Using Exercise 8 of §3.8 we find that

$$
\begin{aligned}
&(\ulcorner true_a \urcorner \times 1_a) \circ \langle !_d, p_2 \circ f \rangle \\
&\quad = \langle \ulcorner true_a \urcorner \circ !_d, 1_a \circ p_2 \circ f \rangle \\
&\quad = \langle p_a \circ f, p_2 \circ f \rangle \\
&\quad = \langle p_a, p_2 \rangle \circ f && \text{(Exercise 3.8.2)} \\
&\quad = 1_{\Omega^a \times a} \circ f && \text{(Exercise 3.8.3)} \\
&\quad = f
\end{aligned}
$$

Thus f factors through $\ulcorner true_a \urcorner \times 1_a$. Since the latter factors through $\in_a$, in $\mathrm{Sub}(\Omega^a \times a)$ we have $f \subseteq \in_a$. Hence (Theorem 7.5.1),

$$\chi_f \Rightarrow \chi_{\in_a} = true_{\Omega^a \times a},$$

which is the desired result.

(2) Exercise – use the diagram given with the definition of $\exists_a$ to show $\in_a \subseteq g$, where $\chi_g = \exists_a \circ p_a$. □

Now in **Set**, if we take the sequence $\langle x_1, \ldots, x_m \rangle$, form $\langle x_1, \ldots, x_m, x_t \rangle$, and then apply T_i^{m+1} we end up with $\langle x_1, \ldots, x_{i-1}, x_t, x_{i+1}, \ldots, x_m \rangle$ – the overall effect is to perform $\delta[i/t]$. Abstracting, we have

THEOREM 5. *Let* $U_t^m : a^m \to a^{m+1}$ *be the product arrow* $\langle 1_{a^m}, \rho_t^m \rangle$, *Then*

(1)

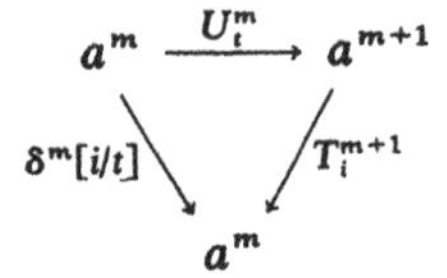

commutes, and

(2)

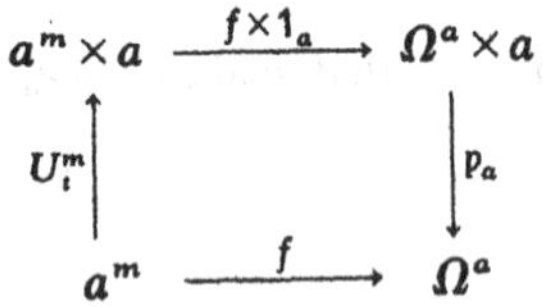

commutes for any f as shown.

PROOF. (1) Exercise – you will need to know $1_{a^m} = \langle pr_1^m, \ldots, pr_m^m \rangle$.

(2) By definition of the product arrow $f \times 1_a$,

$$p_a \circ (f \times 1_a) \circ \langle 1_{a^m}, \rho_t \rangle = f \circ pr \circ \langle 1_{a^m}, \rho_t \rangle .$$

(where $pr: a^m \times a \to a^m$ is projection)

$$= f \circ 1_{a^m}$$

$$= f \qquad \square$$

Part (1) of this theorem, with the Substitution Lemma, gives

$$[\![\varphi(v_i/t)]\!]^m = [\![\varphi]\!]^m \circ T_i^{m+1} \circ U_i^m,$$

and since

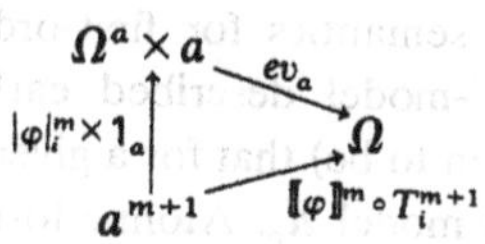

commutes, by definition of $|\varphi|_i^m$ as exponential adjoint to $[\![\varphi]\!]^m \circ T_i^{m+1}$, we get

$$[\![\varphi(v_i/t)]\!]^m = ev_a \circ (|\varphi|_i^m \times 1_a) \circ U_i^m.$$

Moreover by taking $f = |\varphi|_i^m$ in Theorem 5(2), we have

$$|\varphi|_i^m = p_a \circ (|\varphi|_i^m \times 1_a) \circ U_i^m.$$

Using these last two equations, and putting $(|\varphi|_i^m \times 1_a) \circ U_i^m = g$, we calculate

$$\begin{aligned}
[\![\forall v_i \varphi \supset \varphi(v_i/t)]\!]^m &= \Rightarrow \circ \langle [\![\forall v_i \varphi]\!]^m, [\![\varphi(v_i/t)]\!]^m \rangle \\
&= \Rightarrow \circ \langle \forall_a \circ |\varphi|_i^m, ev_a \circ g \rangle \\
&= \Rightarrow \circ \langle \forall_a \circ p_a \circ g, ev_a \circ g \rangle \\
&= \Rightarrow \circ \langle \forall_a \circ p_a, ev_a \rangle \circ g \\
&= (\forall_a \circ p_a \Rightarrow ev_a) \circ g \\
&= true_{\Omega^a \times a} \circ g && \text{(Theorem 4)} \\
&= true_{a^m} && (a^m \xrightarrow{g} \Omega^a \times a)
\end{aligned}$$

Hence the axiom UI is valid. $\square$

EXERCISE 5. Show that EG is valid by an anologous argument using the second part of Theorem 4. $\square$

The soundness of the rules ($\forall$) and ($\exists$) are left for the enthusiastic reader. The details have been worked out in Brockway [76].

11.6. Kripke Models

The algebraic and topological interpretations of intuitionistic propositional logic extend readily to first-order logic. The truth-value of a formula becomes a function $[\![\varphi]\!]^m : A^m \to \mathbf{H}$, where $\mathbf{H}$ is a suitable Heyting algebra, e.g. the lattice of open sets of some topological space. A comprehensive study of this type of model is undertaken by Rasiowa and Sikorski [63] (cf. also its application to intuitionistic analysis by Dana Scott [68].)

In his 1965 paper, Kripke gave a semantics for first-order IL that generalises the classical notion of $\mathscr{L}$-model described earlier in this chapter. The basic idea is (or can be seen to be) that for a given poset $\mathbf{P}$, a model assigns to each $p \in P$ a classical model $\mathfrak{A}_p$. Atomic formulae have their truth value at p determined by their classical truth value in $\mathfrak{A}_p$, and then the connectives can be dealt with as in the propositional case (§8.4). In fact Kripke's theory did not discuss individual constants, or the identity predicate, so in order to do so ourselves we introduce a slightly more general notion of model than that considered previously.

Let $\mathbf{P}$ be a poset. An *$\mathscr{L}$-model based on* $\mathbf{P}$ is defined to be a structure $\mathfrak{A}$ consisting of

(a) for each $p \in P$ a classical $\mathscr{L}$-model $\mathfrak{A}_p = \langle A_p, R_p, c_p \rangle$;

(b) for each arrow $p \sqsubseteq q$ in $\mathbf{P}$, a function $A_{pq} : A_p \to A_q$, such that

(i) if $p \sqsubseteq q$ then $A_{pq}(c_p) = c_q$

(ii) if $p \sqsubseteq q$ then $xR_p y$ only if $A_{pq}(x) R_q A_{pq}(y)$

(iii) A_{pp} is the identity $1 : A_p \to A_p$

(iv) if $p \sqsubseteq q \sqsubseteq r$, then

$$\begin{array}{ccc} A_p & \xrightarrow{A_{pq}} & A_q \\ & {\scriptstyle A_{pr}}\searrow \quad \swarrow{\scriptstyle A_{qr}} & \\ & A_r & \end{array}$$

commutes. Thus (i) requires that A_{pq} take the interpretation of $\mathbf{c}$ at p to its interpretation at q, while by (ii) A_{pq} "preserves" the truth of atomic formulae of the form $t\mathbf{R}u$. Notice that the collection $\{A_p : p \in P\}$ of sets together with the transition maps A_{pq} constitute a functor $A : \mathbf{P} \to \mathbf{Set}$, i.e. an object in the topos $\mathbf{Set}^{\mathbf{P}}$. This is a consequence of the definition, rather than the motivation for it. The reason why $\mathscr{L}$-models are defined as above is that this seems to be the natural way to treat $\approx$ as the relation of identity of individuals. Kripke's definition has in place of (b) the requirement that

$$p \sqsubseteq q \quad \text{implies} \quad A_p \subseteq A_q \quad \text{and} \quad R_p \subseteq R_q.$$

This amounts to putting A_{pq} as the inclusion $A_p \hookrightarrow A_q$. As pointed out by Richmond Thomason in [68], if $\approx$ is interpreted as identity, such a model would validate $(t \approx u) \vee \sim(t \approx u)$, for distinct individuals are left distinct by inclusions, and so remain "distinct forever". Thomason's solution is to interpret $\approx$ as an equivalence relation E_p on A_p, with perhaps $E_p \neq \Delta$. However by introducing the transitions A_{pq} we are able to give $\approx$ its natural interpretation and still not have the above instance of XII come out valid. For it is quite possible to have $x_t \neq x_u$ in A_p, but $A_{pq}(x_t) = A_{pq}(x_u)$. We thus give an account of the notion that things not known to be identical could come to be so known later, and also formalise some of the discussion of §10.1.

Now if φ is an $\mathscr{L}$-formula to which m is appropriate, we may define the relation

$$\mathfrak{A} \models_p \varphi[x_1, \ldots, x_m]$$

for $x_1, \ldots, x_m \in A_p$, of satisfaction of φ in $\mathfrak{A}$ at p.

In the interest of legibility we will abbreviate $A_{pq}(x)$ to x^q.

(1) If φ is atomic, $\mathfrak{A} \models_p \varphi[x_1, \ldots, x_m]$ iff $\mathfrak{A}_p \models \varphi[x_1, \ldots, x_m]$ in the classical sense.

(2) $\mathfrak{A} \models_p \varphi \wedge \psi[x_1, \ldots, x_m]$ iff $\mathfrak{A} \models_p \varphi[x_1, \ldots, x_m]$ and $\mathfrak{A} \models_p \psi[x_1, \ldots, x_m]$.

(3) $\mathfrak{A} \models_p \varphi \vee \psi[x_1, \ldots, x_m]$ iff $\mathfrak{A} \models_p \varphi[x_1, \ldots, x_m]$ or $\mathfrak{A} \models_p \psi[x_1, \ldots, x_m]$.

(4) $\mathfrak{A} \models_p \sim\varphi[x_1, \ldots, x_m]$ iff for all q with $p \sqsubseteq q$, not $\mathfrak{A} \models_q \varphi[x_1^q, \ldots, x_m^q]$.

(5) $\mathfrak{A} \models_p \varphi \supset \psi[x_1, \ldots, x_m]$ iff for all q with $p \sqsubseteq q$, if $\mathfrak{A} \models_q \varphi[x_1^q, \ldots, x_m^q]$ then $\mathfrak{A} \models_q \psi[x_1^q, \ldots, x_m^q]$.

(6) $\mathfrak{A} \models_p \exists v_i \varphi[x_1, \ldots, x_m]$ iff for some $a \in A_p$, $\mathfrak{A} \models_p \varphi[x_1, \ldots, x_{i-1}, a, x_{i+1}, \ldots, x_m]$.

(7) $\mathfrak{A} \models_p \forall v_i \varphi[x_1, \ldots, x_m]$ iff for every q with $p \sqsubseteq q$, and every $a \in A_q$, $\mathfrak{A} \models_q \varphi[x_1^q, \ldots, x_{i-1}^q, a, x_{i+1}^q, \ldots, x_m^q]$.

Thus $\exists v \varphi$ is to be true at stage p iff φ is true of some individual present at stage p, while the truth of $\forall v \varphi$ at p requires φ to be true not only of all individuals present at p but also all that occur at later stages.

If $\varphi(v_{i_1}, \ldots, v_{i_n})$ has index n, we put $\mathfrak{A} \models_p \varphi[x_1, \ldots, x_n]$ iff $\mathfrak{A} \models_p \varphi[y_1, \ldots, y_m]$ for some (hence any) appropriate m and $y_1, \ldots, y_m$ having $y_{i_1} = x_1, \ldots, y_{i_n} = x_n$.

Then we put $\mathfrak{A} \models_p \varphi$ (φ *is true at* p) iff $\mathfrak{A} \models_p \varphi[x_1, \ldots, x_n]$ for all $x_1, \ldots, x_n \in A_p$, and finally $\mathfrak{A} \models \varphi$ ($\mathfrak{A}$ *is a model of* φ) iff for all $p \in P$, $\mathfrak{A} \models_p \varphi$.

EXERCISE. 1. Show that this definition reduces to the classical notion of $\mathscr{L}$-model when P has only one member.

EXERCISE 2. Show that if $\mathfrak{A} \models_p \varphi[x_1, \ldots, x_m]$ and $p \sqsubseteq q$, then $\mathfrak{A} \models_q \varphi[x_1^q, \ldots, x_m^q]$, any φ. □

Now the **P**-model $\mathfrak{A}$ is turned into a $\mathbf{Set}^{\mathbf{P}}$ model $\mathfrak{A}^* = \langle A, r, f_c \rangle$, by taking

(i) $A : \mathbf{P} \to \mathbf{Set}$ as the functor associated with $\mathfrak{A}$ described earlier.

(ii) $r : A \times A \dot{\to} \Omega$ as the natural transformation with components $r_p : A_p \times A_p \to \Omega_p$ given by

$$r_p(\langle x, y \rangle) = \{q : p \sqsubseteq q \text{ and } A_{pq}(x) R_q A_{pq}(y)\}.$$

(iii) $f_c : 1 \dot{\to} A$ as the arrow with components $(f_c)_p : \{0\} \to A_p$ having $(f_c)_p(0) = c_p$.

EXERCISE 3. Show that $r_p(\langle x, y \rangle)$ is an hereditary subset of $[p)$.

EXERCISE 4. Show that xR_py iff $r_p(\langle x, y \rangle) = [p)$ and hence (cf. §10.3)

$$\begin{array}{ccc} R_p & \hookrightarrow & A_p^2 \\ \downarrow & & \downarrow r_p \\ 1 & \xrightarrow{\text{true}_p} & \Omega_p \end{array}$$

is a pullback.

EXERCISE 5. Verify that r and f_c *are* natural transformations. □

The exercises tell us how to reverse the construction. Given a $\mathbf{Set}^{\mathbf{P}}$ model $\langle A, r, f_c \rangle$ we specify $\mathfrak{A}_p$ by *defining* c_p by the equation (iii) in Exercise 2, and *defining* R_p by the equation in Exercise 4. This establishes a bijective correspondence between $\mathscr{L}$-models $\mathfrak{A}$ based on **P** and $\mathbf{Set}^{\mathbf{P}}$-models $\mathfrak{A}^*$ for $\mathscr{L}$.

Undoubtedly the reader has anticipated that corresponding models have the same formulae true in them. Indeed the connection is much finer than that. Let us calculate $[\![\varphi]\!]^m$, relative to $\mathfrak{A}^*$, for φ an atomic formula.

We have

$$\begin{array}{ccc} A^m & \xrightarrow{\langle \rho_t,\, \rho_u\rangle} & A^2 \\ {\scriptstyle [\![t\mathbf{R}u]\!]^m} \searrow & & \swarrow {\scriptstyle r} \\ & \Omega & \end{array}$$

where A^m is the product functor having $A^m_p=(A_p)^m$ etc., and $\rho_t : A^m \to A$ has components

$$(\rho_t)_p : A^m_p \to A_p,$$

where

$$(\rho_t)_p(\langle x_1, \ldots, x_m\rangle) = x_t.$$

From this we see that the component $[\![t\mathbf{R}u]\!]^m_p : A^m_p \to \Omega_p$ assigns to $\langle x_1, \ldots, x_m\rangle$ the set

$$\begin{aligned} r_p(\langle x_t, x_u\rangle) &= \{q : p \sqsubseteq q \text{ and } x^q_t R_q x^q_u\} \\ &= \{q : p \sqsubseteq q \text{ and } \mathfrak{A} \models_q t\mathbf{R}u[x^q_1, \ldots, x^q_m]\}. \end{aligned}$$

This situation is quite typical, as expressed in the:

TRUTH LEMMA. *For any φ, and appropriate m, then relative to $\mathfrak{A}^*$ the* $\mathbf{Set}^{\mathbf{P}}$*-arrow* $[\![\varphi]\!]^m : A^m \to \Omega$ *has p-th component*

$$[\![\varphi]\!]^m_p : A^m_p \to \Omega_p,$$

where

$$[\![\varphi]\!]^m_p(\langle x_1, \ldots, x_m\rangle) = \{q : p \sqsubseteq q \textit{ and } \mathfrak{A} \models_q \varphi[x^q_1, \ldots, x^q_m]\}.$$

Given the analysis of $\mathbf{Set}^{\mathbf{P}}$ in Chapter 10, the proof of the Truth Lemma for the inductive cases of the connectives should be evident. For identities and quantification we need to examine the arrows δ_A, $\forall_A$, and $\exists_A$, for a $\mathbf{Set}^{\mathbf{P}}$-object $A : \mathbf{P} \to \mathbf{Set}$.

THEOREM 1. *$\delta_A : A \times A \to \Omega$ has*

$$(\delta_A)_p : A_p \times A_p \to \Omega_p$$

given by

$$(\delta_A)_p(\langle x, y\rangle) = \{q : p \sqsubseteq q \text{ and } x^q = y^q\}.$$

PROOF. $\Delta_A : A \to A \times A$ has $(\Delta_A)_p$ as the map $\langle 1_{A_p}, 1_{A_p}\rangle : A_p \to A^2_p$. $(\Delta_A)_p$

then can be identified with the identity relation $\Delta_p = \{\langle x, y\rangle : x = y\} \subseteq A_p \times A_p$. The characteristic function of this set is $(\delta_A)_p$, and so (cf. §10.3)

$$(\delta_A)_p(\langle x, y\rangle) = \{q : p \sqsubseteq q \text{ and } \langle A_{pq}(x), A_{pq}(y)\rangle \in \Delta_q\}$$

as required. □

EXERCISE 6. Use Theorem 1 to prove the Truth Lemma for the case that φ has the form $(t \approx u)$. □

The definition of $\forall_A$ uses the operation of exponentiation in $\mathbf{Set}^{\mathbf{P}}$. Given functors F and G from $\mathbf{P}$ to $\mathbf{Set}$, this operation produces a functor $G^F : \mathbf{P} \to \mathbf{Set}$ consisting of a collection $\{(G^F)_p : p \in P\}$ of sets indexed by P, together with transitions $(G^F)_{pq} : (G^F)_p \to (G^F)_q$ whenever $p \sqsubseteq q$. Now for each p we define the *restriction* of F to the category $[p)$ to be the functor $F \restriction p : [p) \to \mathbf{Set}$ that assigns to each object $q \in [p)$ the set F_q, and to each arrow $q \to r$ in $[p)$ (i.e. $p \sqsubseteq q \sqsubseteq r$) the function F_{qr}. Similarly we define the functor $G \restriction p$, and then put

$$(G^F)_p = \{\sigma : F \restriction p \xrightarrow{\sigma} G \restriction p\}$$

to be the set of all natural transformations from $F \restriction p$ to $G \restriction p$. Thus an element σ of $(G^F)_p$ may be directly described as a collection $\{\sigma_q : p \sqsubseteq q\}$ of functions, indexed by the members of $[p)$, with $\sigma_q : F_q \to G_q$, such that

$$\begin{array}{ccc} F_q & \xrightarrow{\sigma_q} & G_q \\ {\scriptstyle F_{qr}}\downarrow & & \downarrow{\scriptstyle G_{qr}} \\ F_r & \xrightarrow{\sigma_r} & G_r \end{array}$$

commutes, whenever $p \sqsubseteq q \sqsubseteq r$.

Now one way of obtaining such a σ would be to take an arrow $\tau : F \to G$ in $\mathbf{Set}^{\mathbf{P}}$ and restrict it to the subcategory $[p)$, i.e. let $\sigma = \{\tau_q : p \sqsubseteq q\}$. This process also yields the transition map $(G^F)_{pq}$ when $p \sqsubseteq q$. For $\sigma \in (G^F)_p$ we put

$$(G^F)_{pq}(\sigma) = \{\sigma_r : q \sqsubseteq r\}.$$

The arrow $ev : G^F \times F \to G$ has p-th component

$$ev_p : (G^F)_p \times F_p \to G_p$$

given by

$$ev_p(\langle \sigma, x\rangle) = \sigma_p(x),$$

for each

$$\sigma \in (G^F)_p \quad \text{and} \quad x \in F_p.$$

EXERCISE 7. Verify that $(G^F)_{pq}(\sigma)$ is a natural transformation $F \restriction q \dashrightarrow G \restriction q$.

EXERCISE 8. Relate this construction to its analogue for $\mathbf{Set}^{\mathscr{C}}$ in Chapter 9. □

Now for an arrow $\tau : H \times F \dashrightarrow G$ the exponential adjoint

$$\hat{\tau} : H \to G^F$$

has as p-th component a function

$$\hat{\tau}_p : H_p \to (G^F)_p.$$

For each y in H_p,

$$\hat{\tau}_p(y) = \{\tau_q^y : p \sqsubseteq q\}$$

is a natural transformation

$$F \restriction p \dashrightarrow G \restriction p.$$

Its q-th component

$$\tau_q^y : F_q \to G_q$$

has, for each $x \in F_q$,

$$\tau_q^y(x) = \tau_q(H_{pq}(x), x).$$

The reader should now take a deep breath and go through that again. Having done so he may test his understanding of the definition in some further exercises:

EXERCISE 9. $true_A \circ pr_A : 1 \times A \to \Omega$ has as p-th component $\{0\} \times A_p \dashrightarrow \Omega_p$ the function assigning $[p)$ to each input $\langle 0, x\rangle$.

EXERCISE 10. The p-th component $\ulcorner true_A \urcorner_p : \{0\} \to (\Omega^A)_p$ of $\ulcorner true_A \urcorner : 1 \dashrightarrow \Omega^A$ may be identified with the natural transformation $\sigma : A \restriction p \to \Omega \restriction p$ that has $\sigma_q : A_q \to \Omega_q$, where $p \sqsubseteq q$, given by $\sigma_q(x) = [q)$, all $x \in A_q$. Thus $\sigma_q = true_q \circ !_{A_q}$, i.e. $\ulcorner true_A \urcorner_p(0) = \{true_q \circ !_{A_q} : p \sqsubseteq q\}$. □

THEOREM 2. *$\forall_A : \Omega^A \to \Omega$ has*

$$(\forall_A)_p : (\Omega^A)_p \to \Omega_p$$

given by

$$(\forall_A)_p(\sigma)=\{q : p\sqsubseteq q, \textit{ and for every } r \textit{ with } q\sqsubseteq r, \textit{ and every } x\in A_r,\ \sigma_r(x)=[r)\}.$$

PROOF. For $\sigma\in(\Omega^A)_p$, since $\forall_A$ is the character of $\ulcorner true_A\urcorner$ we have

$$\begin{aligned}(\forall_A)_p(\sigma)&=\{q: p\sqsubseteq q \text{ and } (\Omega^A)_{pq}(\sigma)=\ulcorner true_A\urcorner_q(0)\}\\ &=\{q: p\sqsubseteq q \text{ and } \{\sigma_r: q\sqsubseteq r\}=\{true_r\circ !_{A_r}: q\sqsubseteq r\}\}\\ &=\{q: p\sqsubseteq q, \text{ and if } q\sqsubseteq r \text{ then } \sigma_r=true_r\circ !_{A_r}\}\end{aligned}$$

from which the theorem follows. □

If, for each p, we define $\in_p\subseteq(\Omega^A)_p\times A_p$ to be the set $\in_p=\{\langle\sigma, x\rangle: \sigma_p(x)=[p)\}$ then

$$\begin{array}{ccc}\in_p & \xrightarrow{(\in_A)_p} & (\Omega^A)_p\times A_p\\ \downarrow & & \downarrow{\scriptstyle (ev_A)_p}\\ 1 & \xrightarrow{true_p} & \Omega_p\end{array}$$

is a pullback, by §10.3, and the description of ev_A given above. Thus the inclusions $(\in_A)_p$ are the components of the "membership relation" on A, i.e. the arrow $\in_A:\in\rightarrowtail\Omega^A\times A$ whose character is ev_A.

EXERCISE 11. The collection $\{\in_p: p\in P\}$ gives rise to a functor (**Set**$^{\mathbf{P}}$-object) $\in$ as just mentioned. What are its transitions $\in_{pq}$?

EXERCISE 12. Show that the component $(p_A\circ\in_A)_p$ of the composite of $\in_A$ and the first projection $p_A:\Omega^A\times A\to\Omega^A$ has $(p_A\circ\in_A)_p(\langle\sigma, x\rangle)=\sigma$.

EXERCISE 13. Let ι be the image arrow of $p_A\circ\in_A$. Show that the p-th component of ι is the inclusion

$$\iota_p\hookrightarrow(\Omega^A)_p,$$

where

$$\iota_p=\{\sigma: \text{for some } x\in A_p, \langle\sigma, x\rangle\in\in_p\}.$$ □

THEOREM 3. $\exists_A:\Omega^A\to\Omega$ *has*

$$(\exists_A)_p:(\Omega^A)_p\to\Omega_p$$

given by

$$(\exists_{\mathbf{A}})_p(\sigma)=\{q\colon p\sqsubseteq q \textit{ and for some } x\in A_q,\ \sigma_q(x)=[q)\}$$

PROOF. $\exists_{\mathbf{A}}$ is the character of the image arrow of $p_{\mathbf{A}}\circ\in_{\mathbf{A}}$. Using Exercise 13 then,

$$(\exists_{\mathbf{A}})_p(\sigma)=\{q\colon p\sqsubseteq q \text{ and } (\Omega^{\mathbf{A}})_{pq}(\sigma)\in\iota_q\}$$

$$=\{q\colon p\sqsubseteq q \text{ and for some } x\in A_q,\ \langle\sigma',x\rangle\in\in_q\}$$

$$(\text{where } \sigma'=\Omega^{\mathbf{A}}_{pq}(\sigma)=\{\sigma_r\colon q\sqsubseteq r\})$$

$$=\{q\colon p\sqsubseteq q \text{ and for some } x\in A_q,\ \sigma'_q(x)=[q)\},$$

and since $\sigma'_q=\sigma_q$, the result follows. □

The descriptions of $\forall_{\mathbf{A}}$ and $\exists_{\mathbf{A}}$ in Theorems 2 and 3 reflect the structure of the satisfaction clauses for $\forall$ and $\exists$ in Kripke models. The explicit link is given by

THEOREM 4. *For each $\mathscr{L}$-formula φ and appropriate m, the* **Set**$^{\mathbf{P}}$*-arrow*

$$|\varphi|^m_i\colon \mathbf{A}^m\to\Omega^{\mathbf{A}}$$

has as p-th component the function

$$f_p\colon A^m_p\to(\Omega^{\mathbf{A}})_p,$$

which assigns to $\langle x_1,\ldots,x_m\rangle\in A^m_p$ the natural transformation

$$f_p(\langle x_1,\ldots,x_m\rangle)=\{\sigma_q\colon p\sqsubseteq q\} \textit{ from } \mathbf{A}\restriction p \textit{ to } \Omega\restriction p,$$

with $\sigma_q\colon A_q\to\Omega_q$ having

$$\sigma_q(x)=[\![\varphi]\!]^m_q(\langle x^q_1,\ldots,x^q_{i-1},x,x^q_{i+1},\ldots,x^q_m\rangle)$$ □

EXERCISE 14. Prove Theorem 4.

EXERCISE 15. Show that $[\![\exists v_i\varphi]\!]^m_p\colon A^m_p\to\Omega_p$ assigns to $\langle x_1,\ldots,x_m\rangle\in A^m_p$ the collection

$$\{q\colon p\sqsubseteq q \text{ and for some } x\in A_q,$$
$$[\![\varphi]\!]^m_q(\langle x^q_1,\ldots,x^q_{i-1},x,x^q_{i+1},\ldots,x^q_m\rangle)=[q)\}.$$

EXERCISE 16. Derive the corresponding description of $[\![\forall v_i\varphi]\!]^m_p$ in terms of the $[\![\varphi]\!]^m_q$'s.

EXERCISE 17. Hence complete the inductive proof of the Truth Lemma. □

11.7. Completeness

Our first application of the Truth Lemma is a description of $[\![\varphi]\!]_{\mathfrak{A}} : A^n \rightarrow \Omega$, in $\mathbf{Set}^{\mathbf{P}}$, where φ has index n.

THEOREM 1. $[\![\varphi]\!]_p : A_p^n \rightarrow \Omega_p$ has

$$[\![\varphi]\!]_p(\langle x_1, \ldots, x_n\rangle) = \{q : p \sqsubseteq q \text{ and } \mathfrak{A} \models_q \varphi[x_1^q, \ldots, x_n^q]\}.$$

PROOF. Exercise – use the fact that there is a commuting triangle

$$\begin{array}{ccc} A_p^n & \longrightarrow & A_p^m \\ & {\scriptstyle [\![\Phi]\!]_p} \searrow \quad \swarrow {\scriptstyle [\![\Phi]\!]_p^m} & \\ & \Omega_p & \end{array}$$

whenever m is appropriate to φ. □

THEOREM 2. *For any $\mathscr{L}$-model $\mathfrak{A}$ based on* $\mathbf{P}$, *and associated* $\mathbf{Set}^{\mathbf{P}}$ *model* $\mathfrak{A}^*$, *we have for all $\mathscr{L}$-formulae φ,*

$$\mathfrak{A} \models \varphi \quad \textit{iff} \quad \mathfrak{A}^* \overset{\mathbf{Set}^{\mathbf{P}}}{\models} \varphi.$$

PROOF. Take any p, and $x_1, \ldots, x_n \in A_p^n$, where n is the index of φ. Then

$$p \in [\![\varphi]\!]_p(\langle x_1, \ldots, x_n\rangle) \quad \text{iff} \quad [\![\varphi]\!]_p(\langle x_1, \ldots, x_n\rangle) = [p)$$

by properties of hereditary sets (§10.2, Exercise 3(ii)). Thus by Theorem 1

$$\mathfrak{A} \models_p \varphi[x_1, \ldots, x_n] \quad \text{iff} \quad [\![\varphi]\!]_p(\langle x_1, \ldots, x_n\rangle) = (true_{A^n})_p (\langle x_1, \ldots, x_n\rangle).$$

Since this is the case for all n-length sequences from A_p, we have

$$\mathfrak{A} \models_p \varphi \quad \text{iff} \quad [\![\varphi]\!]_p = (true_{A^n})_p.$$

Since that is the case for all $p \in P$,

$$\mathfrak{A} \models \varphi \quad \text{iff} \quad [\![\varphi]\!]_{\mathfrak{A}} = true_{A^n}.$$ □

Now by the methods used by Thomason [68] (and also by Fitting [69]), we can construct a canonical poset $\mathbf{P}_{\mathscr{L}}$, and a canonical model $\mathfrak{A}_{\mathscr{L}}$ based on $\mathbf{P}_{\mathscr{L}}$ such that for any φ,

$$\mathfrak{A}_{\mathscr{L}} \models \varphi \quad \textit{iff} \quad \vdash_{\mathrm{IL}} \varphi.$$

(Thomason's models interpret $\approx$ as an equivalence relation E_p on A_p. However by taking A_p instead to be the set of E_p-equivalence classes, and A_{pq} the transition that maps the E_p-equivalence class of x to the E_q-equivalence class of x, we realise $\mathfrak{A}_{\mathscr{L}}$ as a canonical IL-model on which $\approx$ is interpreted as the diagonal relation Δ.)

Now if $\mathfrak{A}^*_{\mathscr{L}}$ is the associated model in the topos $\mathscr{E}_{\mathscr{L}} = \mathbf{Set}^{\mathbf{P}_{\mathscr{L}}}$, by Theorem 2 we have

$$\mathfrak{A}^*_{\mathscr{L}} \overset{\mathscr{E}_{\mathscr{L}}}{\models} \varphi \quad \textit{iff} \quad \vdash_{\mathrm{IL}} \varphi.$$

Hence, with the Soundness Theorem we get

$$\mathscr{E}_{\mathscr{L}} \models \varphi \quad \textit{iff} \quad \vdash_{\mathrm{IL}} \varphi.$$

From this follows a general

COMPLETENESS THEOREM. *If φ is valid in every topos, then $\vdash_{\mathrm{IL}} \varphi$.*

An example of a topos model in which the Law of Excluded Middle fails is now readily given. We take $\mathbf{P}$ as the ordinal poset $\mathbf{2} = \langle\{0, 1\}, \leqslant\rangle$. $\mathfrak{A}$ has

$$\mathfrak{A}_0 = \langle\{b, c\}, R_0, c\rangle$$

$$\mathfrak{A}_1 = \langle\{c\}, R_1, c\rangle,$$

where b and c are two distinct entities, R_0 and R_1 are any relations on $A_0 = \{b, c\}$ and $A_1 = \{c\}$ of the reader's fancy, and $A_{01} : \{b, c\} \to \{c\}$ is the only map it can be. Then if φ is the sentence $(\forall v_1)(v_1 \approx \mathbf{c})$, φ is true at $\mathfrak{A}_1$ but false at $\mathfrak{A}_0$.

Thus we have not $\mathfrak{A} \models_0 \varphi$, but we do have $\mathfrak{A} \models_1 \varphi$, so not $\mathfrak{A} \models_0 \sim\varphi$, hence not $\mathfrak{A} \models_0 \varphi \vee \sim\varphi$.

Now we saw in §7.4 that, for propositional logic, a topos can validate all instances of $\alpha \vee \sim\alpha$ (since Sub(1) is a **BA**) but still not be Boolean (since Sub(Ω) is not a **BA**). This occurs for example in the topos $\mathbf{M}_2$. Similarly we have $\mathbf{M}_2 \models \varphi \vee \sim\varphi$ whenever φ is an $\mathscr{L}$-sentence, since then $[\![\varphi]\!]_{\mathfrak{A}}$ is a truth-value $1 \to \Omega$. However the situation is not the same for open formulae.

THEOREM 3. *If $\mathscr{E} \models \varphi \vee \sim\varphi$ for every $\mathscr{L}$-formula φ, then $\mathscr{E}$ is a Boolean topos.*

PROOF. Let $\mathfrak{A}$ be a $\mathscr{L}$-model of the form $\langle \Omega, r, true \rangle$, i.e. a model in which $\mathbf{c}$ is interpreted as the element $true: 1 \to \Omega$ of Ω. Let $\varphi(v_1)$ be the formula $(v_1 \approx \mathbf{c})$. Then $[\![\varphi]\!]_{\mathfrak{A}}: \Omega \to \Omega$ is $\delta_\Omega \circ \langle 1_\Omega, true_\Omega \rangle$. By Exercise 2 of §5.1, the equaliser of 1_Ω and $true_\Omega$ is $true$ $1 \to \Omega$ so (*Theorem* 2, §11.5) $[\![\varphi]\!]_{\mathfrak{A}} = \chi_{true} = 1_\Omega$. But $\mathfrak{A} \models \varphi \vee \sim\varphi$, so $[\![\varphi \vee \sim\varphi]\!]_{\mathfrak{A}} = true_\Omega$, i.e.

$$[\![\varphi]\!]_{\mathfrak{A}} \cup (\neg \circ [\![\varphi]\!]_{\mathfrak{A}}) = true_\Omega$$

i.e.

$$1_\Omega \cup (\neg \circ 1_\Omega) = true_\Omega$$

which by Theorem 3 of §7.4 implies that Sub(Ω) is a **BA**. □

EXERCISE. The proof of Theorem 3 used the fact that $\mathscr{L}$ had an individual constant. Show that this assumption is not needed, by considering the process of "adjoining" a constant to a language. □

11.8. Existence and free logic

The assumption of non-emptiness, ($\mathscr{E}(1, a) \neq \emptyset$), for $\mathscr{L}$-models in a topos has been needed, not just for interpreting constants, but also for our definition of $[\![\varphi]\!]_{\mathfrak{A}}$ and hence of truth in a model. In **Set** of course the only empty object is the null set $\emptyset$, and if *that* is admitted as a model, then as Andrzej Mostowski [51] observed, the rule of DETACHMENT no longer preserves validity. Informally we regard any universal sentence $\forall v\varphi$, or any open formula $\varphi(v)$, as being true of $\emptyset$, since there is nothing in $\emptyset$ of which φ is false. On the other hand an existential statement $\exists v\varphi$ is false in $\emptyset$ since the latter has no element of which φ is true. More formally, since $2^0 = \{0\}$, $\forall_\emptyset: \{\emptyset\} \to 2$ is simply the map *true*, while $\exists_\emptyset: \{\emptyset\} \to 2$ is the map *false*. Moreover if φ has index $n \geqslant 1$, then $\emptyset^n = \emptyset$, so $[\![\varphi]\!]_{\mathfrak{A}}: \emptyset \to 2$ is the empty map, i.e. the map $true_\emptyset$. Thus, e.g., the open formulae

$$(v_1 \approx v_1) \supset \exists v_1(v_1 \approx v_1)$$

and

$$(v_1 \approx v_1)$$

and true in $\emptyset$, while the sentence

$$\exists v_1(v_1 \approx v_1)$$

is false.

There are two basic methods that have been developed of doing logic when empty models are allowed (so called "free" logic). Mostowski modified the rule of DETACHMENT to read:

> *From φ and $\varphi \supset \psi$ infer ψ, provided that all variables free in φ are free in ψ.*

(Alternatively we allow ψ to be detached only if $\exists v(v \approx v)$ has also been derived for each variable v that is free in φ.)

This approach is used in the topos setting by the Montréal school (cf. Robitaille-Giguère [75], Boileau [75]). The other method is to introduce a special *existence* predicate **E**, with **E**(t) read "t exists", and to modify the definition of satisfaction to accommodate the possibility that "t may not denote anything". This notion has been studied by Dana Scott and Michael Fourman [74], and has a very interesting interpretation for sheaves and bundles, as well as Kripke models.

Let us consider an object $a = (A, f)$ in the topos $\mathbf{Bn}(I)$ of bundles over I. An element $s: 1 \to a$ of a is a *global* section $s: I \to A$ of the bundle, picking one "germ" $s(i)$ out of each stalk A_i. But if the stalk is empty, $A_i = \emptyset$, then no such $s(i)$ exists. So we see that if a has at least one empty stalk (because f is not epic), that is enough to prevent there being any elements $1 \to a$. (We also see that $\mathbf{Bn}(I)$ has many significant and non-isomorphic objects that are empty in the categorial sense). At best we can consider *local* sections $s: D \to A$, with $f \circ s = D \hookrightarrow I$, defined on some subset D of I. This possible if $A_i \neq \emptyset$ for all $i \in D$. Recall (§4.4 Example 6) that the set $D \subseteq I$ can be regarded as a subobject of the terminal object 1 under the isomorphism

$$\mathcal{P}(I) \cong \mathbf{Bn}(I)(1, \Omega) \cong \mathrm{Sub}(1)$$

that obtains for $\mathbf{Bn}(I)$.

A similar situation arises in the context of a $\mathbf{Set}^{\mathbf{P}}$ model $\langle A, r, f_c \rangle$. If the object (functor) A has element $f_c: 1 \to A$, then for each p, $(f_c)_p(0) \in A_p$, so $A_p \neq \emptyset$. So if just one A_p were empty, A would have no elements. However even if A does have elements, it may be undesirable to interpret a constant as an arrow of the form $1 \to A$. We may for instance wish to expand our language $\mathcal{L}$ to include a "name" $\mathbf{c}_0$ for a particular element c_0 of some A_p. $\mathbf{c}_0$ would then be interpreted (as c_0^q) only in those $\mathfrak{A}_q$ for $q \in [p)$. Notice that $[p)$ being hereditary can be identified (Exercise 2, §10.6) with a subobject $D \rightarrowtail 1$ of the terminal object in $\mathbf{Set}^{\mathbf{P}}$. The interpretation of $\mathbf{c}_0$ then yields an arrow $f_{c_0}: D \to A$ with $(f_{c_0})_q: D_q \to A_q$ picking out c_0^q whenever $p \sqsubseteq q$, i.e. $D_q = \{0\}$, and $(f_{c_0})_q = \,!: \emptyset \to A_q$ otherwise.

We are thus lead to replace elements $1 \to a$ of an object a by arrows $d \to a$ whose domains are subobjects $d \rightarrowtail 1$ of the terminal 1. Such things are called *partial elements* of a. This comes from the more general notion of *partial arrow*. In **Set** we say that f is a *partial function* from A to B, written $f: A \rightsquigarrow B$, if f is a function from a subset of A to B, i.e. $\operatorname{dom} f \subseteq A$ and $\operatorname{cod} f = B$. In a general category $\mathscr{C}$ we put $f: a \rightsquigarrow b$ if f is a $\mathscr{C}$-arrow with $\operatorname{cod} f = b$, and there is a $\mathscr{C}$-monic $\operatorname{dom} f \rightarrowtail a$. Thus a partial element of a is an arrow $s: 1 \rightsquigarrow a$.

Now in the **Set** case, if $f: A \rightsquigarrow B$ there may be some elements $x \in A$ with $x \notin \operatorname{dom} f$. This is often expressed as "$f(x)$ is undefined". But if we introduce some new entity $*$, with $* \notin B$, and write "$f(x) = *$" whenever $x \notin \operatorname{dom} f$ then we can regard f as being defined on all of A (we need $* \notin B$, or else "$f(x) = *$" could be compatible with $x \in \operatorname{dom} f$). A convenient choice for $*$ would be the null set $\emptyset$ ($f(x) = \emptyset$ means "x has null denotation"). However it may be that $\emptyset \in B$. We can get around this by replacing each element y of B by the singleton subset $\{y\}$ and replacing B by the collection of these singletons, i.e. we replace B by its isomorphic copy $B' = \{\{y\}: y \in B\}$. Then $\emptyset \notin B'$ so we add $\emptyset$ to B' to form

$$\tilde{B} = \{\{y\}: y \in B\} \cup \{\emptyset\}.$$

Then given $f: D \to B$, with $D \subseteq A$, define $\tilde{f}: A \to \tilde{B}$ by

$$\tilde{f}(x) = \begin{cases} \{f(x)\} & \text{if} \quad x \in \operatorname{dom} f = D \\ \emptyset & \text{otherwise} \end{cases}$$

It is clear then that

$$\begin{array}{ccc} D & \hookrightarrow & A \\ {\scriptstyle f}\downarrow & & \downarrow{\scriptstyle \tilde{f}} \\ B & \xrightarrow{\eta_B} & \tilde{B} \end{array}$$

commutes, where $\eta_B(y) = \{y\}$, all $y \in B$.

Moreover the pullback of η_B and $\tilde{f}$ has domain

$$\begin{aligned} \{\langle y, x\rangle: \{y\} = \tilde{f}(x)\} &= \{\langle y, x\rangle: x \in D \text{ and } y = f(x)\} \\ &= \{\langle f(x), x\rangle: x \in D\} \cong D. \end{aligned}$$

Thus, knowing $\tilde{f}$, we pull it back along η_B to recover f. In fact (exercise) it can be shown that $\tilde{f}$ as defined is the only map $A \to \tilde{B}$ making this diagram a pullback. Thus the arrow $\eta_B: B \to \tilde{B}$ is a generalisation of *true*: $1 \to 2$. It acts as a "partial function classifier", providing a bijective correspondence between (equivalence classes of) partial maps $f:A \rightsquigarrow B$ with codomain B, and "total" maps $\tilde{f}:A \to \tilde{B}$ with codomain $\tilde{B}$.

PARTIAL ARROW CLASSIFIER THEOREM. *If $\mathscr{E}$ is any topos, then for each $\mathscr{E}$-object b there is an $\mathscr{E}$-object $\tilde{b}$ and an arrow $\eta_b : b \rightarrowtail \tilde{b}$ such that given any pair (f, g) of arrows as in the following diagram, there is one and only one arrow $\tilde{f}$ as shown that makes the diagram a pullback.*

$$\begin{array}{ccc} d & \overset{g}{\rightarrowtail} & a \\ {\scriptstyle f}\big\downarrow & & \big\downarrow{\scriptstyle \tilde{f}} \\ b & \underset{}{\overset{\eta_b}{\rightarrowtail}} & \tilde{b} \end{array}$$

□

The proof of this theorem is given in detail by Kock and Wraith [71]. To define η_b, the arrow $\{\cdot\}_b : b \to \Omega^b$ is introduced as the exponential adjoint to $\delta_b : b \times b \to \Omega$ (in **Set** $\{\cdot\}_b$ maps y to $\{y\}$). $\{\cdot\}_b$ proves to be monic, and so is $\langle\{\cdot\}_b, 1_b\rangle : b \to \Omega^b \times b$. The latter has a character $h : \Omega^b \times b \to \Omega$ and this in turn has an exponential adjoint $\hat{h} : \Omega^b \to \Omega^b$ (in **Set** $\hat{h}$ is the identity on singletons and maps all other subsets of b to $\emptyset$). It is then shown that $\hat{h} \circ \{\cdot\}_b = \{\cdot\}_b$, so

$$\begin{array}{ccccc} \tilde{b} & \rightarrowtail & \Omega^b & \overset{1}{\underset{\hat{h}}{\rightrightarrows}} & \Omega^b \\ {\scriptstyle \eta_b}\big\uparrow & \nearrow {\scriptstyle \{\cdot\}_b} & & & \\ b & & & & \end{array}$$

defining $\tilde{b}$ as the (domain of the) equaliser of 1_{Ω^b} and $\hat{h}$, η_b is the unique arrow factoring $\{\cdot\}_b$ through $\tilde{b}$.

EXERCISE 1. Examine the details of this construction in **Set**.

EXERCISE 2. Show that

$$\eta_1 : 1 \to \tilde{1}$$

is a subobject classifier in any topos. □

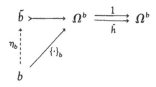

Returning now to free logic, a semantical theory in the classical case may be developed by allowing variables and constants to be interpreted in a model $\mathfrak{A}=\langle A,\ldots\rangle$ as elements of $A\cup\{*\}$. The existence predicate $\mathbf{E}$ is interpreted as the set (one-place relation) A, i.e. for $a\in A\cup\{*\}$

$$\mathfrak{A}\models\mathbf{E}(v)[a] \quad \text{iff} \quad a\in A,$$

while the range of quantification remains A itself, i.e.

$$\mathfrak{A}\models\forall v\varphi \quad \text{iff for all} \quad a\in A, \quad \mathfrak{A}\models\varphi[a].$$

Under this semantics, DETACHMENT preserves validity, while the axioms UI and EG are modified to

$$(\forall v)\varphi\wedge\mathbf{E}(t)\supset\varphi(v/t)$$

and

$$\varphi(v/t)\wedge\mathbf{E}(t)\supset(\exists v)\varphi$$

More details of this type of theory may be found in Scott [67] – where, as is often done, $\mathbf{E}(t)$ is taken to stand for a formula of the form $\exists v(v\approx t)$.

Moving to models $\mathfrak{A}=\langle a,\ldots\rangle$ in a general topos, we see that instead of dealing with partial elements $1\rightsquigarrow a$ as suggested by the examples discussed earlier, we may deal with elements $1\to\tilde{a}$ of the "object of partial elements of a" ($\tilde{a}$ always has elements, since a has at least the partial element $0\rightarrowtail a$). The interpretation of the predicate $\mathbf{E}$ becomes the character $e:\tilde{a}\to\Omega$ of the monic $\eta_a:a\rightarrowtail\tilde{a}$, and each formula φ determines an arrow $[\![\varphi]\!]_{\mathfrak{A}}:(\tilde{a})^n\to\Omega$. Then given a partial element $f_c:1\rightsquigarrow a$,

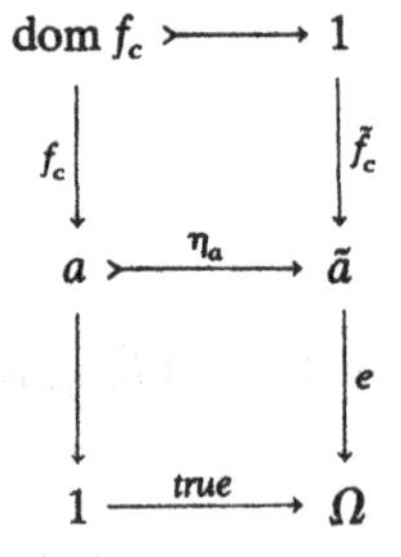

we have $[\![\mathbf{E}(\mathbf{c})]\!]=e\circ\tilde{f}_c$, and so as the diagram indicates,

$$[\![\mathbf{E}(\mathbf{c})]\!] \text{ is the character of } \operatorname{dom} f_c\rightarrowtail 1.$$

Hence

$$\begin{aligned}\mathfrak{A}\models\mathbf{E}(\mathbf{c}) \quad &\text{iff} \quad [\![\mathbf{E}(\mathbf{c})]\!]_{\mathfrak{A}}=\mathit{true}\\ &\text{iff} \quad \operatorname{dom} f_c\rightarrowtail 1\simeq 1_1 \text{ in } \mathrm{Sub}(1)\\ &\text{iff} \quad f_c \text{ is a "total" element of } a.\end{aligned}$$

In the case of a bundle $a = \langle A, f\rangle$, $\tilde{a}$ is a bundle of (disjoint) copies of the sets $\tilde{A}_i$, with η_a acting on the stalk A_i being the map $\eta_{A_i} : A_i \to \tilde{A}_i$. An element $\tilde{f}_c : 1 \to \tilde{a}$ is essentially a partial element $f_c : 1 \rightsquigarrow a$, i.e. a local section $f_c : I \rightsquigarrow A$, with

$$\text{dom } f_c = \{i : \tilde{f}_c(i) \neq \emptyset \text{ in } \tilde{A}_i\}$$

Identifying truth values with subsets of I we may then simply say that

$$[\![\mathbf{E}(\mathbf{c})]\!]_{\mathfrak{A}} = \text{dom } f_c,$$

and

$$\mathfrak{A} \models \mathbf{E}(\mathbf{c}) \quad \text{iff} \quad f_c \textit{ is a global section.}$$

Now the set $\tilde{A}$ is isomorphic in **Set** to $A+1$, the latter being the disjoint union of A and $\{0\}$. The iso arrow in question is the co-product arrow $[\eta_A, \emptyset_A]$, where $\emptyset_A : \{0\} \to \tilde{A}$ has $\emptyset_A(0) = \emptyset$. Thus $\emptyset_A$ "is" the element of $\tilde{A}$ corresponding to the partial element $! : \emptyset \to A$ of A. The obvious question then arises as to whether $\tilde{a}$ is isomorphic to $a+1$ in general. If this were so, we would have in particular $\tilde{1} \cong 1+1$. But (Exercise 2 above) $\tilde{1}$ is an object of truth values, and we know that $\Omega \cong 1+1$ only in *Boolean* topoi.

To formulate the situation precisely, let $\emptyset_a : 1 \to \tilde{a}$, where a is an object of topos $\mathscr{E}$, be the unique $\mathscr{E}$-arrow making

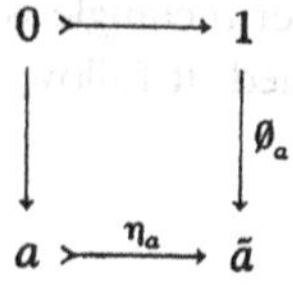

a pullback, and form the co-product arrow

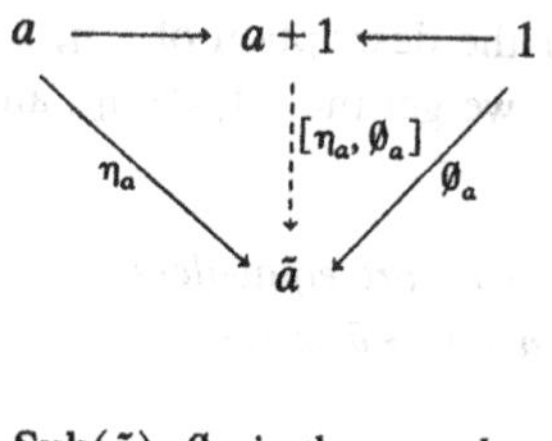

LEMMA. *In* Sub($\tilde{a}$), $\emptyset_a$ *is the pseudo-complement of* η_a.

Proof. If $-\eta_a : -a \rightarrowtail \tilde{a}$ is the pseudo-complement of η_a, then $\eta_a \cap -\eta_a \simeq 0_{\tilde{a}}$ (§7.2) and so

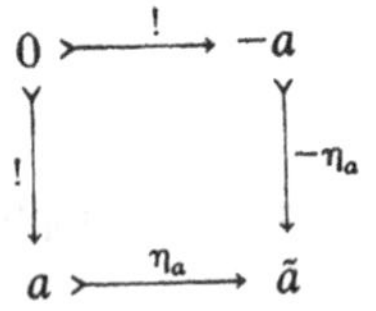

is a pullback. But the Partial Arrow Classifier Theorem then implies that $-\eta_a$ is the only arrow that makes the diagram thus a pullback.

Now consider

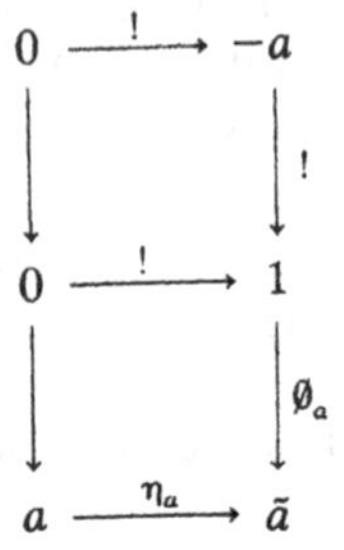

The top square is a pullback (exercise), and the bottom square is the pullback defining $\emptyset_a$. Hence by the PBL the outer rectangle is a pullback. In view of the unique role of $-\eta_a$ just mentioned, it follows that

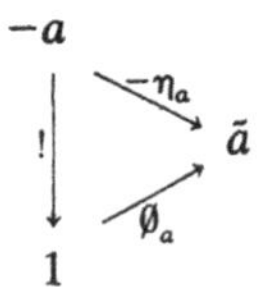

commutes, showing that $-\eta_a \subseteq \emptyset_a$. But the pullback square defining $\emptyset_a$ shows that $\eta_a \cap \emptyset_a \simeq 0_{\tilde{a}}$. In view of the description of $-\eta_a$ as the largest element of $\mathrm{Sub}(\tilde{a})$ disjoint from η_a, we get then $\emptyset_a \subseteq -\eta_a$, and altogether $\emptyset_a \simeq -\eta_a$. □

Theorem. *In any topos $\mathscr{E}$, the following are equivalent*

(1) *For all $\mathscr{E}$-objects a, $[\eta_a, \emptyset_a] : a + 1 \to \tilde{a}$ is iso*

(2) $[\eta_1, \emptyset_1] : 1 + 1 \to \tilde{1}$ *is iso*

(3) $\mathscr{E}$ *is Boolean.*

PROOF. Clearly (1) implies (2). But $\emptyset_1$ is defined by the pullback

$$\begin{array}{ccc} 0 & \xrightarrow{!} & 1 \\ \big\downarrow & & \big\downarrow \phi_1 \\ 1 & \xrightarrow{\eta_1} & \tilde{1} \end{array}$$

which shows that when η_1 is used as subobject classifier, i.e. $\eta_1 = true$, then $\emptyset_1$ is the arrow *false*. Hence (2) asserts that the co-product [*true*, *false*] is iso, which yields Booleanness as we saw in §7.3.

Finally, if (3) holds, then applying the Lemma to any $\mathscr{E}$-object a, we have

$$\eta_a \cup \emptyset_a \simeq \eta_a \cup -\eta_a \simeq 1_{\tilde{a}}.$$

But η_a and $\emptyset_a$ are disjoint monics, so the Lemma following Theorem 3 of §5.4 implies that $[\eta_a, \emptyset_a]$ is monic, and hence is its own epi-monic factorisation, i.e. $\eta_a \cup \emptyset_a \simeq [\eta_a, \emptyset_a]$ in Sub($\tilde{a}$). Thus $1_{\tilde{a}} \simeq [\eta_a, \emptyset_a]$, and so the latter is iso (Exercise, 7.2.1). □

EXERCISE 3. Let $a = f : A \to B$ be an object in the topos $\mathbf{Set}^{\to}$ of set functions. Form the co-product function

$$A + B \xrightarrow{[f, id_B]} B,$$

and let $[f, id_B]^{\sim} : (A+B)^{\sim} \to \tilde{B}$ be defined by the ~-construction in **Set**. Then

$$\begin{array}{ccc} A & \xrightarrow{g} & (A+B)^{\sim} \\ f \big\downarrow & & \big\downarrow [f, id_B]^{\sim} \\ B & \xrightarrow{\eta_B} & \tilde{B} \end{array}$$

commutes, where g is the composite of $i_A : A \to A + B$ and η_{A+B}.

Show that $\eta_a : a \to \tilde{a}$ is a partial arrow classifier with respect to a in $\mathbf{Set}^{\to}$, where $\tilde{a}$ is the function $[f, id_B]^{\sim}$ and η_a is the pair (g, η_B).

Apply the construction just given to the terminal 1 in $\mathbf{Set}^{\to}$ to recover the description of the subobject classifier for $\mathbf{Set}^{\to}$ given in Chapter 4.

11.9. Heyting-valued sets

Building on the ideas of the previous section, we might regard an object in a topos as a "set-like" entity consisting of *potentially* existing (partially defined) elements, only some of which possess *actual* existence (are totally defined). The variables in a formula that are bound by quantifiers are then taken to range over actually existing elements. In the context of this "logic of partial elements" we distinguish two concepts of sameness. The sentence $\exists v(v \approx \mathbf{c})$ is tantamount to the assertion that the individual c exists, in that it asserts that there actually exists an individual that is equal to c. So the sentence

(i) $\mathbf{E}(\mathbf{c}) \equiv \exists v(v \approx \mathbf{c})$

is valid on this account. Here the symbol $\equiv$ is the *biconditional* connective read "if and only if". The expression $\varphi \equiv \psi$ is formally introduced as an abbreviation for the formula

$$(\varphi \supset \psi) \wedge (\psi \supset \varphi).$$

In arriving at (i) we have implicitly invoked the principle that anything equal to an existing entity must itself exist. But more strongly than this we are going to require that elements can only be equal if they exist. Equality implies existence, and we thus have

(ii) $(v \approx w) \supset \mathbf{E}(v) \wedge \mathbf{E}(w)$

The other notion of sameness, for which we use the symbol $\approxeq$, is a weaker concept of *equivalence* which does not differentiate elements in regard to their lack of existence. Thus v and w will be equivalent if neither of them exists, or if they both exist and are equal ($\approx$). We can express this in a positive form as "if either of them exists then they are equal" (and hence the other exists by (ii)). Thus equivalence is characterised by

(iii) $(v \approxeq w) \equiv (\mathbf{E}(v) \vee \mathbf{E}(w) \supset v \approx w).$

But then we see, conversely, that we may describe equality in terms of equivalence, since equal elements are those that exist and are equivalent, i.e.

(iiia) $(v \approx w) \equiv ((v \approxeq w) \wedge \mathbf{E}(v) \wedge \mathbf{E}(w)).$

These notions are simply illustrated in the topos $\mathbf{Bn}(I)$. Let f and g be two partial elements $I \rightsquigarrow A$ of a bundle $A \to I$ over I, and put

$$[\![f \approx g]\!] = \{i \in I : f(i) = g(i)\}$$

Then $[\![f \approx g]\!]$, being a subset of I, is a truth-value in $\mathbf{Bn}(I)$. We regard it as the truth-value of the statement "$f = g$", or alternatively as a measure of the extent to which f and g are equal. The expression "$f = g$" is interpreted to mean that $f(i)$ and $g(i)$ are both defined (i.e. i is a member of the domains of both f and g) and they are the same element of A. In particular we must have

$$[\![f \approx g]\!] \subseteq \operatorname{dom} f \cap \operatorname{dom} g$$

and so by the analysis of 11.8 we can put

$$[\![f \approx g]\!] \subseteq [\![\mathbf{E}(f)]\!] \cap [\![\mathbf{E}(g)]\!]$$

which accords with (ii) above.

Notice that

$$[\![f \approx f]\!] = \{i: f(i) = f(i)\} = \operatorname{dom} f = [\![\mathbf{E}(f)]\!]$$

and so $[\![f \approx f]\!]$ is a measure of the degree of existence of f.

For the weaker concept of sameness, we regard the local sections f and g as equivalent if they agree whenever they are defined. Thus as a measure of the extent of their equivalence we take those i where neither is defined, together with those where they are both defined and agree. Thus

$$\begin{aligned}[\![f \approxeq g]\!] &= -(\operatorname{dom} f \cup \operatorname{dom} g) \cup [\![f \approx g]\!] \\ &= -([\![\mathbf{E}(f)]\!] \cup [\![\mathbf{E}(g)]\!]) \cup [\![f \approx g]\!]\end{aligned}$$

which corresponds to (iii), since $-B \cup C = B \Rightarrow C$ in $\mathcal{P}(I)$.

Analogously, in $\mathbf{Top}(I)$ we define a measure of the degree of equality of partial elements (continuous local sections) of a sheaf of germs by putting

$$[\![f \approx g]\!] = \{i: f(i) = g(i)\}^{0},$$

applying the interior operator $(\ )^{0}$ to ensure that $[\![f \approx g]\!]$ is an open set, i.e. a truth-value. $[\![\mathbf{E}(f)]\!] = [\![f \approx f]\!]$ remains as $\operatorname{dom} f$, since local sections always have open domains. For equivalence we put

$$[\![f \approxeq g]\!] = [\![\mathbf{E}(f)]\!] \cup [\![\mathbf{E}(g)]\!] \Rightarrow [\![f \approx g]\!],$$

where $B \Rightarrow C = (-B \cup C)^{0}$ is the relative pseudo-complementation of open sets in I. Notice that whereas $[\![f \approx f]\!]$ may be a proper subset of I ("$f = f$" is not totally true) we always have $[\![f \approxeq f]\!] = I$.

Emerging from this discussion is a generalised concept of a "set" as consisting of a collection of (partial) elements, with some Heyting-algebra-valued measure of the degree of equality of these elements. This notion admits of an abstract axiomatic development in the following way:

Let $(\Omega, \sqsubseteq)$ be a *complete* Heyting algebra (**CHA**), i.e. an **HA** in which *every* subset $A \subseteq \Omega$ has a least upper bound, denoted $\bigsqcup A$, and a greatest lower bound, denoted $\bigsqcap A$, in Ω. (Recall the definitions of l.u.b. and g.l.b. given in §8.3). An *Ω-valued set* (*Ω-set*) is defined to be an entity $\mathbf{A}$ comprising a set A and a function $A \times A \to \Omega$ assigning to each ordered pair $\langle x, y \rangle$ of elements of A an element $[\![x \approx y]\!]_{\mathbf{A}}$ of Ω, satisfying

$$[\![x \approx y]\!]_{\mathbf{A}} \sqsubseteq [\![y \approx x]\!]_{\mathbf{A}}$$

and

$$[\![x \approx y]\!]_{\mathbf{A}} \sqcap [\![y \approx z]\!]_{\mathbf{A}} \sqsubseteq [\![x \approx z]\!]_{\mathbf{A}}$$

for all $x, y, z \in A$. These two conditions give the Ω-validity of the formulae

$$(x \approx y) \supset (y \approx x)$$

$$(x \approx y) \wedge (y \approx z) \supset (x \approx z)$$

that express the symmetry and transitivity of the equality relation. The element $[\![x \approx x]\!]_{\mathbf{A}}$ will often be denoted $[\![Ex]\!]_{\mathbf{A}}$. We introduce the definition

$$[\![x \approxeq y]\!]_{\mathbf{A}} = ([\![Ex]\!]_{\mathbf{A}} \sqcup [\![Ey]\!]_{\mathbf{A}}) \Rightarrow [\![x \approx y]\!]_{\mathbf{A}}$$

The $\mathbf{A}$-subscripts in these expressions will be deleted whenever the meaning is clear without them.

EXERCISE 1. Prove that the following conditions hold for any Ω-valued set:

$[\![x \approx y]\!] \sqsubseteq [\![Ex]\!]$

$[\![x \approx y]\!] = [\![x \approxeq y]\!] \sqcap [\![Ex]\!] \sqcap [\![Ey]\!]$

$[\![Ex]\!] \sqcap [\![x \approxeq y]\!] \sqsubseteq [\![Ey]\!]$

$[\![x \approxeq x]\!]$ is the unit (greatest element) of Ω

$[\![x \approxeq y]\!] \sqsubseteq [\![y \approxeq x]\!]$

$[\![x \approxeq y]\!] \sqcap [\![y \approxeq z]\!] \sqsubseteq [\![x \approxeq z]\!]$

$p \sqsubseteq [\![x \approxeq y]\!]$ iff $p \sqcap [\![Ex]\!] \sqsubseteq [\![x \approxeq y]\!]$ and $p \sqcap [\![Ey]\!] \sqsubseteq [\![x \approxeq y]\!]$. □

The justification for using the subobject-classifier symbol for our **CHA** is that the Ω-sets form the objects of a category, denoted Ω-**Set**, which is a topos, and in which the object of truth-values is Ω itself! More precisely, this object of truth-values is the Ω-set $\mathbf{\Omega}$ obtained by putting

$$[\![p \approx q]\!]_{\mathbf{\Omega}} = (p \Leftrightarrow q)$$

for each $p, q \in \Omega$, where

$$(p \Leftrightarrow q) = (p \Rightarrow q) \sqcap (q \Rightarrow p)$$

is the Ω-operation that interprets the biconditional connective $\equiv$. Since the members of Ω are going to serve as truth-values we will use the symbols $\bot$ and $\top$ to denote the least (zero) and greatest (unit) elements of Ω respectively.

EXERCISE 2. $[\![p \approx q]\!]_{\mathbf{\Omega}} = \top$ iff $p = q$.

EXERCISE 3. $[\![Ep]\!]_{\mathbf{\Omega}} = \top$.

EXERCISE 4. $[\![p \approx \top]\!]_{\mathbf{\Omega}} = p$.

EXERCISE 5. $[\![p \approx \bot]\!]_{\mathbf{\Omega}} = \neg p$ □

An arrow from **A** to **B** in Ω-**Set** may be thought of in the first instance as a function $f: A \to B$. Its graph would then be a subobject of $A \times B$ and so should correspond to a function of the form $A \times B \to \Omega$. We interpret the latter as assigning to $\langle x, y \rangle$ the truth-value $[\![f(x) \approx y]\!]$, giving the degree of equality of $f(x)$ and y, i.e. a measure of the extent to which y is the f-image of x. With this idea in mind we turn to the formal definition.

An arrow from **A** to **B** in Ω-**Set** is a function $f: A \times B \to \Omega$ satisfying

(iv) $[\![x \approx x']\!]_{\mathbf{A}} \sqcap f(\langle x, y \rangle) \sqsubseteq f(\langle x', y \rangle)$

(v) $f(\langle x, y \rangle) \sqcap [\![y \approx y']\!]_{\mathbf{B}} \sqsubseteq f(\langle x, y' \rangle)$

(vi) $f(\langle x, y \rangle) \sqcap f(\langle x, y' \rangle) \sqsubseteq [\![y \approx y']\!]_{\mathbf{B}}$

(vii) $[\![x \approx x]\!]_{\mathbf{A}} = \bigsqcup \{f(\langle x, y \rangle) : y \in B\}$

The first two conditions are laws of extensionality (indistinguishability of equals) and assert the Ω-validity of the formulae

$$(x \approx x') \wedge (f(x) \approx y) \supset (f(x') \approx y)$$

$$(f(x) \approx y) \wedge (y \approx y') \supset (f(x) \approx y')$$

(which are instances of the axiom $I2$ of §11.3). Condition (vi) gives the validity of the "unique output" property for the arrow f. It can be read "partial elements y and y′ are each the f-image of x only to the extent that they are equal". To understand condition (vii) we note that the completeness of the **HA** Ω can be used to interpret an existential quantifier, by construing the latter as a (possibly infinite) disjunction (l.u.b.). That is, the sentence "there exists a $y \in B$ such that $\varphi(y)$" is construed as "$\varphi(y_1)$ or $\varphi(y_2)$ or $\varphi(y_3)$ or ..." where $y_1, y_2, \ldots$ run through all the members of B, and hence is given the truth-value

$$\bigsqcup\{[\![\varphi(y)]\!]: y \in B\}, \quad \text{or} \quad \bigsqcup_{y \in \mathbf{B}} [\![\varphi(y)]\!].$$

(Dually, construing a universal quantifier as a conjunction, the sentence "for all $y \in B$, $\varphi(y)$" would be interpreted by

$$\bigsqcap\{[\![\varphi(y)]\!]: y \in B\}, \quad \text{or} \quad \bigsqcap_{y \in \mathbf{B}} [\![\varphi(y)]\!].)$$

Thus we see that (vii) gives the validity of the statement that each $x \in A$ has some f-image $y \in B$, i.e. f is a total function. By giving an equation of the form $[\![Ex]\!] = [\![\varphi]\!]$ the suggestive reading "x exists to the extent that φ", we may read (vii) as "each element of **A** exists to the extent that it has an image in **B**".

In summary then, an arrow from **A** to **B** is represented, via its graph, as an extensional, functional and total Ω-valued relation from **A** to **B**. But then it is not hard to see that the equality relation on **A** satisfies these properties, i.e. the function $\langle x, y\rangle \mapsto [\![x \approx y]\!]_{\mathbf{A}}$ is an arrow $\mathbf{A} \to \mathbf{A}$ according to (iv)–(vii). And indeed it will be the identity arrow for **A**, with the truth-value of "$\text{id}(x) = y$" thus being precisely that of "$x = y$", as it should be.

The composite of arrows $f: \mathbf{A} \to \mathbf{B}$ and $g: \mathbf{B} \to \mathbf{C}$ is the function $g \circ f: A \times C \to \Omega$ given by

$$g \circ f(\langle x, z\rangle) = \bigsqcup_{y \in \mathbf{B}} (f(\langle x, y\rangle) \sqcap g(\langle y, z\rangle))$$

(compare this to the statement "for some $y \in B$, $f(x) = y$ and $g(y) = z$").

These definitions complete the description of **Ω-Set** as a category. In order to describe its topos structure we will from now on use the notations $f(\langle x, y\rangle)$ and $[\![f(x) \approx y]\!]$ interchangeably in reference to an arrow $f: \mathbf{A} \to \mathbf{B}$.

Terminal Object: This is the Ω-set **1** comprising the ordinary set $\{0\}$ with

$[\![0 \approx 0]\!] = \top$. The unique arrow $f : \mathbf{A} \to \mathbf{1}$ is given by

$$[\![f(x) \approx 0]\!] = [\![Ex]\!]$$

i.e. "$f(x)$ equals 0 to the extent that x exists".

Products: $\mathbf{A} \times \mathbf{B}$ is the product set $A \times B$ with the Ω-valued equality

$$[\![\langle x, y\rangle \approx \langle x', y'\rangle]\!] = [\![x \approx x']\!]_{\mathbf{A}} \sqcap [\![y \approx y']\!]_{\mathbf{B}}$$

The projection arrow $pr_{\mathbf{A}} : \mathbf{A} \times \mathbf{B} \to \mathbf{A}$ has

$$[\![pr_{\mathbf{A}}(\langle x, y\rangle) \approx z]\!] = [\![x \approx z]\!] \sqcap [\![Ex]\!] \sqcap [\![Ey]\!]$$

i.e. "the $\mathbf{A}$-projection of $\langle x, y\rangle$ equals z to the extent that x and y exist and x equals z".

Pullbacks: To realise the diagram

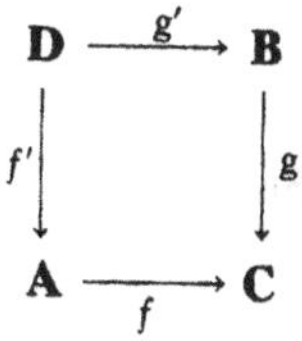

as a pullback we define, for $x \in A$ and $y \in B$

$$E_D(\langle x, y\rangle) = \bigsqcup_{c \in C} ([\![f(x) \approx c]\!] \sqcap [\![g(y) \approx c]\!])$$

(cf. "there exists $c \in C$ with $f(x) = c$ and $g(y) = c$", i.e. "$f(x) = g(y)$").

Then $\mathbf{D}$ is the product set $A \times B$, with

$$[\![\langle x, y\rangle \approx \langle x', y'\rangle]\!]_{\mathbf{D}} = E_D(\langle x, y\rangle) \sqcap E_D(\langle x', y'\rangle) \sqcap [\![x \approx x']\!]_{\mathbf{A}} \sqcap [\![y \approx y']\!]_{\mathbf{B}}$$

Then in fact,

$$[\![E\langle x, y\rangle]\!]_{\mathbf{D}} = E_D(\langle x, y\rangle)$$

i.e. "$\langle x, y\rangle$ exists in D to the extent that $f(x) = g(y)$".

The "projection" f' is given by

$$[\![f'(\langle x, y\rangle) \approx z]\!] = E_D(\langle x, y\rangle) \sqcap [\![x \approx z]\!]_{\mathbf{A}}$$

and similarly for g'.

Subobjects: In **Set**, the pullback is a subset D of $A \times B$ specified by the condition "$f(x) = g(y)$". We have just seen that in Ω-**Set**, **D** is a kind of subobject of $\mathbf{A} \times \mathbf{B}$ that has the same partial elements as the latter but with degrees of existence determined by the pullback condition. This sort of phenomenon is typical of the description of subobjects in Ω-**Set**.

Intuitively, a subset of **A** may be represented by a function of the form $s : A \to \Omega$. Such a function assigns to each $x \in A$ an element $s(x)$ of Ω, which we think of as the truth-value of "$x \in s$", or as a measure of the extent to which x belongs to the "set" s. Thus we also denote $s(x)$ by $[\![x \in s]\!]$. Formally, a subset of an Ω-set **A** is a function $s : A \to \Omega$ that has

(viii) $\quad [\![x \in s]\!] \sqcap [\![x \approx y]\!] \sqsubseteq [\![y \in s]\!]$ (extensional)

and

(ix) $\quad [\![x \in s]\!] \sqsubseteq [\![Ex]\!]$ (strict)

Example. Let $E : A \to \Omega$ be given by

$$E(x) = [\![x \approx x]\!] = [\![Ex]\!].$$

E represents the set of *existing* elements of **A**. Since

$$[\![Ex]\!] = [\![x \in E]\!]$$

we have that "x exists to the extent that it belongs to the set of existing elements of **A**". □

Now an arrow $f : \mathbf{A} \to \mathbf{B}$ can be shown to be monic just in case it satisfies

$$[\![f(x) \approx z]\!] \sqcap [\![f(y) \approx z]\!] \sqsubseteq [\![x \approx y]\!]$$

for all $x, y \in A$ and $z \in B$. Such an arrow corresponds to a subset of **B** (the "f-image" of **A**), and hence to a function $s_f : B \to \Omega$. This is given by

$$s_f(y) = \bigsqcup_{x \in A} [\![f(x) \approx y]\!]$$

i.e. "y belongs to s_f to the extent that it is the f-image of some $x \in A$". Thus $s_f(y)$ is the truth-value of "$y \in f(A)$".

Conversely, a subset $s : B \to \Omega$ of **B** determines a monic arrow $f_s : \mathbf{A}_s \rightarrowtail \mathbf{B}$. $\mathbf{A}_s$ has the same collection B of elements as **B**, but with equality given by

$$[\![x \approx y]\!]_{\mathbf{A}_s} = [\![x \in s]\!] \sqcap [\![y \in s]\!] \sqcap [\![x \approx y]\!]_{\mathbf{B}}$$

i.e. "x and y are equal in $\mathbf{A}_s$ to the extent that they are equal in $\mathbf{B}$ and belong to s". The "inclusion" arrow f_s has

$$[\![f_s(x) \approx y]\!] = [\![x \approx y]\!]_{\mathbf{A}_s}.$$

EXERCISE 6. (I) Prove that $s_{f_s} = s$.

(ii) Let $f_{s_f} : \mathbf{A}_{s_f} \rightarrowtail \mathbf{B}$ be constructed from the set s_f corresponding to a monic $f : \mathbf{A} \rightarrowtail \mathbf{B}$ as above. Then $\mathbf{A}_{s_f}$ has the same collection of elements as $\mathbf{B}$. Define $g : \mathbf{A} \to \mathbf{A}_s$ by

$$[\![g(x) \approx y]\!] = [\![f(x) \approx y]\!].$$

Show that g is iso in Ω-**Set** and that

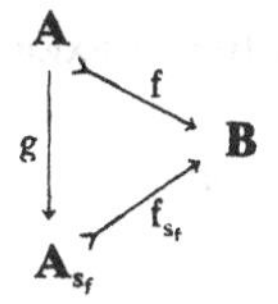

commutes. □

The import of this exercise is that subobjects $\mathbf{A} \rightarrowtail \mathbf{B}$ of $\mathbf{B}$ are uniquely determined by subsets $B \to \Omega$ of $\mathbf{B}$. The latter in fact form the power object $\mathcal{P}(\mathbf{B})$ of $\mathbf{B}$. To define this, let $S(\mathbf{B})$ be the collection of all subsets $s : B \to \Omega$ of $\mathbf{B}$. Then $\mathcal{P}(\mathbf{B})$ comprises $S(\mathbf{B})$ with the equality

$$[\![s \approx t]\!]_{\mathcal{P}(\mathbf{B})} = \bigsqcap_{x \in B} (s(x) \Leftrightarrow t(x))$$

(cf. "for all $x \in B$, $x \in S$ iff $x \in t$").

EXERCISE 7. $[\![s \approx t]\!]_{\mathcal{P}(\mathbf{B})} = \top$ iff $s = t$ (i.e. s and t are the same function).

EXERCISE 8. $[\![Es]\!]_{\mathcal{P}(\mathbf{B})} = \top$

EXERCISE 9. $[\![x \in s]\!] \sqcap [\![s \approx t]\!] \sqsubseteq [\![x \in t]\!]$ □

Now the function $e : A \times S(\mathbf{A}) \to \Omega$ having $e(\langle x, s \rangle) = s(x)$ satisfies (viii) and (ix), and so is a subset of the Ω-set $\mathbf{A} \times \mathcal{P}(\mathbf{A})$. The corresponding subobject f_e is precisely the membership relation $\in_{\mathbf{A}} \rightarrowtail \mathbf{A} \times \mathcal{P}(\mathbf{A})$ on $\mathbf{A}$. The definition of e thus gives that "$\langle x, s \rangle$ belongs to $\in_{\mathbf{A}}$ to the same extent that x belongs to s".

SUBOBJECT CLASSIFIER: The arrow $true: \mathbf{1} \to \mathbf{\Omega}$ has

$$[\![true(0) \approx p]\!] = [\![p \approx \top]\!]_{\mathbf{\Omega}}$$

("p is *true* to the extent that p equals $\top$") and so

$$[\![true(0) \approx p]\!] = (p \Leftrightarrow \top) = p.$$

Now let $f: \mathbf{A} \to \mathbf{D}$ be a monic, with corresponding subset $s_f: \mathbf{D} \to \mathbf{\Omega}$ of $\mathbf{D}$. The character $\chi_f: \mathbf{D} \to \mathbf{\Omega}$ of f has

$$[\![\chi_f(d) \approx p]\!] = [\![Ed]\!]_{\mathbf{D}} \sqcap [\![s_f(d) \approx p]\!]_{\mathbf{\Omega}}$$

i.e. "$\chi_f(d)$ equals p to the extent that d exists and p is the truth-value of "$d \in f(A)$" ".

EXERCISE 10. Show that this construction satisfies the Ω-axiom.

EXERCISE 11. $[\![false(0) \approx p]\!] = [\![p \approx \bot]\!]_{\mathbf{\Omega}} = (p \Leftrightarrow \bot) = \neg p$

EXERCISE 12. The truth arrows $\cap$, $\cup$ have

$$[\![p \cap q \approx r]\!] = [\![(p \sqcap q) \approx r]\!]_{\mathbf{\Omega}}$$

EXERCISE 13. $[\![p \cup q \approx r]\!] = [\![(p \sqcup q) \approx r]\!]_{\mathbf{\Omega}}$

EXERCISE 14. Show that the r.p.c. operation $\Rightarrow: \Omega \times \Omega \to \Omega$ on the **HA** Ω is a subset of $\mathbf{\Omega} \times \mathbf{\Omega}$ in the sense of (viii) and (ix) and that the corresponding subobject is $\circledcirc \rightarrowtail \mathbf{\Omega} \times \mathbf{\Omega}$. Show that the character of the latter, i.e. the implication arrow $\Rightarrow': \mathbf{\Omega} \times \mathbf{\Omega} \to \mathbf{\Omega}$ has

$$[\![p \Rightarrow' q \approx r]\!] = (p \Rightarrow q) \Leftrightarrow r = [\![(p \Rightarrow q) \approx r]\!]_{\mathbf{\Omega}}. \qquad \square$$

Object of partial elements

In **Set**, a "singleton" is a set with exactly one member. In the present context of partial elements we are more interested in sets with *at most* one member. Formally a subset (extensional, strict function) $s: A \to \Omega$ of **A** is a *singleton* if it satisfies

(x) $\qquad [\![x \in s]\!] \sqcap [\![y \in s]\!] \sqsubseteq [\![x \approx y]\!]$

i.e. "elements of A belong to s only to the extent that they are equal".

EXAMPLE 1. If $a \in A$, then the map $\{\mathbf{a}\}: A \to \Omega$ that assigns to $x \in A$ the degree $[\![x \approx a]\!]$ of its equality with a is a singleton in this sense, with $[\![x \in \{\mathbf{a}\}]\!] = [\![x \approx a]\!]$.

EXAMPLE 2. Suppose $\mathbf{A}$ is the Ω-set (with $\Omega=\mathcal{P}(I)$) of all local sections of some bundle over I, as considered earlier. Included in $\mathbf{A}$ is the empty section $\emptyset_{\mathbf{A}}$, the unique section whose domain is the empty subset of I. For any other section x, we have $[\![x\approx\emptyset_{\mathbf{A}}]\!]=\emptyset$. Generalising to an arbitrary Ω and arbitrary Ω-set $\mathbf{A}$, the map $\{\emptyset_{\mathbf{A}}\}:A\to\Omega$ assigning $\bot$ to each $x\in A$ is a singleton, with $[\![x\in\{\emptyset_{\mathbf{A}}\}]\!]=\bot$. □

EXERCISE 15. If s is a singleton

$$[\![x\in s]\!]\sqsubseteq([\![y\in s]\!]\Leftrightarrow[\![y\approx x]\!])$$

EXERCISE 16. $\{\mathbf{a}\}=\{\mathbf{b}\}$ iff $[\![a\approx b]\!]=[\![Ea]\!]=[\![Eb]\!]$.

EXERCISE 17. Let $s\in S(\mathbf{A})$ and $p\in\Omega$. The restriction of s to p is the function $s\restriction p:A\to\Omega$ assigning $s(x)\sqcap p$ to x. Show that $s\restriction p\in S(\mathbf{A})$ and that $s\restriction p$ is a singleton if s is. □

Now the object $\tilde{\mathbf{A}}$ is to be regarded as the Ω-set of all subsets of $\mathbf{A}$ that are singletons in the present sense. Thus $\tilde{\mathbf{A}}$ is to be thought of as itself being a subobject of $\mathcal{P}(\mathbf{A})$ and hence corresponds to a function $sing:S(\mathbf{A})\to\Omega$. The formal definition, for $s\in S(\mathbf{A})$, is

$$[\![s\in sing]\!]=\bigsqcap_{x,y\in A}([\![x\in s]\!]\sqcap[\![y\in s]\!]\Rightarrow[\![x\approx y]\!])$$

(cf. "for all x, $y\in A$, if x and y belong to s then $x=y$".)

The inclusion arrow $\eta_{\mathbf{A}}:\mathbf{A}\rightarrowtail\tilde{\mathbf{A}}$ of $\mathbf{A}$ into $\tilde{\mathbf{A}}$ has

$$[\![\eta_{\mathbf{A}}(a)\approx s]\!]=[\![Ea]\!]_{\mathbf{A}}\sqcap[\![s\approx\{\mathbf{a}\}]\!]_{\mathcal{P}(\mathbf{A})}$$

("$\eta_{\mathbf{A}}(a)$ is s to the extent that a exists and s is $\{\mathbf{a}\}$").

EXERCISE 18. $[\![s\in sing]\!]=\top$ iff s is a singleton.

EXERCISE 19. $[\![\{\mathbf{a}\}\approx s]\!]\sqsubseteq[\![s\in sing]\!]$. □

Now we know that each bundle over I gives rise to an Ω-set, where $\Omega=\mathcal{P}(I)$, whose elements are the partial sections of the bundle. Conversely, given an arbitrary $\mathcal{P}(I)$-set $\mathbf{A}$, each $i\in I$ determines an equivalence relation $\sim_i$ on the set

$$A_i=\{x\in A: i\in[\![Ex]\!]\}$$

that is defined by

$$x \sim_i y \quad \text{iff} \quad i \in [\![x \approx y]\!].$$

We then obtain a bundle over I by taking the quotient set $A_i/{\sim_i}$ as the stalk over the point i. These constructions may be used to establish that the categories **Bn**(I) and $\mathcal{P}(I)$-**Set** are equivalent. They can also be adapted to the case of sheaves of sets of germs, showing that **Top**(I) is equivalent to Θ-**Set**, where Θ is the **CHA** of open subsets of a topological space I. These facts are a special case of a result of D. Higgs [73] to the effect that Ω-**Set**, for any **CHA** Ω, is equivalent to the category of "sheaves over Ω". Precisely what that means will be explained in Chapter 14, where we shall see also that Ω-**Set** is equivalent to a subcategory of itself in which arrows $\mathbf{A} \to \mathbf{B}$ may be identified with actual set-functions $A \to B$.

Elementary Logic in Ω-Set

We have been interpreting the operations $\sqcap$ and $\sqcup$ *informally* as universal and existential quantifiers in order to understand the constructions that define Ω-**Set**. When we come to interpret a first-order language in this topos, these same operations may serve to give meanings to the formal symbols $\forall$ and $\exists$. Moreover, instead of assigning a formula an arrow of the type $\tilde{\mathbf{A}} \to \mathbf{\Omega}$, we may work directly with functions of the form $A \to \Omega$, and take advantage of the presence of the extents $[\![Ea]\!]$ of individuals to formalize the principle that quantifiers are to range over *existing* individuals.

To illustrate this approach, suppose that our language $\mathcal{L}$ has a single two-place relation symbol **R**. Our basic alphabet is presumed to include the existence predicate **E** and the identity (equality) symbol $\approx$. The symbol $\approxeq$ for equivalence is introduced according to clause (iii) at the beginning of this section. Alternatively, $\approx$ may be defined in terms of $\approxeq$ by (iiia).

For this language, a model in Ω-**Set** is a pair $\mathfrak{A} = \langle \mathbf{A}, r \rangle$ comprising an Ω-set **A** and a subset $r: A \times A \to \Omega$ of $\mathbf{A} \times \mathbf{A}$. (By Exercise 6, r corresponds to a unique subobject of $\mathbf{A} \times \mathbf{A}$, hence to a unique *arrow* $\mathbf{A} \times \mathbf{A} \to \mathbf{\Omega}$, and so this approach accords within the definition of "model" in §11.4). We then extend $\mathcal{L}$ by adjoining an individual constant **c** for each element $c \in A$. A truth-value $[\![\varphi]\!]_{\mathfrak{A}} \in \Omega$ can then be calculated for each sentence φ by induction as follows:

Atomic Sentences:

$$[\![\mathbf{c} \approx \mathbf{d}]\!]_{\mathfrak{A}} = [\![c \approx d]\!]_{\mathbf{A}}$$

$$[\![\mathbf{E}(\mathbf{c})]\!]_{\mathfrak{A}} = [\![Ec]\!]_{\mathbf{A}}$$

$$[\![\mathbf{cRd}]\!]_{\mathfrak{A}} = r(\langle c, d\rangle)$$

Propositional Connectives:

$\wedge$, $\vee$, $\supset$, $\sim$ are interpreted by $\sqcap$, $\sqcup$, $\Rightarrow$, $\neg$ in Ω.

Quantifiers:

$$[\![\forall v\varphi]\!]_{\mathfrak{A}} = \bigsqcap_{c \in A} ([\![\mathbf{E}(\mathbf{c}) \supset \varphi(v/\mathbf{c})]\!]_{\mathfrak{A}})$$

("$\varphi(\mathbf{c})$ holds for all existing c")

$$[\![\exists v\varphi]\!]_{\mathfrak{A}} = \bigsqcup_{c \in A} ([\![\mathbf{E}(\mathbf{c}) \wedge \varphi(v/\mathbf{c})]\!]_{\mathfrak{A}})$$

("$\varphi(\mathbf{c})$ holds for some existing c").

Satisfaction: For a formula $\varphi(v_1, \ldots, v_n)$ we define $\mathfrak{A} \models \varphi[c_1, \ldots, c_n]$, where $c_1, \ldots, c_n \in A$, to mean that $[\![\varphi(v_1/\mathbf{c}_1, \ldots, v_n/\mathbf{c}_n)]\!]_{\mathfrak{A}} = \top$. Then truth- $\mathfrak{A} \models \varphi$ - of φ in $\mathfrak{A}$ can then be defined as usual by

$$\mathfrak{A} \models \varphi[c_1, \ldots, c_n] \quad \text{for all} \quad c_1, \ldots, c_n \in A.$$

EXERCISE 20. Show that the following are true in $\mathfrak{A}$:

$$(t \approxeq u) \wedge \varphi(v/u) \supset \varphi(v/t)$$

$$\forall v_i((v_i \approxeq v_j) \equiv (v_i \approxeq v_k)) \supset (v_j \approxeq v_k)$$

$$\forall v\varphi \wedge \mathbf{E}(t) \supset \varphi(v/t)$$

$$\varphi(v/t) \wedge \mathbf{E}(t) \supset \exists v\varphi$$

$$\mathbf{E}(t) \equiv \exists v(v \approx t)$$

$$\exists v(v \approx \mathbf{t}) \equiv \exists v(v \approxeq \mathbf{t})$$

$$\forall v_i \forall v_j((v_i \approxeq v_j) \equiv (v_i \approx v_j))$$

$$\forall v\varphi \equiv \forall v(\mathbf{E}(v) \supset \varphi)$$

$$\exists v\varphi \equiv \exists v(\mathbf{E}(v) \wedge \varphi)$$

$$\forall v\mathbf{E}(v)$$

$$(\mathbf{E}(v_i) \vee \mathbf{E}(v_j) \supset (v_i \approxeq v_j)) \supset (v_i \approxeq v_j)$$

EXERCISE 21. Show that the following rules preserve truth in $\mathfrak{A}$:

$$\text{From } \varphi \wedge \mathbf{E}(v) \supset \psi \quad \text{infer} \quad \varphi \supset \forall v \psi$$

$$\text{From } \psi \wedge \mathbf{E}(v) \supset \varphi \quad \text{infer} \quad \exists v \psi \supset \varphi$$

provided in both cases that v is not free in φ. □

This semantical theory will be used in Chapter 14 to define number-systems in **Ω-Set**. We will find it convenient there to have available the following result, which simplifies the calculation of the truth-value of quantified formulae in some cases by allowing the range of quantification to be further restricted.

We say that a subset C of A *generates* the Ω-set $\mathbf{A}$ if for each $a \in A$,

$$[\![Ea]\!]_{\mathbf{A}} = \bigsqcup_{c \in C} [\![a \approx c]\!]_{\mathbf{A}}$$

EXERCISE 22. If C generates A then

$$[\![\forall v \varphi]\!]_{\mathfrak{A}} = \bigsqcap_{c \in C} ([\![\mathbf{E}(\mathbf{c}) \supset \varphi(v/\mathbf{c})]\!]_{\mathfrak{A}})$$

and

$$[\![\exists v \varphi]\!]_{\mathfrak{A}} = \bigsqcup_{c \in C} ([\![\mathbf{E}(\mathbf{c}) \wedge \varphi(v/\mathbf{c})]\!]_{\mathfrak{A}})$$ □

11.10. Higher-order logic

In closing this chapter on quantificational logic we mention briefly the study that has been made of the relationship between higher order logic and topoi.

Higher order logic has formulae of the form $(\forall X)\varphi$ and $(\exists X)\varphi$, where X may stand for a set, a relation, a set of sets, a set of relations, a set of sets of sets of ..., etc. So for a classical model $\mathfrak{A} = \langle A, \ldots \rangle$ the range of X may be any of $\mathcal{P}(A)$, $\mathcal{P}(A^n)$, $\mathcal{P}(\mathcal{P}(A^n))$, etc. Analogues of these exist in any topos, in the form of Ω^a, Ω^{a^n}, etc., and so higher order logic is interpretable in $\mathscr{E}$. In fact the whole topos becomes a model for a many sorted language, having one sort (infinite list) of individual variables for each $\mathscr{E}$-object. Given a theory Γ (i.e. a consistent set of sentences) in this language, a topos $\mathscr{E}_\Gamma$ can be constructed that is a model of Γ. Conversely given a topos $\mathscr{E}$ a theory $\Gamma_{\mathscr{E}}$ can be defined whose associated topos $\mathscr{E}_{\Gamma_{\mathscr{E}}}$ is categorially equivalent to $\mathscr{E}$. These results were obtained for the logic of partial elements by Fourman [74] and subsequently for the other approach to free logic by Boileau [75]. They amount to a demonstration that the concept of "elementary topos" is co-extensive with that of "model for many-sorted higher-order intuitionistic free logic", and hence provide a full explication of Lawvere's statement in [72] that "the notion of topos summarizes in objective categorical form the essence of 'higher-order logic'." The work of Fourman incorporates a number of interesting and unusual logical features, which we will outline briefly.

Firstly, as already noted in §11.8, variables are to be thought of as ranging over, and constants denoting, *potential* elements of an $\mathscr{E}$-object a. Thus a formula is interpreted by an arrow of the form $[\![\varphi]\!]:(\tilde{a})^n \to \Omega$, corresponding to the subobject of all n-tuples of potential elements that satisfy φ.

Next, the system includes a theory of *definite descriptions* as terms of the formal language. A definite description is an expression of the form $\mathbf{I}v\varphi$, which is read "the unique v such that φ". The expression serves as a name for this unique v whenever it exists. The basic axiom governing this descriptions-operator is

$$\forall u((u \approx \mathbf{I}v\varphi(v)) \equiv \forall v(\varphi(v) \equiv (v \approx u)))$$

which has the reading "an existing element u is equivalent to the element $\mathbf{I}v\varphi(v)$ iff u is the one and only existing element satisfying φ" (recall that quantifiers range over existing elements).

To interpret a definite description semantically in $\mathscr{E}$ suppose, by way of example, that the $\mathscr{E}$-arrow $[\![\varphi]\!]:\tilde{a} \to \Omega$ has been defined, where $\varphi(v)$ has index 1. Let $f:1 \to \Omega^a$ be the *name* of the arrow $[\![\varphi]\!] \circ \eta_a : a \to \Omega$ (cf. §4.1). (In **Set**, f corresponds to the element

$$|\varphi| = \{x \in a: \varphi(x)\}$$

of the powerset of a, i.e. the subset of a defined by φ).

Form in $\mathscr{E}$ the pullback

$$\begin{array}{ccc} b & \rightarrowtail & 1 \\ {\scriptstyle g}\downarrow & & \downarrow{\scriptstyle f} \\ a & \xrightarrow{\{\cdot\}_a} & \Omega^a \end{array}$$

of f along the "singleton arrow" $\{\cdot\}_a$, that was defined in §11.8. (In **Set** we may regard g as the inclusion $b \hookrightarrow a$, with $b = |\varphi|$ if $|\varphi|$ is a non-empty singleton, i.e. if $|\varphi| = \{x\}$ for some $x \in a$, and $b = \emptyset$ otherwise). Notice that $g:1 \rightsquigarrow a$, i.e. g is a partial element of a, and so corresponds to an arrow

$\tilde{g}:1 \to \tilde{a}$. We take this $\tilde{g}$ to be $[\![\mathbf{l}v\varphi]\!]$. (In **Set**, taking $\tilde{a}$ as $a \cup \{*\}$, $\tilde{g}$ corresponds to the element x of a if $|\varphi| = \{x\}$, and is the "null entity" $*$ otherwise.)

Of course the description operator and its semantic interpretation can be developed in the context of first-order logic. In higher order logic it becomes particularly useful, in that is provides simple and straightforward ways of expressing both the Comprehension Principle, and the operation of *functional abstraction*, the latter being the process of defining a term that denotes a function whose graph is specified by a formula.

To consider Comprehension, suppose by way of example that $\varphi(v)$ has a single free variable whose range is a collection of entities of a certain level, or type, in a higher-order structure comprising subsets, sets of subsets, sets of sets of subsets etc. In a higher-order language there will also be variables w that range over the subsets of the range of v. Then the sentence

$$\mathbf{E}\mathbf{l}w\forall v(\varphi(v) \equiv w(v))$$

asserts the actual existence of the unique set whose elements are precisely those entities that satisfy φ.

If instead $\varphi(v, w)$ has two free variables, it defines a relation when interpreted. We denote by $\varphi'(v)$ *the term*

$$\mathbf{l}w\varphi(v, w).$$

If the interpretation of φ is a functional relation (one with the unique output property) then this term will provide a notation for function values. Functional abstraction may now be performed by forming the expression

$$\mathbf{l}u\forall v\forall w(u(v, w) \equiv \varphi'(v) \approx w)$$

(which is abbreviated to $\lambda v \cdot \varphi'(v)$), where u is a variable that ranges over the relations from the range of v to the range of w. The expression $\lambda v \cdot \varphi'(v)$ may be read "the function which for input v gives output $\varphi'(v)$".

The details of this higher-order language and its use in characterising topoi as models of higher-order theories may be found in Fourman's article "The Logic of Topoi" in Barwise [77]. This work is important for a broad understanding of the structural properties of topoi. It offers a different perspective to the one we are dealing with here. Our present concern is to develop the view of a topos as a universe of set-like objects and hence, qua foundation for mathematics, as a model of a first-order theory of set-membership. We take this up in earnest in the next chapter.

CHAPTER 12

CATEGORIAL SET THEORY

> "... *the mathematics of the future, like that of the past, will include developments which are relevant to the philosophy of mathematics .... They may occur in the theory of categories where we see, once again, a largely successful attempt to reduce all of pure mathematics to a single discipline*".
>
> Abraham Robinson

While a topos is in general to be understood as a "generalised universe of sets", there are, as we have seen, many topoi whose structure is markedly different from that of **Set**, the domain of classical set theory. Even within a topos that has classical logic (is Boolean) there may be an infinity of truth-values, non-initial objects that lack elements, distinct arrows not distinguished by elements of their domain etc. So in order to identify those topoi that "look the same" as **Set** we will certainly impose conditions like well-pointedness and (hence) bivalence.

However, in order to say precisely which topoi look like **Set** we have to know precisely what **Set** looks like. Thus far we have talked blithely about *the* category of all sets without even acknowledging that there might be some doubt as to whether, or why, such a unique thing may exist at all. We resolve (sidestep?) this matter by introducing a formal first-order language for set-theory, in which we write down precise versions of set-theoretic principles. Instead of referring to "the universe **Set**", we confine ourselves to discussion of interpretations of this language. The notion of a topos is also amenable to a first-order description, as indicated in the last chapter, and so the relationship between topos theory and set theory can be rigorously analysed in terms of the relationship between models of two elementary theories.

Before looking at the details of this program we need to develop two more fundamental aspects of the category of sets.

12.1 Axioms of choice

Let $f: A \twoheadrightarrow I$ be an epic (onto) set function. Then, construing f as a bundle over I, we may construct a *section* of f, i.e. a function $s: I \to A$ having $f \circ s = \mathrm{id}_I$. The point here is that for each $i \in I$ the stalk A_i over i is non-empty (since f is onto) and so we may choose some element of A_i and take it as $s(i)$ (unless $I = \emptyset$, in which case $A = \emptyset$ so we take s as the empty map $!: \emptyset \to \emptyset$). The section s is sometimes said to "split" the epic f. In sum then we have produced an argument to the effect that in **Set**, all epics split. We lift this now to the categorial statement

ES: *Each epic* $a \overset{f}{\twoheadrightarrow} b$ *has a section* $b \overset{s}{\to} a$ *with* $f \circ s = 1_b$.

EXERCISE 1. Show that a section is always monic. □

The principle ES is a variant of what is known as the *axiom of choice.* The name relates to our making an arbitrary choice of the element $s(i)$ of A_i. The function s, in selecting an element from each A_i is called a *choice function.* Informally, the axiom of choice asserts that it is permissible to make an unlimited number of arbitrary choices. It was first isolated as a principle of mathematical reasoning by Zermelo in 1904 and subsequently has been shown to be implied by, indeed equivalent to, many substantial "theorems" of classical mathematics. To many classically minded mathematicians the axiom of choice is a perfectly acceptable principle. It is difficult for someone so minded to see what could be wrong with the above argument that purports to show that ES is true of **Set**.

Nonetheless the status of the axiom of choice remained in doubt until Paul Cohen [66] proved that it was not derivable from the Zermelo–Fraenkel axioms for set theory (Gödel [40] had earlier shown that it was not refutable by this system). The point would seem to be that the choice function s cannot be explicitly defined in terms of any set-theoretic operations involving $f: A \to I$. In general we are unable to formulate a rule for s of the form "let $s(i)$ be the element of A_i such that φ", where φ is some property that demonstrably is possessed by only one element of A_i. So if we wish to include ES in our account of what **Set** looks like we will simply have to take it as an axiom (unless of course we adopt some equally "unprovable" axiom that implies it).

Now if $f: A \to I$ is a function that is not onto, then f will not have a section. This, as explained in §11.8, is why the $\mathbf{Bn}(I)$-object $a = (A, f)$ is empty, i.e. has no elements $1 \to a$. However f will have a "partial section" $s: I \rightsquigarrow A$. For, taking the epi-monic factorisation

$$\begin{array}{ccc} A & \xrightarrow{f} & I \\ & {}_{f^*}\searrow \quad \nearrow & \\ & f(A) & \end{array}$$

of f we find that a section of the epic f^* is a partial function from I to A.

Now the image $f(A)$ is sometimes known as the *support* of the bundle a. It is the subset of I

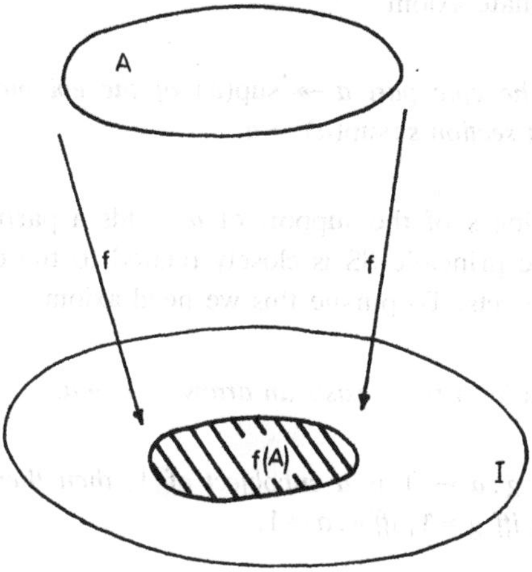

Fig. 12.1.

over which the stalks actually "sit". As a subset of I, $f(A)$ is identifiable with a subobject of $1 = (I, \mathrm{id}_I)$ in $\mathbf{Bn}(I)$. Indeed since

$$\begin{array}{ccc} A & \xrightarrow{f} & I \\ & {}_{f^*}\searrow \quad \nearrow & \\ {}_{f}\searrow & f(A) & \swarrow_{\mathrm{id}_I} \\ & \downarrow & \\ & I & \end{array}$$

commutes in **Set**, so does

$$\begin{array}{ccc} a & \xrightarrow{f} & 1 \\ & {}_{f^*}\searrow \quad \nearrow & \\ & \sup(a) & \end{array}$$

where the object $\sup(a)$ is the function (bundle) $f(A) \hookrightarrow I$.

Lifting this to a general topos $\mathscr{E}$ we define the *support* of an $\mathscr{E}$-object a to be the subobject $\sup(a) \rightarrowtail 1$ of 1 given by the epi-monic factorisation

$$\begin{array}{ccc} a & \longrightarrow & 1 \\ & \searrow \quad \nearrow & \\ & \sup(a) & \end{array}$$

of the unique arrow $!: a \to 1$.

We may now formulate axiom

SS (supports split): *The epic part* $a \twoheadrightarrow \sup(a)$ *of the epi-monic factorisation of* $a \to 1$ *has a section* $s: \sup(a) \to a$.

Notice that a splitting s of the support of a yields a partial element $s: 1 \rightsquigarrow a$ of a, so the principle SS is closely related to the question of (non) emptiness of objects. To pursue this we need axiom

NE: *For every non-initial* a *there exists an arrow* $x: 1 \to a$.

LEMMA. *In any* $\mathscr{E}$, *if* $g: a \rightarrowtail 1$ *is a subobject of* 1, *then there exists an element* $x: 1 \to a$ *of* a *iff* $g \simeq 1_1$ *iff* $g: a \cong 1$.

PROOF. This is the essence of Case 2 in the proof of Theorem 5.4.2. □

CONVENTION. $\mathscr{E}$ is always non-degenerate, i.e. $0 \not\cong 1$.

NOTATION. We write $\mathscr{E} \models \mathrm{NE}$, $\mathscr{E} \models \mathrm{SS}$ etc. to mean that NE (SS etc.) holds for $\mathscr{E}$.

THEOREM 1. *For any topos* $\mathscr{E}$,

$$\mathscr{E} \models \mathrm{NE} \textit{ iff } \mathscr{E} \textit{ is bivalent and } \mathscr{E} \models \mathrm{SS}.$$

PROOF. Suppose $\mathscr{E} \models \mathrm{NE}$, and let $t: 1 \to \Omega$ be a truth-value. Pull t back along $\top$ to get $g: a \rightarrowtail 1$ with $\chi_g = t$. Then if $t \neq \bot$, a is non-initial, so by NE there exists $x: 1 \to a$. But then by the Lemma, $g \simeq 1_1$, so $\chi_g = \chi_{1_1}$, i.e. $t = \top$. Hence $\mathscr{E}$ is bivalent.

To see why supports split, consider

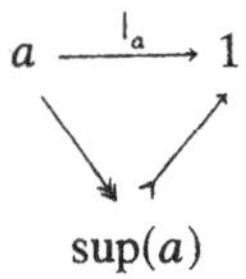

If $\sup(a)\cong 0$, then $a\cong 0$ (Theorem 3.16.1, (2)) and so the unique arrow $\sup(a)\to a$ will split the unique $a\twoheadrightarrow \sup(a)$. If not $\sup(a)\cong 0$, then by NE there is an element $1\to\sup(a)$, from which by the Lemma, $\sup(a)$ is terminal, $\sup(a)\cong 1$, and hence $!_a$ is epic. Then if $a\cong 0$, $!_a$ would be monic (Theorem 3.16.1, (4)), hence altogether iso, making $0\cong a\cong 1$, and thus $\mathscr{E}$ degenerate. So we may invoke NE again to get an element $x:1\to a$. Since $\sup(a)\cong 1$ this yields an arrow $\sup(a)\to a$ which must be a section of the unique $!:a\twoheadrightarrow\sup(a)$.

Conversely if $\mathscr{E}$ is bivalent then in Sub(1), $\sup(a)\rightarrowtail 1$ can only be 0_1 or 1_1. But if $a\not\cong 0$, then $\sup(a)\not\cong 0$ (as above), so it cannot be 0_1. We must then have $\sup(a)\rightarrowtail 1=1_1$, so $\sup(a)\cong 1$. Then if $\mathscr{E}\models$SS, there is an arrow $\sup(a)\to a$, hence an arrow $1\to a$. This establishes NE. □

COROLLARY. *$\mathscr{E}$ is well-pointed iff $\mathscr{E}$ is Boolean (classical), bivalent, and has splitting supports.*

PROOF. Theorem 5.4.5 (proven in §7.6) gives $\mathscr{E}$ well-pointed iff $\mathscr{E}$ is classical and $\mathscr{E}\models$NE. □

Even when there are more than two truth-values, the splitting of epics in a Boolean topos has implications for extensionality. We will say that $\mathscr{E}$ is *weakly extensional* if for every pair $f,g:a\rightrightarrows b$ with $f\neq g$ there is a partial element $x:1 \rightsquigarrow a$ such that $f\circ x\neq g\circ x$. Recall that $x:1\rightsquigarrow a$ means that cod $x=a$ and there is a monic dom $x\rightarrowtail 1$ (hence x could not be $!:0\to a$ if $f\circ x\neq g\circ x$).

Category theorists will recognise "$\mathscr{E}$ is weakly extensional" as "Sub(1) is a set of generators for $\mathscr{E}$".

THEOREM 2. *If $\mathscr{E}$ is Boolean and $\mathscr{E}\models$SS, then $\mathscr{E}$ is weakly extensional.*

PROOF. Let $h:c\rightarrowtail a$ equalise $f,g:a\rightrightarrows b$, and let $-h:-c\rightarrowtail a$ be the complement of h in Sub(a). Then as in §7.6, if $f \neq g$, $-c \not\cong 0$. Now if $y : \sup(-c) \to -c$ is a section of $-c \twoheadrightarrow \sup(-c)$,

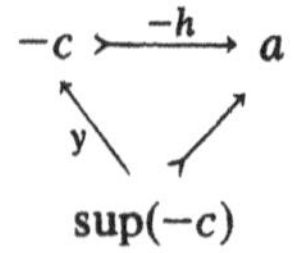

then putting $x = -h \circ y$ gives $x : 1 \rightsquigarrow a$. If $f \circ x = g \circ x$, then reasoning as in §7.6,

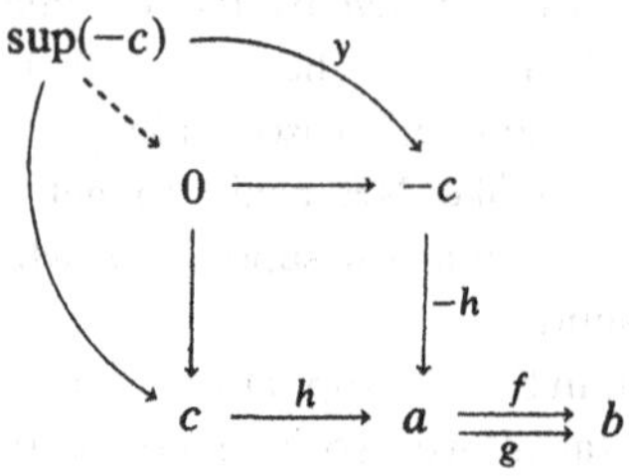

x would factor through h, ultimately making $\sup(-c) \cong 0$ and hence $-c \cong 0$. Therefore x distinguishes f and g. □

Example. In general **Bn**(I), though Boolean, is not extensional (well-pointed), since NE fails. However **Bn**(I) *is* weakly extensional. Given bundles $a = (A, h)$, $b = (B, k)$ and distinct arrows $f, g : a \rightrightarrows b$, then the distinguishing $x : 1 \rightsquigarrow a$, as in Theorem 2, is a local section of a, defined on a subset $-C$ of the support $h(A)$ of a. For each $i \in -C$ (hence $A_i \neq \emptyset$), x selects an element x_i of the stalk A_i that distinguishes f and g, i.e. $f(x_i) \neq g(x_i)$. □

Returning to **Set** once more, let $f : A \to I$ be any function and, invoking ES, let $s : f(A) \to A$ be a section of $f^* : A \twoheadrightarrow f(A)$. Then if $A \neq \emptyset$, by choosing a particular $x_0 \in A$ we can obtain a function $f : I \to A$ by the rule

$$g(y) = \begin{cases} s(y) & \text{if } y \in f(A) \\ x_0 & \text{otherwise.} \end{cases}$$

Of course if there exists $y \notin f(A)$, g will not be a section of f, since $f(g(y)) \in f(A)$. However, starting with $x \in A$ we find that $g(f(x)) = s(f(x))$ lies in the stalk over $f(x)$ so f simply takes $g(f(x))$ to $f(x)$, i.e. $f \circ g \circ f(x) = f(x)$. This yields another version of the axiom of choice, due to Maclane, that has the categorial formulation

AC: *If* $a \not\cong 0$ *then for any arrow* $a \xrightarrow{f} b$ *there exists* $b \xrightarrow{g} a$ *with* $f \circ g \circ f = f$.

THEOREM 3. *If* $\mathcal{E} \models \mathrm{AC}$, *then* $\mathcal{E} \models \mathrm{NE}$, $\mathcal{E} \models \mathrm{ES}$, *and* $\mathcal{E}$ *is bivalent.*

PROOF. If $a \not\cong 0$, apply AC to $! : a \to 1$ to get $g : 1 \to a$. Hence NE holds. To derive ES, observe that if $f : a \twoheadrightarrow b$ is epic, and $a \cong 0$, then f is monic, (Theorem 3.16.1), hence altogether iso, so is split by its inverse. If $a \not\cong 0$, apply AC to get $g : b \to a$, with $f \circ g \circ f = f = 1_b \circ f$. Since f is right cancellable, we get $f \circ g = 1_b$, making g a section of f.

For bivalence, observe that if $g : a \rightarrowtail 1$ has $a \not\cong 0$, then by AC there is an arrow $1 \to a$. Hence, as in Theorem 1, $g \simeq 1_1$. Thus Sub(1) has only the two elements 0_1 and 1_1. □

The argument that yields AC from ES in **Set** will lift to a topos only if that topos is sufficiently "**Set**-like". To see this, consider a set I with at least two elements. Then $\mathbf{Bn}(I)$ has at least four truth-values (subsets of I) so by the last result AC fails (alternatively observe that NE fails). But if epics split in **Set**, they will in $\mathbf{Bn}(I)$ also. For $h : (A, f) \twoheadrightarrow (B, g)$ means that h is an onto function with

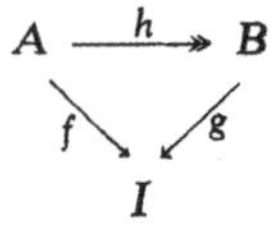

$g \circ h = f$. But then if $s : B \rightarrowtail A$ is a section of h,

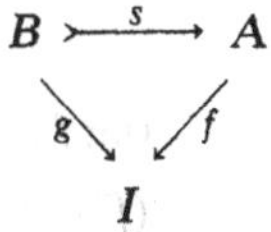

will commute, making s a splitting of h in $\mathbf{Bn}(I)$.

Rather than rely on the assumption that ES holds in **Set**, we can use the result of Gödel that there exist models of formal set theory in which the axiom of choice is true. We may then construct a category of bundles of "sets" from such a model to obtain a topos in which ES holds but AC fails.

THEOREM 4. *If* $\mathcal{E} \models \mathrm{ES}$, *and* $\mathcal{E}$ *is well-pointed, then* $\mathcal{E} \models \mathrm{AC}$.

PROOF. Take $f: a \to b$ and perform the factorisation

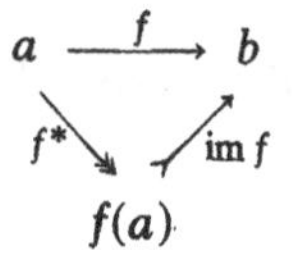

Since $\mathscr{E}$ is well-pointed, it is Boolean, so $\operatorname{im} f$ has a complement $-\operatorname{im} f: -f(a) \rightarrowtail b$, with $\operatorname{im} f \cup -\operatorname{im} f \simeq 1_b$ in $\operatorname{Sub}(b)$. But $\operatorname{im} f$ and $-\operatorname{im} f$ are disjoint monics (Theorem 7.2.3), and so $[\operatorname{im} f, -\operatorname{im} f]: f(a) + -f(a) \to b$ is monic (Lemma, §5.4). But then

$$[\operatorname{im} f, -\operatorname{im} f] \simeq \operatorname{im} f \cup -\operatorname{im} f \simeq 1_b,$$

and so this co-product arrow is iso. This allows us to use b as a co-product object for $f(a)$ and $-f(a)$, with $\operatorname{im} f$ and $-\operatorname{im} f$ serving as the associated injections.

Now suppose $a \not\simeq 0$. Then as well-pointed topoi satisfy NE, we take some $x: 1 \to a$ and let $h: -f(a) \to a$ be the composite $x \circ !: -f(a) \to 1 \to a$. Since $\mathscr{E} \models \mathrm{ES}$, we have also a section $s: f(a) \to a$ of f^*. Then

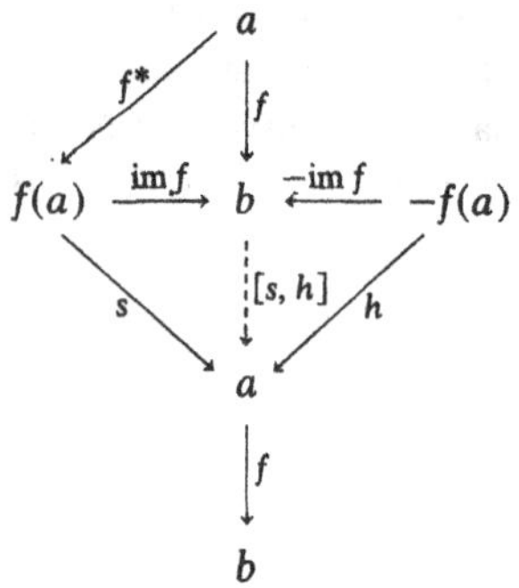

$$\begin{aligned} f \circ [s, h] \circ f &= \operatorname{im} f \circ f^* \circ [s, h] \circ \operatorname{im} f \circ f^* \\ &= \operatorname{im} f \circ f^* \circ s \circ f^* && (\operatorname{im} f \text{ as injection}) \\ &= \operatorname{im} f \circ f^* && (f^* \circ s = 1_{f(a)}) \\ &= f. \end{aligned}$$

Thus $g = [s, h]$ gives the required arrow for AC. □

The hypothesis of Theorem 4, as stated, assumes more than it need do. We know that "well-pointed" = "NE plus Boolean". But in the presence of ES, the last of these conditions can be derived! We have the remarkable fact, discovered by Radu Diaconescu [75], that the axiom of choice implies that the logic of a topos must be classical.

THEOREM 5. *If $\mathscr{E}$ satisfies ES, then $\mathscr{E}$ is Boolean.*

The basis of Diaconescu's result is that if epics with domain $d+d$ have sections, then each subobject $f: a \rightarrowtail d$ of d has a complement in Sub(d). The construction, as described in Boileau [75], is best illustrated in **Set**, where we can see how it produces a categorial characterisation of the complement $-A$ in D of a subset $A \subseteq D$.

(1) Form the co-product $i_1, i_2: d \rightrightarrows d+d$, with injections, i_1, i_2.

In **Set** we take D_1 and D_2 as two disjoint "copies" of D, containing copies A_1 and A_2 respectively of A. $D+D$ is $D_1 \cup D_2$.

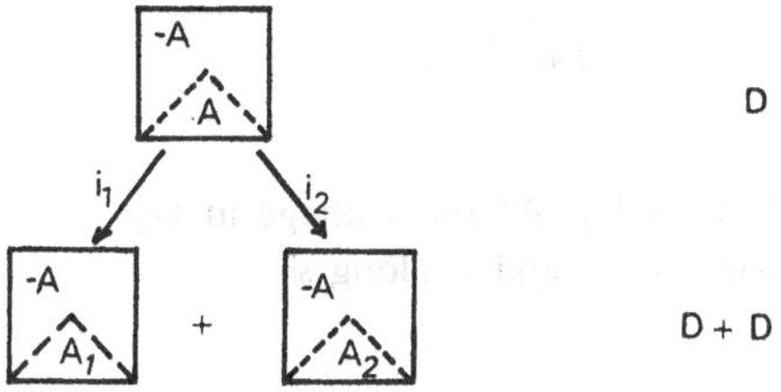

Fig. 12.2.

(2) Let $g: d+d \twoheadrightarrow b$ be the co-equaliser (hence an epic) of $i_1 \circ f: a \rightarrow d+d$ and $i_2 \circ f: a \rightarrow d+d$.

In **Set** f is the inclusion $A \hookrightarrow D$. The effect of g is to amalgamate the two copies A_1 and A_2 of A into a single copy $A' \cong A$, and to leave $-A_1$ and $-A_2$ as they are

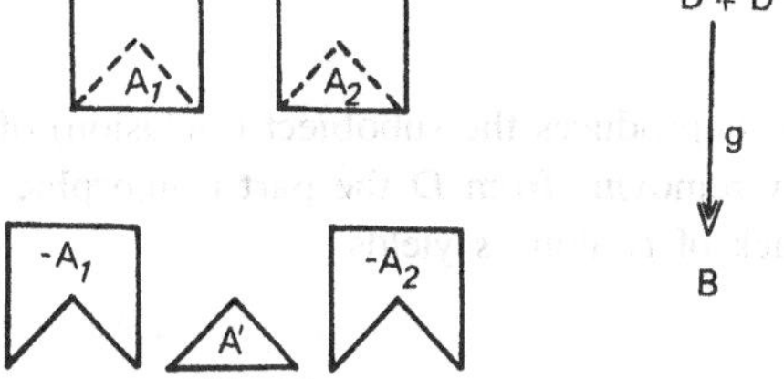

Fig. 12.3.

(3) Let $s: b \rightarrowtail d+d$ be a section of g.

In **Set**, s acts to literally split A' into two pieces, part going into D_1 and part into D_2

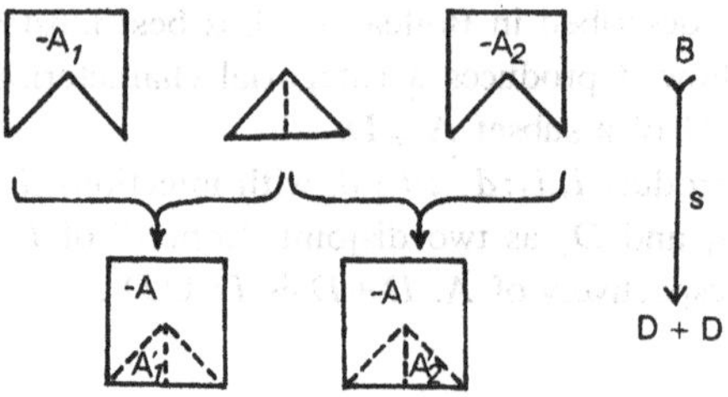

Fig. 12.4.

A'_1 is the s-image of A' in D_1, A'_2 the s-image in D_2.

(4) Form the pullbacks of i_1 and i_2 along s

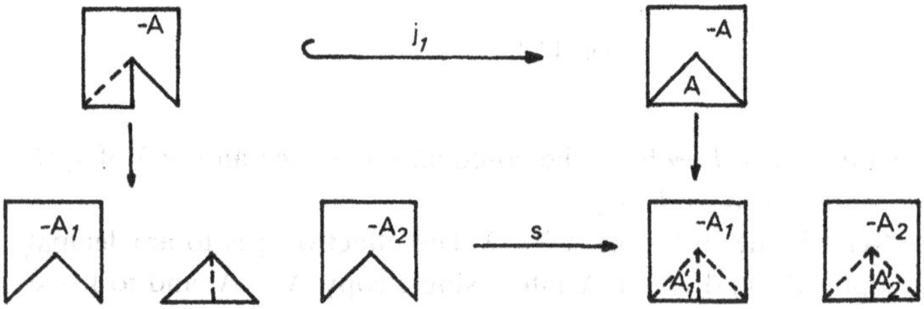

Fig. 12.5.

In **Set** the pullback of i_1 produces the subobject (inclusion) of D whose domain is obtained by removing from D the part isomorphic to A'_2.

Similarly the pullback of i_2 along s yields

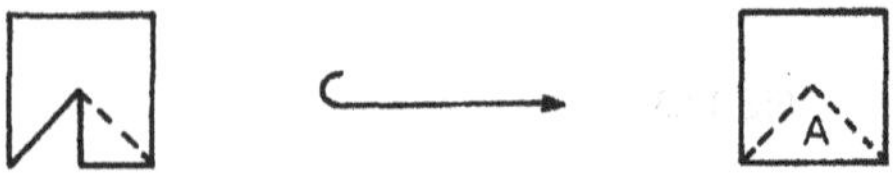

Fig. 12.6.

(5) Form the intersection (pullback) of j_1 and j_2.
In **Set** this gives the intersection

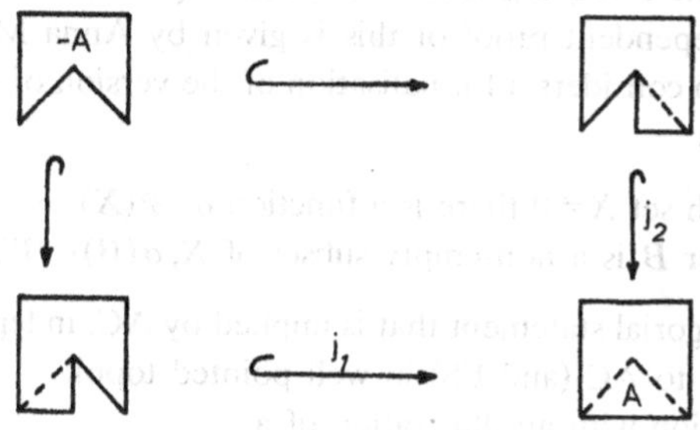

Fig. 12.7.

of the domains of j_1 and j_2, i.e. the subset $-A$.

The five steps of this construction can be carried out in any topos to show that the intersection of the pullbacks of i_1 and i_2 along a section of the co-equaliser of the diagram

$$a \overset{f}{\rightarrowtail} d \underset{i_2}{\overset{i_1}{\rightrightarrows}} d+d$$

is a complement of f in Sub(d). Thus all elements of Sub(d) have complements if $\mathscr{E}\models$ES, and since Sub(d) is a distributive lattice, it must therefore be a Boolean algebra. A detailed proof of Theorem 5, using a modification of this construction, and due to G. M. Kelly, is given by Brook [74]. There is also a proof given in Johnstone [77], Chapter 5.

Note that, by §7.3, for $\mathscr{E}$ to be Boolean it suffices to have a complement for $true: 1 \to \Omega$ in Sub(Ω). Thus a sufficient condition for Booleanness is that the co-equaliser of

$$1 \xrightarrow{true} \Omega \underset{i_2}{\overset{i_1}{\rightrightarrows}} \Omega+\Omega$$

splits.

THEOREM 6. *$\mathscr{E}\models$AC iff $\mathscr{E}\models$ES and $\mathscr{E}\models$NE.* □

We have already noted that topoi, e.g. **Bn**(I), can have splitting epics but not be fully extensional (well-pointed). However in view of Theorem 5, we see from Theorem 2 that if $\mathscr{E}\models$ES, then $\mathscr{E}$ is at least weakly extensional, since then $\mathscr{E}\models$SS and $\mathscr{E}$ is Boolean. Extensionality on the other hand does not imply ES or AC. By Cohen's work [66] there are models of set theory, hence well-pointed topoi, in which the axiom of choice fails.

It follows from the foregoing results that AC implies Booleanness for any topos. An independent proof of this is given by Anna Michaelides Penk [75], who also considers a formalisation of the version of the choice principle that reads

> "for each set $X \neq \emptyset$ there is a function $\sigma : \mathcal{P}(X) \to X$ such that whenever B is a non-empty subset of X, $\sigma(B) \in B$".

This leads to a categorial statement that is implied by AC, independent of ES, and equivalent to AC (and ES) in well-pointed topoi.

We end this section with an illustration of a

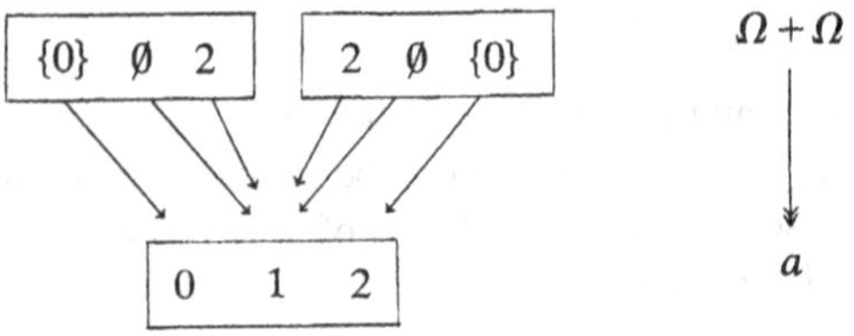

"non-splitting" epic arrow $\Omega + \Omega \twoheadrightarrow a$ in the topos $\mathbf{M_2}$. Here $a = (\{0, 1, 2\}, \lambda)$ has $\lambda(1, x) = x$, and $\lambda(0, x) = 1$, all $x \in \{0, 1, 2\}$.

Exercise 2. Show that λ as defined is an action on $\{0, 1, 2\}$ and that the displayed epic is an $\mathbf{M_2}$-arrow (equivariant). Explain why it has no section.

Exercise 3. Make a similar display of the co-equaliser of

$$1 \underset{i_2 \circ \top}{\overset{i_1 \circ \top}{\rightrightarrows}} \Omega + \Omega$$

in $\mathbf{M_2}$ and explain why it has no section.

Exercise 4. Show that SS holds in $\mathbf{M_2}$, and (hence?) that NE does as well.

Exercise 5. Show that SS and (hence?) NE fail in $\mathbf{Z_2}$-**Set** where $\mathbf{Z_2}$ is the group

+	0	1
0	0	1
1	1	0

of the integers mod 2 under addition. Explain why the situation is typical, i.e. why SS and NE always fail in **M-Set** when **M** is a (non-trivial) group.

EXERCISE 6. Carry out Exercise 3 for the topos $\mathbf{Set}^{\to}$. □

12.2. Natural numbers objects

An obvious difference between **Set**, and the topoi **Finset** and **Finord** is that in the latter, all objects are finite. Various definitions of "finite object" in a topos are explored by Brook [74], and Kock, Lecouturier, and Mikkelsen [75]. Our concern now is with the existence in set theory of infinite objects, the primary example being the set $\omega = \{0, 1, 2, \ldots\}$ of all finite ordinals, whose members are the set-theoretic representatives of the intuitively conceived natural numbers.

ω can be thought of as being generated by starting with 0 and "repeatedly adding 1", to produce the series $1 = 0+1$, $2 = 1+1$, $3 = 2+1, \ldots$. The process of "adding 1" yields the *successor* function $s: \omega \to \omega$ which for each input $n \in \omega$ gives output $n+1$. That is, $s(n) = n+1$.

(Notice that $n = \{0, \ldots, n-1\}$ and $n+1 = \{0, \ldots, n\}$ so that an explicit set-theoretic definition of s is available:- $s(n) = n+1 = n \cup \{n\}$.)

Now the initial ordinal 0 may be identified with an arrow $O: 1 \to \omega$ in the usual way (indeed the arrow is the inclusion $\{0\} \hookrightarrow \omega$). Then we have a diagram

$$1 \xrightarrow{O} \omega \xrightarrow{s} \omega$$

which was observed by Lawvere [64] to enjoy a kind of co-universal property that characterises the natural numbers uniquely up to isomorphism in **Set**. The property that the diagram has is that all diagrams of its type, i.e. of the type

$$1 \xrightarrow{x} A \xrightarrow{f} A$$

factor uniquely throught it. For, given functions x and f as shown we may use f and the element $x(0)$ of A to generate a sequence

$$x(0),\ f(x(0)),\ f(f(x(0))),\ f(f(f(x(0)))), \ldots$$

in A by "repeatedly applying f". Now this sequence can itself be described as a function $h:\omega\to A$ from ω to A, displayed as

$$h(0), h(1), h(2), h(3), \ldots$$

h is defined inductively, or *recursively* in two parts.

(1) We let $h(0)$ be the first term $x(0)$ in the sequence, i.e.

$$(*) \qquad h(0)=x(0),$$

(2) Having defined the n-th term $h(n)$, apply f to it to get the next term $h(n+1)$, i.e.

$$h(n+1)=f(h(n)).$$

Since $n+1=s(n)$, this equation becomes

$$(**) \qquad h\circ s(n)=f\circ h(n).$$

$(*)$ and $(**)$ mean that the diagram

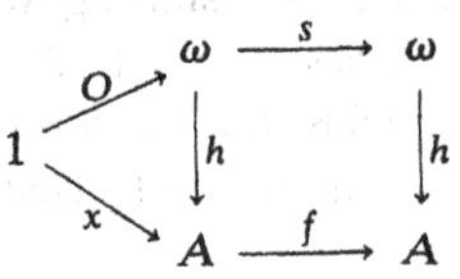

commutes, giving the "factoring" mentioned above. But also we see that the only way for this diagram to commute is for h to obey the equations $(*)$ and $(**)$, so h can only be the function generated in the way we did it. h is said to be *defined recursively* from the data x and f.

Inductive definitions of this type are called definitions by *simple recursion* and would seem to originate with Dedekind [88]. They lead us to the following axiom, which we have seen to be true of **Set**.

NNO: *There exists a natural numbers object* (nno), *i.e. an object* N *with arrows* $1\xrightarrow{O}N\xrightarrow{s}N$ *such that for any object* a, *and arrows* $1\xrightarrow{x}a\xrightarrow{f}a$ *there is exactly one arrow* $h:N\to a$ *making*

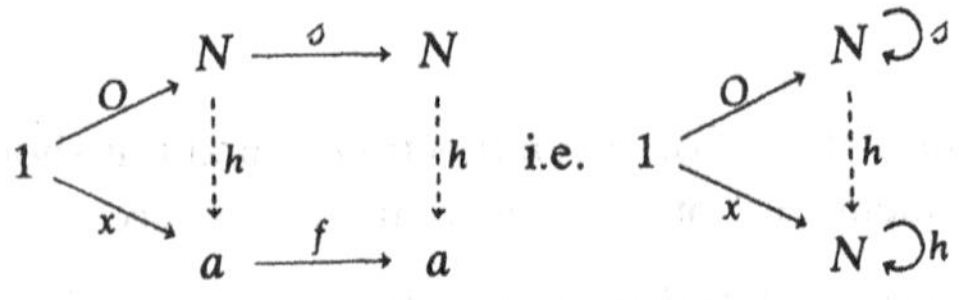

commute.

EXERCISE 1. If $1 \xrightarrow{O} N \xrightarrow{s} N$ and $1 \xrightarrow{O'} N' \xrightarrow{s'} N'$ are nno's, then the unique h in

$$\begin{array}{ccccc} & & N & \xrightarrow{s} & N \\ & \nearrow^{O} & \big\downarrow h & & \big\downarrow h \\ 1 & & & & \\ & \searrow_{O'} & & & \\ & & N' & \xrightarrow{s'} & N' \end{array}$$

is iso. □

This exercise establishes that natural numbers objects are unique up to isomorphism in any category. Arrows $h : N \to a$ with $\text{dom} = N$ will on occasion be called *sequences*.

A multiplicity of examples of nno's is provided by

THEOREM 1. *For any (small) category* $\mathscr{C}$, $\mathbf{Set}^{\mathscr{C}} \models \text{NNO}$.

CONSTRUCTION FOR PROOF. Let $N : \mathscr{C} \to \mathbf{Set}$ be the constant functor having

$$N(a) = \omega, \quad \text{all } \mathscr{C}\text{-objects } a$$

$$N(f) = \text{id}_\omega, \quad \text{all } \mathscr{C}\text{-arrows } f.$$

$s : N \dot{\to} N$ is the constant natural transformation with component $s_a : N(a) \to N(a)$ being the successor function $s : \omega \to \omega$ for each a.

$O : 1 \dot{\to} N$ is the constant transformation with each component $O_a : 1(a) \to N(a)$ being $\{0\} \hookrightarrow \omega$. That this construction satisfies the axiom NNO is left for the reader to establish (the definition of the unique h is obvious, that it is a natural transformation is not). □

EXERCISE 2. Describe the natural numbers objects in $\mathbf{Set}^2$, $\mathbf{Set}^{\to}$, and **M-Set**, in terms appropriate to the way these topoi were originally defined. □

In $\mathbf{Bn}(I)$ as one would expect, N is a bundle of copies of ω. Formally N is $\text{pr}_I : I \times \omega \to I$, so that the stalk N_i over i is

$$\{i\} \times \omega \cong \omega.$$

$s : N \to N$ has $s(\langle i, n \rangle) = \langle i, n+1 \rangle$, i.e. s acts as the successor function on each stalk. $O:1 \to N$ has $O(i) = \langle i, 0\rangle$, so that

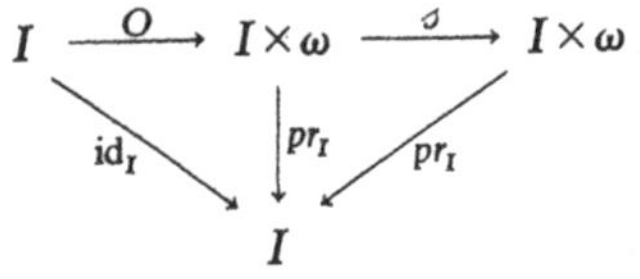

commutes, making O and $\mathfrak{s}$ arrows in $\mathbf{Bn}(I)$.

Given a bundle $a = (A, g)$ and arrows $x:1 \to a$, $f:a \to a$, then

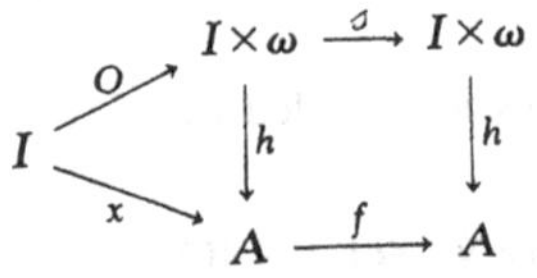

a unique arrow $h: I\times\omega \to A$ may be defined to make the last diagram commute. Fixing attention on the stalk over i, we recursively define h on that stalk by

$$h(i, 0) = x(i)$$
$$h(i, n+1) = f(h(n, i)).$$

This is evidently the only way to make the diagram commute and so h provides the unique arrow from N to a in $\mathbf{Bn}(I)$ defined recursively from the data x and f.

EXERCISE 3. Verify (inductively) that $h: N \to a$, i.e. that $g \circ h = pr_I$.

EXERCISE 4. Show that $\mathfrak{s}$ is the product map $\mathrm{id}_I \times s$, and $O = \langle \mathrm{id}_I, O_I\rangle$, where $O_I : I \to \omega$ has $O_I(i) = 0$, all $i \in I$. □

The spatial topos $\mathbf{Top}(I)$ of sheaves of sets of germs over a topological space I also has a natural numbers object – the same one as $\mathbf{Bn}(I)$. We take the product topology on the stalk space $I \times \omega$, assuming the discrete topology on ω. Thus the basic sets are all those of the form $U \times A$, with U open in I and A any subset of ω. For each point $\langle i, n\rangle$, if U is any open neighbourhood of i in I (e.g. $U = I$), then $U \times \{n\}$ will be an open neighbourhood of $\langle i, n\rangle$ in $I\times\omega$ that projects homeomorphically

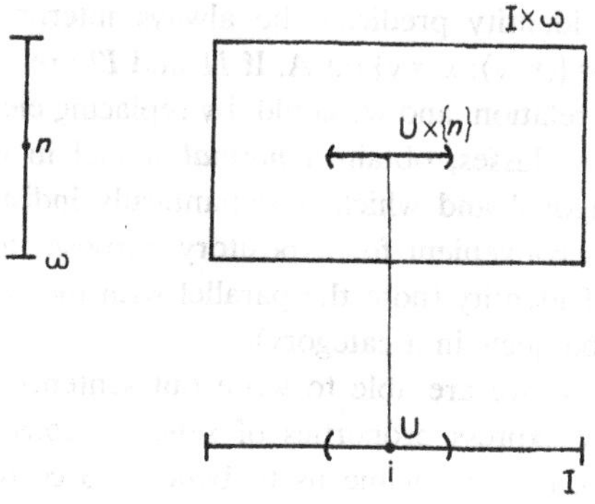

Fig. 12.8.

onto U. Thus pr_I is a local homeomorphism. Moreover each of $\sigma = \mathrm{id}_I \times s$ and $O = \langle \mathrm{id}_I, O_I\rangle$ is a product of continuous maps, hence is continuous, i.e. a **Top**(I)-arrow.

EXERCISE 5. If $x : 1 \to a$ and $f : a \to a$ are **Top**(I)-arrows, so that x and f are continuous, prove (inductively) that the unique h defined recursively from x and f in **Bn**(I) is also continuous, hence a **Top**(I)-arrow. □

We shall reconsider the structure of nno's in **Top**(I) again in Chapter 14, in relation to "locally constant natural-number-valued functions on I".

In any topos satisfying NNO a good deal of the arithmetic of the natural numbers can be developed. This will be considered in the next chapter.

The co-universal property of a natural numbers object will be fully elucidated in Chapter 15.

12.3. Formal set theory

The first-order language $\mathscr{L}$ that we shall use for set-theory has a single binary predicate $\boldsymbol{\varepsilon}$, and no function symbols, or individual constants. Thus $\mathscr{L} = \{\boldsymbol{\varepsilon}\}$.

The definition of $\mathscr{L}$-model that we shall adopt is a little wider than that of §11.2. A model is a structure $\mathfrak{A} = \langle A, E, \simeq\rangle$, where E and $\simeq$ are binary relations on A, such that the identity axioms *I1* and *I2* are valid in $\mathfrak{A}$ when ε is interpreted as E and $\approx$ as $\simeq$. Thus we are giving up the requirement that the identity predicate be always interpreted as the "diagonal" relation $\Delta = \{\langle x, y\rangle : x = y\}$ on A. If *I1* and *I2* are valid then $\simeq$ will be an equivalence relation, and we could, by replacing elements of A by their $\simeq$-equivalence classes, obtain a *normal* model in which $\approx$ is interpreted as the diagonal and which is semantically indistinguishable from $\mathfrak{A}$. However it is convenient for expository purposes to allow the wider interpretation of identity (note the parallel with the way we have treated equality of subobjects in a category).

Using the language $\mathscr{L}$, we are able to write out sentences (strings of symbols) that formally express properties of sets. By considering sentences that our intuitions may incline us to believe to correctly codify ways that sets actually do behave, and by using the precise and rigorous machinery of deduction in elementary logic, we are able to examine the consequences of our intuitively based assumptions about sets. Thus if Σ is a collection of sentences expressing what we take to be truths of set theory, and φ holds in all $\mathscr{L}$-models of Σ, then we would regard φ as a truth of set theory, whatever "the universe of sets" looks like.

Our intention then is to regard an $\mathscr{L}$-structure $\mathfrak{A} = \langle A, E, \simeq\rangle$ as a formal, abstract, model or representation of the intuitively-conceived universe of all sets, from which we developed the idea of the category **Set**. There is a conceptual barrier to this that seems to belong uniquely to the study of set theory. While we have no difficulty in thinking of, say, a Boolean algebra as being any model of a certain group of axioms, since a Boolean algebra is *conceived of* as an abstract set satisfying appropriate laws, it is difficult not to think of a model for set theory as consisting of very particular sorts of things, namely sets. We regard the variables $v_1, v_2, \ldots$ as referring to collections, whereas the individuals in $\mathfrak{A}$ are just that – individuals with no particular presupposed structure. We give the atomic formula $v_1 \varepsilon v_2$ its *intended* reading "v_1 is a *member* of v_2", whereas all we mean is $\mathfrak{A} \models v_1 \varepsilon v_2[x_1, x_2]$, i.e. $x_1 E x_2$.

Having taken pains to spell this out, we should recognise it as being, not a source of pedantry, but rather the very essence of the enterprise itself. By forcing ourselves to regard ε as being an abstract relation between indeterminate things, we force ourselves to stand back from our presuppositions about what "membership" means, and thereby to identify those assumptions and determine what they commit us to.

We must also be careful to distinguish between *metalanguage* and *object-language*, between the language *in* which we speak and the language *about* which we speak. The object language is the first-order language $\mathscr{L}$. The metalanguage is the language we use to talk about $\mathscr{L}$ and about the meanings of $\mathscr{L}$-sentences (interpretations, models). It is the language in which we make statements like "φ is satisfied by every valuation in $\mathfrak{A}$". This metalanguage consists basically of sentences of English and unformalised, intuitive, set theory, which is concerned with actual collections. Thus the $\mathscr{L}$-formulas form a collection, a model $\mathfrak{A}$ is based on a collection A of individuals, the relation E is a collection of ordered pairs, and so on. These collections are described by the metalanguage. They are "metasets", and we continue to use the symbol $\in$ to denote membership of such collections. The individuals in A on the other hand might be called "sets in the sense of $\mathfrak{A}$", or simply "$\mathfrak{A}$-sets".

The distinction between these two levels can perhaps be made, somewhat colloquially, by contrasting our perspective, as we look at $\mathfrak{A}$ "from outside", with that of an imaginary person who lives "inside" $\mathfrak{A}$ and is aware only of the existence of the individuals in A, i.e. of the $\mathfrak{A}$-sets. While to us, A is a set – an individual in our metauniverse of metasets – the $\mathfrak{A}$-person does not see A at all as an individual in his world. Rather, A represents the whole universe for the $\mathfrak{A}$-person. Similarly if B is a subset of A (i.e. $B \subseteq A$), the metaset B may not be an $\mathfrak{A}$-set (if $B \notin A$). However it is possible in some cases that B *corresponds* to an $\mathfrak{A}$-set. This occurs when there is an $\mathfrak{A}$-set b (i.e. $b \in A$) whose E-members are just the $\in$-members of B, i.e. $B = \{x: x \in A \text{ and } xEb\}$. We shall return to this point shortly.

Now if a and b are members of $A(a, b \in A)$, then the statement "a is a member of b" when interpreted on the metalevel means $a \in b$. However when uttered by the $\mathfrak{A}$-person it means aEb. In some models, the *standard* ones, these two interpretations are the same. Thus a model is standard if E is simply the meta-membership relation restricted to A, i.e. the relation

$$\in \restriction A = \{\langle x, y\rangle: x \in A, y \in A, \text{ and } x \in y\}.$$

In a standard model, the metalevel/object-level distinction can be very delicate. If y is an $\mathfrak{A}$-set, and $x \in y$, we cannot then assume that the statement "$x \in y$" makes any sense inside $\mathfrak{A}$. Unless $x \in A$ as well, which is not necessary, the $\mathfrak{A}$-person will be unaware of the existence of x. Thus he may not recognise all the y-members that we do.

We recall now the expression $\varphi \equiv \psi$ as an abbreviation for the $\mathscr{L}$-formula $(\varphi \supset \psi) \wedge (\psi \supset \varphi)$.

Axiom of Extensionality. This is the $\mathscr{L}$-formula

Ext: $\quad (\forall t)(t\varepsilon u \equiv t\varepsilon v) \supset u \approx v,$

which formalises the principle that sets with the same members are equal.

In a model $\mathfrak{A}$, if $x \in A$, let

$$E_x = \{z : z \in A \text{ and } zEx\}.$$

Then $\mathfrak{A} \models \text{Ext}$ iff $E_x = E_y$ implies $x \simeq y$, for all $x, y \in A$.

NULL SET:

$$(\exists t)(\forall u)(\sim(u\varepsilon t))$$

"there exists a set with no members". In $\mathfrak{A}$ this is true when there is some $x \in A$ such that E_x is the empty metaset.

PAIRS:

$$\forall u \forall v \exists t[\forall w(w\varepsilon t \equiv w \approx u \vee w \approx v)]$$

"given sets x and y there exists a set having just x and y as members", i.e. "$\{x, y\}$ exists".

POWERSETS: Let "$v \subseteq u$" abbreviate the formula $\forall w(w\varepsilon v \supset w\varepsilon u)$, i.e. "$v$ is a subset of u".

The axiom of powersets is the sentence

$$\forall u \exists t[\forall v(v\varepsilon t \equiv v \subseteq u)]$$

formalising the statement "for any x, there is a set whose members are just the subsets of x".

UNIONS:

$$\forall u \exists t[\forall v(v\varepsilon t \equiv \exists w(w\varepsilon u \wedge v\varepsilon w)]$$

Intuitively, all individuals in the universe are sets, so the members of x are themselves collections. This axiom states the existence of the union of all the members of x.

SEPARATION: If $\varphi(v)$ is a formula with free v, the following is an instance of the Separation axiom schema

$$\text{Sep}_\varphi: \quad \forall u \exists t[\forall v(v\varepsilon t \equiv v\varepsilon u \wedge \varphi(v))]$$

i.e. "given x, there exists a set consisting just of the members of x satisfying φ". Or, "given x, $\{y : y \in x \;\&\; \varphi(y)\}$ exists". This is a formal statement of the separation principle discussed in Chapter 1.

BOUNDED SEPARATION: A formula φ is *bounded* if all occurrences of $\forall$ in φ are at the front of a subformula of φ of the form $\forall v(v \varepsilon t \supset \psi)$, and all occurrences of $\exists$ are of the form $\exists v(v \varepsilon t \wedge \psi)$. Thus quantifiers in bounded formulae have readings of the form "for all v in t" and "there exists a v in t". The *bounded separation* (Δ_0-separation) schema takes as axioms all the formulae Sep_φ for *bounded* φ. It allows us to "separate out" a subset of x defined by a formula, provided that the quantifiers of that formula are restricted to range over sets.

The system Z_0 of axiomatic set theory has, in addition to the classical axioms for first-order logic with identity (§11.3), the axioms of Extensionality, Null Set, Pairs, Powersets, Unions, and Bounded Separation. From Sep_φ and Ext one can derive in Z_0 the sentence

$$\forall u \exists ! t[\forall v(v \varepsilon t \equiv v \varepsilon u \wedge \varphi(v))]$$

that asserts the existence of a unique set having the property that its members are precisely those members of x for which φ holds. Because of this we introduce expressions of the form $\{u: \varphi\}$, called *class abstracts*, as abbreviations for certain $\mathscr{L}$-formulae. The use of class abstracts is determined by stipulating that we write

$$v \varepsilon \{u: \varphi\} \quad \text{for} \quad \varphi[u/v]$$

$$v \approx \{u: \varphi\} \quad \text{for} \quad \forall t(t \varepsilon v \equiv \varphi[u/t]$$

$$\{u: \varphi\} \varepsilon v \quad \text{for} \quad \exists t(t \varepsilon v \wedge t \approx \{u: \varphi\})$$

Class abstracts play the same sort of role in $\mathscr{L}$ as do the corresponding expressions in the metalanguage. If φ has only the variable u free, then intuitively $\{u: \varphi\}$ denotes a collection, the collection of all sets (individuals in the universe) having the property φ. For a model $\mathfrak{A}$, $\{u: \varphi\}$ will determine a metasubset of A, viz the collection

$$\mathfrak{A}_\varphi = \{x: x \in A \text{ and } \mathfrak{A} \models \varphi[x]\}.$$

In some cases, the metaset $\mathfrak{A}_\varphi$ will correspond to an $\mathfrak{A}$-set, as above. This occurs when there is some $y \in A$ such that $\mathfrak{A}_\varphi = E_y = \{x: x \in A \text{ and } xEy\}$. Thus if φ is $\sim(u \approx u)$, we find that $\mathfrak{A}_\varphi = \emptyset$ (the empty metaset), and $\mathfrak{A}_\varphi$ corresponds to an $\mathfrak{A}$-set iff the Null Set axiom is true in $\mathfrak{A}$.

The formula Sep_φ can now be given in the form

$$\forall u \exists t(t \approx \{v: v \varepsilon u \wedge \varphi(v)\}).$$

This is true in $\mathfrak{A}$ when for each $x \in A$ there is some $y \in A$ such that $E_y = E_x \cap \mathfrak{A}_\varphi$.

Some familiar abstracts, and their abbreviations are

$$\begin{array}{rll} \mathbf{0} & \text{for} & \{u: \sim(u \approx u)\} \\ \{u, v\} & \text{for} & \{t: t \approx u \vee t \approx v\} \\ \{u\} & \text{for} & \{u, u\} \\ u \cap v & \text{for} & \{t: t\varepsilon u \wedge t\varepsilon v\} \\ u \cup v & \text{for} & \{t: t\varepsilon u \vee t\varepsilon v\} \\ u - v & \text{for} & \{t: t\varepsilon u \wedge \sim(t\varepsilon v)\} \\ \bigcup u & \text{for} & \{z: \exists t(t\varepsilon u \wedge z\varepsilon t)\} \\ \bigcap u & \text{for} & \{z: \forall t(t\varepsilon u \supset z\varepsilon t)\} \\ \mathbf{1} & \text{for} & \{\mathbf{0}\} \\ u+1 & \text{for} & u \cup \{u\} \\ \mathcal{P}(u) & \text{for} & \{z: z \subseteq u\} \end{array}$$

EXERCISE 1. Let $\varphi(v)$ be the formula $v \approx \{u: u\varepsilon v\}$. Explain why, for any $x \in A$, $\mathfrak{A} \models \varphi[x]$. Show that $\varphi(v)$ is a theorem of first-order logic.

EXERCISE 2. Let $\varphi(t, u, v)$ be the formula $t \approx \{u, v\}$. Show that $\mathfrak{A} \models \varphi[x, y, z]$ iff $E_x = \{y, z\}$.

EXERCISE 3. Show that the Pairs axiom can be written as

$$\forall u \forall v \exists t(t \approx \{u, v\}).$$

EXERCISE 4. Rewrite the other axioms of Z_0 using class abstracts. □

To formalise the notions of *relation* and *function* we denote by $\langle u, v\rangle$ the abstract $\{\{u\}, \{u, v\}\}$. The point of this definition is simply that it works, i.e. that we can derive in Z_0 the sentence

$$(\langle u, v\rangle \approx \langle t, w\rangle) \equiv (u \approx t \wedge v \approx w)$$

which captures the essential property of ordered pairs. Then we put

$$\begin{array}{rll} \{\langle u, v\rangle: \varphi\} & \text{for} & \{t: \exists u \exists v(t \approx \langle u, v\rangle \wedge \varphi)\} \\ t \times w & \text{for} & \{\langle u, v\rangle: u\varepsilon t \wedge v\varepsilon w\} \\ \mathbf{OP}(u) & \text{for} & \exists t \exists v(u \approx \langle t, v\rangle) \\ \mathbf{Rel}(u) & \text{for} & \forall v(v\varepsilon u \supset \mathbf{OP}(v)) \\ \mathbf{Fn}(u) & \text{for} & \mathbf{Rel}(u) \wedge \forall v \forall t \forall w(\langle v, t\rangle\varepsilon u \wedge \langle v, w\rangle\varepsilon u \supset t \approx w) \\ \mathbf{Dom}(u) & \text{for} & \{t: \exists v(\langle t, v\rangle\varepsilon u)\} \end{array}$$

$$\mathbf{Im}(u) \quad \text{for} \quad \{t: \exists v(\langle v, t\rangle \boldsymbol{\varepsilon} u)\}$$
$$\boldsymbol{\Delta}(u) \quad \text{for} \quad \{\langle v, v\rangle: v \boldsymbol{\varepsilon} u\}$$
$$v \circ u \quad \text{for} \quad \{\langle t, w\rangle: \exists s(\langle t, s\rangle \boldsymbol{\varepsilon} u \wedge \langle s, w\rangle \boldsymbol{\varepsilon} v)\}$$

Using these definitions we can construct from any Z_0-model $\mathfrak{A} = \langle A, E, \approx\rangle$ a category $\mathscr{C}(\mathfrak{A})$ by formalising our definition of the category **Set**. The $\mathscr{C}(\mathfrak{A})$-objects are the $\mathfrak{A}$-sets, i.e. the elements $a \in A$. The $\mathscr{C}(\mathfrak{A})$-arrows are the triples $f = \langle a, k, b\rangle$, where a, k, and b are $\mathfrak{A}$-sets, such that

$$\mathfrak{A} \models \varphi[a, k, b]$$

where $\varphi(t, u, v)$ is the formula

$$\mathbf{Fn}(u) \wedge \mathbf{Dom}(u) \approx t \wedge \mathbf{Im}(u) \subseteq v.$$

We take the domain of arrow f to be a, and the codomain to be b. The composite of $f = \langle a, k, b\rangle$ and $g = \langle b, l, c\rangle$, where $\operatorname{cod} f = \operatorname{dom} g$, is $g \circ f = \langle a, h, c\rangle$, where $h \in A$ has

$$\mathfrak{A} \models \psi[h, k, l],$$

$\psi(t, u, v)$ being the formula $t \approx v \circ u$.

The identity arrow for a is $\mathrm{id}_a = \langle a, k, a\rangle$, where, for $\varphi(t, u)$ the formula $t \approx \boldsymbol{\Delta}(u)$, we have

$$\mathfrak{A} \models \varphi[k, a].$$

THEOREM 1. *If $\mathfrak{A}$ is a model of all the Z_0-axioms, then $\mathscr{C}(\mathfrak{A})$ is a well-pointed topos.*

EXERCISE 5. Verify in detail that Theorem 1 holds, by formalising in $\mathscr{L}$, and interpreting in $\mathfrak{A}$, the descriptions of pullbacks, terminal object, exponentials, and subobject classifier given for **Set**. □

AXIOM OF INFINITY: Let $\mathbf{inf}(u)$ be the formula

$$\mathbf{0} \boldsymbol{\varepsilon} u \wedge \forall v(v \boldsymbol{\varepsilon} u \supset v \cup \{v\} \boldsymbol{\varepsilon} u).$$

Intuitively $\mathbf{inf}(u)$ asserts of a set x that the initial ordinal $\emptyset$ is an element of x, and x is closed under the successor function (recall $n+1 = n \cup \{n\}$ in **Set**). Hence $\omega \subseteq x$, and x has infinitely many members. The axiom of infinity is

Inf: $\exists u(\mathbf{inf}(u))$.

In $Z_0 + \mathrm{Inf}$ one can derive

$$\exists t(\mathbf{inf}(t) \wedge t \approx \bigcap\{u: \mathbf{inf}(u)\})$$

and so in any Z_0 model $\mathfrak{A}$ such that $\mathfrak{A} \models \text{Inf}$, there will be an $\mathfrak{A}$-set that the $\mathfrak{A}$-person thinks is the set of all finite ordinals. By formalising the discussion of §12.2 we can then show that this $\mathfrak{A}$-set produces a natural numbers object for $\mathscr{E}(\mathfrak{A})$, i.e. $\mathscr{E}(\mathfrak{A}) \models \text{NNO}$.

Axiom of Choice: There is some choice about which sentence we use to formalise the choice principle in classical set theory. Perhaps the simplest is

$$\forall u \forall v(\mathbf{Fn}(u) \wedge \sim(\mathbf{Dom}(u) \approx \mathbf{0}) \wedge \mathbf{Im}(u) \subseteq v \supset \exists t(\mathbf{Fn}(t) \wedge \mathbf{Dom}(t) \approx v \wedge \mathbf{Im}(t) \subseteq \mathbf{Dom}(u) \wedge u \circ t \circ u \approx u)$$

which formalises the statement AC of §12.1. For a Z_0-model of this sentence we will have $\mathscr{E}(\mathfrak{A}) \models \text{AC}$.

Axiom of Regularity:

Reg: $\forall u(\sim(u \approx \mathbf{0}) \supset \exists v(v \varepsilon u \wedge v \cap u \approx \mathbf{0}))$.

Intuitively, Reg asserts that if $x \neq \emptyset$ then x has a member $y \in x$ such that y and x have no members in common. The basic viewpoint of set theories of the type that we are developing is that sets are built up "from below" by operations such as union, powerset, separation etc. Reg asserts that if x exists, then its construction must have started somewhere, i.e. we cannot have all members of x consisting of members of x. This axiom proscribes relationships like $x \in x$, $x \in y \in x$, $x \in y \in z \in x$, etc., as well as "infinitely descending" membership chains $x_1 \ni x_2 \ni x_3 \ni \dots$.

Axiom of Replacement: Intuitively, the replacement axiom schema asserts that if the domain of a function is a set (individual in the universe) then so is its range, or image. The type of function it deals with is the functional relation defined by a formula φ with two free variables.

Rep_φ $$\forall u \forall v \forall w(\varphi(u, v) \wedge \varphi(u, w) \supset v \approx w) \supset \forall t \exists s(s \approx \{v : \exists u(u \varepsilon t \wedge \varphi(u, v))\}).$$

This asserts that if the ordered pairs satisfying φ form a relation with the "unique output" property of functions, and if for each $u \in t$, $f(u)$ is the unique individual such that $\langle u, f(u) \rangle$ satisfies φ, then the collection $\{f(u) : u \in t\}$ is a set.

The Zermelo–Fraenkel system of set-theory, ZF, can be defined as $Z_0 + \text{Inf} + \text{Reg} + \text{Replacement}$. We see then that ZF is a much more powerful system than is needed to construct topoi. The description of **Set**, when formalised, turns any model of the weaker system Z_0 into a well-pointed topos. In order to reverse the procedure, and construct models of set theory from topoi, we have to analyse further the arrow-theoretic account of the membership relation.

12.4. Transitive sets

A set B determines a metamembership structure that can be displayed as:

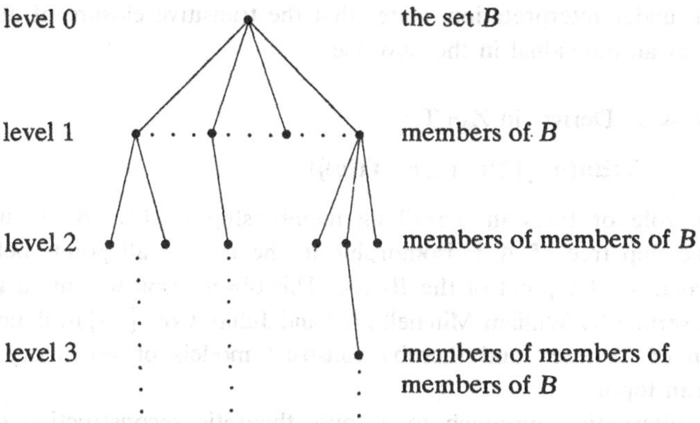

This diagram is called the *membership tree* of B. The tree is in fact upside down – from each point there is a unique path upward towards the root (top point) of the tree. The collection T_B of all points in the tree except the top point B has a special property called *transitivity*. In general a set A is *transitive* if it satisfies the condition

$$x \in A \quad \textit{implies} \quad x \subseteq A,$$

i.e. if x is a member of A then all members of x are themselves members of A. (Notice that if a model $\mathfrak{A}$ is standard, and is based on a transitive A, then for each $\mathfrak{A}$-set x all the metamembers of x will be $\mathfrak{A}$-sets. Thus the $\mathfrak{A}$-person will see the same members of x that we do.)

Now if x appears in T_B at say level n, then all the members of x appear in T_B at level $n+1$. So T_B is transitive. But if A is any transitive set that contains B, it follows that $T_B \subseteq A$. The assumption that $B \subseteq A$ means that all level 1 points of T_B are in A. Then if all level n points are in A, transitivity of A puts all level $n+1$ points in A. Thus by an inductive proof we show that T_B is contained in all transitive sets containing B. It is the "smallest" transitive set containing B, and so is called the *transitive closure* of B.

Axiom of Transitivity: We write $\mathbf{Tr}(u)$ for the formula $\forall v(v \varepsilon u \supset v \subseteq u)$. The axiom of transitivity is

TA: $\forall t \exists u(t \subseteq u \wedge \mathbf{Tr}(u))$

In Z_0+TA we can derive

$$\forall t \exists ! u(t \subseteq u \wedge \mathbf{Tr}(u) \wedge \forall v(t \subseteq v \wedge \mathbf{Tr}(v) \supset u \subseteq v))$$

which, under interpretation, states that the transitive closure of any set exists as an individual in the universe.

Exercise 1. Derive, in Z_0+TA,

$$\forall t \exists u(u \approx \bigcap\{v: t \subseteq v \wedge \mathbf{Tr}(v)\})$$ □

The role of trees in describing membership is this: $A \in B$ iff the membership tree of A is isomorphic to the tree of all points below a particular level 1 point of the B-tree. This observation was lifted to the topos setting by William Mitchell [72] and Julian Cole [73] to define the notion of "$\mathscr{E}$-tree" and thereby construct models of set-theory from Boolean topoi.

An alternative approach to a topos-theoretic reconstruction of set theory was subsequently developed by Gerhard Osius [74], based on a characterisation of those **Set**-objects that are transitive as sets. Transitivity of A simply means that if $x \in A$ then $x \in \mathscr{P}(A)$, i.e. A is transitive iff $A \subseteq \mathscr{P}(A)$. This property gives transitive sets a tractability not enjoyed by sets that are not "closed under $\in$". The relations $E \subseteq A \times A$ on a set A are in bijective correspondence with the functions $r_E : A \to \mathscr{P}(A)$. Given E, then r_E assigns to $y \in A$ the subset

$$r_E(y) = \{x: x \in A \text{ and } xEy\} = E_y \text{ of } A.$$

In the case that E is the membership relation

$$\in \restriction A = \{\langle x, y\rangle : x \in A, y \in A \text{ and } x \in y\},$$

we find that

$$r_{\in}(y) = \{x: x \in A \text{ and } x \in y\}.$$

But if A is *transitive*, this simplifies: $x \in y$ *implies* $x \in A$ for $y \in A$, and so

$$r_{\in}(y) = \{x: x \in y\} = y.$$

Thus we see that for transitive A, the membership relation $\in \restriction A$ on A gives rise to the inclusion $A \hookrightarrow \mathcal{P}(A)$ as $r_\in$, making A a subobject of $\mathcal{P}(A)$.

Now let us consider the problem of defining "membership" in a topos $\mathscr{E}$. We already know what $x \in f$ means if x is an "element" $1 \rightarrow a$ of an $\mathscr{E}$-object a, and $f: b \rightarrowtail a$ is a subobject of a (§4.8). But what about $g \in f$, where $g: c \rightarrowtail a$ is some other subobject of a?

Returning to **Set**, we see that if $g: C \hookrightarrow A$ and $f: B \hookrightarrow A$ are subsets of A, then if C is going to be an element of B, $C \in B$, then since $B \subseteq A$ we will have $C \in A$, so there will be an arrow $\hat{g}: \{0\} \rightarrow A$ with $\hat{g}(0) = C$. But then, knowing that $\hat{g}$ exists, i.e. $C \in A$, deciding whether $C \in B$ is equivalent to deciding whether $\hat{g} \in f$, i.e. whether

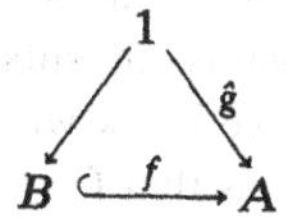

$\hat{g}$ factors through f.

Thus the question of membership of C in B can be resolved in the language of arrows once we know, categorially, whether $\hat{g}$ exists. In the event that A is transitive, the problem can be transferred into $\mathcal{P}(A)$ and restated. In general, $g: C \hookrightarrow A$, as a subset of A, corresponds to an "element" $\ulcorner g \urcorner: 1 \rightarrow \mathcal{P}(A)$ of the powerset of A, where $\ulcorner g \urcorner(0) = C$. Identifying $\mathcal{P}(A)$ with 2^A, we see that $\ulcorner g \urcorner$ becomes $\ulcorner \chi_g \urcorner$, the *name* of $\chi_g: A \rightarrow 2$ as defined in §4.1. Then if there is an inclusion $r_\in: A \hookrightarrow \mathcal{P}(A)$, we have that $C \in A$, i.e. $\hat{g}$ as defined is an arrow from 1 to A, iff $\ulcorner g \urcorner \in r_\in$, that is, $C \in A$ iff $\hat{g}$ exists to make

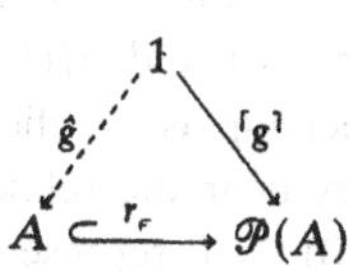

$\ulcorner g \urcorner$ factor (uniquely) through $r_\in$.

Altogether then, for transitive A, we can characterise the "local set theory" of subsets of A. For $f:B \hookrightarrow A$ and $g:C \hookrightarrow A$, we have $g \in f$ iff $C \in B$ iff the *name* of g factors through $r_{\in} \circ f$,

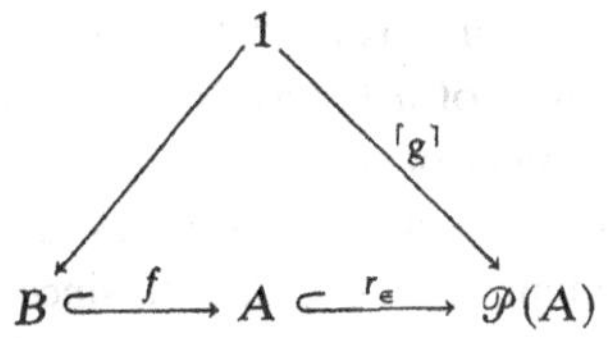

i.e. iff $\ulcorner g \urcorner \in r_{\in} \circ f$.

Characterising the local set theory of an object (set) is, as Osius notes, sufficient for the needs of the "working mathematician", who tends to deal with any given problem within the context of some fixed "universal" set A. But the "global" question of membership for $\mathscr{E}$ can be reduced to the local one. First we need to deal with equality of subobjects. If $f:b \rightarrowtail a$ and $g:c \rightarrowtail a$ have the same codomain, we know what it means for f and g to represent the same "subset" – it means that $f \simeq g$ in Sub(a). But $f:b \rightarrowtail a$ and $g:c \rightarrowtail d$ may still represent the same set, even if they have distinct codomains. In **Set**, the codomains of $f:B \rightarrowtail A$ and $g:C \rightarrowtail D$ may overlap, and indeed we may have $f(B) = g(C) \subseteq A \cap D$, in which case we would want to put $f \simeq g$. But it is clear in this situation that if T is any set that includes both A and D (e.g. $T = A \cup D$), so that there are inclusions $i:A \hookrightarrow T$ and $j:D \hookrightarrow T$, then $f(B) = g(C)$ iff $i(f(B) = j(g(C))$. Thus $f \simeq g$ iff in Sub(T), $i \circ f \simeq j \circ g$.

So the identification of subobjects – the general definition of $f \simeq g$ – is resolved by localising to the set-theory of any object that includes the co-domains of both f and g. The global membership for **Set** can now be described as follows. For $f:B \rightarrowtail A$ and $g:C \rightarrowtail D$ we put

$g \in f$ iff for some *transitive* T including both A and D, in $\mathscr{P}(T)$ we have $[j(g(C) \hookrightarrow T] \in [i(f(B)) \hookrightarrow T]$.

Here i and j are the inclusion as above. For a suitable T we may use the transitive closure of $A \cup D$. Although the arrows f and $i(f(B)) \hookrightarrow T$ are not the same thing, the definition of membership is justified precisely because they are equal as subobjects, i.e. they bear the relation "$\simeq$" to each other. Similarly the arrows g and $j(g(C)) \hookrightarrow T$ represent the same set.

EXERCISE 2. Verify this last statement.

EXERCISE 3. Show that the definition of $g \in f$ does not depend on the choice of appropriate T.

EXERCISE 4. For any sets A, B, show $A \in B$ iff $\mathrm{id}_A \in \mathrm{id}_B$.

EXERCISE 5. Let T_A be the transitive closure of A, so that $A \hookrightarrow T_A$. Show that $g \in f$ iff for some $h : Y \hookrightarrow T_A$, $g \simeq h$ and in $\mathscr{P}(T_A)$, $h \in (f(B) \hookrightarrow T_A)$. Thus $g \underset{\bullet}{\in} f$ iff g is "equal" to a member of $f(B)$ in T_A. □

In lifting these considerations to a topos $\mathscr{E}$, we take an $\mathscr{E}$-object a that is the domain of a subobject $r : a \rightarrowtail \Omega^a$ of its own power object. Then a "membership" relation $\in_r$ can be defined on Sub(a) by putting, for $f : b \rightarrowtail a$ and $g : c \rightarrowtail a$,

$$g \in_r f \quad \textit{iff} \quad \ulcorner g \urcorner \in r \circ f$$

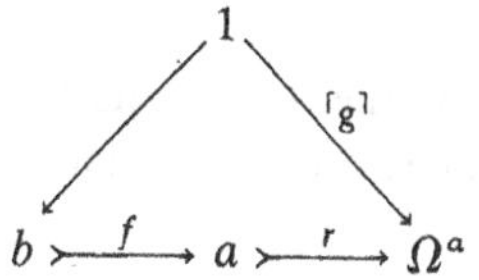

i.e. iff $\ulcorner g \urcorner$ factors through $r \circ f$, where $\ulcorner g \urcorner = \ulcorner \chi_g \urcorner$ is the exponential adjoint of $\chi_g \circ pr_a : 1 \times a \to \Omega$.

Although this definition can be made for any r of this form, the simple requirement that r be monic does not capture the essence of transitivity. Indeed, it does not even capture the fact that for transitive A, $A \hookrightarrow \mathscr{P}(A)$ arises from the *metamembership* relation $\in \restriction A$. For if $\mathfrak{A} = \langle A, E, \Delta \rangle$ is any normal $\mathscr{L}$-model, then since $r_E(y) = \{x : x \in A$ and $xEy\} = E_y$, $r_E : A \to \mathscr{P}(A)$ will be monic if (and only if) $\mathfrak{A} \models \mathrm{Ext}$.

So the problem remains of determining when $r : A \rightarrowtail \mathscr{P}(A)$ represents the membership relation of a *transitive* set.

COLLAPSING LEMMA (Mostowski [49]). *Let $E \subseteq A \times A$ be a relation on A. Then there exists a transitive set B such that*

$$\langle A, E \rangle \cong \langle B, \in \restriction B \rangle$$

iff

(1) *E is extensional, and*
(2) *E is well-founded.* □

Here, (1) means that $r_E : A \to \mathcal{P}(A)$ is monic. Well-foundedness means that every non-empty subset of A has an E-minimal element. That is, if $C \subseteq A$ and $C \neq \emptyset$, there exists $x \in C$ such that $E_x \cap C = \emptyset$, so that if yEx, then $y \notin C$.

The sense of isomorphism in $\langle A, E\rangle \cong \langle B, \in \restriction B\rangle$ is that "E-membership" within A looks exactly like "$\in$-membership" within B. This requires that there be a bijective map $f\colon A \cong B$ such that

$$(*) \qquad xEy \quad \text{iff} \quad f(x) \in f(y), \quad \text{all } x, y \text{ in } A.$$

For such an f, the diagram

$$\begin{array}{ccc} A & \xrightarrow{\ f\ } & B \\ {\scriptstyle r_E}\downarrow & & \downarrow{\scriptstyle r_\in} \\ \mathcal{P}(A) & \xrightarrow{\ \mathcal{P}f\ } & \mathcal{P}(B) \end{array}$$

commutes, where $\mathcal{P}f$ assigns to $C \in \mathcal{P}(A)$ (i.e. $C \subseteq A$) its f-image $f[C] = \{f(y)\colon y \in C\} \in \mathcal{P}(B)$. The diagram requires, for $x \in A$, that

$$f[E_x] = f(x)$$

i.e.

$$\{f(y)\colon yEx\} = \{z\colon z \in f(x)\},$$

which for bijective f is equivalent to $(*)$.

Mostowski's lemma has been stated as a fact about our metaset-theory. It can be expressed as a sentence of the formal language $\mathcal{L}$. "E is a relation on A" would be replaced by "$\mathbf{Rel}(u) \wedge u \subseteq v \times v$", $\in \restriction B$ would be replaced by an abstract of the form $\boldsymbol{\varepsilon} \restriction t = \{\langle u, v\rangle\colon u \boldsymbol{\varepsilon} t \wedge v \boldsymbol{\varepsilon} t \wedge u \boldsymbol{\varepsilon} v\}$, and so on. The resulting formal sentence can then be derived only if we assume the full strength of the ZF axioms. Thus Mostowski's "theorem" is a theorem only if our metaset-theory satisfies all the ZF-axioms.

Note that the lemma implies in particular that $\in \restriction B$ is well-founded on B. This in fact can be deduced if we assume our metaset-theory satisfies the Regularity axiom. For then if $C \subseteq B$ is non-empty there will be some $x \in C$ with $x \cap C = \emptyset$, so that if $y \in x$, $y \notin C$, making x $\in$-minimal in B.

Now a well-founded relation E on A can be used to define functions with domain A by "recursion" in a similar manner to the operation of nno's. The intuitive idea is that in order to define $f(x)$, where $f\colon A \to B$, we make the inductive assumption that $f(y)$ has been defined for all yEx, i.e. f is defined for all "E-members" of x. We then input the collection $\{f(y)\colon yEx\}$ to some other function g and let $f(x)$ be defined to be the resulting output. Thus

$$f(x) = g(\{f(y)\colon yEx\}) = g(f[E_x])$$

i.e.

$$(**) \qquad f(x) = g(\mathcal{P}f(E_x))$$

Since we want $f(x) \in B$, and since $\mathcal{P}f(E_x) \in \mathcal{P}(B)$, g has to be a function from $\mathcal{P}(B)$ to B. Equation $(**)$ states that the diagram

$$\begin{array}{ccc} A & \xrightarrow{\ f\ } & B \\ {\scriptstyle r_E}\big\downarrow & & \big\uparrow{\scriptstyle g} \\ \mathcal{P}(A) & \xrightarrow{\ \mathcal{P}f\ } & \mathcal{P}(B) \end{array}$$

commutes. But, given g, if f exists to make this diagram commute then it is uniquely determined by the equation $(**)$.

THEOREM 1. *E is well-founded on A iff for any set B and function $g\colon\mathcal{P}(B)\to B$ there exists exactly one function $f\colon A\to B$ making the last diagram commute.*

A proof of this result is given by Osius in [74]. Again the statement can be expressed as an $\mathcal{L}$-sentence, but this time it can be derived just using Z_0-axioms. Thus we see that in ZF, transitive sets are essentially extensional (monic) well-founded relations, and that well-foundedness can be characterised, even in Z_0, by an arrow-theoretic property.

This will lead us to a definition of "transitive sets" in a topos, for which we will also appeal to the following description of inclusions between transitive sets.

THEOREM 2. *If A and B are transitive then*

$$\begin{array}{ccc} A & \xrightarrow{\ f\ } & B \\ {\scriptstyle r_\in}\big\downarrow & & \big\downarrow \\ \mathcal{P}(A) & \xrightarrow{\ \mathcal{P}f\ } & \mathcal{P}(B) \end{array}$$

commutes iff $A \subseteq B$ *and* f *is the inclusion* $A \hookrightarrow B$.

Proof. If f is the inclusion, it is clear, for $x \in A$ (hence $x \subseteq A$) that $f[x] = \{y : y \in x\} = x$, so the diagram commutes. On the other hand, if the diagram does commute, then $f(x) = f[x]$, for all $x \in A$. To show that f is the inclusion we have to show that $f(x) = x$, all x in A, or that

$$C = \{x : x \in A \text{ and } f(x) \neq x\} = \emptyset$$

To do this we need to assume $\in \restriction A$ is well-founded.

Then if C were a non-empty subset of A it would have an element x_0 that is $\in$-minimal in C. Thus $x_0 \neq f(x_0)$, but (using transitivity)

$$y \in x_0 \quad \text{implies} \quad y \notin C, \quad \text{and so} \quad f(y) = y.$$

But then $f(x_0) = f[x_0] = \{f(y) : y \in x_0\} = \{y : y \in x_0\} = x_0$, a contradiction. □

Theorem 2 can be expressed as an $\mathscr{L}$-sentence derivable in $Z_0 + \text{Reg}$ (Regularity being used to give well-foundedness of $\in \restriction A$). The proof of the theorem indicates what lies behind Theorem 1, i.e. how inductive definitions and constructions depend on the property of well-foundedness for their validity.

12.5. Set-objects

Images: If $f : a \to b$ is an arrow in topos $\mathscr{E}$, then for each subobject $g : c \rightarrowtail a$ of a we define the image $f[g] : f(g(c)) \rightarrowtail b$ of g under f to be the monic part of the epi-monic factorisation

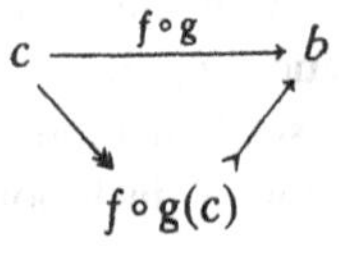

Thus $f[g] = \text{im}(f \circ g)$.

This construction establishes a map from Sub(a) to Sub(b), that in fact has an internal version $\Omega^f : \Omega^a \to \Omega^b$. In **Set** Ω^f is the function $\mathscr{P}f : \mathscr{P}(A) \to \mathscr{P}(B)$ used in the last section.

Now by the identification of subobjects with their characters, the image construction assigns to each $h : a \to \Omega$ an arrow $f[h] : b \to \Omega$. Then, starting with $f : a \to b$ we form $1 \times f : \Omega^a \times a \to \Omega^a \times b$ and then take the image $1_{\Omega^a} \times f[ev_a]$ of $ev_a : \Omega^a \times a \to \Omega$ under $1_{\Omega^a} \times f$.

Ω^f is then defined as the unique arrow making

$$\begin{array}{ccc} \Omega^b \times b & \xrightarrow{\,ev_b\,} & \\ \uparrow{\scriptstyle \Omega^f \times 1_b} & \searrow & \Omega \\ \Omega^a \times b & \xrightarrow{1_{\Omega^a} \times f[ev_a]} & \end{array}$$

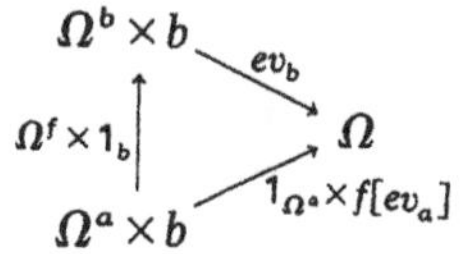

commute, i.e. Ω^f is the exponential adjoint of $1_{\Omega^a} \times f[ev_a]$.

EXERCISE 1. If $f : a \rightarrowtail b$ is monic, then $f[g] \simeq f \circ g$.

EXERCISE 2. Verify that the definition of Ω^f-characterises $\mathscr{P}f$ in **Set**.

EXERCISE 3. Show that $\Omega^{1_a} = 1_{\Omega^a}$, and that if

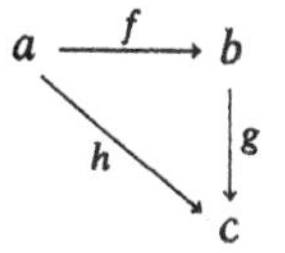

commutes, then so does

$$\begin{array}{ccc} \Omega^a & \xrightarrow{\;\Omega^f\;} & \Omega^b \\ & \searrow{\scriptstyle \Omega^h} & \downarrow{\scriptstyle \Omega^g} \\ & & \Omega^c \end{array}$$

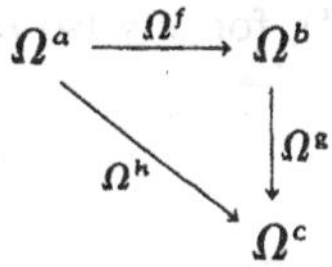

i.e. $\Omega^{g \circ f} = \Omega^g \circ \Omega^f$.

EXERCISE 4. Given $c \xrightarrow{g} a \xrightarrow{f} b$, show

$$\begin{array}{ccc} & \xrightarrow{\ulcorner g \urcorner} & \Omega^a \\ 1 & & \downarrow{\scriptstyle \Omega^f} \\ & \xrightarrow{\ulcorner f[g] \urcorner} & \Omega^b \end{array}$$

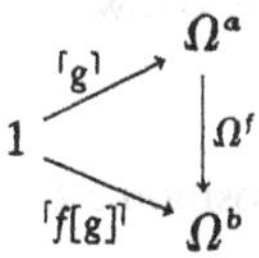

commutes.   □

DEFINITION. A *transitive set object* (tso) is an $\mathscr{E}$-arrow $r : a \rightarrowtail \Omega^a$ that is
(1) *extensional*, i.e. monic, and
(2) *recursive*, i.e. for any $\mathscr{E}$-arrow of the form $g : \Omega^b \to b$ there is

exactly one $\mathscr{E}$-arrow $f:a\to b$ making

$$\begin{array}{ccc} a & \xrightarrow{\ f\ } & b \\ {\scriptstyle r}\big\downarrow & & \big\uparrow{\scriptstyle g} \\ \Omega^a & \xrightarrow{\ \Omega^f\ } & \Omega^b \end{array}$$

commute. (f is said to be defined recursively from g over r:— $f=\mathrm{rec}_r(g)$).

Exercise 5. $0\to\Omega^0$ is a tso.

Exercise 6. $\perp:1\to\Omega^1$ is a tso (why is this so in **Set**?) □

If $r:a\to\Omega^a$ and $s:b\to\Omega^b$ are "relations" then $h:a\to b$ is an *inclusion* from r to s, written $h:r\hookrightarrow s$, iff

$$\begin{array}{ccc} a & \xrightarrow{\ h\ } & b \\ {\scriptstyle r}\big\downarrow & & \big\downarrow{\scriptstyle s} \\ \Omega^a & \xrightarrow{\ \Omega^h\ } & \Omega^b \end{array}$$

commutes. We write $r\subseteq s$ if there exists an inclusion $h:r\hookrightarrow s$.

Exercise 7. Show that $(0\to\Omega^0)\subseteq(r:a\rightarrowtail\Omega^a)$, for any tso r.

Exercise 8. $r\subseteq r$.

Exercise 9. $r\subseteq s\subseteq t$ implies $r\subseteq t$. (cf. Exercise 3) □

An inclusion between transitive set-objects, if it exists, is unique. To see this, we introduce a construction that assigns to each monic $s:b\rightarrowtail\Omega^b$ a unique arrow $\hat{s}:\Omega^{\tilde{b}}\to\tilde{b}$, where $\tilde{b}$ is the codomain of the partial arrow classifier $\eta_b:b\to\tilde{b}$ described in §11.8. The arrow

$$\Omega^{\eta_b}:\Omega^b\to\Omega^{\tilde{b}}$$

will in fact be monic, since η_b is (Osius [74], Proposition 5.8(a)). $\hat{s}$ is then defined as the unique arrow making

$$\begin{array}{ccc} b & \xrightarrow{\ \Omega^{\eta_b}\circ s\ } & \Omega^{\tilde{b}} \\ {\scriptstyle 1_b}\big\downarrow & & \big\downarrow{\scriptstyle \hat{s}} \\ b & \xrightarrow{\ \eta_b\ } & \tilde{b} \end{array}$$

a pullback (note that $\Omega^{\eta_b} \circ s$ is monic iff s is monic).

THEOREM 1. *If $r : a \rightarrow \Omega^a$ is recursive, and $s : b \rightarrowtail \Omega^b$ extensional, then*
(1) *$f : a \rightarrow b$ is an inclusion iff*

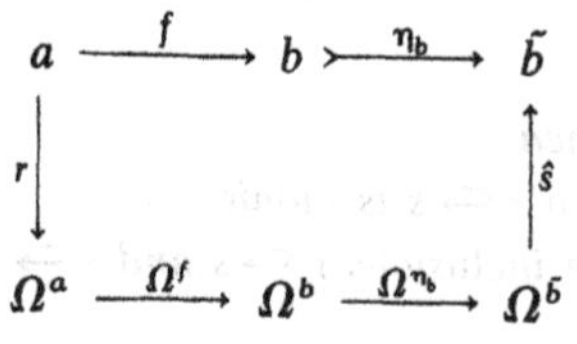

commutes, iff $\eta_b \circ f = \text{rec}_r(\hat{s})$.
(2) *If $r \subseteq s$ then there is a unique inclusion $r \hookrightarrow s$ of r into s.*

PROOF. (1) Consider

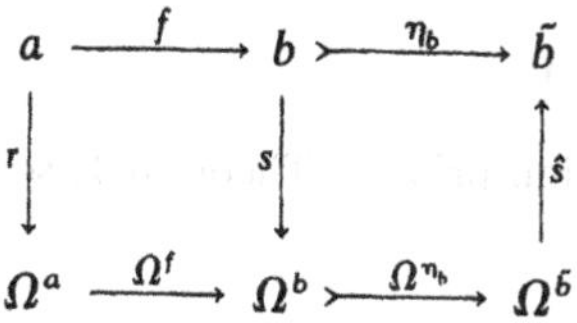

The right hand square always commutes, by the definition of $\hat{s}$. Then if f is an inclusion, the left hand square commutes, hence the whole diagram does. Conversely, if the perimeter of the diagram commutes then this means precisely that the perimeter of the diagram

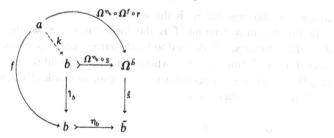

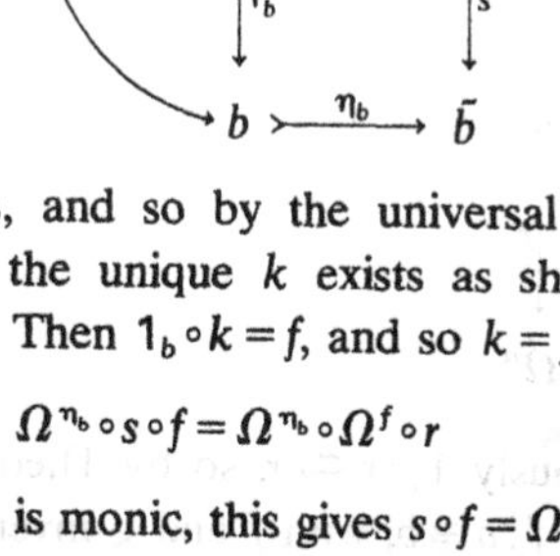

commutes, and so by the universal property of the inner square as pullback, the unique k exists as shown to make the whole diagram commute. Then $1_b \circ k = f$, and so $k = f$. Hence from the upper triangle

$$\Omega^{\eta_b} \circ s \circ f = \Omega^{\eta_b} \circ \Omega^f \circ r$$

Since Ω^{η_b} is monic, this gives $s \circ f = \Omega^f \circ r$, i.e. the left hand square of the previous diagram commutes, making f an inclusion.

To complete part (1), note that since $\Omega^{(\eta_b \circ f)} = \Omega^{\eta_b} \circ \Omega^f$, recursiveness of r implies the diagram commutes precisely when $\eta_b \circ f$ is the unique arrow defined recursively from $\hat{s}$ over r.

(2) If $f_1 : r \hookrightarrow s$ and $f_2 : r \hookrightarrow s$, then by (1), $\eta_b \circ f_1 = \eta_b \circ f_2 = \mathrm{rec}_r(\hat{s})$. Since η_b is monic we get $f_1 = f_2$. □

THEOREM 2. *If r and s are tso's, then*

(1) *If $r \subseteq s$, the (unique) inclusion $r \hookrightarrow s$ is monic.*

(2) *If $r \subseteq s \subseteq r$, then $r \cong s$, i.e. the inclusions $r \hookrightarrow s$ and $s \hookrightarrow r$ are iso.*

PROOF. (1) Consider

$$\begin{array}{ccccc}
a & \xrightarrow{\;f\;} & b & \xrightarrow{\;g\;} & \tilde{a} \\
{\scriptstyle r}\downarrow & & \downarrow{\scriptstyle s} & & \uparrow{\scriptstyle \hat{r}} \\
\Omega^a & \xrightarrow{\;\Omega^f\;} & \Omega^b & \xrightarrow{\;\Omega^g\;} & \Omega^{\tilde{a}}
\end{array}$$

Here $\hat{r}$ is defined by the construction prior to Theorem 1, so $\hat{r} \circ \Omega^{\eta_a} \circ r = \eta_a \circ 1_a = \eta_a$. Hence

$$\begin{array}{ccc}
a & \rightarrowtail^{\eta_a} & \tilde{a} \\
{\scriptstyle r}\downarrow & & \uparrow{\scriptstyle \hat{r}} \\
\Omega^a & \xrightarrow{\;\Omega^{\eta_a}\;} & \Omega^{\tilde{a}}
\end{array}$$

commutes, showing that η_a is the arrow $\mathrm{rec}_r(\hat{r})$.

In the previous diagram, f is the inclusion $r \hookrightarrow s$, so the left hand diagram commutes. g is defined to be the arrow $\mathrm{rec}_s(\hat{r})$ given by recursion from $\hat{r}$ over s. But then the whole diagram commutes, and so $g \circ f = \mathrm{rec}_r(\hat{r}) = \eta_a$. Thus $g \circ f$ is monic, so f itself must be monic (Exercise 3.1.2).

(2) If $r \subseteq s \subseteq r$, then from

$$\begin{array}{ccccc}
a & \overset{f}{\hookrightarrow} & b & \overset{g}{\hookrightarrow} & a \\
{\scriptstyle r}\downarrow & & \downarrow{\scriptstyle s} & & \downarrow{\scriptstyle r} \\
\Omega^a & \xrightarrow{\;\Omega^f\;} & \Omega^b & \xrightarrow{\;\Omega^g\;} & \Omega^a
\end{array}$$

we see that $g \circ f : r \hookrightarrow r$. But obviously $1_a : r \hookrightarrow r$, so by Theorem 1 (2), $g \circ f = 1_a$. Similarly $f \circ g = 1_b$, hence $f : a \cong b$, with f and g inverse to each other. □

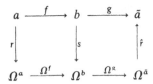

Thus, defining $r \simeq s$ iff $r \subseteq s$ and $s \subseteq r$ leads to a definition of equality of (isomorphism classes of) transitive set-objects, with respect to which the inclusion relation becomes a partial ordering. Osius then gives constructions for

(i) the intersection $r \cap s : a \cap b \to \Omega^{a \cap b}$, which proves to be the greatest lower bound of r and s in the inclusion ordering of tso's; and

(ii) the union $r \cup s : a \cup b \to \Omega^{a \cup b}$, which is the least upper bound of r and s.

For (i), the cube

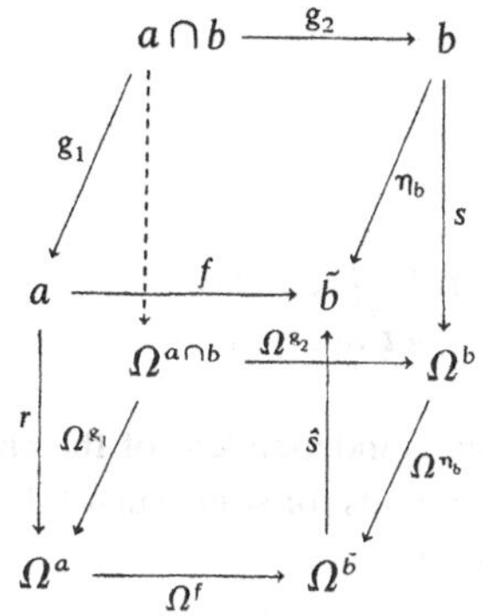

is formed by first defining f to be $\mathrm{rec}_r(\hat{s})$, and obtaining the top face as the pullback of f along η_b. Thus the right-hand face is the square defining $\hat{s}$, the front face the square defining f. The bottom square then proves to be a pullback whose universal property yields the unique arrow $a \cap b \dashrightarrow \Omega^{a \cap b}$ making the whole diagram commute. This arrow is $r \cap s$.

For (ii), $a \cup b$ comes from the pushout

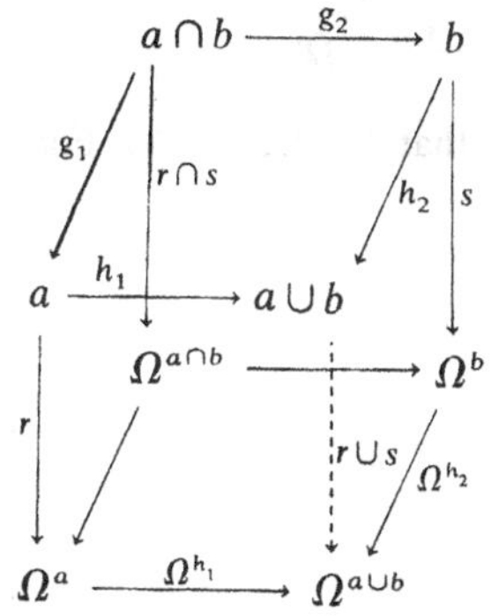

of g_1 and g_2, with $r \cup s$ arising from the co-universal property of pushouts.

DEFINITION. A *set-object* in a topos $\mathscr{E}$ is a pair (f, r) of $\mathscr{E}$-arrows of the form

$$b \overset{f}{\rightarrowtail} a \overset{r}{\rightarrowtail} \Omega^a,$$

where r is a transitive set-object.

Equality of set-objects is defined as follows: $(f, r) \simeq_{\mathscr{E}} (g, s)$ iff for some tso $t: e \to \Omega^e$ such that $r \subseteq t$ and $s \subseteq t$,

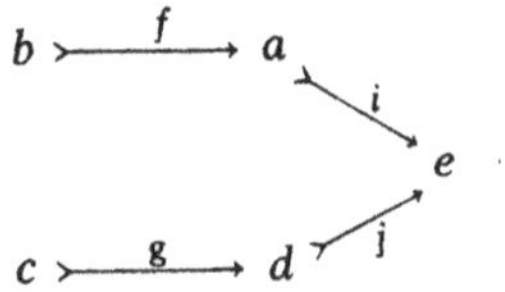

we have $i[f] \simeq j[g]$ in Sub(e), (i.e. $i \circ f \simeq j \circ g$, since i and j are monic) where i and j are the inclusions $i : r \hookrightarrow t$ and $j : s \hookrightarrow t$.

Osius establishes that the definition is independent of the choice of the tso t containing r and s: the condition holds for some such t iff it holds for all such t (hence iff it holds when $t = r \cup s$).

EXERCISE 10. $(f, r) \simeq_{\mathscr{E}} (g, r)$ iff in Sub(a), $f \simeq g$.

EXERCISE 11. Suppose that $\ulcorner f \urcorner \in r$ and $\ulcorner g \urcorner \in s$, i.e. there are commutative diagrams

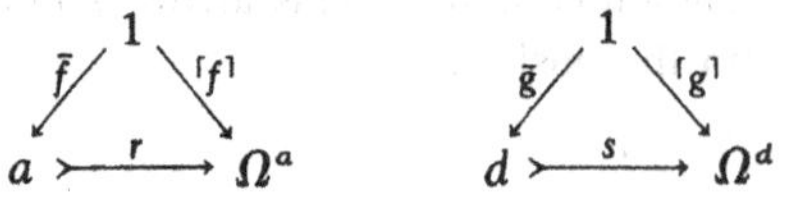

for certain elements $\bar{f}$ and $\bar{g}$. Show that $f \in_r 1_a$, $g \in_s 1_d$. For t such that $r \subseteq t$ and $s \subseteq t$, show

$$(f, r) \simeq_{\mathscr{E}} (g, s) \quad \text{iff} \quad i \circ \bar{f} = j \circ \bar{g},$$

i.e.

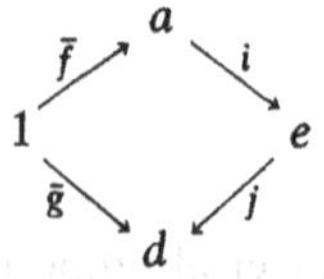

commutes. □

"Membership" for set objects is defined by

$$(g, s)E_{\mathscr{E}}(f, r)$$

iff for some tso $t: e \to \Omega^e$ such that $r \subseteq t$ and $s \subseteq t$, $j \circ g \in_t i \circ f$,

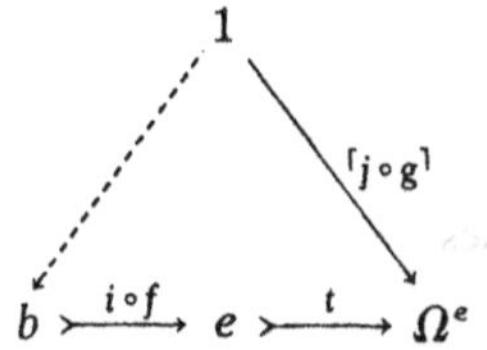

i.e. $\ulcorner j \circ g \urcorner$ factors through $t \circ i \circ f$.

Again the definition is independent of the choice of t, and can be given with $t = r \cup s$.

Equivalent definitions of $(g, s)E_{\mathscr{E}}(f, r)$ are

(i) There exist set objects (g', t) and (f', t) with

$$(g, s) \simeq_{\mathscr{E}} (g', t), \qquad (f, r) \simeq_{\mathscr{E}} (f', t)$$

and

$$g' \in_t f',$$

and

(ii) There exists $g': c' \rightarrowtail a$ such that

$$(g, s) \simeq_{\mathscr{E}} (g', r)$$

and

$$g' \in_r f.$$

EXERCISE 12. For set objects (g, r), (f, r),

$$(g, r)E_{\mathscr{E}}(f, r) \quad \text{iff} \quad g \in_r f. \qquad \square$$

We now have a definition of an $\mathscr{L}$-model

$$\mathfrak{A}(\mathscr{E}) = \langle A_{\mathscr{E}}, E_{\mathscr{E}}, \simeq_{\mathscr{E}} \rangle,$$

where $A_{\mathscr{E}}$ is the collection of all set objects in $\mathscr{E}$. Notice that the definition has been given for any topos $\mathscr{E}$. Osius proves

THEOREM 3. *If $\mathscr{E}$ is well-pointed, then $\mathfrak{A}(\mathscr{E})$ is a model of all of the Z_0-axioms, together with the axiom of Regularity and the Transitivity axiom* (TA). *If* NNO (*respectively* ES) *holds in $\mathscr{E}$ then the Axiom of Infinity* (*respectively Axiom of Choice*) *holds in $\mathfrak{A}(\mathscr{E})$.* $\square$

It is also shown that for each tso $r: a \rightarrowtail \Omega^a$, the set object $(1_a, r)$ is a "transitive set" in the sense of $\mathfrak{A}(\mathscr{E})$, i.e. the $\mathscr{L}$-formula $\mathbf{Tr}(u)$ is satisfied in $\mathfrak{A}(\mathscr{E})$ when u is interpreted as the $\mathfrak{A}(\mathscr{E})$-set $(1_a, r)$.

12.6 Equivalence of models

We now have two construction processes

$$\mathfrak{A} \mapsto \mathscr{E}(\mathfrak{A})$$
$$\mathscr{E} \mapsto \mathfrak{A}(\mathscr{E})$$

of well-pointed topoi from models of Z_0 and conversely. It remains to determine the extent to which these constructions are inverse to each other.

To do this we will need to assume that Mostowski's lemma is true in $\mathfrak{A}$. Rather than confine ourselves to ZF-models, we take the statement of the lemma as a further axiom.

Axiom of Transitive Representation: This is the $\mathscr{L}$-sentence that formally expresses the statement

ATR: Any extensional, well-founded relation $r: A \rightarrow \mathscr{P}(A)$ is isomorphic to the membership relation $r_\in : B \hookrightarrow \mathscr{P}(B)$ of some transitive set B.

B is called the *transitive representative* of r.

The system Z is Z_0+Reg+TA+ATR.

Now let us assume $\mathfrak{A} = \langle A, E, \simeq \rangle$ is a Z-model. If $b \in A$ is an $\mathfrak{A}$-set, then, working "inside" $\mathfrak{A}$, from Z_0+TA there will be an $\mathfrak{A}$-set a that is the transitive closure of b in the sense of $\mathfrak{A}$, and so there will be an $\mathfrak{A}$-inclusion $f: b \hookrightarrow a$. Moreover by Ext and Reg the $\mathfrak{A}$-membership relation $r_a : a \rightarrow \mathscr{P}(a)$ on the $\mathfrak{A}$-transitive object a will be $\mathfrak{A}$-monic and $\mathfrak{A}$-well-founded, hence $\mathfrak{A}$-recursive. But the $\mathfrak{A}$-functions f and r_a will be arrows in the topos $\mathscr{E}(\mathfrak{A})$, and so $\langle f, r_a \rangle$ will be a set-object in $\mathscr{E}(\mathfrak{A})$, i.e. an individual ("set") in the $\mathscr{L}$-model $\mathfrak{A}(\mathscr{E}(\mathfrak{A}))$. Putting $Ob(a) = \langle f, r_a \rangle$ gives a transformation from $\mathscr{L}$-model $\mathfrak{A}$ to $\mathscr{L}$-model $\mathfrak{A}(\mathscr{E}(\mathfrak{A}))$ that satisfies

$$a \simeq c \quad \text{iff} \quad Ob(a) \simeq_{\mathscr{E}(\mathfrak{A})} Ob(c)$$

and

$$a E c \quad \text{iff} \quad Ob(a) E_{\mathscr{E}(\mathfrak{A})} Ob(c).$$

In the opposite direction, given a set-object $X=(f:b \rightarrowtail a,\ r:a \rightarrowtail \Omega^a)$ in $\mathfrak{A}(\mathscr{E}(\mathfrak{A}))$, then r is a monic, recursive arrow in $\mathscr{E}(\mathfrak{A})$, i.e. an extensional, well-founded relation in $\mathfrak{A}$. Since ATR holds in $\mathfrak{A}$ there is some $\mathfrak{A}$-transitive set $c \in A$, and an $\mathfrak{A}$-bijection $g:a \to c$ that makes r $\mathfrak{A}$-isomorphic to the $\mathfrak{A}$-membership relation on c. We let $St(X)$ be the $\mathfrak{A}$-set "$g(f(a))$", i.e. the $\mathfrak{A}$-image of b in c under the $\mathfrak{A}$-function $g \circ f$.

In view of Theorem 2 of §12.4, transitive representatives are unique (in Z) and so this gives us a map St from $\mathfrak{A}(\mathscr{E}(\mathfrak{A}))$ to $\mathfrak{A}$ that can be shown to satisfy

$$X \simeq_{\mathscr{E}(\mathfrak{A})} Y \quad \text{iff} \quad St(X) \simeq St(Y)$$

and

$$X E_{\mathscr{E}(\mathfrak{A})} Y \quad \text{iff} \quad St(X) E St(Y).$$

Moreover Ob, and St are "almost inverse" in the sense that we have

$$a \simeq St(Ob(a))$$

and

$$X \simeq_{\mathscr{E}(\mathfrak{A})} Ob(St(X))$$

Were we to "normalise" $\mathfrak{A}$ and $\mathfrak{A}(\mathscr{E}(\mathfrak{A}))$ by replacing individuals by their $\simeq$-equivalence classes we would obtain two fully isomorphic $\mathscr{L}$-models.

Exercise 1. Show, for any $\mathscr{L}$-formula φ, that

$$\mathfrak{A} \models \varphi[a] \quad \text{iff} \quad \mathfrak{A}(\mathscr{E}(\mathfrak{A})) \models \varphi[Ob(a)]$$

and

$$\mathfrak{A}(\mathscr{E}(\mathfrak{A})) \models \varphi[X] \quad \text{iff} \quad \mathfrak{A} \models \varphi[St(X)].$$

Exercise 2. Show

$$\mathfrak{A} \models \varphi[a] \quad \text{iff} \quad \mathfrak{A} \models \varphi[St(Ob(a))]$$

and

$$\mathfrak{A}(\mathscr{E}(\mathfrak{A})) \models \varphi[X] \quad \text{iff} \quad \mathfrak{A}(\mathscr{E}(\mathfrak{A})) \models \varphi[Ob(St(X))]. \qquad \square$$

Beginning now with a well-pointed topos $\mathscr{E}$, a transformation $F:\mathscr{E}(\mathfrak{A}(\mathscr{E})) \to \mathscr{E}$ is defined as follows. If X is an $\mathscr{E}(\mathfrak{A}(\mathscr{E}))$ object then X is an $\mathfrak{A}(\mathscr{E})$-set, i.e. a set-object (f, r), where $f:b \rightarrowtail a$ and $r:a \to \Omega^a$ are $\mathscr{E}$-arrows. We put $F(X) = \operatorname{dom} f = b$.

Osius shows how to define F on $\mathscr{E}(\mathfrak{A}(\mathscr{E}))$-arrows so that it becomes a functor from $\mathscr{E}(\mathfrak{A}(\mathscr{E}))$ to $\mathscr{E}$. The image of F in $\mathscr{E}$ proves to be a full subcategory of $\mathscr{E}$ containing those $\mathscr{E}$-objects b that are *partially transitive.* Partial transitivity of b means that there exists a tso $r: a \to \Omega^a$ in $\mathscr{E}$, and an $\mathscr{E}$-monic $f: b \rightarrowtail a$ from b to a. This makes (f, r) a set-object, i.e. an object in $\mathscr{E}(\mathfrak{A}(\mathscr{E}))$, with $F(f, r) = b$.

Axiom of partial transitivity:

APT: Every object is partially transitive.

Notice that if $\mathfrak{A}$ is any Z-model, then the topos $\mathscr{E}(\mathfrak{A})$ of $\mathfrak{A}$-sets and $\mathfrak{A}$-functions always satisfies APT. The definition of $Ob(b)$ shows that every b is partially transitive.

Now if $\mathscr{E} \models \text{APT}$, then the functor F described above will be "onto" – its image is the whole of $\mathscr{E}$. Moreover F will then be an equivalence of categories, as defined in Chapter 9. Thus $\mathscr{E}$ and $\mathscr{E}(\mathfrak{A}(\mathscr{E}))$ are equivalent categories. They are "isomorphic up to isomorphism". By identifying isomorphic objects in each we obtain two (skeletal) categories that are isomorphic in the category **Cat** of all small categories. Furthermore if $\mathscr{E}$ is partially transitive (i.e. $\mathscr{E} \models \text{APT}$) then the functor F can be used to show that the axiom ATR of transitive representation holds in $\mathfrak{A}(\mathscr{E})$, and so $\mathfrak{A}(\mathscr{E})$ is a Z-model. For, if R is an extensional well-founded relation on X inside $\mathfrak{A}(\mathscr{E})$ then R corresponds to an $\mathfrak{A}(\mathscr{E})$-function $r: X \to \mathscr{P}(X)$ which becomes a tso in $\mathscr{E}(\mathfrak{A}(\mathscr{E}))$. F transfers this to a tso $t: a \to \Omega^a$ in $\mathscr{E}$. The set object $\langle 1_a, t \rangle$ then proves to be the transitive representative of X in $\mathfrak{A}(\mathscr{E})$.

In summary then, there is an exact correspondence between models of the set theory Z and well-pointed, partially transitive, topoi. The concept of a "well-pointed partially transitive topos" can be expressed in the first-order language of categories, and so we have an exact correspondence between models of two first-order theories. Indeed the whole exercise can be treated as a syntactic one, the set-theoretic definition of "function (arrow)" and the categorial definition of "set-object" providing theorem-preserving interpretations of two formal systems in each other.

The theory as developed may be extended to stronger set theories. A categorial version of the Replacement schema can be defined to characterise those topoi that correspond to models of ZF. Further results of this nature are given in Section 9 of Osius. In the event that epics split in well-pointed $\mathscr{E}$, the axiom APT is redundant. By lifting to $\mathscr{E}$ the set-theoretic proof that any object A has a well-ordering (and hence yields a tso $A \to \mathscr{P}(A)$), it can be shown from ES that all objects are partially transitive. Thus well-pointed topoi satisfying ES correspond exactly to models of ZC (Z+axiom of choice).

A fuller account of the technical details of the theory just described, including proofs of the main results, is to be found in Chapter 9 of Johnstone [77].

CHAPTER 13

ARITHMETIC

"Abstraction is a crucial feature of [rational] knowledge, because in order to compare and to classify the immense variety of shapes, structures and phenomena around us we cannot take all their features into account, but have to select a few significant ones. Thus we construct an intellectual map of reality in which things are reduced to their general outlines."

Fritjof Capra

13.1. Topoi as foundations

Category theory promotes the viewpoint that the concept of "arrow" be taken as fundamental in place of "membership", and the development of topos theory substantiates that position. By imposing natural conditions on a topos (extensionality, sections for epics, natural numbers object), we can make it correspond precisely to a model of classical set theory. Thus, to the extent that set theory provides a foundation for mathematics, so too does topos theory. What then are the attractions of this new system?

The first thing one could point to is that the concepts of topos theory are natural ones to the practising mathematician. Category theory was originally developed as a language for use in the areas of topology and algebra. The alternative account it has subsequently produced of the nature of mathematical structures and their essential features is a most compelling one. Entities are characterised by their universal properties, which specify their role in relation to other entities. Thus it is the universal property that a product has that most effectively conveys its usage and function in relation to the two objects from which it is obtained. Once this "operational" description is known, its internal structure – the way it was constructed – is of lesser importance.

It was suggested in Chapter 1 that the purpose of foundational studies is to provide a rigorous explication of the nature of mathematical concepts and entities. There is of course no single correct way to do this. Set theory offers one approach, topos theory another. As against either one might retort that we really know what such things as whole numbers are, and always have. And yet as long as there are mathematicians, there will be new and different attempts to define and describe them. Contexts and perspectives change in the light of new knowledge. Forms of language change to deal with new perspectives. Whenever this occurs, old ideas are re-examined in a different light. To some people, discovering topoi will constitute a revelation. Just re-expressing familiar ideas in a new language, relating them to different concepts, somehow carries the force of *explanation*, even if the new new concepts themselves ultimately require explaining. It may well be, in the future, that those bought up on a solid diet of "arrow-language" will seek to reappraise what to them will have been standard fare. When that happens, new concepts, and new foundations will emerge.

One of the new analyses of mathematical structure developed by the categorial foundation is an alternative account of what sets are and how they behave. Instead of the "universe of (ZF) sets" we are offered the "category of sets". In formal theories like ZF a set is an entity that has members that have members that have members that have .... The membership structure determined by a set can be very rich indeed (think about the membership tree for example of $\mathscr{P}(\omega)$). The informal picture that the ZF-set-theorist has of his universe is an open-ended cone

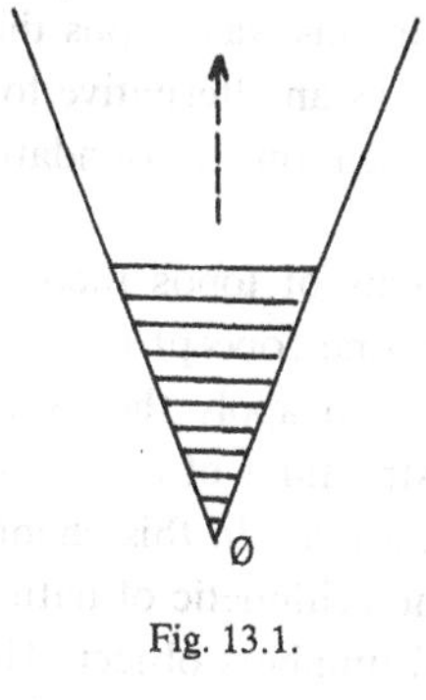

Fig. 13.1.

with the null set at the base point. Starting with $\emptyset$, all the individuals in the universe are built up by repeatedly forming powersets and taking unions. As these operations are iterated, sets of greater and greater complexity appear at higher and higher levels that pile up in the cone ad infinitum.

Now the elements of the collections that are used in mathematics are indeed often sets themselves. Thus a topology is a collection of subsets, as is a powerset, and a Heyting algebra $\mathbf{P}^+$. An analyst deals daily with collections of functions, and with function(al)s whose inputs are themselves functions. Rarely however does one find in practice the need for more than three or four levels of membership. Even then one can distinguish these examples of the use of set theory from the actual conception, the essential idea, of what a set is. As Lawvere [76] puts it, "an abstract set X has elements each of which has no internal structure whatsoever". A set, "naively", is a collection of indeterminate, quite arbitrary, things. Indeed in algebra the word "abstract" is used to convey precisely that sense. One studies abstract group theory when one studies groups as collections that support a certain algebraic structure, the nature of the elements of those collections being immaterial. In general topology, the elements of a topological space are universally called "points", therein a point being, as it was for Euclid, "that which has no parts". Likewise, in the category of sets, a set is an object X that has elements $1 \to X$, these elements being fundamental and indivisible. Topos theory has shown us how to develop foundations for standard mathematical concepts in these terms.

Intuitive set theory is, and will doubtless remain, central to our metalanguage for the doing of mathematics. It is part of the language in which we speak, whether the object of our discourse be geometry, algebra, or foundations, whether the objects about which we speak be topological spaces, groups, or sets. Seen in this way, topos theory stands not so much as a rival to set theory per se as an alternative to formalised set theory in presenting a rigorous explication, a foundation, of our intuitive notion of "set".

One of the most significant achievements of topos theory is to have crystallised the core of basic set theory in one concept that is manifest in such hitherto diverse contexts. Thus we can apply the "set of points" notion and our familiarity with it to the structures of algebraic geometry, intuitionistic logic, and monoid representations. In this chapter we shall look briefly at how the foundations of the arithmetic of natural numbers can be lifted to any topos with a natural numbers object. The power of the axiomatic method, and the ability of abstraction to simplify and get at the heart of things will perhaps be brought home if one reflects that a "natural number", i.e. element $1 \to N$ of N, referred to below might in fact be anything from a continuous function between sheaves of sets of germs (local homeomorphisms) to an equivariant mapping of monoid actions, or a natural transformation between set-valued functors defined on an arbitrary small category.

13.2. Primitive recursion

Throughout this section, $\mathscr{E}$ denotes a topos that has a natural numbers object $1 \xrightarrow{O} N \xrightarrow{s} N$. So for any diagram $1 \xrightarrow{x} a \xrightarrow{f} a$ in $\mathscr{E}$ we have a unique "$\mathscr{E}$-sequence" $h : N \to a$ defined by simple recursion from f and x, i.e. making

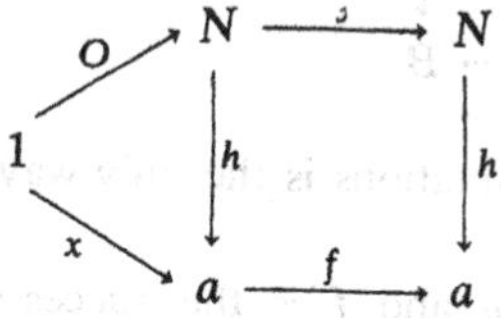

commute.

Now there are many basic arithmetical functions that can be defined inductively by more complex forms of recursion than that captured by the axiom NNO. Consider, for example, the process of forming the sum $m+n$ of two numbers. We may do this by holding m fixed and "repeatedly adding 1 to m" to generate the sequence

$$m, m+1, m+2, \ldots, m+n, \ldots$$

Then $m+n$ is defined by "recursion on n" from the equations

$$m+0=m$$

and

$$m+(n+1)=(m+n)+1$$

i.e.

$$m+s(n)=s(m+n).$$

The form of these equations is the same as those that defined the unique $h : I \times \omega \to A$ used to verify NNO for **Bn**(I) in §12.2, and readily generalises. The "parameter" m is replaced by an element x of an arbitrary set A, and in place of $m+n$ we define a function $h(x, n)$ with inputs from $A \times \omega$, and outputs in some other set B. To start the induction on n we need a function of the form $h_0: A \to B$ so that we can put

(1) $\qquad h(x, 0) = h_0(x).$

Then, assuming a function $f: B \to B$ has been given, repeated application of f will generate h. Thus we put

(2) $\qquad h(x, n+1) = f(h(x, n)).$

By (1) and (2) the diagram

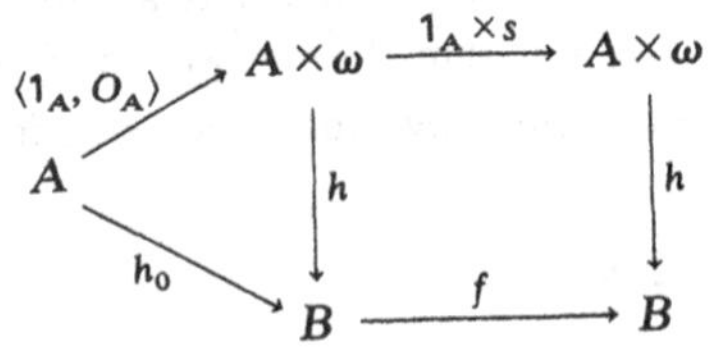

commutes, and defining h by these equations is the only way that it can commute.

In the case that h_0 is $\mathrm{id}_\omega : \omega \to \omega$ and f is the successor function $s: \omega \to \omega$, the unique h defined recursively from h_0 and f by (1) and (2) is the addition function $+: \omega \times \omega \to \omega$.

THEOREM 1. (Freyd [72]). *If* $\mathscr{E} \models \mathrm{NNO}$, *then for any diagram* $a \xrightarrow{h_0} b \xrightarrow{f} b$ *there is exactly one* $\mathscr{E}$*-arrow* $h: a \times N \to b$ *such that*

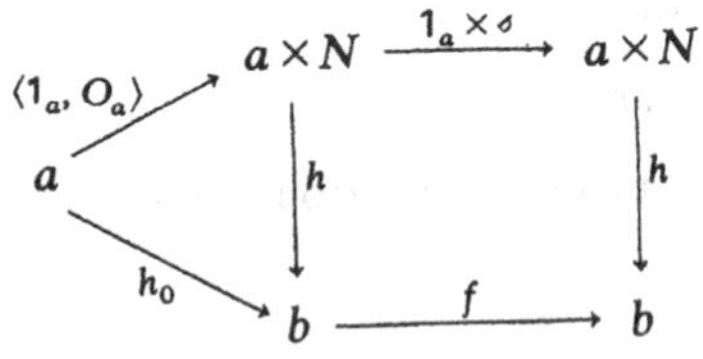

commutes, where O_a *is the composite of* $a \to 1 \xrightarrow{O} N$.

CONSTRUCTION FOR PROOF. h is the "twisted" exponential adjoint of the unique sequence $N \to b^a$ that makes

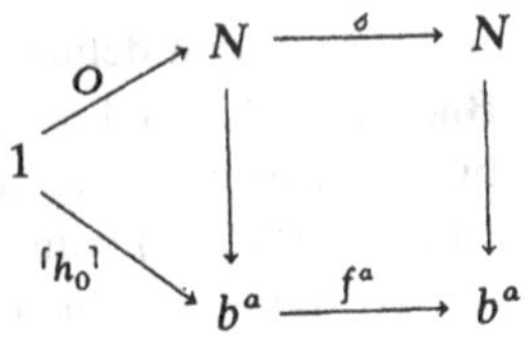

commute. Here f^a is the exponential adjoint of $f \circ ev : b^a \times a \to b$

$$\begin{array}{ccc} b^a \times a & \xrightarrow{ev} & b \\ & {\scriptstyle f \circ ev} \searrow & \downarrow {\scriptstyle f} \\ & & b \end{array}$$

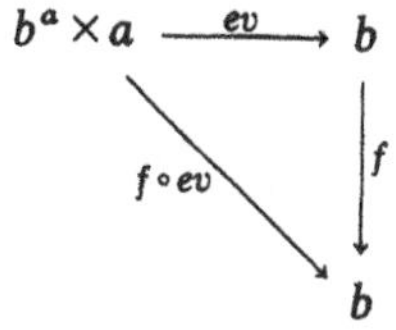

In **Set**, $f \circ ev$ maps $\langle g, x \rangle \in B^A \times A$ to $f(g(x)) \in B$, so that f^A maps $g \in B^A$ to

$$\begin{array}{ccc} A & \xrightarrow{g} & B \\ & {\scriptstyle f \circ g} \searrow & \downarrow {\scriptstyle f} \\ & & B \end{array}$$

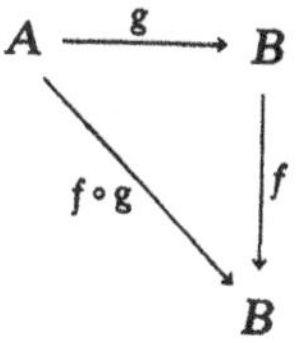

$f \circ g \in B^A$. □

Applying Theorem 1 to a diagram of the form $b \xrightarrow{1_b} b \xrightarrow{f} b$, the unique $h : b \times N \to b$ defined by recursion from $\mathbf{1}_b$ and f has in **Set** the recursive equations

$$h(x, 0) = x$$

$$h(x, n+1) = f(h(x, n)).$$

Thus for fixed x, h generates the sequence

$$x, f(x), f(f(x)), f(f(f(x))), \ldots$$

and so h is called the *iterate* of f.

The iterate of the successor arrow $\mathfrak{s} : N \to N$ is, by definition, the addition arrow $\oplus : N \times N \to N$.

EXERCISE 1. What does $\oplus$ look like in $\mathbf{Set}^{\mathscr{C}}$ and $\mathbf{Bn}(I)$?

EXERCISE 2. Let $i(f)$ be the iterate of f. Show that

$$\begin{array}{ccc} b \times N \times N & \xrightarrow{i(f) \times 1_N} & b \times N \\ {\scriptstyle 1_b \times \oplus} \downarrow & & \downarrow {\scriptstyle i(f)} \\ b \times N & \xrightarrow{i(f)} & b \end{array}$$

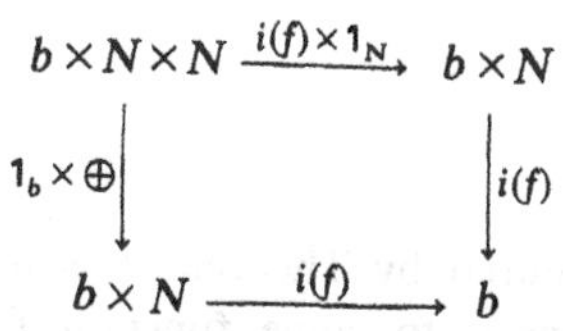

commutes.

EXERCISE 3. Explain why Exercise 2, in the case $f = s$, gives the "associative law for addition".

EXERCISE 4. Show that

$$\begin{array}{ccc} b\times N & \xrightarrow{f\times 1_N} & b\times N \\ {\scriptstyle i(f)}\downarrow & & \downarrow{\scriptstyle i(f)} \\ b & \xrightarrow{f} & b \end{array}$$

and

$$\begin{array}{ccc} b\times N & \xrightarrow{f\times 1_N} & b\times N \\ {\scriptstyle 1_b\times s}\downarrow & & \downarrow{\scriptstyle i(f)} \\ b\times N & \xrightarrow{i(f)} & b \end{array}$$

commute.

EXERCISE 5. Show that $\langle O_N, 1_N\rangle\circ O = \langle 1_N, O_N\rangle\circ O = \langle O, O\rangle$.

EXERCISE 6. $\oplus\circ\langle O, O\rangle = O$.

EXERCISE 7. $(0+m=m)$. Show that

$$\begin{array}{ccc} N & \xrightarrow{\langle O_N, 1_N\rangle} & N\times N \\ & {\scriptstyle 1_N}\searrow \quad \swarrow{\scriptstyle \oplus} & \\ & N & \end{array}$$

commutes.

EXERCISE 8. (Commutativity of Addition)

$$\begin{array}{ccc} N\times N & \xrightarrow{\langle pr_2, pr_1\rangle} & N\times N \\ & {\scriptstyle \oplus}\searrow \quad \swarrow{\scriptstyle \oplus} & \\ & N & \end{array}$$

commutes. □

The basic idea of recursion captured by Theorem 1 is that $h(x, n)$, having been defined, serves as input to some function f to obtain $h(x, n+1)$ as output. But there are some functions with natural inductive definitions in which $h(x, n+1)$ depends, not just on $h(x, n)$, but also on x and n in a very direct way, i.e. we need to input one or both of x and n as well as $h(x, n)$ to get $h(x, n+1)$. Take for example the multiplication $x \times n$ of x by n, i.e. "x added to itself n times". This is given by the equations

$$x \times 0 = 0$$

$$x \times (n+1) = x + (x \times n)$$

i.e.

$$x \times s(n) = f(x, x \times n)$$

where f is the addition function.

For an example in which $h(x, n+1)$ depends directly on n consider the predecessor function $\rho : \omega \to \omega$ that has $\rho(n) = n - 1$ (unless $n = 0$, in which case we put $\rho(n) = 0$). Recursively ρ is specified by

$$\rho(0) = 0$$

$$\rho(n+1) = n.$$

These two considerations may be combined into one: given functions $h_0 : A \to B$ and $f : A \times \omega \times B \to B$ we define $h : A \times \omega \to B$, by "primitive recursion", through the equations

$$h(x, 0) = h_0(x)$$

$$h(x, n+1) = f(x, n, h(x, n)).$$

By putting h_0 as $O_\omega : \omega \to \omega$ and f as the "2nd projection" $pr_2^3 : \omega^3 \to \omega$, the resulting h is the predecessor function ρ. Using the same h_0, but with f the composite of

$$\omega^3 \xrightarrow{\langle pr_1, pr_3 \rangle} \omega^2 \xrightarrow{+} \omega$$

we recover the multiplication function as h.

Primitive Recursion Theorem (Freyd [72]). *If $\mathscr{E} \models$ NNO, then for any $\mathscr{E}$-arrows $h_0 : a \to b$ and $f : a \times N \times b \to b$ there is a unique $\mathscr{E}$-arrow $h : a \times N \to b$ making*

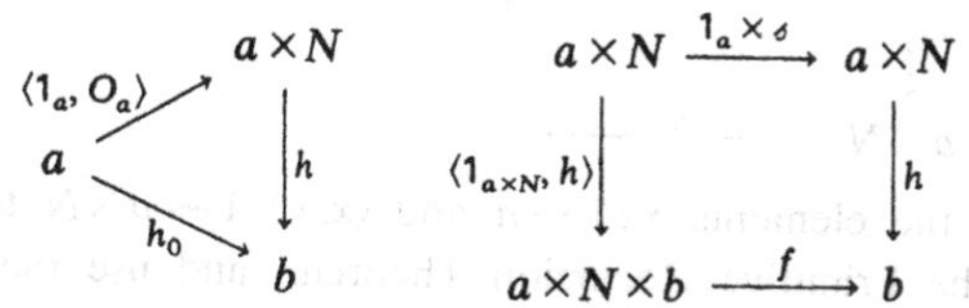

commute.

Construction for Proof. By Theorem 1, there is a unique h' such that

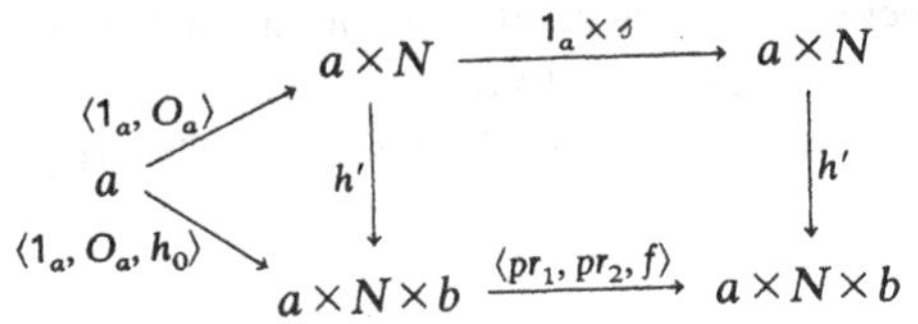

commutes.

In **Set** $\langle pr_1, pr_2, f\rangle$ takes $\langle x, n, y\rangle$ to $\langle x, n, f(x, n, y)\rangle$. Hence h' has the equations

$$h'(x, 0) = \langle x, 0, h_0(x)\rangle$$

$$h'(x, n+1) = \langle x, n, f(x, n, h'(x, n))\rangle.$$

The desired $\mathscr{E}$-arrow h is the composite

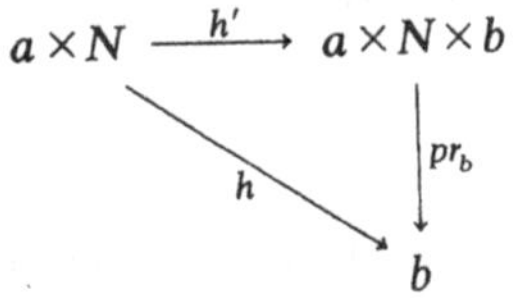

of h' and the projection to b. □

Corollary. *If h is defined recursively from h_0 and f as in the Theorem, then for any elements $x: 1 \to a$ and $y: 1 \to N$ of a and N we have*

(i) $h \circ \langle x, 0\rangle = h_0 \circ x$

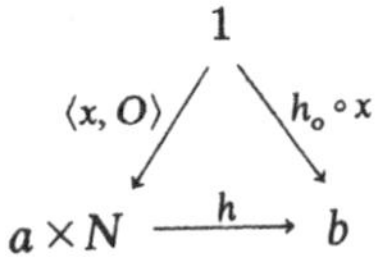

(ii) $h \circ \langle x, s \circ y\rangle = f \circ \langle x, y, h \circ \langle x, y\rangle\rangle$

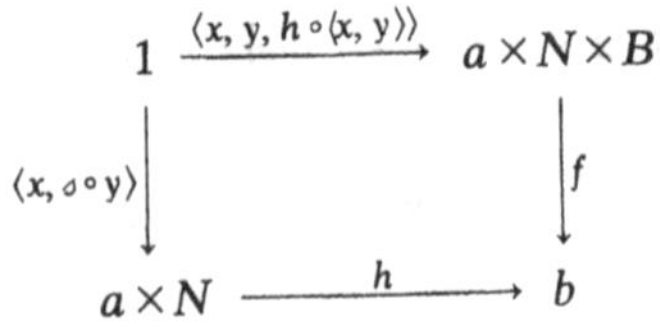

Proof. Apply the elements $x: 1 \to a$ and $\langle x, y\rangle: 1 \to a \times N$ to the two diagrams of the Primitive Recursion Theorem, and use the rules for product arrows given in the Exercises of §3.8.

The original formulation of the Primitive Recursion Theorem, in the context of well-pointed categories, is due to Lawvere [64] and states that there is a unique h satisfying the two conditions of the corollary. A full proof of this is given by Hatcher [68], wherein extensionality is invoked to show uniqueness of h.

Some special cases

(1) (*Independence of n*). Given $h_0: a \to b$ and $f: a \times b \to b$, there is a unique $h: a \times N \to b$ making

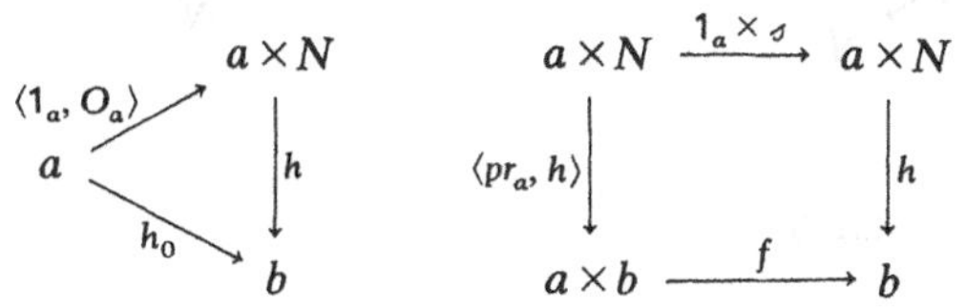

commute. (h is obtained by primitive recursion from h_0 and $f \circ \langle pr_a, pr_b \rangle : a \times N \times b \to b$, using $\mathbf{1}_{a \times N} = \langle pr_a, pr_N \rangle$.)

(2) (*Independence of x*). Given $h_0: a \to b$ and $f: N \times b \to b$ there is a unique $h: a \times N \to b$ making

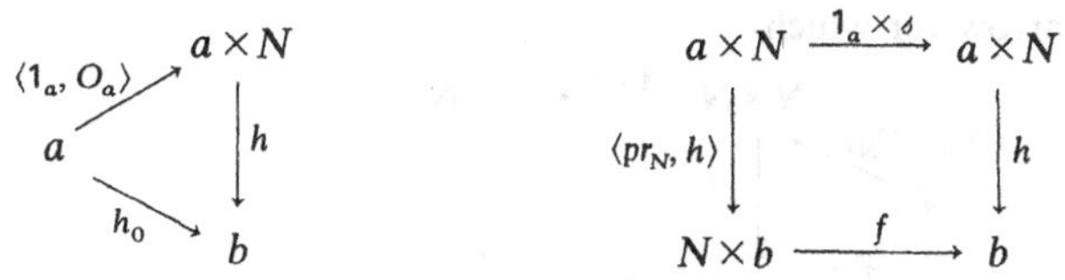

commute.

(3) (*Dependence only on n*). Given $h_0: 1 \to b$ and $f: N \to b$ there is a unique $h: N \to b$ such that

commute (this comes from Case (2), defining $h': 1 \times N \to b$ from h_0 and $f \circ pr_N : N \times b \to b$ and using the isomorphism $1 \times N \cong N$).

(4) (*Iteration*). Theorem 1 is itself a special case: given $h_0: a \to b$, $f: b \to b$, the unique $h: a \times N \to b$ is defined by primitive recursion from h_0 and $f \circ pr_b : a \times N \times b \to b$.

Using the Primitive Recursion Theorem and its special cases, we can define in any topos with a natural numbers object analogues of many arithmetical operations.

DEFINITION (*Predecessor*). $\rho : N \to N$ is defined by recursion from O: $1 \to N$ and $1_N : N \to N$ (Case (3)) as the unique arrow that exists to make

commute.

COROLLARY. σ *is monic.*

PROOF. If $\sigma \circ f = \sigma \circ g$, $\rho \circ \sigma \circ f = \rho \circ \sigma \circ g$, i.e. $1_N \circ f = 1_N \circ g$.

EXERCISE 9. Show that ρ is epic.   □

DEFINITION (*Subtraction*). $\dot- : N \times N \to N$ is the iterate of ρ, i.e. the unique arrow for which

$$
\begin{array}{ccc}
& N \times N \xrightarrow{\ 1 \times \sigma\ } N \times N & \\
{}^{\langle 1_N,\, O_N\rangle}\nearrow & \Big\downarrow {\scriptstyle \dot-} \qquad \Big\downarrow {\scriptstyle \dot-} & \\
N & & \\
{}_{1_N}\searrow & N \xrightarrow{\quad \rho \quad} N &
\end{array}
$$

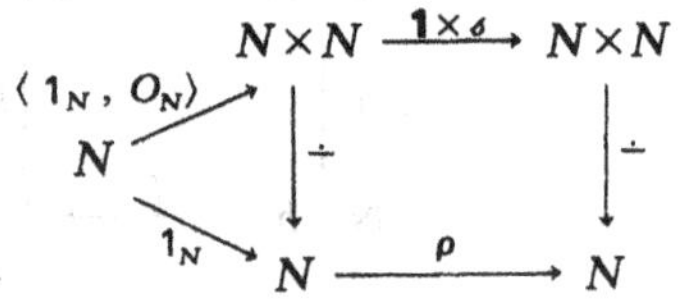

commutes.

EXERCISE 10. Verify that in **Set**

$$
m \dot- n = \begin{cases} m - n & \text{if } m \geqslant n \\ 0 & \text{otherwise.} \end{cases}
$$

EXERCISE 11.

$$
\begin{array}{ccc}
N \times N & \xrightarrow{\ 1_N \times \sigma\ } & N \times N \\
{\scriptstyle \langle \oplus,\, pr_2\rangle}\Big\downarrow & & \Big\downarrow {\scriptstyle \langle \oplus,\, pr_2\rangle} \\
N \times N & \xrightarrow{\ \sigma \times \sigma\ } & N \times N
\end{array}
$$

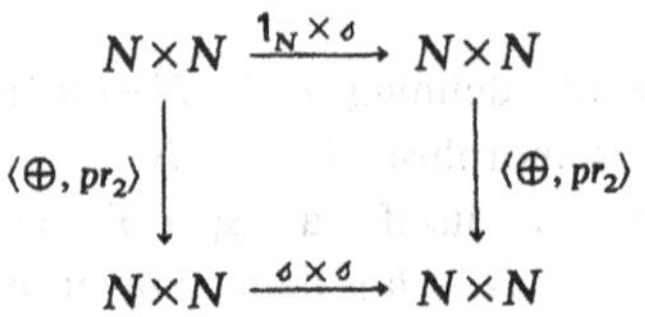

commutes.

EXERCISE 12. $((n+1)\dot{-}1=n)$. The diagram

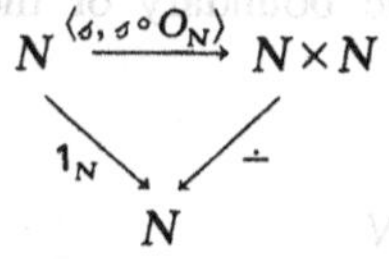

commutes. □

THEOREM 2. (1) $[(m+1)\dot{-}(n+1)=m\dot{-}n]$

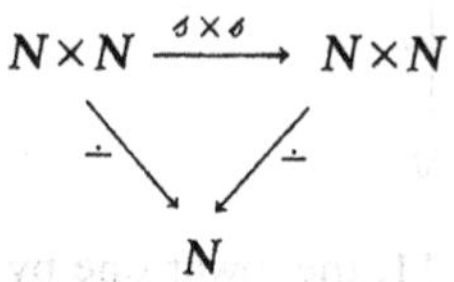

commutes.

(2) $[(m+n)\dot{-}n=m]$

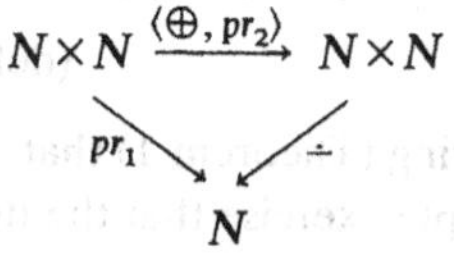

commutes.

PROOF. (1) Consider

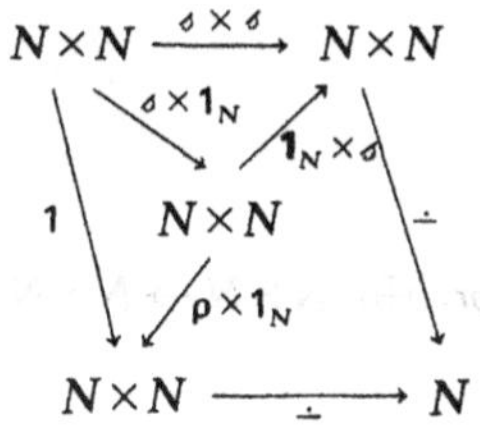

That the upper triangle commutes is a standard exercise (3.8.8) in product arrows. For the other triangle we have

$$\rho\times 1_N\circ\sigma\times 1_N=\rho\circ\sigma\times 1_N\circ 1_N \qquad (3.8.8)$$
$$=1_N\times 1_N$$
$$=1_{N\times N}.$$

But the lower part of the diagram commutes by the second diagram of Exercise 4 above (tipped over). Hence the boundary of the diagram commutes as required.

(2) Consider

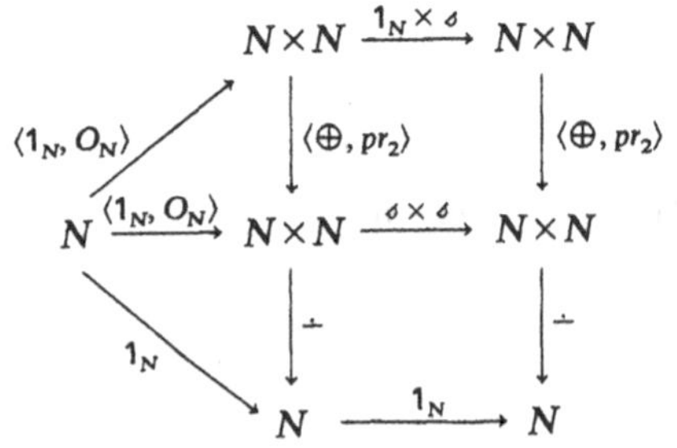

The upper square commutes by Exercise 11, the lower one by part (1) of this theorem. The lower triangle is part of the definition of $\dot{-}$, and for the upper triangle we have

$$\begin{aligned}\langle\oplus, pr_2\rangle\circ\langle 1_N, O_N\rangle &= \langle\oplus\circ\langle 1_N, O_N\rangle, pr_2\circ\langle 1_N, O_N\rangle\rangle\\ &= \langle 1_N, O_N\rangle \qquad\qquad \text{(definition } \oplus).\end{aligned}$$

Thus the whole diagram commutes, showing (Theorem 1) that $\dot{-}\circ\langle\oplus, pr_2\rangle$ is the unique iterate of 1_N. But it is a simple exercise that the iterate of 1_N is $pr_1: N\times N \to N$. □

Corollary.

(1)

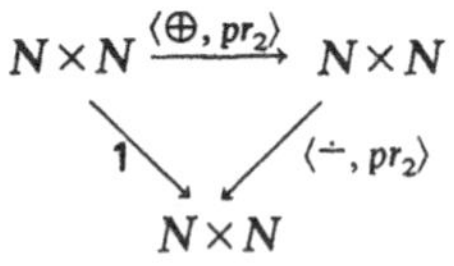

commutes.

(2) $\langle\oplus, pr_2\rangle: N\times N\to N\times N$ and $\langle pr_1, \oplus\rangle: N\times N \to N\times N$ *are both monic.*

Proof. (1)

$$\begin{aligned}&\langle\dot{-}, pr_2\rangle\circ\langle\oplus, pr_2\rangle\\ &= \langle\dot{-}\circ\langle\oplus, pr_2\rangle, pr_2\circ\langle\oplus, pr_2\rangle\rangle\\ &= \langle pr_1, pr_2\rangle \qquad\qquad \text{(Theorem, part (2))}\\ &= 1_{N\times N}.\end{aligned}$$

(2) From (1) (as in the proof that s is monic), we get $\langle \oplus, pr_2\rangle$ monic. But then, since

$$\begin{array}{ccc} N\times N & \xrightarrow{\langle pr_2,\, pr_1\rangle} & N\times N \\ & {\scriptstyle\langle \oplus,\, pr_1\rangle}\searrow \quad \swarrow{\scriptstyle\langle \oplus,\, pr_2\rangle} & \\ & N\times N & \end{array}$$

commutes, using Exercise 8, and so too does

$$\begin{array}{ccc} N\times N & \xrightarrow{\langle \oplus,\, pr_1\rangle} & N\times N \\ & {\scriptstyle\langle pr_1,\, \oplus\rangle}\searrow \quad \swarrow{\scriptstyle\langle pr_2,\, pr_1\rangle} & \\ & N\times N & \end{array}$$

the fact that the twist arrow $\langle pr_2, pr_1\rangle$ is iso means that $\langle pr_1, \oplus\rangle$ is monic.

□

Order relations

The standard ordering $\leqslant$ on ω yields the relation

$$L=\{\langle m, n\rangle: m\leqslant n\}.$$

Since, in general, $m\leqslant n$ iff for some $p\in\omega$, $m+p=n$, we have

$$L=\{\langle m, m+p\rangle: m, p\in\omega\}.$$

But $\langle m, m+p\rangle$ is the output of the function $\langle pr_1, \oplus\rangle:\omega\times\omega\to\omega\times\omega$, for input $\langle m, p\rangle$, so we have the epi-monic factorisation

$$\begin{array}{ccc} \omega\times\omega & \xrightarrow{\langle pr_1,\, \oplus\rangle} & \omega\times\omega \\ & {\scriptstyle\langle pr_1,\, \oplus\rangle^*}\searrow \quad \nearrow & \\ & L & \end{array}$$

Thus in $\mathscr{C}$ we may define the order relation on N to be that subobject of $N\times N$ that arises from the epi-monic factorisation of $\langle pr_1, \oplus\rangle$. Since, as we have just seen, this arrow is monic already, we may take *it* to represent the order on N.

The strict order $<$ on ω is given from $\leqslant$ by the condition

$$m<n \quad \text{iff} \quad m+1\leqslant n.$$

Thus in $\mathscr{C}$ we define $\textcircled{<} : N\times N \rightarrowtail N\times N$ by the diagram

$$\begin{array}{ccc} N\times N & \xrightarrow{s\times 1_N} & N\times N \\ & {\scriptstyle \textcircled{<}}\searrow \quad \swarrow{\scriptstyle \langle pr_1,\oplus\rangle} & \\ & N\times N & \end{array}$$

$s\times 1_N$ is monic, being a product of monics, and so $\textcircled{<}$ is indeed a subobject of $N\times N$.

Exercise 13. Define the $\mathscr{C}$-arrows corresponding to the relations

$$\{\langle m, n\rangle: m \geqslant n\}$$

and

$$\{\langle m, n\rangle: m > n\}$$

on ω. □

Definition (*Multiplication*). $\otimes: N\times N \to N$ is defined recursively from O_N and $\oplus$ (Special Case (1)) as the unique arrow making

$$\begin{array}{ccc} & & N\times N \\ & \nearrow{\scriptstyle \langle 1_N, O_N\rangle} & \downarrow{\scriptstyle \otimes} \\ N & \xrightarrow{O_N} & N \end{array} \qquad \begin{array}{ccc} N\times N & \xrightarrow{1_N\times s} & N\times N \\ {\scriptstyle \langle pr_1,\otimes\rangle}\downarrow & & \downarrow{\scriptstyle \otimes} \\ N\times N & \xrightarrow{\oplus} & N \end{array}$$

commute.

Exercise 14. Show that, for $x: 1\to N$ and $y: 1 \to N$

$$\begin{array}{ccc} N\times N & \xrightarrow{\langle pr_1,\oplus\rangle} & N\times N \\ & \nwarrow\cdots \quad \nearrow{\scriptstyle \langle x, y\rangle} & \\ & 1 & \end{array}$$

$\langle x, y\rangle \in \langle pr_1, \oplus\rangle$ iff for some $z: 1 \to N$, $\oplus\circ\langle x, z\rangle = y$.

Exercise 15. Show that $\langle x, y\rangle \in \textcircled{<}$ iff for some z, $\oplus\circ\langle s\circ x, z\rangle = y$.

Exercise 16. Show for any $x: 1 \to N$, that

$$\otimes\circ\langle x, s\circ O\rangle = x.$$

EXERCISE 17. Define in $\mathscr{E}$ analogues of the following arithmetical arrows in **Set**

(i) $\exp(m, n) = m^n$

(ii) $|m-n| = \begin{cases} m-n & \text{if } m \geqslant n \\ n-m & \text{otherwise} \end{cases}$

(iii) $\max(m, n) =$ maximum of m and n

(iv) $\min(m, n) =$ minimum of m and n. □

Further information about recursion on natural numbers objects in topoi is given by Brook [74], on which much of this section has been based.

13.3. Peano postulates

In **Set** one can prove of the system $1 \xrightarrow{O} \omega \xrightarrow{s} \omega$ that

(1) $s(x) \neq 0$, all $x \in \omega$.

(2) $s(x) = s(y)$ only if $x = y$, all $x, y \in \omega$.

(3) if $A \subseteq \omega$ satisfies

(i) $0 \in A$, and

(ii) whenever $x \in A$ then $s(x) \in A$,

then $A = \omega$.

Statement (3) formalises the principle of Finite Mathematical Induction. Any natural number is obtainable from 0 by repeatedly adding 1 a finite number of times. (i) and (ii) tell us that this process always results in a member of A.

The three statements (1), (2), (3), known as the Peano Postulates, provide the basis for an axiomatic development of classical number theory. They characterise ω in **Set**, in the sense that if $1 \xrightarrow{O'} \omega' \xrightarrow{s'} \omega'$ was any other system satisfying the analogues of (1), (2), (3), then the unique $h : \omega \to \omega'$ for which

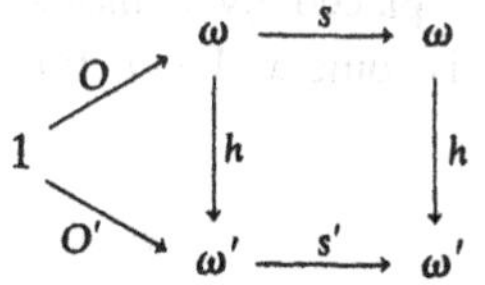

commutes would be iso (i.e. a bijection) in **Set**. (1)′ and (2)′ are used to show that h is injective, and (3)′ applied to $h(\omega) \subseteq \omega'$ shows that $h(\omega) = \omega'$, i.e. h is surjective.

In this section we show that an nno in any topos satisfies analogues of (1), (2), (3). We will then appeal to some deep results of Freyd [72] to show that the notion of a natural numbers object is exactly characterised by categorial Peano Postulates.

It should be clear to the reader how the condition "$s(x) \neq 0$" abstracts to

PO: $\qquad O \notin s$, i.e.

$$\begin{array}{ccc} & 1 & \\ {}^{x}\swarrow & & \searrow^{O} \\ N & \xrightarrow{\;s\;} & N \end{array}$$

does not commute for any "natural number" $x:1 \to N$.

Alternatively, Postulate (1) asserts that

$$s^{-1}(\{0\}) = \emptyset,$$

where $s^{-1}(\{0\}) = \{x \in \omega : s(x) = 0\}$ is the inverse image of $\{0\}$ under s. According to §3.13, the inverse image of a subset of the codomain arises by pulling the inclusion of that subset back along the function in question. Hence we contemplate another abstraction of Postulate (1)

P1:

$$\begin{array}{ccc} 0 & \longrightarrow & 1 \\ \downarrow & & \downarrow O \\ N & \xrightarrow{\;s\;} & N \end{array}$$

is a pullback.

Postulate (2) states precisely that the successor function is injective, and so becomes

P2: $\qquad N \xrightarrow{s} N$ *is monic.*

In Postulate (3), the subset $A \subseteq \omega$ is replaced by a monic $f: a \rightarrowtail N$. Hypothesis (i) becomes $O \in f$, i.e. there is some $x:1 \to a$ for which

$$\begin{array}{ccc} & 1 & \\ {}^{x}\swarrow & & \searrow^{O} \\ a & \overset{f}{\rightarrowtail} & N \end{array}$$

commutes. Hypotheses (ii) states that $s(A) \subseteq A$, where $s(A) = \{s(x) : x \in A\}$ is the image of A under s. Recalling the discussion of images at the beginning of §12.6, $s(A)$ generalises to $s[f]=\text{im}(s\circ f)$, and since s and f are monic, $s[f]\simeq s\circ f$. Thus (ii) becomes the statement that in Sub(N), $s\circ f\subseteq f$, i.e.

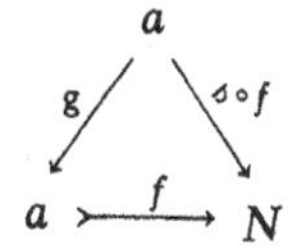

commutes for some g.

Altogether then Postulate (3) becomes

P3:

For any subobject $a \overset{f}{\rightarrowtail} N$ of N, *if*

(i) $O\in f$, *and*
(ii) $s\circ f\subseteq f$

then $f\simeq 1_N$.

THEOREM 1. *Any natural numbers object* $1\overset{O}{\to}N\overset{s}{\to}N$ *satisfies P0, P2, and P3.*

PROOF. *P0:* If $s\circ x=O$ for some $x:1\to N$, then

$$\rho\circ s\circ x=\rho\circ O$$

and so

$$1_N\circ x=O$$

i.e.

$$x=O \qquad \text{(by definition of } \rho)$$

But then we have $s\circ O=s\circ x=O$, and so if h is defined by recursion

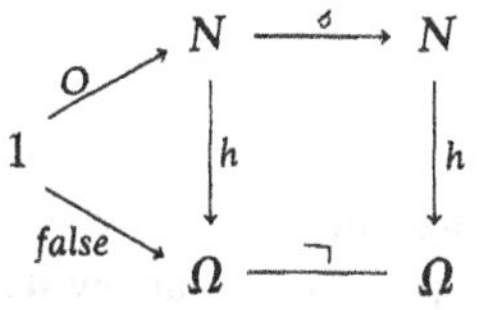

from *false* and $\neg$ we would have

$$\begin{aligned} true &= \neg\circ false \\ &= h\circ s\circ O \\ &= h\circ O \\ &= false \end{aligned}$$

which would make $\mathscr{E}$ degenerate.

P2: $\mathscr{s}$ was shown to be monic in the last section.

P3: Suppose $f \subseteq 1_N$, and there are commuting diagrams

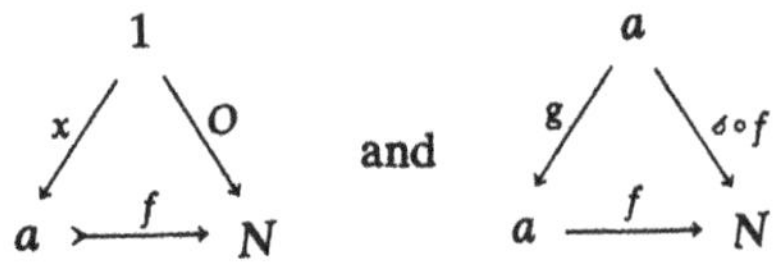

Let h be defined from x and g by simple recursion and consider

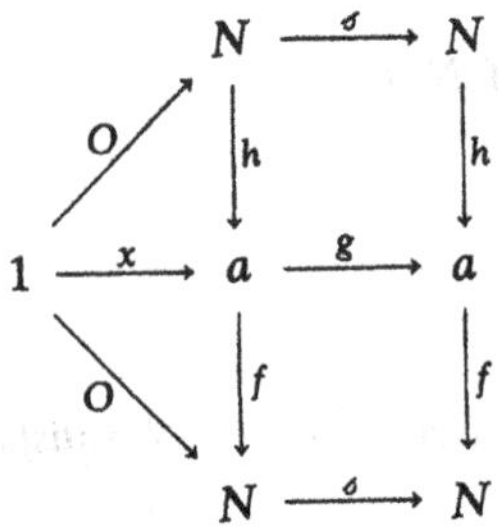

The upper triangle and square commute by definition of h, the lower two by the previous diagrams. Hence the whole diagram commutes, revealing $f \circ h$ as the unique arrow defined by recursion from O and $\mathscr{s}$. But obviously these last two arrows recursively define 1_N. Hence

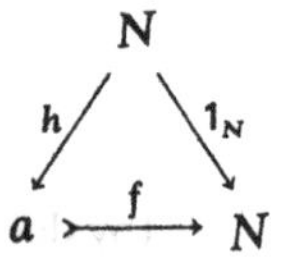

commutes, showing that $1_N \subseteq f$, and so $1_N \simeq f$. □

Exercise 1. Derive *P0* from *P1*. □

The elements of N in **Set** are of course just the finite ordinals $n \in \omega$. Correspondingly, in $\mathscr{E}$ we define, for each $n \in \omega$, an arrow $\mathbf{n}: 1 \to N$ by

$$\mathbf{n} = \underbrace{\mathscr{s} \circ \mathscr{s} \circ \ldots \circ \mathscr{s}}_{n \text{ times}} \circ O$$

The arrows $\mathbf{n}$ will be called the *finite ordinals of* $\mathscr{E}$. Using these, and the more general natural numbers $x: 1 \to N$ of $\mathscr{E}$, we can formulate two variants of the third Peano postulate.

P3A:
For any $a \overset{f}{\rightarrowtail} N$, *if*

(i) $O \in f$, *and*

(ii) $x \in f$ *implies* $s \circ x \in f$, *all* $1 \xrightarrow{x} N$

then $f \simeq 1_N$.

P3B:
For any $a \overset{f}{\rightarrowtail} N$, *if*

(i) $O \in f$, *and*

(ii) $\mathbf{n} \in f$ *implies* $s \circ \mathbf{n} \in f$, *all* $n \in \omega$

then $f \simeq 1_N$.

EXERCISE 1. Show that in $\mathbf{Bn}(I)$, $\mathbf{n}$ is the section of $pr_I : I \times \omega \to I$ that has $\mathbf{n}(i) = \langle i, n \rangle$, all i.

EXERCISE 2. Show that in $\mathbf{Bn}(\omega)$, the diagonal map $\Delta : \omega \to \omega \times \omega$ is a natural number $\Delta : 1 \to N$, with $\Delta \neq \mathbf{n}$, all n.

EXERCISE 3. Show that *P3B* implies *P3A* and *P3A* implies *P3* in general.

EXERCISE 4. Show that *P3B* holds in $\mathbf{Set}^{\mathbf{2}}$ and in $\mathbf{Bn}(I)$ and $\mathbf{Top}(I)$.

EXERCISE 5. Use Theorem 7.7.2 to show that in a well-pointed topos *P3* implies *P3A*. □

Before examining *P1*, we look at two further properties of ω in **Set**. First we observe that

$$\omega \underset{s}{\overset{\mathrm{id}}{\rightrightarrows}} \omega \xrightarrow{!} \{0\}$$

is a co-equaliser diagram in **Set**. For if

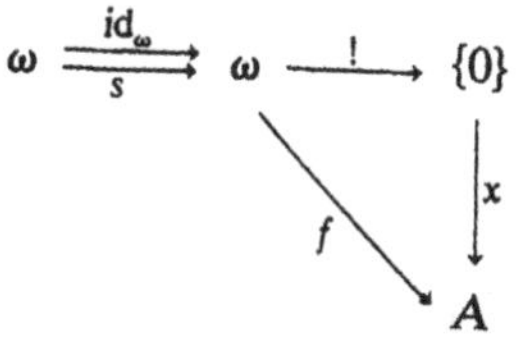

$f \circ s = f \circ \mathrm{id}_\omega = f$, then for each $n \in \omega$, $f(n+1) = f(n)$, and hence (by induction) $f(n) = f(0)$ all n. Thus f is a constant function with $f(0)$ its sole output. Putting $x(0)=f(0)$ then makes the last diagram commute, and clearly x is uniquely defined and exists iff $f\circ s=f$.

Thus we formulate

F1: $N \xrightarrow{!} 1$ *is the co-equaliser of* s *and* 1_N.

Exercise 6. According to §3.12 the codomain of the co-equaliser of id_ω and s is the quotient set ω/R, where R is the smallest equivalence relation on ω having $nRs(n)$, all $n\in\omega$. Show that there is only one such R, namely the universal relation $R=\omega\times\omega$, having $\omega/R=\{\omega\}$, a terminal object in **Set**. □

Since, in **Set**, $\mathrm{Im}\, s=\{1,2,3,\ldots\}$, we have $\{0\}\cup\mathrm{Im}\, s=\omega$. But (Postulate (1)) $\{0\}\cap\mathrm{Im}\, s=\emptyset$, and so the union is a disjoint one – $\{0\}+\mathrm{Im}\, s\cong\{0\}\cup\mathrm{Im}\, s=\omega$. Identifying $\{0\}$ with $O:1\to\omega$ and $\mathrm{Im}\, s$ with the monic s we have

$$[0,s]\colon 1+\omega\cong\omega,$$

and thus we formulate

F2: *The co-product arrow* $[O, s]:1+N\to N$ *is iso.*

Theorem 2. *F1 and F2 hold for any natural numbers object.*

Proof. *F1:* Suppose that

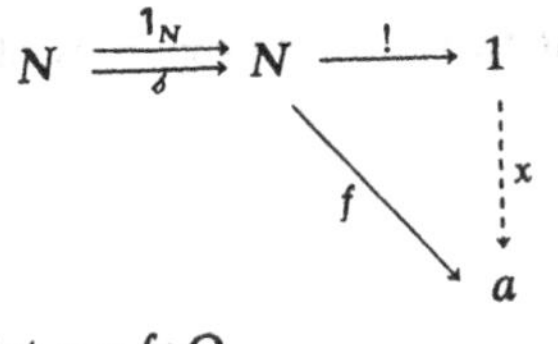

$f\circ s=f$. Put $x=f\circ O$,

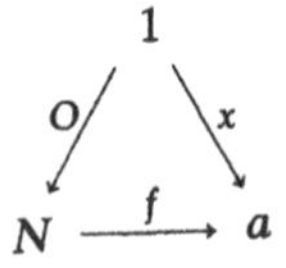

so that

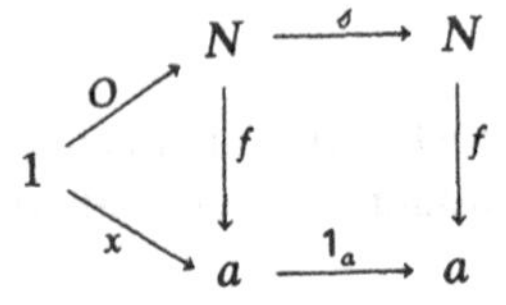

commutes. But

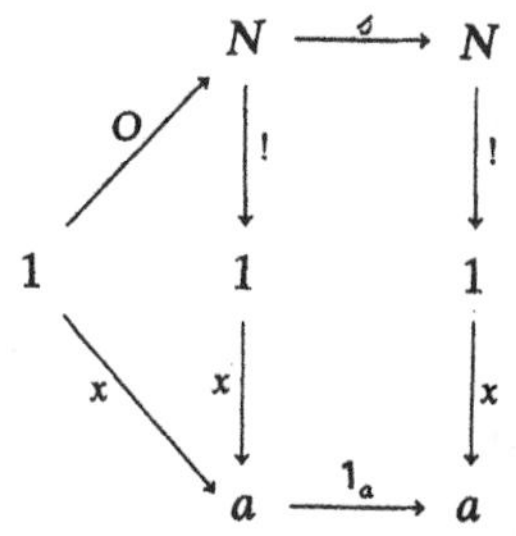

commutes, and so by the axiom NNO,

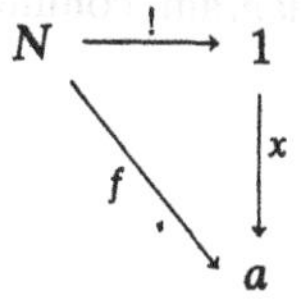

commutes as required. That there can be only one such x making this diagram commute follows from the fact that $!: N \to 1$ is epic. To see why, observe that

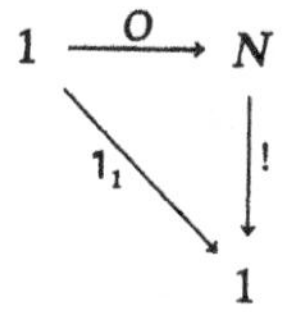

commutes, and use the fact that 1_1 is epic (or derive the result directly).

F2: Let $t: 1+N \to 1+N$ be the arrow $j \circ [O, s]$

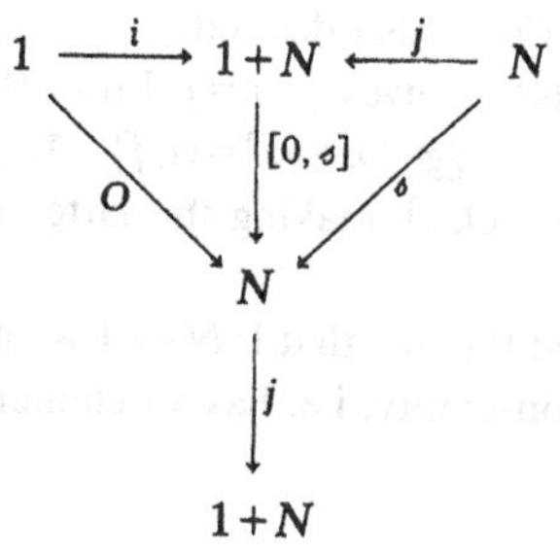

where i and j are the injections. Let g be defined by recursion from i and t, and consider

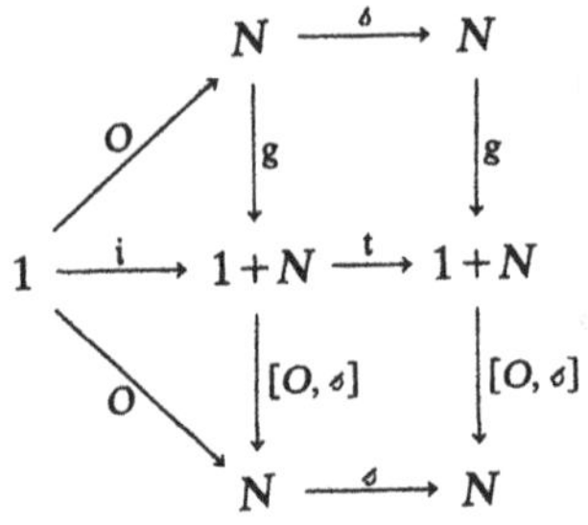

Since i is an injection, $[O, s]\circ i = O$. Since j is an injection, $[O, s]\circ t = [O, s]\circ j\circ[O, s] = s\circ[O, s]$. Hence the whole diagram commutes. NNO then gives $[O, s]\circ g = 1_N$.

Now the diagrams

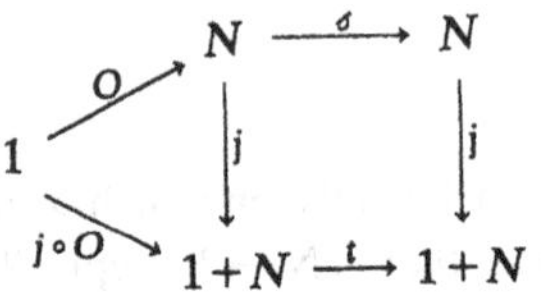

and

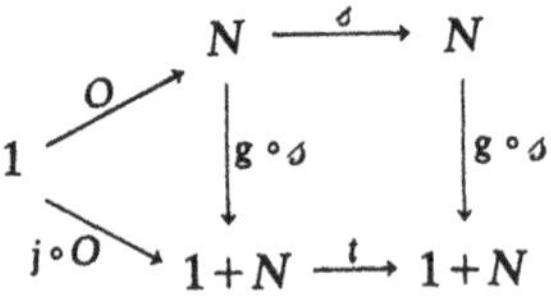

both commute. The first is left as an exercise. For the second, observe from the previous diagram that $t\circ g = g\circ s$. This yields $t\circ g\circ s = g\circ s\circ s$ as desired, and also $t\circ g\circ O = g\circ s\circ O$. But $t\circ g\circ O = j\circ[O, s]\circ g\circ O = j\circ 1_N\circ O = j\circ O$, hence $j\circ O = g\circ s\circ O$, as also desired.

From these last two diagrams, NNO gives $g\circ s = j$. From the previous one we have $g\circ O = i$. Thus $g\circ[O, s] = [g\circ O, g\circ s] = [i, j] = 1_{1\times N}$. Thus we have shown that g is an inverse to $[O, s]$, making the latter iso. □

EXERCISE 7. In deriving *F1* we used the fact that $! : N \to 1$ is epic. Show in any category with 1, that if a is non-empty, i.e. has an element $x : 1 \to a$, then $! : a \to 1$ is epic. □

LEMMA. *In any topos, if*

$$
\begin{array}{ccc}
d & \xrightarrow{\ g\ } & a \\
f\downarrow & & \downarrow k \\
b & \xrightarrow{\ h\ } & c
\end{array}
$$

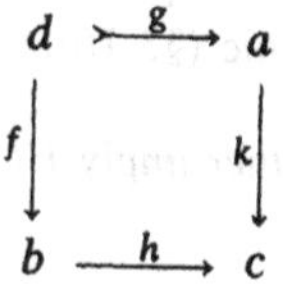

is a pushout with g monic, then h is monic and the square is a pullback.

PROOF. By the Partial Arrow Classifier Theorem (§11.8), using the classifier $\eta_b : b \to \tilde{b}$ associated with b, we have a diagram

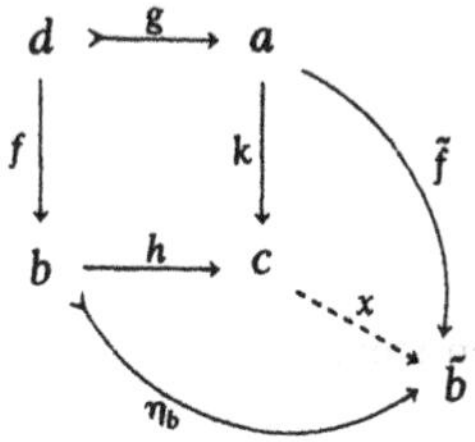

whose boundary is a pullback. The co-universal property of pushouts then implies the existence of the unique x as shown to make the whole diagram commute. That the original square is also a pullback is then a straightforward exercise. Finally, since $x \circ h = \eta_b$ is monic, h must be too. $\square$

THEOREM 3. *Any natural numbers object satisfies*

P1:

$$
\begin{array}{ccc}
0 & \longrightarrow & 1 \\
\downarrow & & \downarrow O \\
N & \xrightarrow{\ \sigma\ } & N
\end{array}
$$

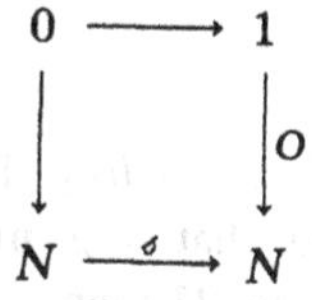

is a pullback.

PROOF. Since, by F2 we have an isomorphism $[O, \sigma]: 1 + N \to N$, it is readily established that

$$1 \xrightarrow{O} N \xleftarrow{\sigma} N$$

is a co-product diagram in $\mathscr{E}$. The co-universal property of co-products then makes it immediate that the diagram for *P1* is a pushout, and so the result follows by the Lemma, since $0 \rightarrow 1$ is monic (§3.16). □

THEOREM 4. *The conditions P1, P2, and P3 together imply F1 and F2 for any diagram*

$$1 \xrightarrow{O} N \xrightarrow{s} N$$

in a topos.

PROOF. *F1:* Suppose that $f \circ s = f$, and let $g : b \rightarrowtail N$ be the equaliser

$$\begin{array}{ccccc} b & \overset{g}{\rightarrowtail} & N & \xrightarrow{!_N} & 1 \\ & & & \searrow^{f} & \downarrow{\scriptstyle f \circ O} \\ & & & & a \end{array}$$

of f and $f \circ O \circ !_N$. Let $f \circ O \circ !_N = h$. Then since

$$\begin{array}{ccc} 1 & \xrightarrow{O} & N \\ & {\scriptstyle 1_1}\searrow \quad \swarrow{\scriptstyle !_N} & \\ & 1 & \end{array}$$

$!_N \circ O = 1_1$, it follows that $h \circ O = f \circ O$. Since g equalises f and h, O must then factor through g, hence $O \in g$.

Next, observe that

$$\begin{array}{ccc} N & \xrightarrow{s} & N \\ & {\scriptstyle !_N}\searrow \quad \swarrow{\scriptstyle !_N} & \\ & 1 & \end{array}$$

$!_N \circ s = !_N$, from which it follows readily that $h \circ s \circ g = h \circ g$. But $h \circ g = f \circ g = f \circ s \circ g$. Thus $h \circ (s \circ g) = f \circ (s \circ g)$, implying that $s \circ g$ must factor through the equaliser g, i.e. $s \circ g \subseteq g$. The postulate *P3* then gives $g \simeq 1_N$, so that g is iso, in particular epic, the latter being enough to give $f = h = f \circ O \circ !_N$. Hence

$$\begin{array}{ccccc} N & \underset{s}{\overset{1_N}{\rightrightarrows}} & N & \xrightarrow{!} & 1 \\ & & & \searrow^{f} & \downarrow{\scriptstyle f \circ O} \\ & & & & a \end{array}$$

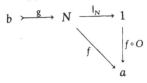

commutes. But $!: N \to 1$ is epic, since N has the element $O: 1 \to N$, and so $f \circ O$ is the only element of a that will make this diagram commute. This establishes *F1*.

F2: By *P2* and *P1*, s and O are disjoint monics, so the Lemma associated with Theorem 5.4.3 gives $[O, s]$ as monic. To prove *F2* then, it suffices to show that $[O, s]$ is epic.

Suppose then that $f \circ [O, s] = g \circ [O, s]$

$$\begin{array}{ccccc} 1 & \xrightarrow{i} & 1+N & \xleftarrow{j} & N \\ & {\scriptstyle O}\searrow & \downarrow{\scriptstyle [O,s]} & \swarrow{\scriptstyle s} & \\ & & N & & \\ & & {\scriptstyle f}\downarrow\downarrow{\scriptstyle g} & & \\ & & a & & \end{array}$$

From the diagram we see that $f \circ O = g \circ O$ and $f \circ s = g \circ s$. Then if $h: b \rightarrowtail N$ equalises f and g we must have $O \in h$, and since then $f \circ s \circ h = g \circ s \circ h$, $s \circ h$ factors through h, i.e. $s \circ h \subseteq h$. Postulate *P3* then gives $h \simeq 1_N$, from which $f = g$ follows. Thus $[0, s]$ is right cancellable.

Corollary. *In any topos $\mathscr{E}$, the following are equivalent for a diagram of the form $1 \xrightarrow{O} N \xrightarrow{s} N$.*

(1) *The diagram is a natural numbers object.*

(2) *The diagram satisfies the Peano Postulates P1, P2, and P3.*

(3) *The diagram satisfies the Freyd Postulates F1 and F2.*

Proof. (1) implies (2): Theorems 1 and 3.

(2) implies (3): Theorem 4.

(3) implies (1): Freyd [72], Theorem 5.4.3. □

The equivalence of (1) and (3) established by Freyd requires techniques beyond our present scope. Freyd also establishes the equivalence in any topos of

(a) *there exists a natural numbers object,*

(b) *there exists a monic $f: a \rightarrowtail a$ and an element $x: 1 \to a$ of its domain for which*

$$\begin{array}{ccc} 0 & \longrightarrow & 1 \\ \downarrow & & \downarrow{\scriptstyle x} \\ a & \overset{f}{\rightarrowtail} & a \end{array}$$

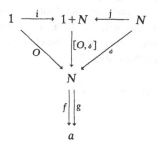

is a pullback, and

(c) *there exists an isomorphism of the form* $1+a \cong a$.

With regard to (c), observe that in **Finset**, where there is no nno, isomorphic objects are finite sets with the same number of elements, and $1+a$ has one more element than a.

The intuitive import of (b) is that the sequence $x, f(x), f(f(x)), \ldots$ has all terms distinct and so forms a subset $\{x, f(x), \ldots\}$ of a isomorphic to ω. The natural numbers object then arises as the "intersection" of all subobjects $g: b \rightarrowtail a$ that contain this set, i.e. have $x \in g$ and $f \circ g \subseteq g$. These ideas are formalised in another approach to the characterisation of natural numbers objects developed by Osius [75].

EXERCISE 8. Derive *P1* and *P2* directly from *F2*. □

CHAPTER 14

LOCAL TRUTH

"a Grothendieck topology appears most naturally as a modal operator, of the nature 'it is locally the case that'"

F. W. Lawvere

The notion of a topological bundle represents but one side of the coin of sheaf theory. The other involves the conception of a sheaf as a functor defined on the category of open sets in a topological space. Our aim now is to trace the development of ideas that leads from this notion, via Grothendieck's generalisation, to the notion of a "topology" on a category and its attendant sheaf concept, and from there to the first-order concept of a topology on a topos and the resultant axiomatic sheaf theory of Lawvere and Tierney. The chapter is basically a survey, and its intention is to direct the reader to the appropriate literature.

14.1. Stacks and sheaves

Let I be a topological space, with Θ its set of open subsets. Θ becomes a poset category under the set inclusion ordering, in which the arrows are just the inclusions $U \hookrightarrow V$.

A *stack* or *pre-sheaf* over I is a *contravariant* functor from Θ to **Set**. Thus a stack F assigns to each open V a set $F(V)$, and to each inclusion $U \hookrightarrow V$ a function $F_U^V : F(V) \to F(U)$ (note the contravariance – reversal of arrow direction), such that

(i) $F_U^U = \mathrm{id}_U$, and

(ii) if $U \subseteq V \subseteq W$, then

$$\begin{array}{ccc} & F(W) & \\ {\scriptstyle F_V^W} \swarrow & & \searrow {\scriptstyle F_U^W} \\ F(V) & \xrightarrow[F_U^V]{} & F(U) \end{array}$$

commutes, i.e. $F_U^W = F_U^V \circ F_V^W$.

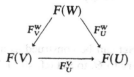

EXAMPLE. Let $f: A \to I$ be a sheaf of sets of germs over I, as in Chapter 4. Define a stack $F_f: \Theta \to \mathbf{Set}$ as follows.

$$F_f(V) = \text{the set of local sections of } f \text{ defined on } V$$

$$= \{V \xrightarrow{s} A : s \text{ is continuous and } f \circ s = V \hookrightarrow I\}$$

For an inclusion $U \hookrightarrow V$, F_{fU}^{V} is the "restricting" or "localising" map that assigns to each section $s: V \to A$ over V its restriction $s \restriction U : U \to A$ to U. Identifying sections s with their images $s(I) \subseteq A$ we have the picture

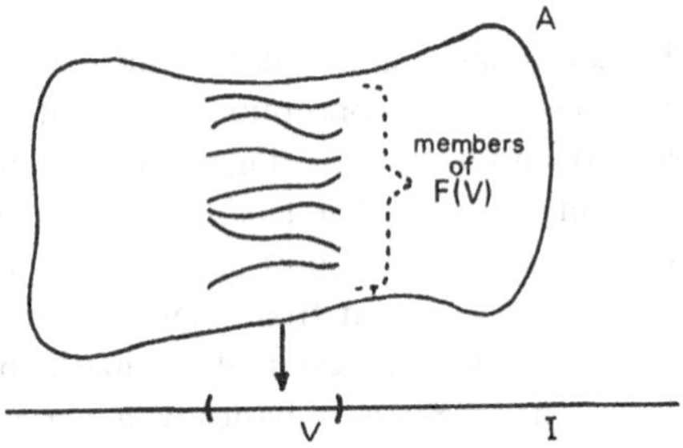

Fig. 14.1.

which indicates the origin of the word "stack". F_f is the *stack of sections over I*. The category $\mathbf{St}(I)$ has as objects the stacks $F: \Theta \to \mathbf{Set}$ and as arrows $\tau: F \dot{\to} G$ the natural transformations, i.e. collections $\{\tau_U : U \in \Theta\}$ of functions $\tau_U: F(U) \to G(U)$ such that

$$\begin{array}{ccccc} V & \quad & F(V) & \xrightarrow{\tau_V} & G(V) \\ \uparrow & & \Big\downarrow {\scriptstyle F_U^V} & & \Big\downarrow {\scriptstyle G_U^V} \\ U & & F(U) & \xrightarrow{\tau_U} & G(U) \end{array}$$

commutes whenever $U \subseteq V$.

Now a contravariant functor $\Theta \to \mathbf{Set}$ can be construed as a covariant functor from Θ^{op}, the opposite category to Θ, to $\mathbf{Set}$ (cf. §9.1). Thus $\mathbf{St}(I)$ is equivalent to the topos $\mathbf{Set}^{\Theta^{\mathrm{op}}}$

EXERCISE 1. Let $h:(A, f) \to (B, g)$ be an arrow in the spatial topos **Top**(I) of sheaves of sets of germs over I. For each open V, define $h_V: F_f(V) \to F_g(V)$ to be the function that maps a section $s \in F_f(V)$ to

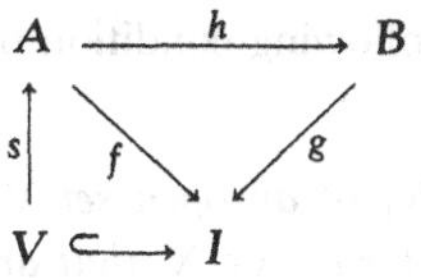

$h \circ s$, i.e $h_V(s) = h \circ s$. Verify that $h \circ s$ *is* a section of g, i.e. $h_V(s) \in F_g(V)$, and that the h_V's, for all $V \in \Theta$, form the components of an arrow $\tau_h: F_f \to F_g$ in **St**(I). Show that the assignments $f \mapsto F_f$ and $h \mapsto \tau_h$ constitute a functor $\mathcal{S}$ from **Top**(I) to **St**(I). □

Now given a stack F, the question arises as to when F is (isomorphic to) a stack of sections, i.e. when is there a sheaf of germs f such that in **St**(I), $F \cong F_f$. The answer is to be found in the answer to another question about the behavior of local sections of $f: A \to I$. Suppose that $\{V_x \xrightarrow{s_x} A : x \in X\}$ is a collection of local sections of f, indexed by some set X, and that each of their domains V_x is a subset of some open set V. Thus, for all x, $s_x \in F_f(V_x)$ and $V_x \subseteq V$. The question is – when can we "paste" together all of the sections s_x to form a single section $s: V \to A \in F_f(V)$. The rule defining the desired s is this: if $i \in V$, choose some V_x that has $i \in V_x$, and put $s(i) = s_x(i)$. In order to have dom $s = V$ we require that each $i \in V$ be a member of at least one V_x. This means that V is the union of the collection of V_x's, i.e. $V = \bigcup\{V_x : x \in X\} = \{i$: for some $x \in X$, $i \in V_x\}$. In general a collection of open sets whose union is V will be called an *open cover* of V.

In order for s to satisfy the "unique output" property of functions, the definition of $s(i)$ should be independent of the choice of V_x containing i. Thus if $i \in V_x$ and $i \in V_y$, we require $s_x(i) = s_y(i)$. So any two of our local sections s_x and s_y must agree on the part $V_x \cap V_y$ of their domains that they have in common. In symbols –

$$\text{for all } x, y \in X, \quad s_x \restriction V_x \cap V_y = s_y \restriction V_x \cap V_y.$$

Under this "compatibility" condition, s will be a well-defined member of $F_f(V)$, with $s \restriction V_x = s_x$, all x. Moreover s is the only section over V whose restriction to V_x is always s_x. For, if $t: V \to A$ has $t \restriction V_x = s_x$, all $x \in X$, then $t = s$.

Now the compatibility condition on the s_x's can be expressed in terms of the restricting maps F_U^V of a functor F. We let $F_y^x : F(V_x) \to F(V_x \cap V_y)$ and $F_x^y : F(V_y) \to F(V_x \cap V_y)$ be the F-images of the inclusions $V_x \cap V_y \hookrightarrow V_x$ and $V_x \cap V_y \hookrightarrow V_y$, and $F_x : F(V) \to F(V_x)$ the image of $V_x \hookrightarrow V$. Then what we have shown is that the following condition obtains for the case that F is of the form F_f.

COM: *Given any open cover* $\{V_x : x \in X\}$ *of an open set* V, *and any selection of elements* $s_x \in F(V_x)$, *for all* $x \in X$, *that are pairwise compatible, i.e.* $F_y^x(s_x) = F_x^y(s_y)$ *all* $x, y \in X$, *then there is exactly one* $s \in F(V)$ *such that* $F_x(s) = s_x$ *all* $x \in X$.

Notice that COM is a statement that can be made about any stack $F : \Theta \to \mathbf{Set}$. Any F satisfying COM will be called a *sheaf of sections over* I, and the full subcategory of $\mathbf{St}(I)$ generated by those objects that satisfy COM will be denoted $\mathbf{Sh}(I)$.

EXERCISE 2. Show that the constant stack $1 : \Theta \to \mathbf{Set}$, where $1(U) = \{0\}$, is a sheaf.

EXERCISE 3. Consider the space $I = \{0, 1\}$, with $\Theta = \mathcal{P}(I)$ (the discrete topology). Let $F(U) = \{0, 1\}$, all $U \in \Theta$ and $F_U^V = f$, all $U \subseteq V$, where $f(0) = f(1) = 0$. By considering the cover $\{\{0\}, \{1\}\}$ of I, show that F is not a sheaf, i.e. COM fails.

EXERCISE 4. Why must $F(\emptyset)$, *for any sheaf* F, be a one-element set?

EXERCISE 5. Show that

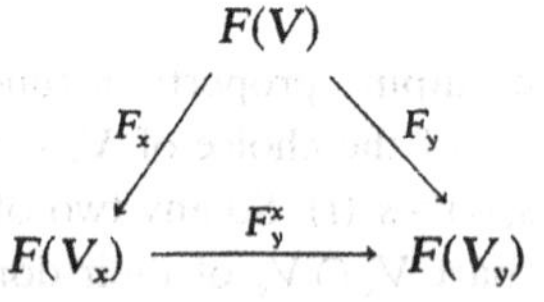

commutes whenever $V_y \subseteq V_x$, and so $F(V)$, together with the maps F_x, for all $x \in X$, forms a cone for the diagram consisting of the objects $F(V_x)$ and the arrows F_y^x. Show that COM is equivalent to the condition that this cone be universal for that diagram, i.e. that $F(V)$ be the limit of the diagram, denoted $F(V) = \varprojlim_{x \in X} F(V_x)$ (cf. §3.11). □

Now given an arbitrary stack $F : \Theta \to \mathbf{Set}$, a corresponding sheaf of germs $p_F : A_F \to I$ may be defined. For each $i \in I$ the collection

$\{F(V): i \in V\}$ of F-images of neighbourhoods of i, together with their associated restricting maps, forms a diagram in **Set**. The stalk over i in A_F is defined to be the co-limit, denoted $\varinjlim_{i \in V} F(V)$ of this diagram. Explicitly, an equivalence relation $\sim_i$ is defined on $\cup \{F(V): i \in V\}$ thus: if $s_x \in F(V_x)$ and $s_y \in F(V_y)$ (where V_x and V_y are i-neighbourhoods), we put

$$s_x \sim_i s_y \quad \text{iff} \quad F_z^x(s_x) = F_z^y(s_y), \quad \text{for some } i\text{-neighbourhood} \\ V_z \subseteq V_x \cap V_y.$$

Intuitively, $F_z^x(s_x)$ is the "localisation" of the element $s_x \in F(V_x)$ to V_z. Thus $s_x \sim_i s_y$ when they are "locally equal", that is when they have the same localisation to some i-neighbourhood. The equivalence class $[s]_i$ of $s \in F(V)$ under $\sim_i$, i.e. the set $[s]_i = \{t: s \sim_i t\}$, is called the *germ of s at i*. The stalk for p_F over i is then the set $F_i = \{\langle i, [s]_i \rangle: s \in \cup \{F(V): i \in V\}\}$. The stalk space is the union $A_F = \cup \{F_i: i \in I\}$, and p_F is the projection of A_F onto I. For each open $V \in \Theta$ and $s \in F(V)$, let $N(s, V) = \{\langle i, [s]_i \rangle: i \in V\}$. The collection of all $N(s, V)$'s generates a topology on A_F making p_F a local homeomorphism.

EXERCISE 6. Verify that $\sim_i$ is an equivalence relation.

EXERCISE 7. Define $p_V^i : F(V) \to F_i$ by

$$p_V^i(s) = \langle i, [s]_i \rangle, \quad \text{all} \quad s \in F(V).$$

Show that

$$
\begin{array}{ccc}
F(V) & \xrightarrow{\;\;F_U^V\;\;} & F(U) \\
& & \\
\;_{p_V^i}\searrow & \swarrow_{p_U^i} & \\
& F_i &
\end{array}
$$

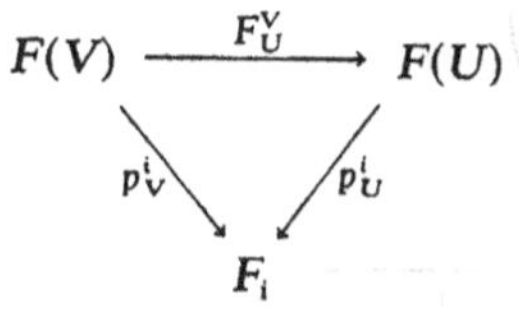

commutes when $U \subseteq V$, so that the p_V^i's form a co-cone for the diagram based on $\{F(V): i \in V\}$. Prove that this co-cone is co-universal for the diagram, so that F_i is its co-limit, $F_i = \varinjlim_{i \in V} F(V)$. (cf. §3.11).

EXERCISE 8. If $s \in F(V)$, define $s_V : V \to A_F$ by putting $s_V(i) = \langle i, [s]_i \rangle = p_V^i(s)$, for all $i \in V$. Show that s_V is a section of the sheaf $p_F : A_F \to I$.

EXERCISE 9. Let F_{p_F} be the sheaf (stack) of sections of the sheaf of germs p_F. For each V, define $\sigma_V : F(V) \to F_{p_F}(V)$ by putting, for $s \in F(V)$, $\sigma_V(s) = s_V$, where s_V is the section of p_F defined in Exercise 8. Show that the σ_V's form the components of an arrow $\sigma : F \to F_{p_F}$ in **St**(I).

Exercise 10. Let $\tau : F \dashrightarrow G$ be an arrow in **St**(I). Define $h_\tau : A_F \to A_G$ as follows: if $\langle i, [s]_i \rangle$ is a germ at i in A_F, with $s \in F(V)$ say, let $h_\tau(i)$ be the germ $\langle i, [\tau_V(s)]_i \rangle$ in A_G, where τ_V is the component $F(V) \to G(V)$ of τ. Show that

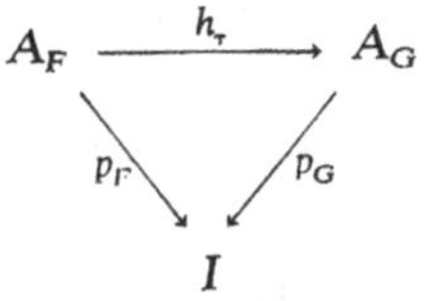

commutes, and that h_τ is a **Top**(I)-arrow from p_F to p_G.

Exercise 11. Verify that the constructions $F \mapsto p_F$, $\tau \mapsto h_\tau$ constitute a functor $\mathcal{G}$ from **St**(I) to **Top**(I).

Exercise 12. Let $f : A \to I$ be any sheaf of germs over I, F_f its stack of sections, and $p_{F_f} : A_{F_f} \to I$ the associated sheaf of germs. Define a map $k : A \to A_{F_f}$ as follows. If $a \in A$, use the local homeomorphism property of f to show that f has a local section $s : V \to A$ through a, i.e. $a \in s(V)$. Let $k(a) = \langle f(a), [s]_{f(a)} \rangle$ be the germ of s at $f(a)$.

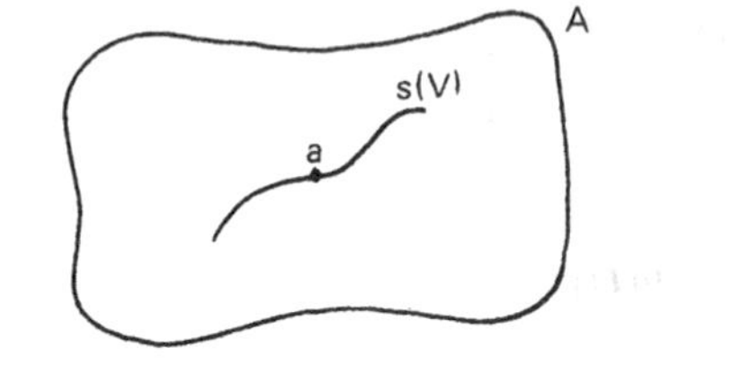

Fig. 14.2.

Check that the definition of $k(a)$ does not depend on which section through a is chosen. Show that

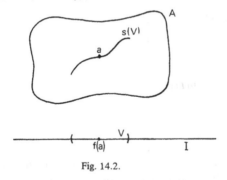

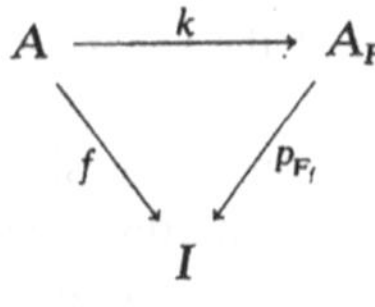

commutes, so that k is a **Top**(I)-arrow from f to p_{F_f}.

EXERCISE 13. Prove that the map k of the last exercise is a bijection, and hence is an iso arrow in **Top**(I), making $f \cong p_{F_f}$.

EXERCISE 14. Let $\sigma_V : F(V) \to F_{p_F}(V)$ be the component of $\sigma : F \dot{\to} F_{p_F}$ defined in Exercise 9. Show that σ_V is a bijection iff the condition COM holds for open covers of V. Hence show that σ is iso iff the stack F is a sheaf, i.e. that $F \cong F_{p_F}$ iff F belongs to **Sh**(I). □

Exercises 1 and 11 provide us with functors $\mathcal{S}: \mathbf{Top}(I) \to \mathbf{St}(I)$ and $\mathcal{G}: \mathbf{St}(I) \to \mathbf{Top}(I)$, with the image of $\mathcal{S}$ being (contained in) **Sh**(I). By Exercise 13,

$$f \cong \mathcal{G}(\mathcal{S}(f)), \quad \text{all} \quad \mathbf{Top}(I)\text{-objects } f.$$

However by Exercise 14, for $F \in \mathbf{St}(I)$, we have

$$F \cong \mathcal{S}(\mathcal{G}(F)) \quad \text{iff} \quad F \in \mathbf{Sh}(I).$$

Thus $\mathcal{S}$, and the restriction of $\mathcal{G}$ to **Sh**(I) are equivalences of categories (§9.2). They establish that the category of sheaves of sections over I is equivalent to the topos of sheaves of germs over I.

We conclude this brief introduction to stacks and sheaves with two major illustrations of the behaviour of **Sh**(I)-objects.

I. NNO in Sh(I)

The category **Sh**(I) has a natural numbers object – the sheaf of *locally constant natural-number-valued functions on I*. Specifically $N: \Theta \to \mathbf{Set}$ is the sheaf that has

$$N(V) = \{V \xrightarrow{g} \omega : g \text{ is continuous}\},$$

where ω is presumed to have the discrete topology, and $N^V_U(g) = g \restriction U$ whenever $U \subseteq V$.

The requirement that g be continuous for the discrete topology on ω means precisely that g is locally constant, i.e. that for each $i \in V$ there is a neighbourhood U_i of i, with $i \in U_i \subseteq V$, such that $g \restriction U_i$ is a constant function. Thus there is a number $g_i \in \omega$ such that $g(i) = g_i$ for all $i \in U_i$. This condition on g can be interpreted as saying that the statement "g is constant" is locally true of its domain V, i.e. true of some neighbourhood of each point of V.

The arrow $O:1 \to N$ has component $O_V:\{0\} \to N(V)$ picking out the constantly zero function $V \to \omega$ on V. The V-th component $\mathfrak{s}_V:N(V) \to N(V)$ of the successor arrow for $\mathbf{Sh}(I)$ has $\mathfrak{s}_V(g) = s \circ g$, where $s:\omega \to \omega$ is the successor function on ω. (Note that $s \circ g$ is locally constant if g is).

Exercise 15. Verify the axiom NNO for this construction.

Exercise 16. For $n \in \omega$, let $n_V: V \to \omega$ have $n_V(i) = n$, all $i \in V$. Explain how the n_V's provide the components for the ordinal arrow $\mathbf{n}:1 \to N$ in $\mathbf{Sh}(I)$.

Exercise 17. If $g \in N(V)$ show that V has an open cover $\{V_x : x \in X\}$ of pairwise disjoint sets, i.e. $V_x \cap V_y = \emptyset$ if $x \neq y$, on each of which g is actually constant. □

Now let $pr_I: I \times \omega \to I$ be the sheaf of germs that is the nno for $\mathbf{Top}(I)$, as described in §12.2. For each continuous $g: V \to \omega$, the product map $\langle \mathrm{id}_V, g\rangle : V \to I \times \omega$ is readily seen to be a section of pr_I, i.e. an element of the stack $F_{pr_I}(V)$ of sections over V. Indeed this construction gives a bijection $N(V) \cong F_{pr_I}(V)$ for each $V \in \Theta$, hence in $\mathbf{Sh}(I)$ we have $N \cong F_{pr_I} = \mathcal{S}(pr_I)$, so that in $\mathbf{Top}(I)$, $\mathcal{G}(N) \cong \mathcal{G}(\mathcal{S}(pr_I)) \cong pr_I$.

Exercise 18. Let $p_N: A_N \to I = \mathcal{G}(N)$ be the sheaf of germs of locally constant ω-valued functions. Define $f: I \times \omega \to A_N$ by $f(\langle i, n\rangle) = \langle i, [n_I]_i\rangle$, where $n_I \in N(I)$ is the "constantly n" function defined in Exercise 16. Show directly that f is a bijection, giving $pr_I \cong \mathcal{G}(N)$. □

II. $\mathbf{Set}^{\mathbf{P}}$ and Sh(P)

If $\mathbf{P}$ is a poset then the collection $\mathbf{P}^+$ of $\mathbf{P}$-hereditary sets is a topology on P, in terms of which we have the category (topos) $\mathbf{Sh}(\mathbf{P})$ of sheaves of the form $F:\mathbf{P}^+ \to \mathbf{Set}$. Given such a functor we can define a Kripke-model (variable set) $F^*:\mathbf{P} \to \mathbf{Set}$ as follows. F^* is to be a collection $\{F^*_p : p \in P\}$ of sets, indexed by P, with transitions $F^*_{pq}: F^*_p \to F^*_q$ whenever $p \sqsubseteq q$. We put

$$F^*_p = F([p)) \qquad \text{(note } [p) \in \mathbf{P}^+\text{)}$$

Whenever $p \sqsubseteq q$, we have $[q) \subseteq [p)$, so we take $F^*_{pq}: F([p)) \to F([q))$ to be the image of the inclusion $[q) \hookrightarrow [p)$ under the contra-variant functor F.

Since F is a sheaf, and since for each $V \in \mathbf{P}^+$, $\{[p) : p \in V\}$ covers V (cf. §10.2), by Exercise 5 above we have

$$(*) \qquad F(V) = \varprojlim_{p \in V} F([p)) = \varprojlim_{p \in V} (F_p^*)$$

This shows us how to *define* a sheaf F of sections over the topology $\mathbf{P}^+$ from a variable set $F^* : \mathbf{P} \to \mathbf{Set}$. In **Set**, all diagrams have limits, and so we can define $F(V)$ from $\{F_p^* : p \in V\}$ by the equation $(*)$. Moreover, if $U \subseteq V$, then $\{F_p^* : p \in U\} \subseteq \{F_p^* : p \in V\}$, so the universal cone $F(V)$ for the latter diagram will be a cone for the former, and so F_U^V may be

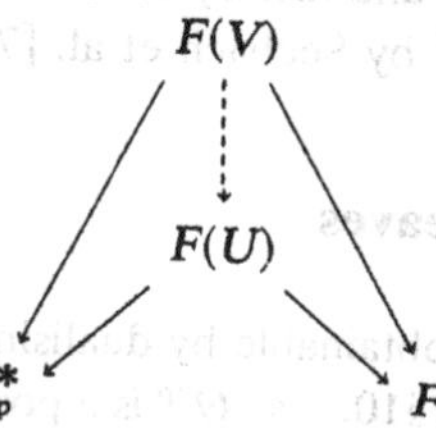

defined as the unique factoring arrow as shown.

Thus we obtain an exact correspondence between objects in $\mathbf{Set}^{\mathbf{P}}$ and $\mathbf{Sh}(\mathbf{P})$. If we pass via the functor $\mathscr{G}$ from the sheaf of sections $F : \mathbf{P}^+ \to \mathbf{Set}$ to the sheaf of germs $p_F : A_F \to P$ we find that the stalk in A_F over a point $p \in P$ turns out to be an isomorphic copy of the original set $F_p^* = F([p))$. The bijection $F([p)) \cong F_p$ (F_p = stalk over p) is given by the function $\rho_{[p)}^p$ (defined in Exercise 7) having

$$\rho_{[p)}^p(s) = \langle p, [s]_p \rangle, \quad \text{all} \quad s \in F([p)).$$

The reason why this is so is that the p-neighbourhood $[p)$ lies inside all other p-neighbourhoods (Exercise 10.2.3), so that the germ of any $s' \in F(V)$ at p is the same as the germ at p of its localisation $s = F_{[p)}^V(s')$ to $[p)$.

In view of the description (Exercise 7) of the stalk F_p as a co-limit we then have that if F is related to F^* by the equation $(*)$, then for each $p \in P$,

$$F_p^* \cong \varinjlim_{p \in V} F(V)$$

EXERCISE 19. Verify that $\rho_{[p)}^p$ is a bijection

EXERCISE 20. Show that

$$\begin{array}{ccc} F(V) & \xrightarrow{F^V_U} & F(U) \\ & {\scriptstyle F^V_{[p)}}\searrow \quad \swarrow{\scriptstyle F^U_{[p)}} & \\ & F^*_p & \end{array}$$

commutes whenever $p \in U \subseteq V$, and so F^*_p is the apex of a co-cone for $\{F(V)\colon p \in V\}$. Verify the co-universal property for this co-cone. □

The reader interested in the origins and history of sheaf theory should consult the paper "What is a Sheaf?" by Seebach et al. [70].

14.2. Classifying stacks and sheaves

The object of truth-values in $\mathbf{St}(I)$ is obtainable by dualising the description of that given for $\mathbf{Set}^{\Theta^{op}}$ in §9.3 (or §10.3, as Θ^{op} is a poset category).

If $V \in \Theta$, let $\Theta_V = \Theta \cap \mathcal{P}(V) = \{U \in \Theta\colon U \subseteq V\}$ be the collection of open subsets of V. (Since V is open, Θ_V is in fact the relative (subspace) topology on V.) A collection $C \subseteq \Theta_V$ of V-open sets is called a *V-crible* when it is closed under the taking of open subsets, i.e. when we have that

if $U \in C$, and $W \subseteq U$ has $W \in \Theta$ (i.e. $W \in \Theta_U$), then $W \in C$.

The stack $\Omega : \Theta \to \mathbf{Set}$ has

$$\Omega(V) = \{C\colon C \text{ is a } V\text{-crible}\}$$

and $\Omega^V_U(C) = C \cap \Theta_U = \{W\colon W \in C \text{ and } W \subseteq U\}$ whenever $U \hookrightarrow V$.

EXERCISE 1. $\Theta_V = \Theta_U$ iff $V = U$.

EXERCISE 2. In the opposite to the inclusion ordering, $V \sqsubseteq U$ iff $U \subseteq V$, of Θ, $\Theta_V = [V)$.

EXERCISE 3. If $U \subseteq V$, then Θ_U is a V-crible, with $\bigcup \Theta_U = U$.

EXERCISE 4. The poset $(\Omega(V), \subseteq)$ of V-cribles under the inclusion relation is a Heyting algebra with the meet and join of V-cribles C, D being their intersection $C \cap D$ and union $C \cup D$. What are $\neg C$ and $C \Rightarrow D$? □

The arrow $true: 1 \to \Omega$ has components $true_V: \{0\} \to \Omega(V)$ given by

$$true_V(0) = \Theta_V, \quad \text{the largest } V\text{-crible.}$$

Given a monic arrow $\tau: F \rightarrowtail G$ of stacks, with each τ_V being the inclusion $F(V) \hookrightarrow G(V)$, the character $\chi_\tau: G \to \Omega$ has V-component $(\chi_\tau)_V: G(V) \to \Omega(V)$, where for $x \in G(V)$,

$$(\chi_\tau)_V(x) = \{U \subseteq V: G^V_U(x) \in F(U)\}$$

EXERCISE 5. Verify that $(\chi_\tau)_V(x)$ is a V-crible. □

In the category **Sh**(I) of sheaves of sections over I there is a subobject classifier, which is not the same as that for **St**(I). This time the object of truth-values is the contravariant functor $\Omega_j: \Theta \to \mathbf{Set}$ that has

$$\Omega_j(V) = \Theta_V, \quad \text{the collection of open subsets of } V,$$

while Ω_j assigns to each inclusion $U \hookrightarrow V$ the restricting map $\Omega_j(V) \to \Omega_j(U)$ that takes $W \in \Theta_V$ to $W \cap U \in \Theta_U$.

The arrow $true_j: 1 \to \Omega_j$ has V-th component $true_{jV}: \{0\} \to \Theta_V$ given by

$$true_{jV}(0) = V, \quad \text{the largest } V\text{-open set.}$$

If $\tau: F \rightarrowtail G$ is a monic arrow in **Sh**(I) its character $\chi^j_\tau: G \to \Omega_j$ has component

$$(\chi^j_\tau)_V: G(V) \to \Omega_j(V),$$

where

$$\begin{aligned}(\chi^j_\tau)_V(x) &= \bigcup\{U \subseteq V: G^V_U(x) \in F(U)\}\\ &= \bigcup(\chi_\tau)_V(x)\end{aligned}$$

(Θ, being a topology, is closed under unions of arbitrary sub-collections).

EXERCISE 6. Show that Ω_j is a sheaf, i.e. satisfies COM.

EXERCISE 7. Verify that the construction just given shows that the Ω-axiom holds in **Sh**(I) for $true_j: 1 \to \Omega_j$, identifying the point at which the condition COM is needed. □

Notice that if the $F \mapsto p_F$ construction (the functor $\mathcal{G}$) is applied to Ω_j, the result is the sheaf of germs of open subsets of I, which is precisely the subobject classifier for the spatial topos **Top**(I) as described in Chapter 4.

In order now to describe the relationship between Ω_j and Ω in categorial terms we define a function $j_V : \Omega(V) \to \Omega(V)$, by putting for each V-crible $C \subseteq \Theta_V$,

$$j_V(C) = \{U \in \Theta : U \subseteq \bigcup C\} = \Theta_{\bigcup C}$$

EXERCISE 8. $j_V(\Theta_U) = \Theta_U$, for $U \in \Theta_V$ (cf. Exercise 3), and so $j_V(true_V(0)) = true_V(0)$.

EXERCISE 9. $C \subseteq j_V(C)$, i.e. $C \cap j_V(C) = C$.

EXERCISE 10. $j_V(j_V(C)) = j_V(C)$.

EXERCISE 11. $j_V(C \cap D) = j_V(C) \cap j_V(D)$, and hence

EXERCISE 12. if $C \subseteq D$ then $j_V(D) \subseteq j_V(D)$, for any $C, D \in \Omega(V)$.

EXERCISE 13. A V-crible of the form Θ_U, for $U \in \Theta_V$, is called a *principal* V-crible. Noting that if $C \subseteq \Theta_V$, then $\bigcup C \in \Theta_V$, show that

$$j_V(C) = C \quad \text{iff} \quad C \text{ is a principal } V\text{-crible}$$

(cf. Exercise 8).

EXERCISE 14. $j_V(C) = \Theta_V$ iff C covers V (i.e. iff $\bigcup C = V$).

EXERCISE 15. If C is any V-crible, and $U \subseteq V$, let

$$C_U = \{W \cap U : W \in C\}.$$

Show that $C_U \subseteq \Omega_U^V(C)$.

EXERCISE 16. Prove, in Exercise 15, that

$$\begin{aligned} U \subseteq \bigcup C \quad &\text{iff} \quad C_U \text{ is an open cover of } U \\ &\text{iff} \quad U = \bigcup C_U = \bigcup\{W \cap U : W \in C\}. \end{aligned}$$

EXERCISE 17. Show, using the last two exercises, that

$$\begin{aligned} U \in j_V(C) \quad &\text{iff} \quad U \text{ has an open cover } D \\ &\qquad \text{with } D \subseteq \Omega_U^V(C). \end{aligned}$$

EXERCISE 18. If $U \subseteq V$, show that

$$\begin{array}{ccc} \Omega(V) & \xrightarrow{j_V} & \Omega(V) \\ {\scriptstyle \Omega_U^V}\downarrow & & \downarrow{\scriptstyle \Omega_U^V} \\ \Omega(U) & \xrightarrow{j_U} & \Omega(U) \end{array}$$

commutes. □

Now if $E \hookrightarrow \Omega(V)$ is the equaliser of id and $j_V : \Omega(V) \to \Omega(V)$, then by what we know of equalisers in **Set**, and by Exercise 13, we have

$$E = \{C \in \Omega(V) : j_V(C) = C\} = \{\Theta_U : U \in \Theta_V\}$$

But the map $e_V : \Theta_V \to \Omega(V)$ having $e_V(U) = \Theta_U$ is monic, by Exercise 1, and so gives a bijection between $\Theta_V = \Omega_j(V)$ and E. Thus we find that

$$\Omega_j(V) \overset{e_V}{\rightarrowtail} \Omega(V) \underset{j_V}{\overset{id}{\rightrightarrows}} \Omega(V)$$

is an equaliser diagram in **Set**. But the import of Exercise 18 is that the j_V's form the components of an arrow $j_\Theta : \Omega \to \Omega$ in **St**(I). The e_V's are also components of a monic $e : \Omega_j \rightarrowtail \Omega$, and we find that

$$\Omega_j \overset{e}{\rightarrowtail} \Omega \underset{j_\Theta}{\overset{1_\Omega}{\rightrightarrows}} \Omega$$

is an equaliser diagram in **St**(I). Thus in **St**(I), Ω_j arises as that subobject of Ω obtained by equalising j_Θ and 1_Ω. Moreover since by Exercise 8 we have $j_\Theta \circ true = true$, there is a unique arrow τ making

$$\begin{array}{ccc} \Omega_j & \overset{e}{\rightarrowtail} & \Omega \\ & {\scriptstyle \tau}\nwarrow \quad \nearrow{\scriptstyle true} & \\ & 1 & \end{array}$$

commute. Clearly τ is in fact the arrow $true_j$.

Not only does the arrow j_Θ give a characterisation of Ω_j, it also characterises, by a property expressible in the first-order language of categories, those stacks over I that are sheaves, i.e. satisfy COM. To see how this works we first observe that j_Θ induces an operator $J : \text{Sub}(G) \to \text{Sub}(G)$ on the **HA** of subobjects of each **St**(I)-object G.

J assigns to the subobject $\tau: F \rightarrowtail G$ the subobject $J(\tau): J(F) \rightarrowtail G$ obtained by pulling *true* back

$$\begin{array}{ccc} J(F) & \xrightarrow{J(\tau)} & G \\ \downarrow & & \downarrow j \circ \chi_\tau \\ 1 & \xrightarrow{true} & \Omega \end{array}$$

along $j \circ \chi_\tau$, so that $\chi_{J(\tau)} = j \circ \chi_\tau$.

EXERCISE 19. In Sub(G) we have

(i) $\tau \subseteq J(\tau)$, i.e. $\tau \cap J(\tau) = \tau$

(ii) $J(J(\tau)) = J(\tau)$

(iii) $J(\tau \cap \sigma) = J(\tau) \cap J(\sigma)$, hence

(iv) if $\tau \subseteq \sigma$, then $J(\tau) \subseteq J(\sigma)$

(cf. Exercises 9–12). □

In general, an operator on a lattice that satisfies (i), (ii), and (iv) (corresponding to Exercises 9, 10, 12) is known as a *closure operator*. An example is the operator on the **BA** $\mathcal{P}(I)$, where I is a topological space, that assigns to each subset $X \subseteq I$ its topological closure (smallest closed superset) $\mathrm{cl}(X)$ in I. If $\mathrm{cl}(X) = I$, then X is said to be *dense* in the space I. By analogy then we say that a monic $\tau: F \rightarrowtail G$ in $\mathbf{St}(I)$ is *dense* iff $J(\tau) \simeq 1_G$ in Sub(G).

EXERCISE 20. Show that $J(F): \Theta \to Set$ assigns to $V \in \Theta$ the subset

$$\{x: (\chi_\tau)_V(x) \text{ covers } V\}$$

of $G(V)$, and that the components of $J(\tau)$ are the corresponding inclusions.

EXERCISE 21. Show that $\tau: F \rightarrowtail G$ is dense iff for all $V \in \Theta$, if $x \in G(V)$, then

$$(\chi_\tau)_V(x) = \{U: G_U^V(x) \in F(U)\} \text{ covers } V. \qquad \square$$

Now the statement "$G_U^V(x) \in F(U)$" can be construed as the localisation to $U \subseteq V$ of the statement "$x \in F(V)$". Thus if $(\chi_\tau)_V(x)$ covers V, the statement "$x \in F(V)$" is locally true of V, i.e. true at some neighbourhood of each point in V. Hence τ is dense when every element of $G(V)$ is *locally* an element of $F(V)$.

THEOREM (Lawvere). *A stack H is a sheaf (satisfies* COM) *iff for every* **St**(I)*-arrow $\sigma : F \dashrightarrow H$ with codomain H, and every dense arrow $\tau : F \rightarrowtail G$, there is exactly one $\sigma' : G \dashrightarrow H$ such that*

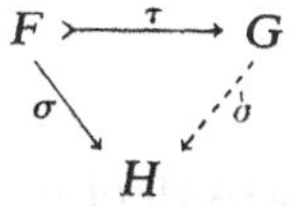

commutes. □

Thus H is a sheaf iff every arrow ending at H can be "lifted" in one and only one way from its domain to any object in which that domain is dense.

It can be shown (Tierney) that the proof of this characterisation can be derived entirely from the fact that the diagrams

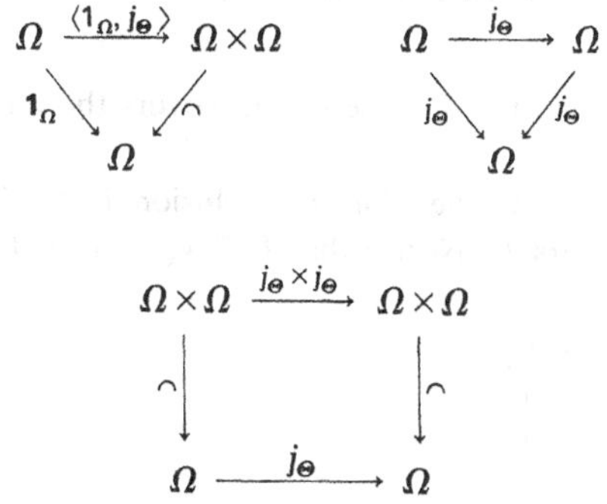

all commute in **St**(I). These diagrams correspond to Exercises 9, 10, 11, and hence to conditions (i)–(iii) of Exercise 19. We shall reserve the name *local operator* for any operator on a lattice that satisfies (i)–(iii) of Exercise 19.

EXERCISE 22. Let **St**(F, H) be the collection of all **St**(I)-arrows from F to H. Given $\tau : F \dashrightarrow G$, let $\tau_0 : \mathbf{St}(G, H) \to \mathbf{St}(F, H)$ be given by $\tau_0(\sigma) = \sigma \circ \tau$

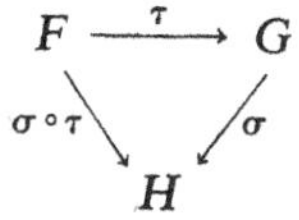

Show that H is a sheaf iff for every dense monic τ, τ_0 is a bijection.

EXERCISE 23. Let $\mathbf{H}=(H,\sqsubseteq)$ be any Heyting algebra. Show that the assignment to each $a\in H$ of its "double pseudo-complement" $\neg\neg a$ is a local operator on $\mathbf{H}$. □

14.3. Grothendieck topoi

Grothendieck's generalisation (cf. Artin et al. [SGA4]) of the functorial notion of a sheaf over a topological space is based on the observation that the axiom COM is expressible in terms of categorial properties of open covers $\{V_x : x\in X\}$, or $\{V_x \hookrightarrow V : x\in X\}$, of objects V in the category Θ. The essential properties of covers needed are

(1) The singleton set $\{V\}$ is a cover of V.

(2) If $\{V_x : x\in X\}$ covers V, and if, for each $x\in X$, $C_x=\{V^x_y : y\in Y_x\}$ is an open cover of V_x, then

$$\bigcup\{C_x : x\in X\}=\{V^x_y : x\in X \text{ and } y\in Y_x\}$$

is an open cover of V.

Thus the union of covers for open sets itself covers the union of those open sets.

(3) If $\{V_x : x\in X\}$ covers V, then for any inclusion $U\hookrightarrow V$, the collection $\{U\cap V_x : x\in X\}$ covers U. Notice that $U\cap V_x\hookrightarrow U$ is the pullback

$$\begin{array}{ccc} U\cap V_x & \hookrightarrow & V_x \\ \downarrow & & \downarrow \\ U & \hookrightarrow & V \end{array}$$

of $V_x\hookrightarrow V$ along $U\hookrightarrow V$.

A *pretopology* on a category $\mathscr{C}$ is an assignment to each $\mathscr{C}$-object a of a collection $Cov(a)$ of sets of $\mathscr{C}$-arrows with codomain a, called *covers* of a, such that

(1) The singleton $\{1_a : a\to a\}\in Cov(a)$.

(2) If $\{a_x \xrightarrow{f_x} a : x\in X\}\in Cov(a)$, and for each $x\in X$, we have an a_x-cover

$$\{a^x_y \xrightarrow{f^x_y} a_x : y\in Y_x\}\in Cov(a_x),$$

then

$$\{a^x_y \xrightarrow{f_x\circ f^x_y} a : x\in X \text{ and } y\in Y_x\}\in Cov(a).$$

(3) If $\{a_x \xrightarrow{f_x} a : x \in X\} \in Cov(a)$, and $g: b \to a$ is any $\mathscr{C}$-arrow, then for each $x \in X$ the pullback

$$\begin{array}{ccc} b \times_a a_x & \longrightarrow & a_x \\ \big\downarrow g_x & & \big\downarrow f_x \\ b & \xrightarrow{\ g\ } & a \end{array}$$

of f_x along g exists, and

$$\{b \times_a a_x \xrightarrow{g_x} b : x \in X\} \in Cov(b).$$

The pair $(\mathscr{C}, Cov)$ of the category $\mathscr{C}$ with the pretopology Cov is called a *site*.

Examples of sites

EXERCISE 1. (Θ, Cov_Θ), where, for open $V \in \Theta$, $Cov_\Theta(V) = \{C : C \subseteq \Theta$ and $\bigcup C = V\}$ is the collection of open covers of V.

EXERCISE 2. $(\mathscr{C}, {}^{\mathrm{i}}Cov)$, $\mathscr{C}$ any category, where ${}^{\mathrm{i}}Cov(a) = \{\{1_a : a \to a\}\}$, all $\mathscr{C}$-objects a.

EXERCISE 3. $(\mathscr{C}, {}^{\mathrm{d}}Cov)$, $\mathscr{C}$ any category, where ${}^{\mathrm{d}}Cov(a) = \mathscr{P}(\{f : \mathrm{cod}\, f = a\})$ is the collection of all sets of $\mathscr{C}$-arrows with codomain a.

EXERCISE 4. Let **2** be the poset category $(\{0, 1\}, \leqslant)$ with $! : 0 \to 1$ the only non-identity arrow. Can you find ten different pretopologies on **2**? □

A *stack*, or *presheaf*, of sets over a category $\mathscr{C}$ is by definition a contravariant functor $F : \mathscr{C} \to \mathbf{Set}$. The category $\mathbf{St}(\mathscr{C})$ of all stacks over $\mathscr{C}$ is thus equivalent to the topos $\mathbf{Set}^{\mathscr{C}^{op}}$.

If Cov is a pretopology on $\mathscr{C}$, and $\{a_x \xrightarrow{f_x} a : x \in X\} \in Cov(a)$, let

$$\begin{array}{ccc} a_x \times_a a_y & \longrightarrow & a_y \\ \big\downarrow & & \big\downarrow f_y \\ a_x & \xrightarrow{\ f_x\ } & a \end{array}$$

be the pullback of f_x and f_y, for each $x, y \in X$.

If F is a stack over $\mathscr{C}$ then F gives rise to the functions $F^x_y: F(a_x) \to F(a_x \times_a a_y)$ and $F^y_x: F(a_y) \to F(a_x \times_a a_y)$ as the F-images of the two new arrows obtained by forming this pullback. We denote also by F_x the arrow $F(f_x): F(a) \to F(a_x)$.

A stack F is a *sheaf* over the site $(\mathscr{C}, Cov)$ iff it satisfies

COM: *Given any cover $\{a_x \xrightarrow{f_x} a : x \in X\} \in Cov(a)$ of a $\mathscr{C}$-object a, and any selection of elements $s_x \in F(a_x)$, for all $x \in X$, that are pairwise compatible, i.e. $F^x_y(s_x) = F^y_x(s_y)$ all $x, y \in X$, then there is exactly one $s \in F(a)$ such that $F_x(s) = s_x$ all $x \in X$.*

The full subcategory of **St**($\mathscr{C}$) generated by those objects that are sheaves over the site $(\mathscr{C}, Cov)$ will be denoted **Sh**(Cov). A *Grothendieck topos* is, by definition, any category that is equivalent to one of the form **Sh**(Cov).

EXERCISE 5. If ${}^iCov$ is the "indiscrete" pretopology on $\mathscr{C}$ of Exercise 2, then **Sh**(${}^iCov$) = **St**($\mathscr{C}$).

EXERCISE 6. Let $F: \mathbf{2} \to \mathbf{Set}$ be a stack over **2**, and choose $s_0 \in F(0)$, $s_1 \in F(1)$. Assuming that $\{1_1, !: 0 \to 1\} \in Cov(1)$, show that s_0 and s_1 are compatible iff $F^1_0(s_1) = s_0$, where F^1_0 is the F-image of $!: 0 \to 1$.

EXERCISE 7. Use the last exercise to show that **St**(**2**) = **Sh**(Cov) if $Cov(0) = \{\{1_0\}\}$ and $Cov(1) = \{\{1_1\}, \{1_1, !\}\}$. □

An *a-crible* (dual to a-sieve) is a collection C of arrows with codomain a that is closed under right composition, i.e. if $f: b \to a \in C$ then $f \circ g: c \to a \in C$ for any $\mathscr{C}$-arrow $g: c \to b$. The stack $\Omega: \mathscr{C} \to \mathbf{Set}$ has

$$\Omega(a) = \{C: C \text{ is an } a\text{-crible}\}$$

while for each $\mathscr{C}$-arrow $f: b \to a$, $\Omega_f: \Omega(a) \to \Omega(b)$ has

$$\Omega_f(C) = \{c \xrightarrow{g} b: f \circ g \in C\}.$$

The **St**($\mathscr{C}$)-arrow *true*: $1 \to \Omega$ has component $\top_a: \{0\} \to \Omega(a)$ given by

$$\top_a(0) = C_a = \{f: \operatorname{cod} f = a\}, \text{ the largest } a\text{-crible.}$$

EXERCISE 8. Show that if $C \in \Omega(a)$, then $\Omega_f(C) \in \Omega(b)$, and that if $f: b \to a \in C$ then $\Omega_f(C) = C_b$. □

The sheaves over a site $(\mathscr{C}, Cov)$ can be described by an arrow $j_{Cov}: \Omega \rightarrow \Omega$ exactly as in the classical case $Cov = Cov_{\Theta}$. The a-th component $j_{Cov\,a}: \Omega(a) \rightarrow \Omega(a)$ is defined, for each a-crible $C \in \Omega(a)$, by

$$j_{Cov\,a}(C) = \{ b \xrightarrow{f} a : \text{ there exists a cover } C_f \in Cov(b) \text{ with } C_f \subseteq \Omega_f(C)\}.$$

This is a direct generalisation of the description of $j_V(C)$ for $V \in \Theta$ given in Exercise 17 of the last section.

EXERCISE 9. Verify that j_{Cov} as defined is a $\mathbf{St}(\mathscr{C})$-arrow.

EXERCISE 10. Show that

(1) $\cap \circ \langle 1_\Omega, j_{Cov}\rangle = 1_\Omega$

(2) $j_{Cov} \circ j_{Cov} = j_{Cov}$

(3) $\cap \circ (j_{Cov} \times j_{Cov}) = j_{Cov} \circ \cap$ □

The characterisation theorem of the last section for sheaves in $\mathbf{St}(I)$ holds for Cov-sheaves in $\mathbf{St}(\mathscr{C})$ when j_Θ is replaced by j_{Cov}. The properties of j_{Cov} needed to prove this are precisely (1)–(3) of Exercise 10, as in the classical topological case.

Notice that, by Exercises 5 and 7, it is possible to have different pretopologies Cov_1 and Cov_2 on the same category that lead to the one category of sheaves, i.e.

$$\mathbf{Sh}(Cov_1) = \mathbf{Sh}(Cov_2).$$

However, it can be shown that this last equation holds iff $j_{Cov_1} = j_{Cov_2}$. Thus the arrow j_{Cov} corresponds to a unique Grothendieck topos.

The notion of pretopology has been further refined by Verdier [SGA4] to yield a notion of "topology" on a category such that distinct topologies yield distinct categories of sheaves. The Verdier topologies are subfunctors of Ω–precisely those whose characters satisfy (1)–(3) of Exercise 10. A detailed introductory account of this theory is given by Shlomiuk [74].

An extensive discussion of sites and logical operations on related categories can be found in the article [74] by Gonzalo Reyes.

14.4. Elementary sites

A *topology* on an elementary topos $\mathscr{E}$ is by definition any arrow $j:\Omega\to\Omega$ that satisfies

(1) $j\circ true = true$

(2) $j\circ j = j$

(3) $\cap\circ(j\times j) = j\circ\cap$.

The pair $\mathscr{E}_j = (\mathscr{E}, j)$ is called an *elementary site*. Notice that condition (1) of Exercise 14.3.10 has been replaced by the simpler $j\circ\top = \top$. This is justified by the following result, for which we need

EXERCISE 1. In any category with 1, a square of the form

$$\begin{array}{ccc} 1 & \xrightarrow{f} & b \\ \downarrow & & \downarrow{\scriptstyle\langle 1_b,\, g\rangle} \\ 1 & \xrightarrow{\langle f,\, h\rangle} & b\times b \end{array}$$

commutes, i.e. $g\circ f = h$, only if it is a pullback. □

THEOREM 1. *For any arrow* $j:\Omega\to\Omega$ *in a topos,*

$$\cap\circ\langle 1_\Omega, j\rangle = 1_\Omega \quad \textit{iff} \quad j\circ true = true.$$

PROOF. Consider the diagram

$$\begin{array}{ccc} 1 & \xrightarrow{\top} & \Omega \\ \downarrow & & \downarrow{\scriptstyle\langle 1_\Omega,\, j\rangle} \\ 1 & \xrightarrow{\langle \top,\, \top\rangle} & \Omega\times\Omega \\ \downarrow & & \downarrow{\scriptstyle\cap} \\ 1 & \xrightarrow{\top} & \Omega \end{array}$$

If $\cap\circ\langle 1_\Omega, j\rangle = 1_\Omega$ then the boundary commutes. But the bottom square is the pullback defining $\cap$, so its universal property implies that $!:1\to 1$ is the unique arrow making the top square commute. But then $\langle\top,\top\rangle = \langle 1_\Omega, j\rangle\circ\top = \langle\top, j\circ\top\rangle$, and so $\top = j\circ\top$.

Conversely if $j\circ\top = \top$, then the top square commutes and is (Exercise 1) a pullback. The PBL then gives the boundary as a pullback and so by the Ω-axiom, $\cap\circ\langle 1_\Omega, j\rangle = \chi_\top = 1_\Omega$. □

Examples of elementary sites

EXERCISE 2. For any site $(\mathscr{C}, Cov)$, $(\mathbf{St}(\mathscr{C}), j_{Cov})$ is an elementary site.

EXERCISE 3. $\mathbf{1}_\Omega : \Omega \to \Omega$ is a topology for any $\mathscr{E}$.

EXERCISE 4. $true_\Omega : \Omega \to \Omega$ is a topology.

EXERCISE 5. $\neg\circ\neg : \Omega \to \Omega$ is a topology, the *double negation* topology, on any topos $\mathscr{E}$ (cf. Exercise 14.2.23).

A topology $j:\Omega \to \Omega$ induces a local operator J on the **HA** Sub(d) for each $\mathscr{E}$-object d, exactly as in the case $j = j_\Theta$. J assigns to $f: a \rightarrowtail d$ the subobject $J(f): J(a) \rightarrowtail d$ having

$$\chi_{J(f)} = j \circ \chi_f.$$

An $\mathscr{E}$-monic $f: a \rightarrowtail d$ is *j-dense* iff $J(f) \simeq \mathbf{1}_d$ in Sub(d).

EXERCISE 6. f is j-dense iff $j \circ \chi_f = true_d$.

EXERCISE 7. In any $\mathscr{E}$, $[\top, \bot]: 1+1 \rightarrowtail \Omega$ is $\neg\circ\neg$-dense. (Hint: show that $\chi_{[\top,\bot]} = \mathbf{1}_\Omega \cup \neg$ and use Exercise 8.3.27).

EXERCISE 8. For $j = \mathbf{1}_\Omega$, $J(f) \simeq f$ and f is j-dense iff $f \simeq \mathbf{1}_d$.

EXERCISE 9. In the site $(\mathscr{E}, true_\Omega)$, $\chi_{J(f)} = true_d$, and every monic is dense.

EXERCISE 10. In the elementary site $\mathscr{E}_{\neg\neg} = (\mathscr{E}, \neg\circ\neg)$, f is dense iff $-(-f) \simeq \mathbf{1}_d$ in Sub(d). Use this to give a different proof of Exercise 7.

EXERCISE 11. Show that for any monic $f: a \rightarrowtail d$, $f \cup -f$ is $\neg\circ\neg$-dense.

□

A *j-sheaf* is, by definition, an $\mathscr{E}$-object b with the property that for any $\mathscr{E}$-arrow $g: a \to b$ and any j-dense $f: a \rightarrowtail d$ there is exactly one $\mathscr{E}$-arrow $g': d \to b$ such that

$$\begin{array}{ccc} a & \stackrel{f}{\rightarrowtail} & d \\ & {\scriptstyle g}\searrow \quad \swarrow{\scriptstyle g'} & \\ & b & \end{array}$$

$g' \circ f = g$.

EXERCISE 12. 1 is a j-sheaf.

EXERCISE 13. If

$$a \rightarrowtail b \rightrightarrows c$$

is an equaliser diagram in $\mathscr{E}$ and b and c are j-sheaves, then so is a.

EXERCISE 14. If a and b are j-sheaves, so is $a \times b$.

EXERCISE 15. If

$$\begin{array}{ccc} a & \longrightarrow & b \\ \downarrow & & \downarrow \\ c & \longrightarrow & d \end{array}$$

is a pullback, and c, d, b are j-sheaves, then so is a.

EXERCISE 16. If $j = 1_\Omega$, *every* $\mathscr{E}$-object is a j-sheaf.

EXERCISE 17. In the site $(\mathscr{E}, true_\Omega)$ the only sheaves are the terminal objects of $\mathscr{E}$. Hint: Consider the diagram

$$\begin{array}{ccc} 0 & \rightarrowtail & a \\ & \searrow & \\ & & b \end{array}$$

EXERCISE 18. b is a j-sheaf iff each j-dense $f: a \rightarrowtail d$ induces a bijection $- \circ f : \mathscr{E}(d, b) \cong \mathscr{E}(a, b)$. □

THEOREM (Lawvere–Tierney). *The full subcategory $sh_j(\mathscr{E})$ of an elementary site $\mathscr{E}_j$ generated by the j-sheaves is an elementary topos. Moreover there is a "sheafication" functor $\mathcal{S}\!\mathscr{h}_j : \mathscr{E} \to sh_j(\mathscr{E})$ that has $\mathcal{S}\!\mathscr{h}_j(b) \cong b$ for each j-sheaf b, and that preserves all finite limits.*

From this result it follows that any Grothendieck topos is an elementary topos. In the case of the elementary site $(\mathbf{St}(I), j_\Theta)$, $\mathcal{S}\!\mathscr{h} : \mathbf{St}(I) \to \mathbf{Sh}(I)$ is the composite $\mathcal{S} \circ \mathcal{G}$, taking stack F to sheaf F_{p_F}.

A proof of this theorem may be found in Freyd [72] or Kock and Wraith [72]. That $sh_j(\mathscr{E})$ has all finite limits is indicated by Exercises 12–15. That it has exponentials is proven by showing that b^a is a j-sheaf whenever b is. Its subobject classifier $true_j : 1 \to \Omega_j$ is formed as the equaliser

$$\begin{array}{ccccc} \Omega_j & \rightarrowtail & \Omega & \overset{1_\Omega}{\underset{j}{\rightrightarrows}} & \Omega \\ & {\scriptstyle true_j}\nwarrow & \nearrow{\scriptstyle true} & & \\ & & 1 & & \end{array}$$

of j and 1_Ω.

An important application of the sheaf construction occurs in the case $j = \neg \circ \neg$. *The topos $sh_{\neg\neg}(\mathscr{E})$ of "double-negation sheaves" in $\mathscr{E}$ is always a Boolean topos*! This is established by showing that in $sh_j(\mathscr{E})$, $[\top_j, \bot_j]$ is $\mathscr{S}\mathscr{h}_j([\top, \bot])$ and that $\mathscr{S}\mathscr{h}_j$ maps a j-dense monic to an iso in $sh_j(\mathscr{E})$. The result then follows by Exercise 7, and can be seen as an analogue of the fact that the *regular* ($\neg\neg a = a$) elements of a Heyting algebra **H** form a *Boolean* subalgebra of **H**.

Thus from any topos $\mathscr{E}$ we can pass via the functor $\mathscr{S}\mathscr{h}_{\neg\neg}$ to a classical subtopos $sh_{\neg\neg}(\mathscr{E})$. This process is used by Tierney [72] to develop a categorial proof of the independence of the Continuum Hypothesis that parallels Cohen's proof for classical set theory. This work reveals that Cohen's "weak-forcing" technique is a version of the technique of passing from a pre-sheaf to its associated sheaf. More recently the method has been used by Marta Bunge [74] to give a topos-theoretic proof of the independence of Souslin's hypothesis.

EXERCISE 19. Let $\mathscr{E} = \mathbf{Set}^{\mathbf{P}}$ and $j = \neg \circ \neg$. Show that $\Omega_{\neg\neg}: \mathbf{P} \to \mathbf{Set}$, the classifier for $\neg\circ\neg$-sheaves, has

$$\Omega_{\neg\neg}(p) = \text{the set of regular members of } [p)^+,$$

where $S \in [p)^+$ is regular iff $\neg_p(\neg_p S) = S$. Show that $\Omega_{\neg\neg}(p)$ is a Boolean subalgebra of the **HA** of hereditary subsets of $[p)$.

EXERCISE 20. Show that in $\mathbf{Top}(I)$, the stalk of $\Omega_{\neg\neg}$ over i is a *Boolean* subalgebra of the **HA** of germs of open sets at i.

EXERCISE 21. Show that in $\mathbf{M_2}$-$\mathbf{Set}$, $\Omega_{\neg\neg} = \{M, \emptyset\}$. □

14.5. Geometric modality

Modal logic is concerned with the study of a one-place connective on sentences that has a variety of meanings, including "it is necessarily true that" (alethic modality), "it is known that" (epistemic modality), "it is believed that" (doxastic), and "it ought to be the case that" (deontic). The quotation that heads this chapter invites us to consider what we might call *geometric* modality. Semantically the modal connective corresponds to an arrow of the form $\Omega \to \Omega$, just as the one-place negation connective corresponds to the arrow $\neg$ of this form. Lawvere suggests that when the arrow is a topology $j : \Omega \to \Omega$ on a topos then the modal connective has the "natural" reading "it is locally the case that."

Let us now extend the sentential language PL of Chapter 6 by the inclusion of a new connective ∇ and the formation rule

if α is a sentence, then so is $\nabla\alpha$

($\nabla\alpha$ is to be read "it is locally the case that α").

Let Ψ be the class of all sentences generated from propositional letters π_i by the connectives $\wedge$, $\vee$, $\sim$, $\supset$, ∇. If $\mathcal{E}_j = (\mathcal{E}, j)$ is any elementary site, then an $\mathcal{E}_j$-valuation $V : \Phi_0 \to \mathcal{E}(1, \Omega)$ extends uniquely to the whole of Ψ, using the semantic rules of §6.7, together with

$$V(\nabla\alpha) = j \circ V(\alpha)$$

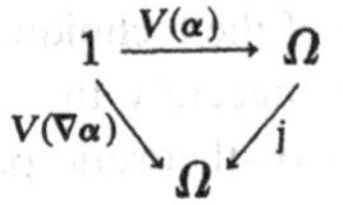

We may then define the validity of any $\alpha \in \Psi$ on the site $\mathcal{E}_j$, denoted $\mathcal{E}_j \vDash \alpha$, to mean that $V(\alpha) = true$ for all $\mathcal{E}_j$-valuations.

Let $\mathcal{J}$ be the axiom system that has Detachment as its sole inference rule, and as axioms the forms I–XI of IL together with the schemata

$$\nabla(\alpha \supset \beta) \supset (\nabla\alpha \supset \nabla\beta)$$

$$\alpha \supset \nabla\alpha$$

$$\nabla\nabla\alpha \supset \nabla\alpha.$$

(Alternatively $\mathcal{J}$ can be defined by replacing the first two of these schemata by

$$(\alpha \supset \beta) \supset (\nabla\alpha \supset \nabla\beta)$$

and

$$\nabla(\alpha \supset \alpha).)$$

Then we have the following characterisation of validity on elementary sites: for any $\alpha \in \Psi$

$$\vdash_{\mathcal{J}} \alpha \quad \text{iff} \quad \text{for all sites } \mathcal{E}_j, \ \mathcal{E}_j \vDash \alpha.$$

The proof of this (described in Goldblatt [77]) uses a Kripke-style model theory for the language Ψ, developed from an analysis of the notion of "local truth". There are in fact two senses in which we have used this idea, one relating to sheaves of germs, the other to sheaves of sections.

(I) Recall the definition of the equivalence relation $\sim_i$ that defines the germ $[U]_i$ of an open set $U \in \Theta$ at i in the sheaf Ω for **Top**(I) (Chapter 4). We have $U \sim_i V$ iff U and V have the same intersection with some i-neighbourhood. We interpret this to mean that the statement "$U = V$" or "$x \in U$ iff $x \in V$" is locally true at i, i.e. true throughout some neighbourhood of i. This in turn represents the intuitive notion that the statement holds for all points "close" to i. The same interpretation was given to the description of germs of sections $s \in F(V)$ in the stalk space A_F of §14.1.

Thus the statement "α is locally true at p" may be rendered as

(i) "α is true at all points close to p",

(ii) "α is true through some neighbourhood of p".

Intuitively (i) and (ii) are equivalent. A p-neighbourhood is any set containing all points that are close to p, while a point is close to p when it belongs to all p-neighbourhoods. Of course in most significant classical topological spaces (any that is at least T_1) there are no points close to p in this sense – other than p itself. The notion can however be given substance by Abraham Robinson's theory of non-standard topology, wherein a space is enlarged to include points "infinitely close" to the original ones. Indeed in his article [69], the germ of U at p is literally a subset of U, namely the intersection of U with the *monad* of p (the set of points infinitely close to p).

Given now a poset **P** we introduce a binary relation $p \prec q$ on P, with the reading "q is close to p". Then given a model $\mathcal{M} = (\mathbf{P}, V)$ based on **P** the connective ∇ can be semantically interpreted as

$$\mathcal{M} \models_p \nabla\alpha \quad \textit{iff} \quad p \prec q \quad \textit{implies} \quad \mathcal{M} \models_q \alpha$$

thereby formalising condition (i).

Writing $\mu(p) = \{q : p \prec q\}$ for the "monad" of p, this clause becomes

$$\mathcal{M} \models_p \nabla\alpha \quad \textit{iff} \quad \mu(p) \subseteq \mathcal{M}(\alpha),$$

where, as in §8.4,

$$\mathcal{M}(\alpha) = \{q : \mathcal{M} \models_q \alpha\}$$

In order for the structure $(P, \sqsubseteq, <)$ to validate the logic $\mathcal{J}$ it suffices that it satisfy

(a) $p<q$ implies $p \sqsubseteq q$, i.e. $\mu(p) \subseteq [p)$ all p,

(b) $<$ is dense, i.e. if $p<q$, then $p<r<q$ for some r, and

(c) $p \sqsubseteq q$ implies $\mu(q) \subseteq \mu(p)$ (this is needed to ensure that $\mathcal{M}(\alpha)$ is $\sqsubseteq$-hereditary).

Notice that we do not require that $p<p$, i.e. $p \in \mu(p)$. Indeed were this to hold for all p, we would have $p<q$ iff $p \sqsubseteq q$. Thus "q is close to p" really means "q is close to but not the same as p", which is akin to the topological notion of "p approximates to q" as formalised by "p is a limit point of $\{q\}$".

To formalise the condition (ii) we could introduce a collection N_p of subsets of P (the p-neighbourhoods) and put

$$\mathcal{M} \models_p \nabla\alpha \quad \textit{iff} \quad \textit{for some} \quad C \in \mathrm{N}_p, \quad C \subseteq \mathcal{M}(\alpha).$$

One possible construction of an N_p, would be to take a relation $<$ and put

$$\mathrm{N}_p^{<} = \{C : \mu(p) \subseteq C\}, \quad \text{for each } p$$

Exercise 1. Show that the structures $(\mathbf{P}, <)$ and $(\mathbf{P}, \mathrm{N}^{<})$ validate the same sentences.

Exercise 2. Given any poset $(P, \sqsubseteq)$ define

$p<q$ iff p is a limit point of $\{q\}$ in the topology $\mathbf{P}^+$ (in which "open" = "hereditary").

Show that

$$p<q \quad \text{iff} \quad p \sqsubset q \quad (\text{i.e. } p \sqsubseteq q \text{ and } p \neq q). \qquad \square$$

(II) The sense of "local truth" that applies to stacks of sections refers to a property holding locally of an open set, or an object of a site, rather than at a point. Thus for example a classical topological space is said to be locally connected if each open set is covered by connected open sets.

In this Chapter a function has been described as "locally constant" on its open domain when that domain is covered by open sets, on each of which the function is constant (Exercise 14.1.17).

In the context of a stack F, if s, $t \in F(V)$ have

$$F_x(s) = F_x(t), \quad \text{all} \quad x \in X,$$

given some cover $\{V_x : x \in X\}$ of V, we can take this to mean that "$s = t$" is locally true of V, i.e. true at all members of some cover of V. It follows from COM then that if F is a sheaf, locally equal sections of F are actually equal.

This same sense of a statement being locally true of an open set V when true of all members of a cover of V appears in the interpretation of j_Θ-density of monics given in §14.2.

If we think now of a poset $\mathbf{P} = (P, \sqsubseteq)$ as being the category of open sets of a topology, with $\sqsubseteq$ the opposite to the inclusion ordering, then we may formalise the foregoing discussion by contemplating structures $(\mathbf{P}, Cov)$, where Cov assigns to each $p \in P$ a collection $Cov(p) \subseteq \mathcal{P}(P)$, the "covers" of p. We define, for $\mathcal{M} = (\mathbf{P}, V)$

$$(\ast)\qquad \mathcal{M} \models_p \nabla\alpha \quad \textit{iff} \quad \textit{for some} \quad C \in Cov(p), \qquad C \subseteq \mathcal{M}(\alpha)$$

(Note that, formally, this is the same as the "neighbourhood system" approach described above.)

In order to guarantee that $\mathcal{M}(\alpha)$ be hereditary the operator Cov must satisfy

$$p \sqsubseteq q \quad \textit{only if} \quad Cov(p) \subseteq Cov(q),$$

i.e., every p-cover is a q-cover.

Example 1. *Grothendieck topology*: Let $(\mathbf{P}, Cov)$ be the site (Θ, Cov_Θ) as defined in Exercise 14.3.1.

If j_V is the V-th component of j_Θ, then $j_V(C) = \Theta_V = true_V(0)$ iff C covers V, for C a V-crible (Exercise 14.2.14). If $\mathcal{M}$ is any model based on Θ, then $\mathcal{M}(\alpha)_V = \mathcal{M}(\alpha) \cap \Theta_V$ is always a V-crible, and we find that, using the above definition $(\ast)$,

$$\mathcal{M} \models_V \nabla\alpha \quad \textit{iff} \quad j_V(\mathcal{M}(\alpha)_V) = true_V(0).$$

Identifying the element $\mathcal{M}(\alpha)_V$ of $\Omega(V)$ with an arrow $\{0\} \to \Omega(V)$ we obtain a $\mathbf{St}(Cov_\Theta)$-arrow $\mathcal{M}(\alpha)\colon 1 \to \Omega$. The role of j_Θ as a modal operator is then given explicitly, as

$$\mathcal{M} \models \nabla\alpha$$

iff

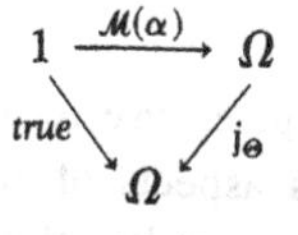

commutes.

EXAMPLE 2. *Cofinality:* Lawvere suggests [70] that the double negation topology $\neg\circ\neg$ is "more appropriately put into words as 'it is cofinally the case that' ".

In general, if S and T are subsets of a poset $\mathbf{P}$, then S is said to be *cofinal* in T if

$$\textit{for all } p \in T \textit{ there is some } q \in S \textit{ such that } p \sqsubseteq q,$$

i.e. every member of T has a member of S "coming after" it.

If we define

$$Cov(p) = \{S \subseteq P: S \text{ is cofinal in } [p)\}$$

then for any model $\mathcal{M}$ on $\mathbf{P}$ we find that

$$\mathcal{M} \models_p \nabla\alpha \quad \text{iff} \quad \mathcal{M} \models_p {\sim}{\sim}\alpha.$$

This is based on the fact that

$$\mathcal{M} \models_p {\sim}{\sim}\alpha \quad \text{iff} \quad \mathcal{M}(\alpha) \text{ is cofinal in } [p).$$

□

By adapting the techniques described in §8.4, a canonical structure $\mathbf{P}_{\mathcal{J}} = (P, \sqsubseteq, \prec)$ is definable for which

$$\mathbf{P}_{\mathcal{J}} \models \alpha \quad \textit{iff} \quad \vdash_{\mathcal{J}} \alpha.$$

On the topos $\mathbf{Set}^{\mathbf{P}_{\mathcal{J}}}$ a topology $j: \Omega \to \Omega$ is then obtained by defining the component $j_p: \Omega_p \to \Omega_p$ to satisfy

$$j_p(S) = \{q: p \sqsubseteq q \text{ and } \mu(q) \subseteq S\},$$

for each $S \in [p)^+$.

We thus obtain the *canonical site* $\mathscr{C}_{\mathcal{J}}$ for $\mathcal{J}$, for which it may be shown that

$$\mathscr{C}_{\mathcal{J}} \models \alpha \quad \textit{iff} \quad \mathbf{P}_{\mathcal{J}} \models \alpha$$

and from this follows the completeness theorem for $\mathcal{J}$ mentioned earlier.

14.6. Kripke–Joyal semantics

The "local character" of properties of sheaves gives rise to a semantical theory, due to André Joyal, that incorporates aspects of Kripke's IL-semantics, together with the principle that the truth-value of a sentence is determined by its local truth-values.

We have already noted that an equality "$s = t$" of sections of a sheaf is true on some open set V iff it is locally true of V, i.e. true throughout some cover of V. Indeed the very essence of the sheaf concept is that an arrow $s : V \to A$ is a section of $f : A \to I$ iff it is locally a section over V. In other words, $s \in F_f(V)$ iff there is a cover $\{V_x : x \in X\}$ with $F_x(s) \in F_f(V_x)$ – "$s \in F_f(V)$" is true when localised to V_x – for all $x \in X$.

To take an example from Lawvere [76] involving existential quantification, suppose that

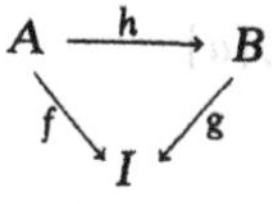

is a map of sheaves of germs and $t \in F_g(V)$ is a section of g over V. We ask – when does there exist a section $s \in F_f(V)$ of f over V with $h \circ s = t$? Answer – precisely when there is a cover $\{V_x : x \in X\}$ of V with for each x a section $s_x \in F_f(V_x)$ such that $h \circ s_x = t \restriction V_x$. Thus the statement $\exists s(h \circ s = t)$ is true of V precisely when it is locally true of V.

Briefly, the basis of Joyal's semantics is this. We consider interpretations of formulae $\varphi(v_1, v_2)$ in a site $(\mathscr{C}, Cov)$. Given arrows $f : a \to b$, $g : a \to c$, suppose we know what it means for $\langle f, g\rangle$ to satisfy φ at a, denoted $a \models \varphi[f, g]$. Then for a particular $f : a \to b$ we put

$a \models \exists v_2 \varphi[f]$ iff there is an a-cover $\{a_x \xrightarrow{f_x} a : x \in X\}$ and arrows $\{a_x \xrightarrow{g_x} c : x \in X\}$ such that $a_x \models \varphi[f \circ f_x, g_x]$, all $x \in X$.

The disjunction connective gets a similar interpretation:

$a \models \varphi \vee \psi[f]$ iff for some $\{a_x \xrightarrow{f_x} a : x \in X\} \in Cov(a)$, we have for each $x \in X$ that $a_x \models \varphi[f \circ f_x]$ or $a_x \models \psi[f \circ f_x]$.

The other connectives, and the universal quantifier, are interpreted by analogues of Kripke's rules, e.g.

$a \models \varphi \supset \psi[f]$ iff for any $a_x \xrightarrow{f_x} a$, if $a_x \models \varphi[f \circ f_x]$ then $a_x \models \psi[f \circ f_x]$

$a \models \forall v_2 \varphi[f]$ iff for all $a_x \xrightarrow{f_x} a$ and all $a_x \xrightarrow{g} c$, $a_x \models \varphi[f \circ f_x, g]$.

The "local character of truth" is then embodied in the consequence that for any formula $\varphi(v)$,

$a \models \varphi[f]$ iff for some $\{a_x \xrightarrow{f_x} a : x \in X\} \in Cov(a)$, $a_x \models \varphi[f \circ f_x]$, all $x \in X$.

The details of the Kripke–Joyal semantics are given by Reyes [76] for sites, Boileau [75] for general topoi and Osius [75(i)] for categorial set theory. For applications of it cf. Kock [76]. In as much as it gives a non-classical interpretation to $\vee$ and $\exists$ it is more analogous to Beth models, and "Beth–Joyal semantics" would perhaps be a more appropriate name. A Beth model for first-order logic has a single set A of individuals, rather than one for each state $p \in P$ as in the structures of §11.6. The universal quantifier has the standard interpretation

$$\mathcal{M} \models_p \forall v \varphi \quad \text{iff} \quad \text{for all} \quad a \in A, \qquad \mathcal{M} \models_p \varphi[a]$$

while the clause for $\exists$ reads

$$\mathcal{M} \models_p \exists v \varphi \quad \text{iff} \quad \text{there is a bar } B \text{ for } p \text{ such that for each } q \in B$$
$$\text{there is some } a \in A \text{ with } \mathcal{M} \models_q \varphi[a].$$

An application of this modelling to intuitionistic metamathematics and an indication of its relation to topological interpretations may be found in van Dalen [78].

14.7. Sheaves as complete Ω-sets

Let $\mathbf{A}$ be an Ω-set, where Ω is a **CHA**. Then, as defined in §11.9, a singleton for $\mathbf{A}$ is a function $s: A \to \Omega$ that satisfies

(i) $s(x) \sqcap [\![x \approx y]\!] \sqsubseteq s(y)$

(ii) $s(x) \sqsubseteq [\![Ex]\!]$

(iii) $s(x) \sqcap s(y) \sqsubseteq [\![x \approx y]\!]$

for all $x, y \in A$. (These are conditions (viii)–(x) of §11.9. Note that (ii) is a consequence of (iii) by putting $x = y$.) Each element $a \in A$ yields the singleton $\{\mathbf{a}\}$ that assigns $[\![x \approx a]\!]$ to each $x \in A$. $\mathbf{A}$ is called a *complete* Ω-set if each of its singletons is of the form $\{\mathbf{a}\}$ for a *unique* $a \in A$.

EXAMPLE 1. Let $\Omega = \Theta$, the **CHA** of open subsets of a space I. Then for any topological space X we have a corresponding Θ-set $\mathbf{C}_X$, which is the set of continuous X-valued partial functions on I. C_X is the set of all continuous functions of the form $f: V \to X$, for all $V \in \Theta$, with degrees of equality measured as

$$[\![f \approx g]\!] = \{i: f(i) = g(i)\}^0.$$

Now suppose that $s: C_X \to \Theta$ is a singleton. For each $f \in C_X$, let $f_s = f \restriction s(f)$ be the restriction of f to the open set $s(f)$. By condition (ii) above, $s(f)$ is a subset of $[\![f \approx f]\!]$, i.e. of the domain of f, and so dom f_s is just $s(f)$ itself. But then by (iii) we see that the f_s's form a compatible family of functions, since if i belongs to dom f_s and dom g_s it must belong to $[\![f \approx g]\!]$ and so $f_s(i) = f(i) = g(i) = g_s(i)$.

Thus we may "patch" together the f_s's to obtain a single element a_s of C_X whose restriction to each $s(f)$ is just f_s. In other words, a_s agrees with f on the set $s(f)$, giving

$$s(f) \subseteq [\![f \approx a_s]\!].$$

For the converse inclusion, if $f(i) = a_s(i)$, then $f(i) = g_s(i) = g(i)$ for some g with $i \in \text{dom}\, g_s = s(g)$. But then the extensionality condition (i) implies that $i \in s(f)$.

Thus we see that our original singleton s is the function $\{\mathbf{a}_s\}$. To see that a_s is unique with this property, observe that whenever $\{\mathbf{f}\} = \{\mathbf{g}\}$, i.e. f and g agree with all members of $\mathbf{C}_X$ to the same extent, then f and g agree with each other to the same extent that they agree with themselves, and so

$$[\![f \approx g]\!] = [\![f \approx f]\!] = [\![g \approx g]\!]$$

(cf. Ex. 16 of §11.9). Thus f and g have the same domain, and they agree on that domain, which means that $f = g$.

This establishes the completeness of $\mathbf{C}_X$.

EXAMPLE 2. Analogously, given a continuous function $k: A \to I$, we obtain the Θ-set $\mathbf{C}_k$ whose elements are the local (partial) continuous sections $I \rightsquigarrow A$ of k. The completeness of $\mathbf{C}_k$ is established exactly as above. In particular, this assigns a complete Θ-set to each object of $\mathbf{Top}(I)$. $\mathbf{C}_X$ itself can be identified with the set of local sections of the projection function $X \times I \to X$, by identifying $f: V \to X$ with $\langle f, V \hookrightarrow I\rangle : V \to X \times I$. (Note that the projection need not be a local homeomorphism, and hence not a $\mathbf{Top}(I)$-object.) □

The completeness property for an Ω-set allows a very elegant abstract treatment of the idea of the restriction of a function to an open set. The development of this theory is due to Dana Scott and Michael Fourman. Given $a \in A$ and $p \in \Omega$, the function $\{\mathbf{a}\} \restriction p$ that assigns $[\![x \approx a]\!] \sqcap p$ to x is a singleton (§11.9, Exercise 17). If $\mathbf{A}$ is complete, then there is exactly one $b \in A$ with $\{\mathbf{b}\} = \{\mathbf{a}\} \restriction p$. We call b the *restriction of a to p*, and denote it $a \restriction p$. (From now on we will often abbreviate the extent $[\![Ea]\!] = [\![a \approx a]\!]$ of a to Ea).

EXERCISE 1. $(a \restriction p) \restriction q = a \restriction (p \sqcap q)$

EXERCISE 2. $a \restriction Ea = a$

EXERCISE 3. $E(a \restriction p) = Ea \sqcap p$

EXERCISE 4. $a \restriction [\![a \approx b]\!] = b \restriction [\![a \approx b]\!]$

EXERCISE 5. $a \restriction (Ea \sqcap Eb) = a \restriction Eb$

EXERCISE 6. Write $a \between b$ to mean that

$$a \restriction Eb = b \restriction Ea$$

i.e. that a and b are *compatible*. Prove that

$$a \between b \quad \text{iff} \quad Ea \sqcap Eb \sqsubseteq [\![a \approx b]\!].$$

EXERCISE 7. Show that $a \between b$, as defined in the last exercise, iff

$$[\![x \in \{\mathbf{a}\}]\!] \sqcap [\![y \in \{\mathbf{b}\}]\!] \sqsubseteq [\![x \approx y]\!]$$

holds for all $x, y \in A$.

EXERCISE 8. Show that the relation $\leqslant$, where

$$a \leqslant b \quad \text{iff} \quad a = b \restriction Ea$$

is a partial ordering on A that satisfies

(i) $a \restriction p \leqslant a$,
(ii) $a \leqslant b$ implies $Ea \sqsubseteq Eb$ and $a \restriction p \leqslant b \restriction p$,
(iii) if $a \leqslant c$ and $b \leqslant c$, for some c, then $a \between b$,
(iv) $a \leqslant b$ iff $Ea \sqsubseteq [\![a \approx b]\!]$,
(v) $a \leqslant b \restriction p$ iff $a \leqslant b$ and $Ea \sqsubseteq p$,
(vi) $a \restriction p = a$ iff $Ea \sqsubseteq p$,
(vii) $a \leqslant b$ iff $a \between b$ and $Ea \sqsubseteq Eb$.

EXERCISE 9. Define $a \in A$ to be the *join* of $B \subseteq A$, written $a = \bigvee B$, iff
(i) $b \leqslant a$ for all $b \in B$, and
(ii) $Ea = \bigsqcup \{Eb : b \in B\}$.

Show that a complete Ω-set $\mathbf{A}$ satisfies the following abstract version of COM:

Every subset $B \subseteq A$ whose elements are pairwise compatible has a unique join.

Prove in fact that

$$s(x) = \bigsqcup_{b \in B} [\![x \approx b]\!]$$

defines a singleton when B has pairwise compatible elements (use Ex. 6) and that the corresponding element of $\mathbf{A}$ to s is $\bigvee B$.

N.B.: To do this exercise, and many of those to follow, you will need to know that a **CHA** satisfies the following law of distribution of $\sqcap$ over $\bigsqcup$:

$$x \sqcap (\bigsqcup C) = \bigsqcup_{c \in C} (x \sqcap c), \quad \text{all} \quad C \subseteq \Omega$$

EXERCISE 10. (i) Prove that $\bigvee B$, when it exists, is the l.u.b. of B for the ordering $\leqslant$, and that in general a set B has a join iff it has a l.u.b. for this ordering.

(ii) $(\bigvee B) \restriction p = \bigvee\{b \restriction p : b \in B\}$. □

A presheaf $F_{\mathbf{A}} : \Omega \to \mathbf{Set}$ over the poset category Ω is defined for *complete* $\mathbf{A}$ by putting

$$F_{\mathbf{A}}(p) = \{x \in A : Ex = p\}$$

for each $p \in \Omega$. Whenever $p \sqsubseteq q$, the assignment of $x \restriction p$ to x is a function from $F_{\mathbf{A}}(q)$ to $F_{\mathbf{A}}(p)$ (Exercise 3). We take this function as the $F_{\mathbf{A}}$-image of the Ω-arrow $p \to q$.

In order to discuss sheaves over the category Ω we define $Cov_\Omega(p)$ to be the collection of all subsets C of Ω that have $\bigsqcup C = p$. This is an obvious generalisation of the definition of Cov_Θ given in §14.3 and (Ω, Cov_Ω) can be shown to be a site. The corresponding category of *sheaves over Ω* is denoted $\mathbf{Sh}(\Omega)$.

EXERCISE 11. Let $C \in Cov_\Omega(p)$ and consider a selection of elements $x_q \in F_{\mathbf{A}}(q)$, all $q \in C$, that are pairwise compatible (in the sense given in COM, or in this section – they mean the same thing). Use the definition of join given in Exercise 9 to construct a unique $x \in F_{\mathbf{A}}(p)$ with $x \restriction q = x_q$, all $q \in C$. Hence verify that $F_{\mathbf{A}}$ is a sheaf (satisfies COM). □

In the converse direction, given a sheaf F over Ω we construct a corresponding Ω-set $\mathbf{A}_F$. We let

$$A_F = \{\langle x, q\rangle : x \in F(q)\}$$

be the disjoint union of the sets $F(q)$ for all $q \in \Omega$. For $a = \langle x, q\rangle$, put $E(a) = q$. Then for any p, we define

$$a \restriction p = \langle F^q_{p\sqcap q}(x), p \sqcap q\rangle,$$

and this allows us to put

$$[\![a \approxeq b]\!]_{\mathbf{A}_F} = \bigsqcup\{p \in \Omega : a \restriction p = b \restriction p\}.$$

Equality can now be given by

$$[\![a \approx b]\!]_{\mathbf{A}_F} = [\![a \approxeq b]\!]_{\mathbf{A}_F} \sqcap E(a) \sqcap E(b).$$

EXERCISE 12. $[\![a \approx a]\!]_{\mathbf{A}_F} = E(a)$

EXERCISE 13. $[\![a \approxeq b]\!]_{\mathbf{A}_F} = (E(a) \sqcup E(b)) \Rightarrow [\![a \approx b]\!]_{\mathbf{A}_F}$

EXERCISE 14. Verify that $\mathbf{A}_F$ is an Ω-set.

EXERCISE 15. Let $s : A_F \to \Omega$ be a singleton. Generalise the argument of Example 1 by showing that the elements $a \restriction s(a)$, for all $a \in A_F$ are pairwise compatible, and use the property COM as it applies to F to show that $s = \{\mathbf{a}\}$, for a unique a (which will in fact be the join of the elements $a \restriction s(a)$). Hence show that $\mathbf{A}_F$ is complete. □

EXAMPLE 3. Let $\Omega = \Theta$ and X be a topological space as in Example 1. The sheaf $F_X : \Theta \to \mathbf{Set}$ of continuous X-valued (partial) functions on I has

$$F_X(V) = \{V \xrightarrow{f} X : f \text{ is continuous}\}$$

with each inclusion $V \hookrightarrow W$ being assigned the usual restriction operator by F_X. In this case $F_X(V)$ and $F_X(W)$ are already disjoint when $V \neq W$, as they consist of sets of functions with distinct domains, and so in forming the associated Θ-set we may simply take the union of the $F_X(V)$'s. We see then that $\mathbf{A}_{F_X}$ is none other than the Θ-set $\mathbf{C}_X$ of Example 1. □

EXERCISE 16. Develop a truly axiomatic theory of "restrictions of elements over a **CHA**" by *defining* a presheaf over Ω to be a set A together with a pair of functions

$$\restriction : A \times \Omega \to A$$

$$E : A \to \Omega$$

that satisfy the laws of Exercises 1–3. Define *compatibility*, the *restriction ordering* $\leqslant$, and *join* for such a structure and call it a *sheaf* if it satisfies the version of COM given in Exercise 9. Use the definition of equality for $\mathbf{A}_F$ to show that such a sheaf carries a complete-Ω-set structure whose $a \restriction p$ operation (defined via singletons) and extent function $[\![Ea]\!]$ are the original $\restriction$ and E you started with.

EXERCISE 17. Let $\mathbf{A} \restriction p$ be the Ω-set based on

$$A \restriction p = \{a \in A : Ea \sqsubseteq p\}$$

with equality as for $\mathbf{A}$. Show that if $B \subseteq A \restriction p$ has a join in $\mathbf{A}$ then this join belongs to $A \restriction p$. Hence show that $\mathbf{A} \restriction p$ is complete if $\mathbf{A}$ is. □

The constructions $\mathbf{A} \mapsto F_{\mathbf{A}}$ and $F \mapsto \mathbf{A}_F$ can be extended to arrows to give an equivalence between $\mathbf{Sh}(\Omega)$ and the sub-category of Ω-$\mathbf{Set}$ generated by the complete objects. In fact $\mathbf{Sh}(\Omega)$ is equivalent to the larger category Ω-$\mathbf{Set}$ itself, a result due originally to D. Higgs [73]. This is because each Ω-set $\mathbf{A}$ is isomorphic in Ω-$\mathbf{Set}$ to a complete Ω-set $\mathbf{A}^*$. We take A^* as the set of all singletons $s : A \to \Omega$ of $\mathbf{A}$, with

$$[\![s \approx t]\!]_{\mathbf{A}^*} = \bigsqcup_{x \in A} (s(x) \sqcap t(x))$$

("there is an x belonging to both s and t". In $\mathbf{Set}$, overlapping singletons are identical).

EXERCISE 18. $[\![Es]\!]_{\mathbf{A}^*} = \bigsqcup_{x \in A} [\![x \in s]\!] = \bigsqcup_{a \in A} [\![s \approx \{\mathbf{a}\}]\!]_{\mathbf{A}^*}$

EXERCISE 19. $[\![\{\mathbf{a}\} \approx s]\!]_{\mathbf{A}^*} = s(a)$

EXERCISE 20. $[\![\{\mathbf{a}\} \approx \{\mathbf{b}\}]\!]_{\mathbf{A}^*} = [\![a \approx b]\!]_{\mathbf{A}}$

EXERCISE 21. $[\![E\{\mathbf{a}\}]\!]_{\mathbf{A}^*} = [\![Ea]\!]_{\mathbf{A}}$

EXERCISE 22. (cf. Example 1). Let $s : A^* \to \Omega$ be a singleton of $\mathbf{A}^*$. For each $f \in A^*$, let f_s be the singleton $f \restriction s(f)$ as defined in Exercise 17 of §11.9, so that $f_s(x) = f(x) \sqcap s(f)$.

Define

$$a_s(x) = \bigsqcup_{f \in A^*} f_s(x).$$

(i) Prove $[\![Ef_s]\!]_{A^*} = s(f)$.

(ii) Show that the f_s's are pairwise compatible in the sense (of Exercise 7) that they satisfy

$$f_s(x) \sqcap g_s(y) \sqsubseteq [\![x \approx y]\!]_{\mathbf{A}}.$$

(iii) Show that a_s is a singleton of $\mathbf{A}$, with $[\![f \approx a_s]\!]_{A^*} = s(f)$, all $f \in A^*$.

(iv) Suppose that $h \in A^*$ has $[\![f \approx h]\!] = s(f)$, all f. Show that $h \restriction s(f) = f_s$ for all f (i.e. $h(x) \sqcap s(f) = f_s(x)$), and hence that $h = a_s$.

Thus prove that $\mathbf{A}^*$ is complete.

Exercise 23. Since $\mathbf{A}^*$ is complete, each element $s \in A^*$ has, for each $p \in \Omega$, a restriction $s \restriction p$ defined as the unique element $t \in A^*$ corresponding to the singleton $\{\mathbf{s}\} \restriction p$ of $\mathbf{A}^*$ (i.e. t is defined by the equation

$$[\![x \approx t]\!]_{A^*} = [\![x \approx s]\!]_{A^*} \sqcap p).$$

Show that this t is precisely the singleton $s \restriction p$ of Exercise 17 of §11.9 (i.e. that $t(x) = s(x) \sqcap p$).

Exercise 24. Show that in $\mathbf{A}^*$,

$$s \restriction s(a) = \{\mathbf{a}\} \restriction s(a)$$

all $s \in A^*$, $a \in A$.

Exercise 25. In view of Exercise 23, use the ideas of Exercise 16 to develop an alternative proof that $\mathbf{A}^*$ is complete.

Exercise 26. Prove that in Ω-**Set**, an arrow $f: A \times B \to \Omega$ from $\mathbf{A}$ to $\mathbf{B}$ is

(i) monic iff it satisfies

$$f(\langle x, y\rangle) \sqcap f(\langle z, y\rangle) \sqsubseteq [\![x \approx z]\!]$$

(ii) epic iff it satisfies

$$[\![Ey]\!]_{\mathbf{B}} = \bigsqcup_{x \in A} f(\langle x, y\rangle)$$

("y exists in $\mathbf{B}$ to the extent that it is the f-image of some x in $\mathbf{A}$").

Exercise 27. Define $i_{\mathbf{A}}: A \times A^* \to \Omega$ by $i_{\mathbf{A}}(\langle x, s\rangle) = s(x)$. Use the last exercise to show that $i_{\mathbf{A}}: \mathbf{A} \to \mathbf{A}^*$ is iso in Ω-**Set**. □

The topos CΩ-Set

The last exercise implies that as far as categorial constructions are concerned, we may confine our attention to complete Ω-sets (also called *Ω-sheaves*). In this context we can take a different approach to arrows, by taking the $[\![f(x)\approx y]\!]$ notation for $f(\langle x, y\rangle)$ literally.

Let **A** and **B** be Ω-sets, and $g: A \to B$ a function from set A to set B satisfying

(i) $[\![x \approx y]\!]_{\mathbf{A}} \sqsubseteq [\![g(x) \approx g(y)]\!]_{\mathbf{B}}$

(ii) $[\![Eg(x)]\!]_{\mathbf{B}} \sqsubseteq [\![Ex]\!]_{\mathbf{A}}$.

Define $\bar{g}: A \times B \to \Omega$ by

$$\bar{g}(\langle x, y\rangle) = [\![g(x) \approx y]\!]_{\mathbf{B}}$$

EXERCISE 28. Prove $[\![Eg(x)]\!] = [\![Ex]\!]$.

EXERCISE 29. Show that $\bar{g}$ is an arrow from **A** to **B** in **Ω-Set**, i.e. an extensional, functional, total Ω-valued relation from A to B (conditions (iv)–(vii) of §11.9). □

To avoid confusion, a function g satisfying (i) and (ii) will be called a *strong* arrow, while the **Ω-Set** arrows will be referred to as *weak*. The two notions are equivalent in the case of a complete codomain. If $f: A \times B \to \Omega$ is a weak arrow, for given $a \in A$ define $s_a : B \to \Omega$ by

$$s_a(y) = f(\langle a, y\rangle).$$

EXERCISE 30. Use the weak arrow properties of f to show that s_a is a singleton of **B**. □

If **B** is complete, there will then be a unique $b \in B$ that has $\{\mathbf{b}\} = s_a$. Put $g_f(a) = b$.

EXERCISE 31. Show that $g_f: A \to B$ is a strong arrow from **A** to **B**, with $\bar{g}_f = f$.

EXERCISE 32. If g is strong, with cod g complete, prove that $g_{\bar{g}} = g$.

EXERCISE 33. Show that for complete **A**, $g_{1_{\mathbf{A}}} = \mathrm{id}_{\mathbf{A}}$.

EXERCISE 34. If $f: \mathbf{A} \to \mathbf{B}$ and $h: \mathbf{B} \to \mathbf{C}$ are weak arrows, with **B** and **C** complete, then $g_{h \circ f}$ is the functional composition $g_h \circ g_f$.

EXERCISE 35. Suppose every weak arrow with codomain **B** is of the form $\bar{g}$ for exactly one strong arrow g. Show that **B** is complete (Hint: consider the one-element Ω-set $\{0\}$ with $[\![0 \approx 0]\!] = [\![Es]\!]_{\mathbf{B}^*}$, and the weak arrow $f:\{0\}\times B \to \Omega$ with $f(\langle 0, y\rangle) = s(y)$).

EXERCISE 36. If **A** and **B** are complete, show that $g: A \to B$ is strong iff it preserves extents and restrictions, i.e. has

$$Eg(a) = Ea$$

$$g(a \upharpoonright p) = g(a) \upharpoonright p$$

all $a \in A$, $p \in \Omega$. □

The category **CΩ-Set** is defined to be that which consists of the complete Ω-sets with strong arrows between them, identities and composites being as in **Set**.

EXERCISE 37. Let $f:\mathbf{A}\to\mathbf{B}$ be a weak arrow, and $s \in A^*$. Define $f_s: B \to \Omega$ by

$$f_s(y) = \bigsqcup_{x \in A} (f(x, y) \sqcap s(x)).$$

Show that f_s is a singleton of **B** ("y belongs to f_s to the extent that it is the f-image of some member of s"). Show that putting $f^*(s) = f_s$ defines a strong arrow $f^*: \mathbf{A}^* \to \mathbf{B}^*$ for which

$$[\![f^*(\{\mathbf{a}\}) \approx \{\mathbf{b}\}]\!]_{\mathbf{B}^*} = f(a, b)$$

all $a \in A$, $b \in B$.

EXERCISE 38. Let the functor F:**CΩ-Set**$\to$**Ω-Set** be the identity on objects, and have $F(g) = \bar{g}$. Let F^*:**Ω-Set**$\to$**CΩ-Set** have $F^*(\mathbf{A}) = \mathbf{A}^*$ and $F^*(f) = f^*$.

Show that F and F^* establish the equivalence of the two categories.

EXERCISE 39. Let $g:\mathbf{A}\to\mathbf{B}$ be a strong arrow, with **A** and **B** complete. Show that in addition to preserving E and $\upharpoonright$, g preserves $\leqslant$ and $\bigvee$, i.e.

(i) $x \leqslant y$ only if $g(x) \leqslant g(y)$

(ii) $g(\bigvee C) = \bigvee\{g(c): c \in C\}$, all $C \subseteq A$.

EXERCISE 40. Let s be a singleton of **A**. Show that in $\mathbf{A}^*$, the elements $\{\mathbf{a}\} \upharpoonright s(a)$ are pairwise compatible, and their join is s, i.e.

$$s = \bigvee\{\{\mathbf{a}\} \upharpoonright s(a): a \in A\}.$$

EXERCISE 41. (i) Let $f: A \to \Omega$ be a subset (extensional and strict) of $\mathbf{A}$ in the sense of §11.9. Define $f^*: A^* \to \Omega$ by putting

$$f^*(s) = \bigsqcup_{a \in A} (f(a) \sqcap s(a)).$$

Show that f^* is a subset of $\mathbf{A}^*$ that has $f^*(\{\mathbf{a}\}) = f(a)$, all $a \in A$.

(ii) Given a subset $g: A^* \to \Omega$ of $\mathbf{A}^*$ show that

$$g_*(a) = g(\{\mathbf{a}\})$$

defines a subset g_* of $\mathbf{A}$, and that

$$g(s) = \bigsqcup_{a \in A} (g_*(a) \sqcap s(a)).$$

Thus show that subsets of $\mathbf{A}^*$ correspond uniquely to subsets of $\mathbf{A}$.

EXERCISE 42. Let $i_{\mathbf{A}}: A \times A^* \to \Omega$ be the weak iso arrow of Exercise 27. Show that the corresponding strong arrow (which we also denote $i_{\mathbf{A}}$) assigns $\{\mathbf{a}\}$ to a.

EXERCISE 43. Let $g: \mathbf{A} \to \mathbf{C}$ be a strong arrow, with $\mathbf{C}$ complete. Show that there exists exactly one strong arrow $h: \mathbf{A}^* \to \mathbf{C}$ for which the diagram

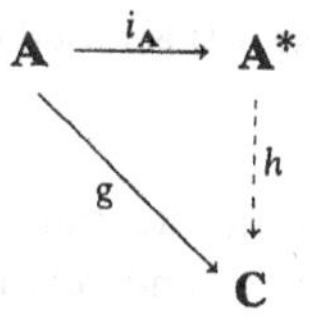

commutes. (Hint: Consider the elements $g(a) \restriction s(a)$, all $a \in A$, for $s \in A^*$. Use Exercises 39, 40).

EXERCISE 44. Show either directly or via the last Exercise that the function f^* of Exercise 37 is uniquely determined by the fact that it has $[\![f^*(\{\mathbf{a}\}) \approx \{\mathbf{b}\}]\!] = f(a, b)$, all $a \in A$ and $b \in B$. □

The topos-structure of **CΩ-Set** could be obtained by applying the completing functor F^* to **Ω-Set**. The relevant constructions admit however of simplified descriptions, which we now outline.

Terminal object

1 is the set Ω, with $[\![p \approx q]\!]_{\mathbf{1}} = p \sqcap q$.

We have $Ep=p$, $p \restriction q = p \sqcap q$, and $[\![p \approx q]\!] = (p \Leftrightarrow q)$.

Notice that in this case $\leqslant$ is the lattice ordering $\sqsubseteq$ on the **CHA** Ω, and $\bigvee$ is the lattice join (l.u.b.) $\bigsqcup$.

The unique arrow $\mathbf{A} \to \mathbf{1}$ is the extent function $a \mapsto [\![a \approx a]\!]_{\mathbf{A}}$.

Exercise 45. Let $\Omega = \Theta$, and X be a topological space. If $f: I \to X$ is continuous, let $f_o: \Theta \to C_X$ assign to each V the restriction $f \restriction V$ of f to V. Interpret the two conditions that define strong arrows to show that in **CΘ-Set** we have $f_o: \mathbf{1} \to \mathbf{C}_X$. Conversely, given an "element" $g: \mathbf{1} \to \mathbf{C}_X$ of $\mathbf{C}_X$, show that $g(V)$ has domain V, and that the $g(V)$'s are pairwise compatible. Hence show that there is a unique $f \in C_X$ that has $f_o = g$.

Thus establish that there is a bijective correspondence between elements $\mathbf{1} \to \mathbf{C}_X$ of $\mathbf{C}_X$ in **CΘ-Set** and *globally defined* continuous functions $I \to X$.

Exercise 46. In view of the last exercise, we say that a is a *global element* of $\mathbf{A}$ in **CΩ-Set** if $Ea = \top$. For such an element, define $f_a: \Omega \to A$ by $f_a(p) = a \restriction p$ and show that $f_a: \mathbf{1} \to \mathbf{A}$.

Conversely, given $h: \mathbf{1} \to \mathbf{A}$, use Exercise 6 to show that the $h(p)$'s are compatible, and hence prove that there is a unique global element a of $\mathbf{A}$ with $f_a = h$.

Exercise 47. The complete Ω-set $\mathbf{1} \restriction e$ (Exercise 17) is based on the set

$$\Omega \restriction e = \{q : q \sqsubseteq e\}.$$

Show that this is a **CHA** in its own right with the same $\sqcup$ and $\sqcap$ operations as Ω, but with pseudo-complement $\neg_e$ and relative pseudo-complement $\Rightarrow_e$ given by

$$\neg_e q = \neg q \restriction e = \neg q \sqcap e$$

and

$$q \Rightarrow_e r = (q \Rightarrow r) \restriction e = (q \Rightarrow r) \sqcap e,$$

all $q, r \in \Omega \restriction e$. □

Initial object

Recall from §11.9 that the function from A to Ω that assigns $\bot$ to every $a \in A$ is a singleton, and so for complete $\mathbf{A}$ corresponds to a unique element $\emptyset_{\mathbf{A}} \in A$. We have $[\![x \approx \emptyset_{\mathbf{A}}]\!] = \bot$, all x.

The initial object **0** for **CΩ-Set** is the one-element set $\{\perp\}$, with $[\![\perp \approx \perp]\!] = \perp$. The unique arrow $\mathbf{0} \to \mathbf{A}$ assigns $\emptyset_{\mathbf{A}}$ to $\perp$.

EXERCISE 48. (i) $\emptyset_{\mathbf{A}}$ is the join of the empty subset of A.
(ii) If $Ea = \perp$, then $a = \emptyset_{\mathbf{A}}$. □

Products

$\mathbf{A} \times \mathbf{B}$ is the set

$$A \cdot B = \{\langle a, b\rangle \in A \times B : Ea = Eb\}, \text{ with}$$
$$[\![\langle a, b\rangle \approx \langle c, d\rangle]\!] = [\![a \approx c]\!] \sqcap [\![b \approx d]\!].$$

We have

$$E\langle a, b\rangle = Ea \sqcap Eb$$

and

$$\langle a, b\rangle \restriction p = \langle a \restriction p, b \restriction p\rangle$$

Projection arrows, and products of arrows are defined just as in **Set**.

Coproducts

$\mathbf{A} + \mathbf{B}$ is the set

$$A + B = \{\langle a, b\rangle \in A \times B : Ea \sqcap Eb = \perp\}, \text{ with}$$
$$[\![\langle a, b\rangle \approx \langle c, d\rangle]\!] = [\![a \approx c]\!] \sqcup [\![b \approx d]\!],$$

giving

$$E\langle a, b\rangle = Ea \sqcup Eb$$

and

$$\langle a, b\rangle \restriction p = \langle a \restriction p, b \restriction p\rangle.$$

The injection $i_{\mathbf{A}} : \mathbf{A} \to \mathbf{A} + \mathbf{B}$ takes $a \in A$ to $\langle a, \emptyset_{\mathbf{B}}\rangle$, while $i_{\mathbf{B}}(b) = \langle \emptyset_{\mathbf{A}}, b\rangle$, all $b \in B$.

The coproduct $[f, g] : \mathbf{A} + \mathbf{B} \to \mathbf{C}$ of two strong arrows f and g assigns to $\langle a, b\rangle$ the join in $\mathbf{C}$ of $f(a)$ and $g(b)$.

EXERCISE 49. Verify that $f(a)$ and $g(b)$ are compatible in $\mathbf{C}$ when $\langle a, b\rangle \in A + B$ and f and g are strong. □

Pullback

The domain of the pullback

$$\begin{array}{ccc} \mathbf{D} & \xrightarrow{g'} & \mathbf{B} \\ {\scriptstyle f'}\downarrow & & \downarrow{\scriptstyle g} \\ \mathbf{A} & \xrightarrow[f]{} & \mathbf{C} \end{array}$$

of f and g as shown has $D=\{\langle x, y\rangle \in A \cdot B: f(x)=g(y)\}$ with its Ω-equality inherited from $\mathbf{A}\times\mathbf{B}$. f' and g' are the evident projections.

EXERCISE 50. Show that $a \in D$ only if $a \restriction p \in D$ all $p \in \Omega$. Prove that if a subset of D has a join in $A \cdot B$ then this join belongs to D. Hence verify that $\mathbf{D}$ is complete. □

Subobject classifier

The object of truth-values is $\dot{\mathbf{\Omega}}$, where

$$\dot{\Omega}=\{\langle p, e\rangle \in \Omega \times \Omega : p \sqsubseteq e\}$$

and

$$[\![\langle p, e\rangle \approx \langle q, e'\rangle]\!]=(p \Leftrightarrow q) \sqcap e \sqcap e'$$

giving

$$E\langle p, e\rangle = e$$

and

$$\langle p, e\rangle \restriction q = \langle p \sqcap q, e \sqcap q\rangle.$$

The arrow $true: \mathbf{1} \to \dot{\mathbf{\Omega}}$ has

$$true(p)=\langle p, p\rangle, \quad \text{all} \quad p \in \Omega.$$

If $f: \mathbf{A} \to \mathbf{B}$ is monic in **CΩ-Set** (which just means that it is injective as a set function-exercise) then for each $b \in B$ we define the truth-value of "$b \in f(A)$" as

$$[\![b \in f(A)]\!] = \bigsqcup_{a \in A} [\![f(a) \approx b]\!]$$

("b belongs to $f(A)$ to the extent that it is equal to the f-image of some $a \in A$").

The character $\chi_f : \mathbf{B} \to \dot{\boldsymbol{\Omega}}$ of f is then given by

$$\chi_f(b) = \langle [\![b \in f(A)]\!], Eb \rangle.$$

EXERCISE 51. Prove that

$$[\![b \in f(A)]\!] = \bigsqcup \{Ea : a \in A \text{ and } f(a) \leqslant b\}$$

("b belongs to $f(A)$ to the extent that there exists a restriction of b in $f(A)$"), and show that the image $f(A)$ of A under f is precisely the set

$$\{b \in B : [\![b \in f(A)]\!] = Eb\}.$$

Hence show how to define the subobject of $\mathbf{B}$ that is classified by a given arrow $\mathbf{B} \to \dot{\boldsymbol{\Omega}}$.

EXERCISE 52. Show that the propositional logic of **CΩ-Set** is as follows:

(i) $false(p) = \langle \bot, p \rangle$ defines $false : \mathbf{1} \to \dot{\boldsymbol{\Omega}}$

(ii) The negation arrow $\neg : \dot{\boldsymbol{\Omega}} \to \dot{\boldsymbol{\Omega}}$ has

$$\neg(\langle p, e \rangle) = \langle \neg_e p, e \rangle$$
$$= \langle \neg p \sqcap e, e \rangle \qquad \text{(cf. Ex. 47)}$$

(iii) Conjunction, disjunction, and implication as arrows $\dot{\Omega} \cdot \dot{\Omega} \to \dot{\Omega}$ have

$$\langle p, e \rangle \cap \langle q, e \rangle = \langle p \sqcap q, e \rangle$$
$$\langle p, e \rangle \cup \langle q, e \rangle = \langle p \sqcup q, e \rangle$$
$$\langle p, e \rangle \Rightarrow \langle q, e \rangle = \langle p \Rightarrow_e q, e \rangle$$

□

Exponentials

$\mathbf{B}^{\mathbf{A}}$ is the set $[A \to B]$ of all pairs of the form $\langle f, e \rangle$ such that $e \in \Omega$ and $f : A \to B$ is a set function satisfying

$$f(a \restriction p) = f(a) \restriction p$$

and

$$Ef(a) = Ea \sqcap e.$$

Equality is defined by

$$[\![\langle f, e \rangle \approx \langle g, e' \rangle]\!] = \bigsqcap_{x \in A} (Ex \Rightarrow [\![f(x) \approx g(x)]\!]) \sqcap e \sqcap e'$$

giving

$$E\langle f, e\rangle = e$$

and

$$\langle f, e\rangle \restriction p = \langle f \restriction p, e \sqcap p\rangle$$

(where $f \restriction p$ is the function $x \mapsto f(x) \sqcap p$ as usual).

The evaluation arrow $ev : A \cdot [A \to B] \to B$ is given by

$$ev(\langle x, \langle f, e\rangle\rangle) = f(x)$$

EXERCISE 53. Given $g : \mathbf{C} \times \mathbf{A} \to \mathbf{B}$, show that the exponential adjoint to g assigns to $c \in C$ the pair $\langle g_c, Ec\rangle$, where $g_c : A \to B$ has

$$g_c(a) = g(\langle c \restriction Ea, a \restriction Ec\rangle).$$

EXERCISE 54. Show that a global element $\langle f, \top\rangle$ of $\mathbf{B}^{\mathbf{A}}$ is essentially a function f that preserves $\restriction$ and E. Hence establish that the global elements of $\mathbf{B}^{\mathbf{A}}$ are essentially the strong $\mathbf{C\Omega}$-**Set** arrows $\mathbf{A} \to \mathbf{B}$. □

Power objects

A simpler description than $\dot{\mathbf{\Omega}}^{\mathbf{A}}$ is available. $\mathcal{P}(\mathbf{A})$ is the set of pairs $\langle f, e\rangle$, where $f : A \to \Omega$ has

$$f(a) \sqsubseteq e$$

and

(†) $$f(a \restriction p) = f(a) \sqcap p$$

all $a \in A$, $p \in \Omega$.

Equality, E, and $\restriction$ are as in the exponential case.

EXERCISE 55. Show that assigning to $\langle f, e\rangle$ the pair $\langle g, e\rangle$, where $g : A \to \dot{\Omega}$ has

$$g(a) = \langle f(a), Ea \sqcap e\rangle$$

establishes the isomorphism of $\mathcal{P}(\mathbf{A})$ and $\dot{\mathbf{\Omega}}^{\mathbf{A}}$.

EXERCISE 56. A global element of $\mathcal{P}(\mathbf{A})$ is essentially a function $f : A \to \Omega$ that satisfies (†) above. Show that such a function is extensional and strict, i.e. satisfies

$$[\![x \approx y]\!] \sqcap f(x) \sqsubseteq f(y)$$

and

$$f(x) \sqsubseteq Ex.$$

Conversely show that any extensional strict function satisfies (†).

In other words, prove that the global elements of the power object $\mathcal{P}(\mathbf{A})$ in **CΩ-Set** are essentially the *subsets of* **A**, *i.e.* the elements of the "weak" power object for **A** in **Ω-Set** described in §11.9.

EXERCISE 57. Prove that $\mathcal{P}(\mathbf{A})$ is "flabby", which means that each of its elements can be extended to (i.e. is a restriction of) some global element.

EXERCISE 58. Prove that the "singleton arrow"

$$\{\cdot\}_{\mathbf{A}}: \mathbf{A} \to \mathcal{P}(\mathbf{A})$$

(cf. §11.8) assigns $\langle\{\mathbf{a}\}, Ea\rangle$ to $a \in A$. □

Object of partial elements

$\tilde{\mathbf{A}}$ has

$$\tilde{A} = \{\langle a, e\rangle: a \in A, e \in \Omega, \text{ and } Ea \sqsubseteq e\}$$

with

$$[\![\langle a, e\rangle \approx \langle a', e'\rangle]\!] = [\![a \approx a']\!] \sqcap e \sqcap e'.$$

As usual $E\langle a, e\rangle = e$, and

$$\langle a, e\rangle \restriction p = \langle a \restriction p, e \sqcap p\rangle.$$

The imbedding $\eta_{\mathbf{A}}: \mathbf{A} \rightarrowtail \tilde{\mathbf{A}}$ has

$$\eta_{\mathbf{A}}(a) = \langle a, Ea\rangle.$$

Notice that $\tilde{\mathbf{1}} = \dot{\Omega}$ explicitly.

EXERCISE 59. If g is a partial arrow

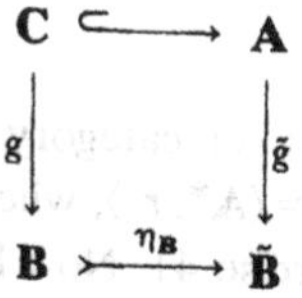

from $\mathbf{A}$ to $\mathbf{B}$ with dom $g \subseteq A$ as shown, show that its character $\tilde{g}: \mathbf{A} \to \tilde{\mathbf{B}}$ has

$$\tilde{g}(a) = \langle g_a, Ea \rangle$$

where

$$g_a = \bigvee \{g(x): x \leqslant a\}$$

□

Formal logic in CΩ-Set

We shall use the same formal semantics for quantificational languages in **CΩ-Set** as that developed for **Ω-Set** in §11.9. A model $\mathfrak{A}$ for our sample language $\mathscr{L} = \{\mathbf{R}\}$ should assign to $\mathbf{R}$ a strong arrow $r: \mathbf{A} \times \mathbf{A} \to \dot{\boldsymbol{\Omega}}$ in **CΩ-Set**. By Exercises 55 and 54 such an r corresponds to a unique global element of $\mathscr{P}(\mathbf{A} \times \mathbf{A})$, and hence by Exercise 56 we can identify it with a subset of $\mathbf{A} \times \mathbf{A}$ (extensional strict function $A \times A \to \Omega$), allowing the theory of §11.9 to proceed unchanged.

There is one notable advantage in working with *complete* Ω-sets as far as formal logic is concerned, and that is that they allow a natural interpretation of definite-description terms of the form $\mathbf{I}v\varphi(v)$ (as described in §11.10). If $\mathfrak{A} = \langle \mathbf{A}, r \rangle$ is an $\mathscr{L}$-model in **CΩ-Set**, and $\varphi(v)$ is a formula with one free variable, define a function $f_\varphi: A \to \Omega$ by

$$f_\varphi(c) = [\![\mathbf{E}(\mathbf{c}) \wedge \forall v(\varphi(v) \equiv (v \approx \mathbf{c}))]\!]$$

Exercise 60. Show that f_φ is a singleton of $\mathbf{A}$, either by a direct calculation, or by expressing this fact in terms of the $\mathfrak{A}$-truth of formulae which you can derive from the $\mathfrak{A}$-true ones of Exercise 20 of §11.9, using $\mathfrak{A}$-truth-preserving rules of inference. □

Since $\mathbf{A}$ is complete, there is a unique $a_\varphi \in A$ that has $\{\mathbf{a}_\varphi\} = f_\varphi$. We take this element as the interpretation of the term $\mathbf{I}v\varphi$.

Exercise 61. Verify the $\mathfrak{A}$-truth of

$$v_i \approx \mathbf{I}v_j(v_i \approx v_j)$$

$$\mathbf{I}v_i\varphi \approx \mathbf{I}v_i(\mathbf{E}(v_i) \wedge \varphi)$$

$$\mathbf{E}(\mathbf{I}v_i\varphi) \equiv \exists v_j \forall v_i((v_i \approx v_j) \equiv \varphi)$$

□

Now, if $\mathfrak{A} = \langle \mathbf{A}, r \rangle$ is an $\mathscr{L}$-model in the weaker category **Ω-Set**, we define the associated complete model to be $\mathfrak{A}^* = \langle \mathbf{A}^*, r^* \rangle$, where r^* is the subset of $\mathbf{A}^* \times \mathbf{A}^*$ corresponding to r as in Exercise 41. Now Exercise 18 of this Section states that in $\mathbf{A}^*$

$$[\![Es]\!] = \bigsqcup_{a \in A} [\![s \approx \{\mathbf{a}\}]\!]$$

which means that the set $\{\{\mathbf{a}\}: a \in A\}$ generates $\mathbf{A}^*$ in the sense of Exercise 22 of §11.9. The latter may then be used to carry through the next result.

EXERCISE 62. For any sentence φ whose closed terms denote only elements of A,

$$[\![\varphi]\!]_{\mathfrak{A}} = [\![\varphi]\!]_{\mathfrak{A}^*}.$$

EXERCISE 63. Suppose $\mathbf{A}$ is complete. Prove that for any formula $\varphi(v_1, \ldots, v_n)$,

$$[\![\varphi(\mathbf{c}_1, \ldots, \mathbf{c}_n)]\!]_{\mathfrak{A}} \sqcap p = [\![\varphi(\mathbf{c}_1 \restriction \mathbf{p}, \ldots \mathbf{c}_n \restriction \mathbf{p})]\!]_{\mathfrak{A}} \sqcap p$$

for all $c_1, \ldots, c_n \in A$ and $p \in \Omega$. □

Comprehension

Given a model $\mathfrak{A}$ based on an Ω-set $\mathbf{A}$, a formula $\varphi(v)$ with one free variable determines a subobject $\mathbf{A}_\varphi \hookrightarrow \mathbf{A}$ of $\mathbf{A}$, namely the Ω-set of A-elements having the "property" φ. In the light of the realisation (§4.8) of the Ω-axiom as a form of the Comprehension principle, $\mathbf{A}_\varphi$ should be constructible by pulling *true* back along an arrow

$$[\![\varphi]\!]: \mathbf{A} \to \dot{\boldsymbol{\Omega}}$$

that semantically interprets φ.

In **CΩ-Set** the appropriate definition is to let $[\![\varphi]\!]$ be the function which assigns to each $c \in A$ the pair $\langle [\![\varphi(\mathbf{c})]\!]_{\mathfrak{A}} \sqcap Ec, Ec\rangle$.

EXERCISE 64. Prove that $[\![\varphi]\!]$ is a strong arrow.

EXERCISE 65. (Exhausting). Describe the arrows $true_{\mathbf{A}}$, $\forall_{\mathbf{A}}$, $\exists_{\mathbf{A}}$ in **CΩ-Set**, and then verify that $[\![\varphi]\!]$ as just defined is precisely the same interpretation $\mathbf{A} \to \dot{\boldsymbol{\Omega}}$ of φ as that that is produced by the $\mathscr{E}$-semantics of §11.4. □

Since $\mathbf{A}_\varphi$ is to be the pullback of *true* along $[\![\varphi]\!]$, we are lead by Exercise 51 to conclude that

$$c \in A_\varphi \quad \text{iff} \quad [\![\varphi(\mathbf{c})]\!]_{\mathfrak{A}} \sqcap Ec = Ec$$

and so

$$A_\varphi = \{c \in A : Ec \sqsubseteq [\![\varphi(\mathbf{c})]\!]\}$$

with Ω-equality in $\mathbf{A}_\varphi$ being as for $\mathbf{A}$.

In the notation introduced for subobject classifiers, we have

$$[\![d \in A_\varphi]\!] = [\![\mathbf{E}(\mathbf{d}) \wedge \varphi(\mathbf{d})]\!]_{\mathbf{A}}$$

EXERCISE 66. $[\![d \in A_\varphi]\!] = [\![\exists v(\varphi(v) \wedge v \approx \mathbf{d})]\!]_{\mathbf{A}}$ □

Simple sheaves

Any set X can be made into an Ω-set by providing it with the *rigid* equality

$$[\![x \approx y]\!] = \begin{cases} \top & \text{if } x = y \\ \bot & \text{if } x \neq y \end{cases},$$

yielding the *rigid Ω-set* $\bar{\mathbf{X}}$. The completion (set of singletons) of $\bar{\mathbf{X}}$ will simply be denoted $\mathbf{X}^*$. A **CΩ-Set**-object obtained in this way is called a *simple sheaf.*

In the case $\Omega = \Theta$, $\mathbf{X}^*$ has a natural representation as the sheaf $\mathbf{C}_X$ of continuous X-valued partial functions on I (Example 1), where here we take the *discrete* topology on X for which singleton subsets $\{x\} \subseteq X$ are open.

Given continuous $f: V \to X$, define

$$s_f(x) = f^{-1}(\{x\}) \in \Theta$$

for all $x \in X$. The $s_f(x)$'s are disjoint for distinct x's, and since $Ex = \top = I$ in $\bar{\mathbf{X}}$, it follows that s_f is a singleton of $\bar{\mathbf{X}}$.

Conversely, if $s: X \to \Theta$ is an element of $\mathbf{X}^*$, for $x \neq y$ we have $s(x) \cap s(y) = \emptyset$, so with $V = \bigcup\{s(x): x \in X\}$ we may define a function $f_s: V \to X$ which corresponds uniquely to s by the construction just given. The rule is that for input $i \in V$,

$$f_s(i) = \text{the unique } x \in X \text{ such that } i \in s(x).$$

Then $f_s^{-1}(\{x\}) = s(x)$ is open, making f_s continuous for the discrete topology.

EXERCISE 67. Verify that the operations $s \mapsto f_s$ and $f \mapsto s_f$ are mutually inverse. □

We noted in §14.1, in discussing natural-numbers objects for $\mathbf{Sh}(I)$, that continuous functions for a discrete codomain are precisely those that are *locally constant* (constant throughout some neighbourhood of each point of their domain). Thus in the topological case, the simple Θ-sheaf $\mathbf{X}^*$ may be thought of as the sheaf of locally constant X-valued functions on I. Its global ($Ea = \top$) elements are of course just those functions that are globally defined (have domain I). We may identify members of the original set X with such functions by associating with $a \in X$ the function $f_a : I \to X$ that has $f_a(i) = a$ for *all* $i \in I$ (f_a corresponds to the singleton $\{\mathbf{a}\}$ in $\mathbf{X}^*$ by the above construction). Thus X is identified with the set of globally defined totally constant functions. There may however be other global elements of $\mathbf{X}^*$. If I is made up of a number of disjoint open pieces then there may well be globally defined but only locally constant functions which assign different constant values to each of these disjoint pieces.

In general, a rigid Ω-set $\bar{\mathbf{X}}$ is *reduced*, which means that it has

$$\{\mathbf{a}\} = \{\mathbf{b}\} \quad \text{only if} \quad a = b$$

for all $a, b \in X$. This implies that the assignment of $\{\mathbf{a}\}$ to a is an injection of X into X^*, and since (Exercise 20)

$$[\![\{\mathbf{a}\} \approx \{\mathbf{b}\}]\!]_{\mathbf{X}^*} = [\![a \approx b]\!]_{\mathbf{X}},$$

we may simply identify a and $\{\mathbf{a}\}$ and regard X as a subset of X^*, i.e. $X \subseteq X^*$. Then by Exercise 18 we find that

$$Es = \bigsqcup_{a \in X} [\![s \approx a]\!]_{\mathbf{X}^*}$$

for all $s \in X^*$, which means that X generates $\mathbf{X}^*$. This greatly simplifies the computation of formal truth-values for a model $\mathfrak{A}$ based on the simple sheaf $\mathbf{X}^*$, since by Exercise 22 of §11.9 we have

$$[\![\forall v\varphi]\!]_{\mathfrak{A}} = \bigsqcap_{c \in X} [\![\mathbf{E}(\mathbf{c}) \supset \varphi(\mathbf{c})]\!]_{\mathfrak{A}}$$

and

$$[\![\exists v\varphi]\!]_{\mathfrak{A}} = \bigsqcup_{c \in X} [\![\mathbf{E}(\mathbf{c}) \wedge \varphi(\mathbf{c})]\!]_{\mathfrak{A}},$$

so that we can confine the range of quantification to the elements of the original (rigid) set X. But the latter elements are all global in $\mathbf{X}^*$, so these equations reduce (via $\top \Rightarrow \mathrm{p} = \mathrm{p}$) to

$$(\dagger) \qquad \begin{aligned} [\![\forall v\varphi]\!]_{\mathfrak{A}} &= \bigsqcap_{c \in X} [\![\varphi(\mathbf{c})]\!]_{\mathfrak{A}} \\ [\![\exists v\varphi]\!]_{\mathfrak{A}} &= \bigsqcup_{c \in X} [\![\varphi(\mathbf{c})]\!]_{\mathfrak{A}} \end{aligned}$$

We shall use these facts later when we come to construct number systems in categories of Ω-sheaves.

Topoi as sheaf-categories

When may an elementary topos $\mathscr{E}$ be construed as the category of sheaves over some **CHA**? To answer that question we shall examine below some of the properties enjoyed by topoi of the form $\mathbf{C\Omega}$-**Set**.

The Heyting algebra most naturally associated with $\mathscr{E}$ is the algebra $\Omega_{\mathscr{E}} = \mathrm{Sub}(1)$ of subobjects $u \rightarrowtail 1$ of $\mathscr{E}$'s terminal object ($\Omega_{\mathscr{E}}$ can alternatively be thought of as the **HA** $\mathscr{E}(1, \Omega)$ of global truth-values of $\mathscr{E}$). But in order to develop sheaf theory over Ω we need the latter to be complete as a lattice. For this it suffices that $\mathscr{E}$ have *arbitrary coproducts of subobjects of* 1, i.e. that for any set $\{u_x : x \in X\}$ of $\mathscr{E}$-objects whose unique arrow $u_x \to 1$ is monic there is an associated co-product object, which we denote $\varinjlim_{x \in X} u_x$. The lattice join $\bigsqcup_{x \in X} u_x$ may then be obtained as the epi-monic factorisation

$$\begin{array}{ccc} \varinjlim_{x \in X} u_x & \longrightarrow & 1 \\ & \searrow \quad \nearrow & \\ & \bigsqcup_{x \in X} u_x & \end{array}$$

of the coproduct of the u_x's (cf. the construction of unions in §7.1).

The existence of coproducts of arbitrary sub-collections of $\Omega_{\mathscr{E}}$ is also necessary for $\mathscr{E}$ to be a sheaf-category since $\mathbf{C\Omega}$-**Set**, for any **CHA** Ω, has coproducts of *all* sets of objects. Given a set $\{\mathbf{A}_x : x \in X\}$ of Ω-sheaves, the coproduct $\varinjlim_{x \in X} \mathbf{A}_x$ is defined, by generalisation of the above definition of $\mathbf{A}+\mathbf{B}$, to be the Ω-set of all *disjoint selections* of the $\mathbf{A}_x$'s. A member of this coproduct is a selection $a = \{a_x : x \in X\}$ of an element $a_x \in A_x$ for each $x \in X$ such that

$$Ea_x \sqcap Ea_y = \bot \quad \text{whenever} \quad x \neq y.$$

Equality of selections is given by

$$[\![a \approx b]\!] = \bigsqcup_{x \in X} [\![a_x \approx b_x]\!]$$

and the injection $\mathbf{A}_y \to \varinjlim_{x \in X} \mathbf{A}x$ assigns to $a_y \in A_y$ the selection a that has

$$a_x = \begin{cases} a_y & \text{if} \quad x = y \\ \emptyset_{\mathbf{A}_x} & \text{otherwise} \end{cases}$$

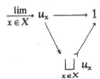

EXERCISE 68. Given a collection $\{\mathbf{A}_x \xrightarrow{f_x} \mathbf{C}: x \in X\}$ of arrows in $\mathbf{C\Omega}$**-Set**, describe their coproduct arrow $\varinjlim_{x\in X} \mathbf{A}_x \to \mathbf{C}$. □

The other property enjoyed by $\mathbf{C\Omega}$**-Set** that will be used to answer our question is *weak extensionality*. In general an $\mathscr{E}$-object a will be called *weakly extensional* if for any two distinct parallel arrows $f,g : a \to b$ with domain a there is a *partial* element $x : 1 \rightsquigarrow a$ that distinguishes them, i.e. has $f \circ x \neq g \circ x$. Thus the whole topos is weakly extensional, as defined in §12.1, just in case each $\mathscr{E}$-object is weakly extensional as just defined.

To see how this property obtains in $\mathbf{C\Omega}$**-Set**, suppose $f,g : \mathbf{A} \to \mathbf{B}$ are distinct strong arrows. Thus we have $f(a) \neq g(a)$ for some $a \in A$. But then assigning $a \restriction q$ to each $q \sqsubseteq Ea$ gives a strong arrow $x : \mathbf{1} \restriction Ea \to \mathbf{A}$ that distinguishes f and g (since $x(Ea) = a$). Here $\mathbf{1} \restriction Ea \hookrightarrow \mathbf{1}$ is the subobject of $\mathbf{1}$ based on the set $\Omega \restriction Ea = \{q \in \Omega: q \sqsubseteq Ea\}$ (*Ex.* 47), so that we have $x : \mathbf{1} \rightsquigarrow \mathbf{A}$.

EXERCISE 69. In any $\mathscr{E}$, given $u \rightarrowtail 1$ and $f : u \to \tilde{a}$ take the pullback g

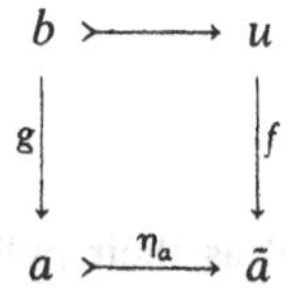

of f along η_a and let $\tilde{g}$ be the unique arrow making the boundary of

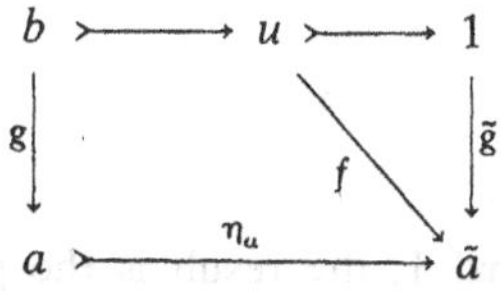

a pullback. Prove that the right-hand triangle of this last diagram commutes.

EXERCISE 70. Use the last exercise to show that if $\tilde{a}$ is weakly extensional then two distinct arrows with domain $\tilde{a}$ are distinguishable by a *global* element of $\tilde{a}$. □

We shall call an $\mathscr{E}$-object a *extensional* if any parallel pair $a \rightrightarrows b$ of distinct arrows with domain a are distinguished by a global element $1 \to a$. (Thus $\mathscr{E}$ is well-pointed precisely when all of its objects are extensional). The last exercise implies that an object $\tilde{a}$ of partial elements is always extensional whenever it is weakly so. Thus in a sheaf-category each object **A** is *sub-extensional*, i.e. is a subobject of an extensional object, (since we have a monic arrow $\mathbf{A} \rightarrowtail \tilde{\mathbf{A}}$).

The two conditions that $\mathscr{E}$ be weakly extensional and have coproducts of subobjects of 1 suffice to make $\mathscr{E}$ a sheaf-category, and indeed to make it equivalent to $\mathbf{C}\Omega_{\mathscr{E}}$**-Set**. The original proof of this result used a great deal of heavy machinery in the form of "geometric morphisms" $\mathscr{E} \to \mathbf{Set}$ and such-like "abstract nonsense". However recent work by Michael Brockway has provided a proof that is much more accessible and has the conceptual advantage of making it possible to see just how an object becomes a sheaf and vice-versa.

The $\Omega_{\mathscr{E}}$-sheaf $\mathbf{A}_a$ corresponding to an $\mathscr{E}$-object a has A_a as the set of partial $\mathscr{E}$-elements of a, with the degree of equality of $x, y : 1 \rightsquigarrow a$ being obtained as the equaliser of the diagram

$$\begin{array}{ccccc} & & \text{dom } x & & \\ & \nearrow & & \overset{x}{\searrow} & \\ \text{dom } x \cap \text{dom } y & & & & a \\ & \searrow & & \underset{y}{\nearrow} & \\ & & \text{dom } y & & \end{array}$$

Here the intersection of domains is given as usual as their pullback

$$\begin{array}{ccc} \text{dom } x \cap \text{dom } y & \rightarrowtail & \text{dom } y \\ \downarrow & & \downarrow \\ \text{dom } x & \longrightarrow & 1 \end{array}$$

but since this is done over the terminal 1, the result is the product of dom x and dom y. Now equalising a product is one way to obtain a pullback, so $[\![x \approx y]\!]$ is alternatively characterised as the pullback

$$\begin{array}{ccc} [\![x \approx y]\!] & \rightarrowtail & \text{dom } y \\ \downarrow & & \downarrow y \\ \text{dom } x & \xrightarrow{\;x\;} & a \end{array}$$

of x and y.

In the case $\mathscr{E} = \mathbf{Top}(I)$, this construction produces the now familiar sheaf of local sections of a topological bundle (Example 2 of this section).

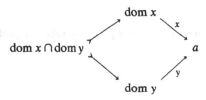

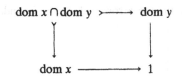

Even in the case $\mathscr{E} = \mathbf{Set}$ it has some interest in assigning to each set X its set $X \cup \{*\}$ of partial elements. Here of course Ω is the 2-element Boolean algebra consisting of $\top = 1$ and $\bot = \emptyset$.

When considered as a 2-sheaf, $X \cup \{*\}$ is actually the simple sheaf $\mathbf{X}^*$ obtained by completing the rigid 2-set $\bar{\mathbf{X}}$. It is not hard to see that the only singletons $s : \bar{\mathbf{X}} \to 2$ are those corresponding to elements of X, together with the unique singleton with empty extent ($Es = \bot$). The latter serves as the null entity $*$. In fact every object of **C2-Set** arises as a simple sheaf, for if $\mathbf{Y}$ is a 2-sheaf and $X = Y - \{\emptyset_{\mathbf{Y}}\}$ then the equality relation of $\mathbf{Y}$ is rigid on X (i.e. makes all elements of X global and distinct elements have $[\![x \approx y]\!] = \bot$.) Thus $\mathbf{X}^* = \mathbf{Y}$.

EXERCISE 71. Prove this last statement. □

In order to categorially recover the **Set**-object X from the **C2-Set** object $\mathbf{X}^* = X \cup \{*\}$ we form the coproduct (disjoint union) of the extents Ex for all $x \in X^*$. If $x \in X$, then $Ex = \top = \{0\}$, so we identify Ex with $\{x\}$. If $x = *$, then $Ex = \bot = \emptyset$, so that the coproduct becomes the union

$$\bigcup \{\{x\} : x \in X\} \cup \emptyset = X.$$

Thus we reconstruct X by representing each element of $\mathbf{X}^*$ by a disjoint copy of its own extent (which is a subobject of 1) and then putting these extents together. The reason why this procedure does faithfully reproduce X is that in $\mathbf{X}^*$ all elements are rigidly separated (disjoint). The same construction will not however work if elements of the sheaf overlap and so are to some extent equal. Consider for example the case $\mathscr{E} = \mathbf{Top}(I)$, where $\Omega_{\mathscr{E}} = \Theta$, the set of open subsets of I. Identifying local sections with their images in the stalk space we have the following sort of picture of the sheaf of partial elements of a bundle $A \to I$.

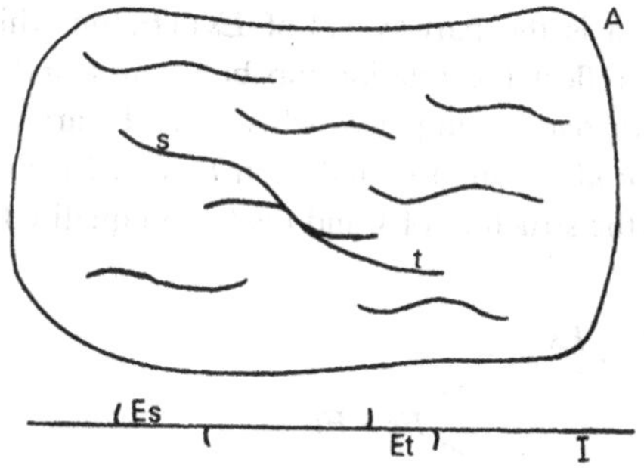

Fig. 14.3.

If we now identify each s with its extent Es and take the coproduct of these Es's we will construct a stalk space larger than A. The two displayed elements s and t will have disjoint copies in the new space, with

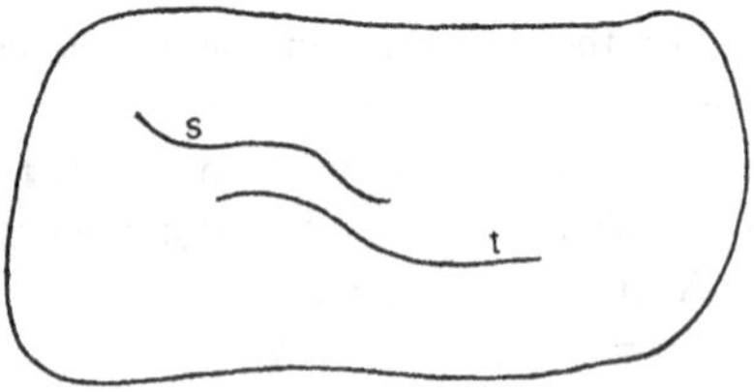

Fig. 14.4.

part of s being duplicated in the copy of t, and vice versa. To recover A we must "reduce" the coproduct by glueing together copies of s and t to the extent that they originally coincided.

Notice that the extents of s and t are arranged thus

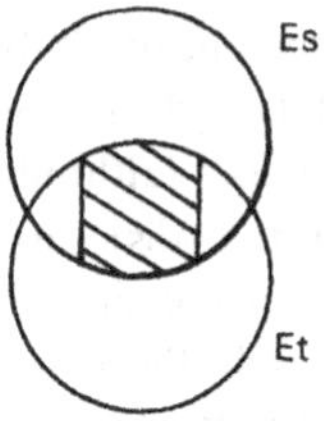

Fig. 14.5.

where the shaded area is the part $[\![s \approx t]\!]$ of $Es \cap Et$ on which s and t agree. This does not reflect the relationship between s and t faithfully either, since Es and Et overlap in places where s and t are distinct. The way to reduce the coproduct, and to build from Es and Et, an object that accurately represents the structure of s and t is to co-equalise the diagram

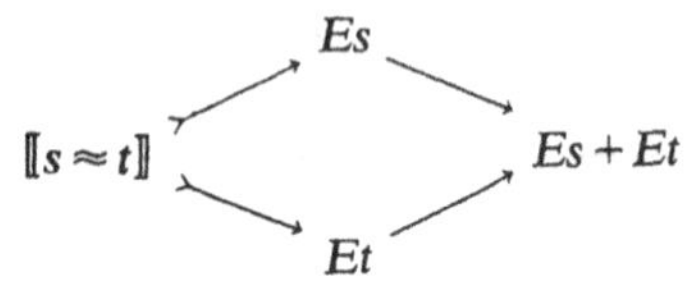

giving

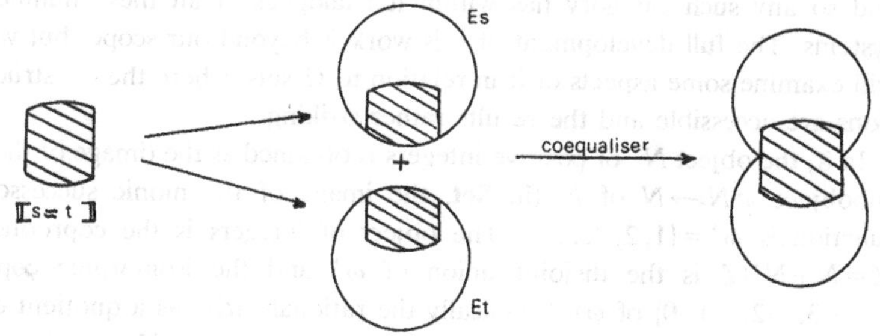

In the case of a general $\Omega_{\mathscr{E}}$-sheaf **A** we take the above diagram for each pair $s, t \in A$ and put them all together by the coproduct construction, yielding a pair of arrows

$$\xrightarrow[s,t \in A]{\lim} [\![s \approx t]\!] \rightrightarrows \xrightarrow[s \in A]{\lim} Es$$

Co-equalising this diagram gives an $\mathscr{E}$-object that has the original **A** as its sheaf of partial elements.

Brockway has developed these ideas to provide functors between $\mathscr{E}$ and **C**$\Omega_{\mathscr{E}}$**-Set** that establish that the latter is equivalent to the full subcategory of $\mathscr{E}$ consisting of the *sub-extensional* $\mathscr{E}$-objects (as defined above). This requires only that $\mathscr{E}$ have all coproducts of subobjects of 1, so that these constructions can be carried out at all. But if $\mathscr{E}$ is also weakly extensional then, by Exercise 70, each $\tilde{a}$ is extensional, so all objects a are sub-extensional, making **C**$\Omega_{\mathscr{E}}$**-Set** equivalent to $\mathscr{E}$ itself.

It can be shown that in order to have coproducts of all subsets of $\Omega_{\mathscr{E}}$ in a topos it suffices to have *arbitrary copowers of* 1, i.e. a coproduct for any set of terminal objects. Thus to put this characterisation in its strongest form (weakest hypothesis), in order to know that $\mathscr{E}$ is the category of sheaves over its subobjects of 1 (global truth-values) it suffices to know that $\mathscr{E}$ has arbitrary copowers of 1 and that each object of partial elements $\tilde{a}$ is weakly extensional.

14.8. Number systems as sheaves

In **Set**, the classical number systems have representations that are built up from the set ω of natural numbers to obtain the integers $\mathbb{Z}$, the rationals $\mathbb{Q}$, the reals $\mathbb{R}$, and finally the complex numbers $\mathbb{C}$. These constructions can be "internalised" to any topos $\mathscr{E}$ that has a natural-numbers object, and so any such category has within it analogues of all these number systems. The full development of this work is beyond our scope, but we will examine some aspects of it in relation to Ω-sets, where the constructions are accessible and the results rather striking.

In $\mathscr{E}$, the object N^+ of positive integers is obtained as the (image of the) subobject $\mathscr{s}: N \rightarrowtail N$ of N (in **Set**, the image of the monic successor function is $\omega^+ = \{1, 2, 3, \ldots\}$). The object of integers is the coproduct $Z = N + N^+$ ($\mathbb{Z}$ is the disjoint union of ω^+ and the isomorphic copy $\{\ldots -3, -2, -1, 0\}$ of ω). Classically the rationals arise as a quotient of $\mathbb{Z} \times \omega^+$, where, thinking of $\langle m, n\rangle$ as the rational m/n, we identify $\langle m, n\rangle$ and $\langle m', n'\rangle$ when $m \cdot n' = m' \cdot n$. Developing this within $\mathscr{E}$ produces the rational-numbers object Q.

In Ω-**Set**, these objects turn out to be the rigid structures $\bar{\omega}$, $\bar{\omega}^+$, $\bar{\mathbb{Z}}$ and $\bar{\mathbb{Q}}$, while in **CΩ-Set** they are the corresponding simple sheaves ω^*, ω^{+*}, $\mathbb{Z}^*$ and $\mathbb{Q}^*$. In particular for **CΘ-Set** we may take them to be the appropriate sheaves of locally constant functions on I.

EXERCISE 1. Define the (rigid) weak successor arrow $\mathscr{s}: \bar{\omega} \to \bar{\omega}$ by

$$\mathscr{s}(\langle m, n\rangle) = \begin{cases} \top & \text{if} \quad n = m+1 \\ \bot & \text{otherwise.} \end{cases}$$

Define the weak "zero arrow" $O: \mathbf{1} \to \bar{\omega}$ in Ω-**Set** analogously. Verify that Ω-**Set**$\models$NNO.

EXERCISE 2. Define $O: \mathbf{1} \to \omega^*$ and $\mathscr{s}: \omega^* \to \omega^*$ in **CΩ-Set** and verify NNO for that category. □

When we come to the reals, the situation is not so clear cut. Classically the two most familiar methods of defining real numbers are as equivalence classes of Cauchy-sequences of rationals, and on the other hand as Dedekind cuts of $\mathbb{Q}$. When carried out in $\mathscr{E}$, these approaches produce an object R_c of "Cauchy-reals" and an object R_d of "Dedekind-reals" *which in general are not isomorphic*! What we do have in general is that $R_c \rightarrowtail R_d$.

Now in Ω-**Set** the construction of Cauchy sequences $N \to Q$ of rationals uses basically the same entities as in **Set** and leads to the same conclusion: R_c is the rigid set $\bar{\mathbb{R}}$. The definition of R_d however, which also proceeds by analogy with the classical case, uses subsets of $\bar{\mathbb{Q}}$, i.e. functions $\mathbb{Q} \to \Omega$, and there may be many more of these than members of $\mathbb{R}$.

In **Set**, a real number $r \in \mathbb{R}$ is uniquely determined by the sets

$$U_r = \{c \in \mathbb{Q}: r < c\}$$
$$L_r = \{c \in \mathbb{Q}: r > c\},$$

called the *upper* and *lower cut* of r. In general an ordered pair $\langle U, L\rangle \in \mathcal{P}(\mathbb{Q}) \times \mathcal{P}(\mathbb{Q})$ of subsets of $\mathbb{Q}$ is called a *Dedekind real number* if it satisfies the sentences

($\delta 1$)	$\exists v \exists w (v \varepsilon \mathbf{U} \wedge w \varepsilon \mathbf{L})$	"non-empty"
($\delta 2$)	$\forall v \sim (v \varepsilon \mathbf{U} \wedge v \varepsilon \mathbf{L})$	"disjoint"
($\delta 3$)	$\forall v (v \varepsilon \mathbf{L} \equiv \exists w (w \varepsilon \mathbf{L} \wedge w > v))$	"open lower cut"
($\delta 4$)	$\forall v (v \varepsilon \mathbf{U} \equiv \exists w (w \varepsilon \mathbf{U} \wedge w < v))$	"open upper cut"
($\delta 5$)	$\forall v \forall w (v > w \supset v \varepsilon \mathbf{U} \vee w \varepsilon \mathbf{L}))$	"close together"

where the symbols $\mathbf{U}$ and $\mathbf{L}$ denote the subsets U and L, ε denotes the standard membership relation, and the variables v and w range over the members of $\mathbb{Q}$. For such a pair $\langle U, L\rangle$ there is one and only one real number $r \in \mathbb{R}$ with $U = U_r$ and $L = L_r$.

Now the conjunction of the sentences ($\delta 1$)–($\delta 5$) may be thought of as a sentence $\delta(\mathbf{r})$, where $\mathbf{r} = \langle \mathbf{U}, \mathbf{L}\rangle$ is an "ordered-pairs symbol" denoting members $r = \langle U, L\rangle$ of $(\mathcal{P}(\mathbb{Q}))^2$. Thus in an axiomatic development of classical set theory the Dedekind real-number system is defined by the Comprehension principle as the set

$$\mathbb{R}_d = \{r: \delta(\mathbf{r}) \text{ is true}\} \subseteq \mathcal{P}(\mathbb{Q})^2.$$

By analogy then, in **Ω-Set** we obtain R_d as the subobject of $\mathcal{P}(\bar{\mathbb{Q}}) \times \mathcal{P}(\bar{\mathbb{Q}})$ defined by $\delta(\mathbf{r})$. According to our earlier discussion of Comprehension, this is the set

$$R_d = \{r: Er \sqsubseteq [\![\delta(\mathbf{r})]\!]\} \subseteq \mathcal{P}(\bar{\mathbb{Q}}) \times \mathcal{P}(\bar{\mathbb{Q}}).$$

Now in **Ω-Set**, power objects, and hence their products, have only global elements (§11.9, Ex. 8), so this simplifies to

$$R_d = \{r: [\![\delta(\mathbf{r})]\!] = \top\}.$$

In order to compute the truth-value $[\![\delta(\mathbf{r})]\!]$ for a given r, we observe that the quanitfied variables v, w in δ range over the rigid set $\bar{\mathbb{Q}}$, i.e. over standard rationals, so that we need to know the "atomic" truth-values $[\![\mathbf{c} < \mathbf{d}]\!]$, $[\![\mathbf{c} > \mathbf{d}]\!]$, $[\![\mathbf{c} \varepsilon \mathbf{U}]\!]$, $[\![\mathbf{c} \varepsilon \mathbf{L}]\!]$, for c, $d \in \mathbb{Q}$. The numerical orderings are interpreted as the standard (rigid) ones

$$[\![\mathbf{c} < \mathbf{d}]\!] = \begin{cases} \top & \text{if } c < d \\ \bot & \text{otherwise,} \end{cases}$$

and similarly for $>$. Moreover $r=\langle U, L\rangle$ is a pair of subsets of $\bar{\mathbb{Q}}$, i.e. strict extensional functions $\mathbb{Q} \to \Omega$, so we put

$$[\![\mathbf{c}\ \varepsilon\ \mathbf{U}]\!] = U(c)$$
$$[\![\mathbf{c}\ \varepsilon\ \mathbf{L}]\!] = L(c)$$

in accordance with our interpretation of subsets developed in §11.9.

Next we notice that since $\bar{\mathbb{Q}}$ is rigid, *every* function $\mathbb{Q} \to \Omega$ is strict and extensional. Thus, putting all of these pieces together with the semantical rules of §11.9 and general lattice-theoretic properties of Ω, it follows that a Dedekind-real number in **Ω-Set** is a pair $r=\langle U, L\rangle$ of functions $\mathbb{Q} \to \Omega$ such that

(δi) $\bigsqcup\{U(c) \sqcap L(d)\colon c, d \in \mathbb{Q}\} = \top$

(δii) $U(c) \sqcap L(c) = \bot$, all $c \in \mathbb{Q}$

(δiii) $L(c) = \bigsqcup\{L(d)\colon d > c\}$, all $c \in \mathbb{Q}$

(δiv) $U(c) = \bigsqcup\{U(d)\colon d < c\}$, all $c \in \mathbb{Q}$

(δv) $U(c) \sqcup L(d) = \top$, all $c > d \in \mathbb{Q}$

(remember $Ec = \top$, all $c \in \bar{\mathbb{Q}}$).

Now in the case $\Omega = \Theta$, we can obtain such a pair by starting with a real-valued function $f\colon I \to \mathbb{R}$ on I and defining

$$\begin{aligned} U_f(c) &= [\![\mathbf{c}\varepsilon\mathbf{U_f}]\!] = \{i\colon c \in U_{f(i)}\} \\ &= \{i\colon f(i) < c\} = f^{-1}(-\infty, c) \end{aligned}$$

and

$$\begin{aligned} L_f(c) &= [\![\mathbf{c}\varepsilon\mathbf{L_f}]\!] = \{i\colon c \in L_{f(i)}\} \\ &= \{i\colon f(i) > c\} = f^{-1}(c, \infty) \end{aligned}$$

where

$$(-\infty, c) = \{x \in \mathbb{R}\colon c > x\}$$

and

$$(c, \infty) = \{x \in \mathbb{R}\colon c < x\}.$$

Now if f is continuous (which means precisely that the inverse images of open sets are open) with respect to the usual topology on $\mathbb{R}$, then $r_f = \langle U_f, L_f\rangle$ will be a pair of functions from $\mathbb{Q}$ to Θ satisfying (δi)–(δv).

Conversely, given a Dedekind-real $r = \langle U_r, L_r\rangle$ in **Θ-Set**, and an element $i \in I$, we put

$$U_i = \{c \in \mathbb{Q}\colon i \in [\![\mathbf{c}\varepsilon\mathbf{U_r}]\!]\}$$

and

$$L_i = \{c \in \mathbb{Q} : i \in [\![c \varepsilon \mathbf{L}_r]\!]\}.$$

Then $\langle U_i, L_i \rangle$ proves to be a classical Dedekind cut in $\mathbb{Q}$, determining a unique real number $r_i \in \mathbb{R}$. Putting $f_r(i) = r_i$ defines a function $f_r : I \to \mathbb{R}$.

EXERCISE 3. (Compulsory). Verify that

(i) r_f satisfies (δi)–(δv).

(ii) $\langle U_i, L_i \rangle$ is a classical Dedekind cut.

(iii) The operations $f \mapsto r_f$ and $r \mapsto f_r$ are mutually inverse.

(iv) f_r is continuous (remember that the sets (c, ∞), $(-\infty, c)$ generate the usual topology on $\mathbb{R}$). □

Thus we have established that in Θ-**Set**, R_d can be represented as the set of all *globally defined continuous real-valued functions on I.* A "Dedekind-real" is a continuous function of the form $I \to \mathbb{R}$, which we envisage as a standard real number "varying continuously" (through the stalks of a bundle) over I. In particular, these "global reals" include, for each $a \in \mathbb{R}$, the totally constant function with output a, and in this way we determine that $R_c \rightarrowtail R_d$.

The analysis just given adapts immediately (in fact reverses) to give a representation of continuous *partial* functions on I. If $V \in \Theta$ then, as defined in §14.2, the set

$$\Theta_V = \{W \in \Theta : W \subseteq V\}$$

is the subspace topology on V, making $\langle V, \Theta_V \rangle$ a topological space in its own right. Notice that, in the terminology of Exercise 47, Θ_V is the **CHA** $\Theta \restriction V$ of all elements "below V" in the **CHA** Θ. We shall also introduce the symbol $\Theta_{\mathbb{R}}$ to denote the open subsets of $\mathbb{R}$ for the usual topology.

Now in saying that $f : V \to \mathbb{R}$ is a *continuous* partial function on I, where I has topology Θ_I, we have meant that for each $W \in \Theta_{\mathbb{R}}$ we have

$$f^{-1}(W) = \{i \in V : f(i) \in W\} \in \Theta_I.$$

But in fact this last condition is equivalent to

$$f^{-1}(W) \in \Theta_V,$$

and so the partial continous $\mathbb{R}$-valued functions on (I, Θ_I) that have domain V are precisely the global continuous $\mathbb{R}$-valued functions on (V, Θ_V). But the latter, by the above construction, correspond precisely to the Dedekind-reals in the topos Θ_V-**Set**!

In other words, if we take a partial continuous $f: I \rightsquigarrow \mathbb{R}$ and relativise to $\Theta \restriction Ef$, within which context f is global, we find that f becomes a Dedekind-real in $\Theta \restriction Ef$-**Set**. And all members of R_d in the latter category arise in this way.

Now let us move to the topos **CΩ-Set** of complete Ω-sets. Here the object R_c is the simple sheaf $\mathbb{R}^*$, so that Cauchy-reals are locally-constant $\mathbb{R}$-valued functions. R_d is again defined by our axioms for Dedekind cuts as the subobject

$$\{r: Er \sqsubseteq [\![\delta(\mathbf{r})]\!]\} \subseteq \mathcal{P}(Q) \times \mathcal{P}(Q).$$

This time Q is the simple sheaf $\mathbb{Q}^*$, and we saw, in analysing models on simple sheaves that we have $\mathbb{Q} \subseteq \mathbb{Q}^*$ as a generating set for $\mathbb{Q}^*$. This means that in determining the truth-value of $\delta(\mathbf{r})$ we can confine the quantifiers to range over the (global) elements of $\mathbb{Q}$ (cf. the equations (†) given earlier).

A typical element r of $\mathcal{P}(\mathbb{Q}^*)^2$ is now a pair $\langle U, L\rangle \in \mathcal{P}(\mathbb{Q}^*) \cdot \mathcal{P}(\mathbb{Q}^*)$ of elements of $\mathcal{P}(Q)$ *with the same extent*, i.e. $Er = EU = EL = e$, say. U itself will be a pair $\langle U_r, e\rangle$, where $U_r: \mathbb{Q}^* \to \Omega$ satisfies

(i) $U_r(a) \sqsubseteq e$,

(ii) $U_r(a \restriction p) = U_r(a) \sqcap p$

all $a \in \mathbb{Q}^*$, $p \in \Omega$.

We put

$$[\![c\varepsilon\mathbf{U}]\!] = U_r(c), \quad \text{all} \quad c \in \mathbb{Q}.$$

Similarly, we have $L = \langle L_r, e\rangle$, and put

$$[\![c\varepsilon\mathbf{L}]\!] = L_r(c).$$

In fact the condition (ii) is immaterial to our purposes, since we observed in Exercise 14.7.56 that it means precisely that U_r is a strict extensional function on $\mathbb{Q}^*$, and in Exercise 14.7.41 that such functions correspond uniquely to strict extensional functions on $\bar{\mathbb{Q}}$. But we know that the latter are simply *all* Ω-valued functions on $\mathbb{Q}$, and anyway we are only interested in the U_r-values of members of the generating set $\mathbb{Q}$. Thus for the present exercise we may simply regard $\mathcal{P}(\mathbb{Q}^*)$ as the set of all pairs $\langle f, e\rangle$, where $f: \mathbb{Q} \to \Omega$ has $f(c) \sqsubseteq e$, all $c \in \mathbb{Q}$.

Having determined the truth-values of atomic sentences, we can establish that the defining condition

$$Er \sqsubseteq [\![\delta(\mathbf{r})]\!]$$

for R_d is equivalent to the satisfaction of the following (remembering that the values of U_r and L_r are "bounded above" by $e = Er$).

(δi_e) $\bigsqcup\{U_r(c) \sqcap L_r(d): c, d \in \mathbb{Q}\} = e$

(δii_e) $U_r(c) \sqcap L_r(d) = \bot$

(δiii_e) $L_r(c) = \bigsqcup\{L_r(d): d > c\}$, all $c \in \mathbb{Q}$

(δiv_e) $U_r(c) = \bigsqcup\{U_r(d): d < c\}$, all $c \in \mathbb{Q}$

(δv_e) $U_r(c) \sqcup L_r(d) = e$, all $c > d \in \mathbb{Q}$

EXERCISE 4. Verify (δi_e)–(δv_e). (The Heyting-algebra involved is a little more complex than for (δi)–(δv).) □

The correspondence between (δi)–(δv) and (δi_e)–(δv_e) is apparent. The only difference is that in (δi) and (δv) we have the unit $\top$ of Ω where in (δi_e) and (δv_e) we have e. But by passing to the **CHA**

$$\Omega \upharpoonright e = \{q : q \sqsubseteq e\} \qquad \text{(Ex. 14.7.47)}$$

we relativise to an algebra in which e *is* the unit. So we see that what (δi_e)–(δv_e) means is precisely that *the pair* $\langle U_r, L_r\rangle$ *is a Dedekind-real in the topos* $\Omega \upharpoonright e$-**Set**.

Conversely, if a pair $\langle U_r, L_r\rangle$ of functions $\mathbb{Q} \to \Omega \upharpoonright e$ is a Dedekind-real in $\Omega \upharpoonright e$-**Set**, then they satisfy (δi_e)–(δv_e), and of course (i), so that $r = \langle\langle U_r, e\rangle, \langle L_r, e\rangle\rangle$ is a Dedekind-real in **C**Ω-**Set**, with $Er = e$.

In summary then we have established that for any given **CHA** Ω, and given $e \in \Omega$,

> *the set of Dedekind-reals in* **C**Ω-**Set** *that have extent e can be identified with the set of* all *Dedekind reals in* $\Omega \upharpoonright e$-**Set**.

Returning to the topological case $\Omega = \Theta_I$ again, an element $r \in R_d$ in **C**Θ_I-**Set** that has $Er = V$, say, is essentially a Dedekind real in $\Theta \upharpoonright V$-**Set**, i.e. a continuous $\mathbb{R}$-valued function defined on all of (V, Θ_V). Thus

> *the Dedekind-reals in* **C**Θ_I-**Set** *with extent V are precisely the continuous* $\mathbb{R}$*-valued partial functions on I that have domain V.*

Putting these "local-reals" together for all $V \in \Theta$ allows us to conclude that

> *in* **C**Θ_I-**Set**, R_d *is the sheaf* $\mathbf{C}_{\mathbb{R}}$ *of all* $\mathbb{R}$*-valued continuous partial functions on I.*

Amongst the continuous $\mathbb{R}$-valued functions on I are the locally constant ones of course, and that observation confirms that we have $R_c \rightarrowtail R_d$ in **CΘ-Set**.

For an arbitrary **CHA**, there is also a representation of R_d available relating to the classical reals $\mathbb{R}$, for which we need the notion of an $\sqcap$-$\bigsqcup$ *map*. This as a function between two **CHA**'s that preserves the operations $\sqcap$ and $\bigsqcup$, i.e. has

$$f(x \sqcap y) = f(x) \sqcap f(y)$$

and

$$f(\bigsqcup B) = \bigsqcup\{f(b) \colon b \in B\}.$$

Such functions are natural objects of study in this generalised topological context, since a topology on a set I is precisely a subset of the lattice $\langle \mathcal{P}(I), \subseteq \rangle$ that is closed under $\sqcap$ and $\bigsqcup$.

EXERCISE 5. Prove that the restriction operator $g_e : \Omega \to \Omega \restriction e$, where $g_e(p) = p \sqcap e$ is a (surjective) $\sqcap$-$\bigsqcup$ map.

EXERCISE 6. Let $f : I \to \mathbb{R}$ be continuous. Define $g_f : \Theta_{\mathbb{R}} \to \Theta_I$ by

$$g_f(W) = f^{-1}(W), \quad \text{all} \quad W \in \Theta_{\mathbb{R}}$$

Show that g_f is an $\sqcap$-$\bigsqcup$ map, and that for any $V \in \Theta_I$,

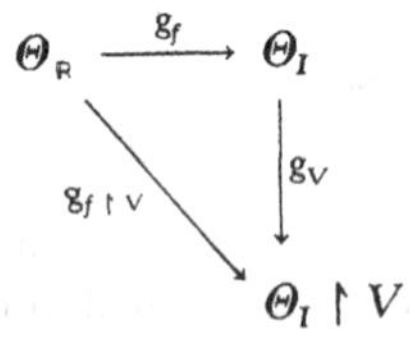

commutes (which may be written $g_{f \restriction V} = g_f \restriction V$). □

In the light of the last exercise, and the earlier representation of R_d in **Θ-Set** we make the following definition:

if $g : \Theta_{\mathbb{R}} \to \Omega$ (Ω an arbitrary **CHA**) is an $\sqcap$-$\bigsqcup$ map ("and-Or" map), put

$$[\![c \varepsilon \mathbf{U}_g]\!] = g(-\infty, c)$$

and

$$[\![\mathbf{c}\varepsilon\mathbf{L_g}]\!] = g(c, \infty).$$

Then the pair $r_g = \langle U_g, L_g \rangle$ satisfies (δi)–(δv) and is a Dedekind-real in **Ω-Set**.

Conversely, given $r = \langle U_r, L_r \rangle$ satisfying (δi)–(δv) we define an $\sqcap$-$\bigsqcup$ map $g_r : \Theta_{\mathbb{R}} \to \Omega$. Intuitively, g_r assigns to each open subset $W \subseteq \mathbb{R}$ the truth-value of "$r \in W$", and the definition uses the fact that each such W is a union of intervals (c, d) with rational end points. For $c,d \in \mathbb{Q}$ we put

$$[\![\mathbf{r}\varepsilon(\mathbf{c}, \mathbf{d})]\!] = [\![\mathbf{c}\varepsilon\mathbf{L_r}]\!] \sqcap [\![\mathbf{d}\varepsilon\mathbf{U_r}]\!]$$

(since, classically,

$$r \in (c, d) \quad \text{iff} \quad r > c \quad \text{and} \quad r < d$$
$$\text{iff} \quad c \in L_r \quad \text{and} \quad d \in U_r).$$

Then the general definition of g_r is

$$g_r(W)(=[\![\mathbf{r}\varepsilon\mathbf{W}]\!]) = \bigsqcup\{[\![\mathbf{r}\varepsilon(\mathbf{c}, \mathbf{d})]\!] : c,d \in \mathbb{Q} \text{ and } (c, d) \subseteq W\}.$$

g_r can be shown to be an $\sqcap$-$\bigsqcup$ map by an argument that uses the compactness of closed intervals $[c, d]$ in $\mathbb{R}$. The constructions $g \mapsto r_g$ and $r \mapsto g_r$ are, as always, mutually inverse, and so we have the presentation of R_d in **Ω-Set** as the set of all $\sqcap$-$\bigsqcup$ maps of the form $\Theta_{\mathbb{R}} \to \Omega$.

EXERCISE 7. You should by now be able to guess what this exercise says.

EXERCISE 8. Prove that in **CΩ-Set**, the members of R_d with extent e are precisely the $\sqcap$-$\bigsqcup$ maps $\Theta_{\mathbb{R}} \to \Omega \restriction e$.

It should be emphasised that it is by no means determinate what object the term "the real-number continuum" denotes in a topos $\mathscr{E}$. One classical property that may fail for R_d is *order-completeness*, i.e. the property

> *every non-empty set of reals with a $\leqslant$-upper-bound has a* least *$\leqslant$-upper-bound.*

A counter-example to this (from Stout [76]) is available in Θ_I-**Set**, where I is the unit interval $[0, 1] \subseteq \mathbb{R}$. The basic idea is conveyed by the following picture.

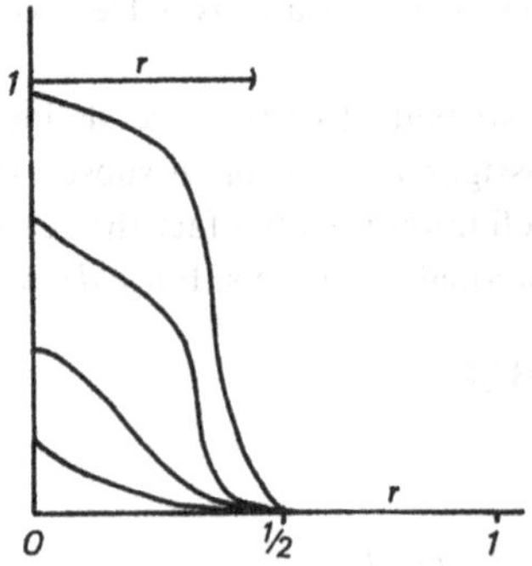

Fig. 14.7.

The ordering $\leqslant$ in R_d (=global continuous functions $I \to \mathbb{R}$) is the Θ-valued relation

$$[\![f \leqslant g]\!] = \{i : f(i) \leqslant g(i)\}^{\circ}.$$

In this model, since everything is global, we will have that "$f \leqslant g$" is true, i.e. $[\![f \leqslant g]\!] = \top$, just in case

(iii) $\qquad f(i) \leqslant g(i) \quad$ for all $\quad i \in I.$

In the picture, $r : I \to \mathbb{R}$ is the characteristic function of $[0, \frac{1}{2})$, i.e.

$$r(i) = \begin{cases} 1 & \text{if} \quad 0 \leqslant i < \frac{1}{2} \\ 0 & \text{if} \quad \frac{1}{2} \leqslant i \leqslant 1. \end{cases}$$

Now consider the set $B \subseteq R_d$ of all continuous functions that are $\leqslant$-below r in the sense of (iii). B has $\leqslant$-upper-bounds (e.g. the function with constant output 1). But it is evident that r can be approximated "arbitrarily closely from below" by members of B, and so the only possible l.u.b. for B is r itself. But r has a "jump discontinuity" at $i = \frac{1}{2}$, and so does not exist at all in R_d.

Exercise 9. Write out a formal sentence $\varphi(\mathbf{B})$ that expresses "B has a $\leqslant$-l.u.b." and show that $[\![\varphi(\mathbf{B})]\!] = I - \{\frac{1}{2}\} \neq \top$. □

It is patent that the counter-example applies to the sheaf R_d in $\mathbf{C}\Theta_I$-**Set**, since what we have been dealing with is just the set of global elements of the latter.

It is possible in fact to "plug the holes" in R_d and expand it to its *order completion* $^*R$ which satisfies the least-upper-bounds principle. A Dedekind-cuts-style definition of an order-complete extension of the rationals within constructive analysis seems to have first been given by John Staples [71]. Christopher Mulvey has modified this approach to show that $^*R$ can be obtained by replacing axiom (δ5) by the sentences

$$(\delta 6) \qquad \forall v(v\varepsilon \mathbf{L} \equiv \exists w(v<w \wedge \forall u(u\varepsilon \mathbf{U} \supset w<u)))$$

$$(\delta 7) \qquad \forall v(v\varepsilon \mathbf{U} \equiv \exists w(v>w \wedge \forall u(u\varepsilon \mathbf{L} \supset w>u))).$$

The object $^*R$ has been used by Charles Burden to derive a version of the Hahn–Banach Theorem in categories of sheaves and more general topoi like **Ω-Set**. A characterisation of those topoi in which the Dedekind-reals, as we have defined them, are order-complete has been given by Peter Johnstone, in a way that graphically illustrates how the underlying logic determines the structure of number systems. This result is that $R_d = {}^*R$ iff the internal logic of the topos validates De Morgan's law

$$\sim(\alpha \wedge \beta) \equiv (\sim\alpha \vee \sim\beta).$$

We will return to the subject of Dedekind cuts and order-completeness below, where the δ-axioms will be put into a rather more perspicacious form.

As for *complex* numbers, they are represented classically as ordered pairs $\langle x, y\rangle = x + iy$ of real numbers, and so given a real-numbers object R, we define an associated complex-numbers object $C = R \times R$. Since in particular a pair of continuous $\mathbb{R}$-valued functions with the same domain can be construed as a single $\mathbb{C}$-valued function on that domain, it transpires that in **$\mathbf{C}\Theta$-Set**, $C_d = R_d \times R_d$ is the sheaf of continuous complex-valued functions on I.

Complex analysis in a topos has been developed by Christiane Rousseau, who derives in $\mathscr{E}$ a version of the Weierstrass Division Theorem for functions of a single complex variable, and establishes that when interpreted in the topos of sheaves over $\mathbb{C}^{n-1}$ the result is equivalent to the classical theorem for functions of n variables. She has also observed that the concept of "holomorphic function" gives rise to an object H that has

$$R_c^2 \rightarrowtail H \rightarrowtail R_d^2,$$

and that H is a suitable "object of complex numbers" upon which to develop complex analysis, although it cannot itself be written as R^2 for any $R \rightarrowtail R_d$.

To close this particular segment about the use of formal logic to construct number systems, we return to the natural numbers once more and record the following version of the Peano Postulates, taken from Fourman [74]:

$$\mathbf{E}(0)$$

$$\forall v \mathbf{E}(\mathfrak{s}(v))$$

$$\forall v \sim (\mathfrak{s}(v) \approx 0)$$

$$\forall v \forall w (\mathfrak{s}(v) \approx \mathfrak{s}(w) \supset v \approx w)$$

$$\forall S((0 \varepsilon S \wedge \forall v (v \varepsilon S \supset \mathfrak{s}(v) \varepsilon S)) \supset \forall v (v \varepsilon S)),$$

Here the symbol S is a second-order variable whose range is the set of all subsets of the set of natural numbers, i.e. the range of S is $\mathcal{P}(N)$.

Exercise 10. Let μ be the conjunction of the above sentences. Show that in **CΩ-Set**, $\langle \mathbf{A}, \mathfrak{s}_{\mathbf{A}}, O_{\mathbf{A}} \rangle$ is a model of μ iff it is isomorphic in a unique way to $\langle \omega^*, \mathfrak{s}_{\omega^*}, O_{\omega^*} \rangle$. □

Ordering the continuum

The standard orderings $\neq$, $<$, $\leqslant$, $>$, $\geqslant$, can be lifted to Θ-valued relations on a Θ-set whose elements are functions $r: I \to \mathbb{R}$ by putting

$$[\![r \neq s]\!] = \{i : r(i) \neq s(i)\}^o$$

$$[\![r < s]\!] = \{i : r(i) < s(i)\}^o$$

$$[\![r \leqslant s]\!] = \{i : r(i) \leqslant s(i)\}^o$$

and similarly for $>$, $\geqslant$ (if both r and s continuous then the interior operator in the definitions of $\neq$, $<$, $>$, are redundant, as the set within the brackets is already open).

We will identify each rational $c \in \mathbb{Q}$ with the constant continuous function $I \to \mathbb{R}$ having c as its sole output. It will simplify matters if we commit a series of abuses of language by using letters c, r, . . . , indiscriminately as informal symbols to refer to elements, as individual constants in formal sentences, and even as variables in such sentences. c, d, b, e, refer always to rationals, and r, s, t, to general reals.

Exercise 11. Show that the above definitions yield

(i) $[\![c < d]\!] = [\![\sim\sim(c < d)]\!]$

(ii) $[\![(c < d]\!] = \top$ or $[\![\sim(c < d)]\!] = \top$

(iii) $[\![(c < d)]\!] = \top$ or $[\![c \approx d]\!] = \top$ or $[\![c > d]\!] = \top$

(iv) $[\![c \leqslant d]\!] = [\![(c < d) \vee (c \approx d)]\!]$. □

The import of these facts is that the structure of the rationals is rigidly determined (constant). This is often expressed by saying that the theory of the ordering of rationals is *decidable*. In other words we may reason with them as if we were working in $\mathbb{Q}$ and applying classical logic (e.g. (ii) above gives the law of excluded middle).

The relation $\neq$ is called *apartness*, a word that comes from intuitionistic mathematics, where it denotes a relation that conveys a positive, constructive sense of difference between elements. (i.e. to be apart is to have been constructively demonstrated to be different). Indeed structures of this kind were first devised by Dana Scott [68, 70] to provide models of intuitionistic theories of the real-number continuum, and in particular to obtain a model that validates Brouwer's theorem on continuity that states: *all* functions $\boldsymbol{R} \to \boldsymbol{R}$ are uniformly continuous on closed intervals.

EXERCISE 12. Show that the following formulae are *true* (i.e. are assigned truth-value $\top = I$).

(i) $\sim(r<s \wedge s<r)$

(ii) $(r<s)\wedge(s<t)\supset(r<t)$

(iii) $((r<s)\vee(s<r))\supset(r\neq s)$

(iv) $(r\leqslant s)\equiv\sim(s<r)$

(v) $(r\approx s)\equiv\sim(r\neq s)$

(vi) $\sim(r\neq r)$

(vii) $(r<s)\equiv((r\leqslant s)\wedge(r\neq s))$

(viii) $(r\approx s)\equiv((r\leqslant s)\wedge(s\leqslant r))$

(ix) $(r\leqslant s\leqslant t)\supset(r\leqslant t)$

(x) $(r\leqslant s<t)\supset(r<t)$

(xi) $(r<s\leqslant t)\supset(r<t)$.

EXERCISE 13. If r and s are continuous, then

$$[\![(r\neq s)\supset((r<s)\vee(s<r))]\!]=\top$$

(converse to (iii) above).

EXERCISE 14. If r, s, t are continuous, then

(i) $[\![(r<s)\supset(r<t)\vee(t<s)]\!]=\top$

(ii) $[\![(r<s)\supset\exists c(r<c<s)]\!]=\top$.

EXERCISE 15. If r is continuous then the following are true

(i) $\exists c,d(c<r<d)$

(ii) $(c<r)\equiv\exists d(c<d<r)$

(iii) $(c\leqslant r)\equiv\forall d(r<d\supset c<d)$

(iv) $(c\leqslant r)\equiv\forall d(d<c\supset d<r)$

(where $\exists c,d$ abbreviates $\exists c\ \exists d$ etc.) □

A word of caution: we are dealing throughout this exposition with *global* elements. For local reals some of these statements must be modified. 12.(vi) is true whether or nor r is global, since invariably $[\![r\not\approx r]\!]=\emptyset$. But (v) and (vi) together yield $[\![r\approx r]\!]=\top$, which is, by definition, false for non-global elements. What we do have in place of (v) is

$$Er\sqcap Es\sqsubseteq[\![(r\approx s)\equiv\sim(r\not\approx s)]\!],$$

the point being that for local elements we need to take account of their extents, and so what is true is the *universal closure*

$$\forall r\forall s((r\approx s)\equiv\sim(r\not\approx s)).$$

EXERCISE 16. Check out the rest of Exercises 12–15 in regard to local elements. □

The principles

$$(r<s)\vee(r\approx s)\vee(r>s)$$

and

$$(r\leqslant s)\equiv((r<s)\vee(r\approx s))$$

both fail in general. A counter example to both is provided by taking the two displayed continuous functions on $I=[0,1]$

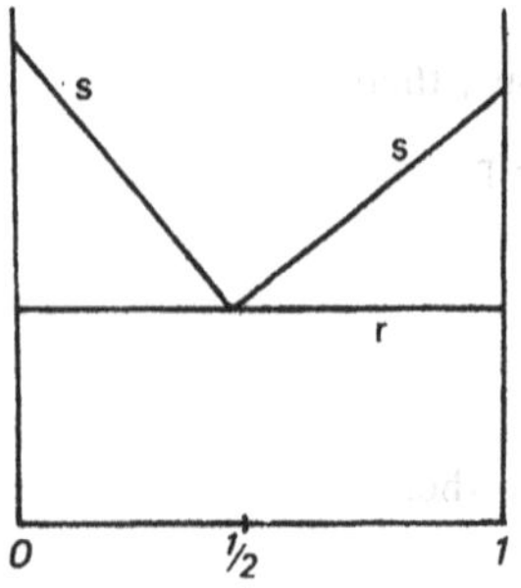

Fig. 14.8.

We have $[\![r \leqslant s]\!] = I$, $[\![r < s]\!] = I - \{\frac{1}{2}\}$, $[\![r \approx s]\!] = \{\frac{1}{2}\}^0 = \emptyset$ and $[\![r > s]\!] = \emptyset$.

Let us now return to the axioms (δ1)–(δ5) that characterise those pairs $\langle U, L\rangle$ of sets of rationals that are the pair $\langle U_r, L_r\rangle$ of cuts for a unique real number r. By invoking the definitions of U_r and L_r we can rewrite these after appropriate conversions as

01: $\exists c, d(c < r < d)$

02: $\forall c \sim (c < r < c)$

03: $\forall c((c < r) \equiv \exists d(c < d < r))$

04: $\forall c((r < c) \equiv \exists d(r < d < c))$

05: $\forall c, d((c < d) \supset (c < r \vee r < d))$.

We have in fact observed in the above Exercises that 01–05 hold for any continuous $r: I \to \mathbb{R}$. As one would expect from our previous work, these axioms characterise the continuous functions.

Exercise 17. Suppose $r: I \to \mathbb{R}$ satisfies 01–05. Prove that for all i,

$$r(i) = \text{g.l.b.}\{c : i \in [\![r < c]\!]\}$$
$$= \text{l.u.b.}\{d : i \in [\![d < r]\!]\}.$$

Hence show that

$$r^{-1}(-\infty, c) = [\![r < c]\!]$$

and

$$r^{-1}(d, \infty) = [\![d < r]\!]$$

and so r is continuous. □

In sum then, if $0(r)$ is the conjunction of 01–05, we find that the subset of the Θ-set $\mathbf{A}_\mathbb{R}$ of *all* $\mathbb{R}$-valued functions on I that is defined by $0(r)$ is precisely the object R_d of Dedekind-reals for **Θ-Set**.

The necessity of continuity for 05 is illustrated by our earlier example that showed R_d was not order-complete for $I = [0, 1]$. With r the characteristic function of $[0, \frac{1}{2})$ we have $[\![\frac{1}{2} < 1]\!] = I$, while $[\![\frac{1}{2} < r]\!] \cup [\![r < 1]\!] = [0, \frac{1}{2}) \cup (\frac{1}{2}, 1] = I - \{\frac{1}{2}\}$. Indeed for any rational d that has $0 \leqslant d \leqslant 1$ we find that

$$[\![(d < r) \vee (r < d)]\!] \neq \top,$$

so that d is not rigidly determined to belong to either $L_r = \{c : c < r\}$ or $U_r = \{c : r < c\}$. There is a big gap between these two cuts.

Notice also in this example that if d is *strictly* between 0 and 1 we have $[\![d \not\approx r]\!] = \top$, and this shows that the continuity assumption in Exercise 13 is essential. This means that in considering non-continuous reals we move away from the intuitionistic theory of the continuum. The latter has the "close together" property 05, and introduces apartness *by definition* as meaning $(r<s) \vee (s<r)$.

Let us now move to a more abstract axiomatic level and explore the order properties that are implicit in the 0-axioms. We shall assume only that we are dealing with an extension R of the rationals that has a binary relation $<$ on it that satisfies 01–04, and when restricted to Q is identical with the classical decidable theory of order for the rationals. The point will be to see what properties of $<$ can be derived using only principles of intuitionistic logic.

Axiom 01 implies that the sets L_r and U_r are not empty. The word *inhabited* is often used here, an intuitionistic term conveying a positive sense of membership. To know that A is inhabited is to have constructively proven $\exists a(a \in A)$, whereas to know that A is non-empty is to have proven only $\sim(A = \emptyset)$ i.e. $\sim\forall a(a \notin A)$, which is equivalent to

$$\sim\sim\exists a(a \in A).$$

03 implies two things about L_r. First it gives

$$(c<d<r) \supset (c<r),$$

which means that L_r is unbounded on the left (anything to the left of a member of L_r is also in L_r). Secondly, from

$$(c<r) \supset \exists d(c<d<r)$$

it implies that L_r has no end-point to the right, and so must be all of Q, or else look like

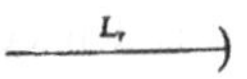

04 gives a dual description of U_r, and 02 implies that the two sets are disjoint, hence neither can be Q and we must have

The linear picture is perhaps misleading, in that we do not have the trichotomy law

$$(c<r) \vee (c \approx r) \vee (r<c).$$

Indeed, we shall take the gap in the line between L_r and U_r to consist only of the points that we know positively to be between the two cuts, i.e.

those that we know to be less than every member of U_r and greater than every member of L_r. The gap then is defined by the conjunction of the sentences

$$\forall d(r<d \supset c<d)$$

and

$$\forall d(d<r \supset d<c).$$

To consider negative membership ($\sim(c \in U_r)$) we introduce the symbol $\leqslant$ by stipulating that $s \leqslant t$ is an abbreviation for $\sim(t<s)$ (cf. Exercise 12 (iv)), and then define

$$L_{\leqslant r}=\{c\colon c \leqslant r\}, \qquad U_{\geqslant r}=\{c\colon r \leqslant c\}.$$

By 02 we get

$$(c<r) \supset (\sim(r<c)),$$

which implies $L_r \subseteq L_{\leqslant r}$. Similarly $U_r \subseteq U_{\geqslant r}$. Since $L_{\leqslant r}$ is defined negatively ($\sim(r<c)$) in terms of the members of U_r, its order properties depend on the axiom 04. From the latter we obtain

$$(d<c) \supset ((r<d) \supset (r<c)),$$

which by contraposition gives

$$(d<c) \supset (\sim(r<c) \supset \sim(r<d)),$$

leading to the transitivity law

$$(d<c \leqslant r) \supset d \leqslant r.$$

This states that $L_{\leqslant r}$ is unbounded on the left. The dual property for $U_{\geqslant r}$ is given by the derivation of

$$(r \leqslant c<d) \supset (r \leqslant d)$$

from 03. Thus far, the picture is

It is easy to see that all members of $L_{\leqslant r}$ are positively to the left of U_r. For if we have $c \leqslant r$, and $r<d$ (hence $d \neq c$) but not $c<d$, we get $d<c$ (Q is decidable), and so $r<c$ by 04. But this contradicts $\sim(r<c)$. Thus we have proven

$$(c \leqslant r) \supset \forall d((r<d) \supset (c<d)).$$

But conversely, if c is less than every member of U_r, assuming $r<c$ would lead to the contradiction $c<c$. Therefore we have $\sim(r<c)$. This establishes

$$(c\leqslant r)\equiv\forall d((r<d)\supset(c<d)),$$

and dually

$$(r\leqslant c)\equiv\forall d((d<r)\supset(d<c)).$$

The picture is now

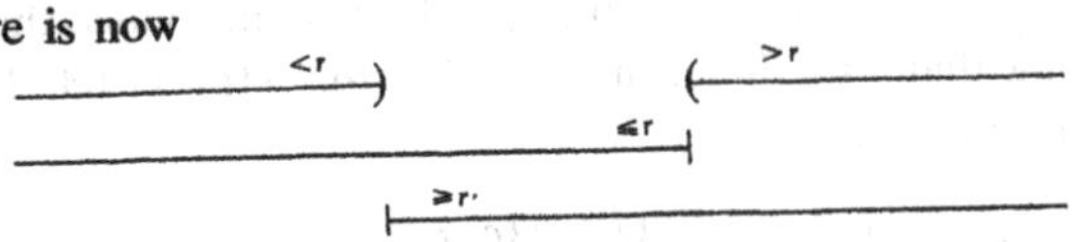

Recall the axioms $\delta 6$ and $\delta 7$ given earlier for the order-complete reals $^*R$. In the present notation $\delta 6$ becomes

06: $\forall c((c<r)\equiv\exists d(c<d\wedge\forall b((r<b)\supset(d<b))))$

and dually for 07. But by the above, $\forall b((r<b)\supset(d<b))$ is equivalent to $d\leqslant r$ so we have

06: $\forall c((c<r)\equiv\exists d(c<d\leqslant r))$

07: $\forall c((r<c)\equiv\exists d(r\leqslant d<c))$.

Then from 06 we obtain

$$(d\leqslant r)\supset((c<d)\supset(c<r)),$$

which means that any member of $L_{\leqslant r}$ has the property that everything to the left of it is in L_r. This has the effect of reducing the gap between L_r and $L_{\leqslant r}$ to (at most) a single point

and so closes the gap in the line. Alternatively by contraposition on the last formula we get

$$(d\leqslant r\leqslant c)\supset(d\leqslant c)$$

which we can interpret as reducing the overlap of $L_{\leqslant r}$ and $U_{\geqslant r}$ to a point.

Exercise 18. Assuming only decidability of Q, prove that 02 is implied by

02′ $\forall c,d(c<r<d\supset c<d)$.

Show that 02 together with either 03 or 04 implies 02′.

EXERCISE 19. Forget the negative definition of $\leqslant$, and assume only that $c \leqslant r$ and $r \leqslant c$ have their positive meanings

$$\forall d(r<d \supset c<d)$$

and

$$\forall d(d<r \supset d<c)$$

respectively. Prove

(i) 02′ is equivalent to each of

$$\forall c(c<r \supset c \leqslant r)$$

and

$$\forall c(r<c \supset r \leqslant c)$$

(ii) 06 and 02′ together imply 03
(iii) 07 and 02′ together imply 04
(iv) Each of 06 and 07 implies 02′.

EXERCISE 20. Show that
(i) 02′, 03 and 05 together give 06.
(ii) 02′, 04 and 05 together give 07. □

The discussion preceding these exercises could be summarised by saying that the axioms for $^*R$ ensure that there is no *positive* gap at the cut determined by a real number. To see how these axioms lead also to order-completeness we continue the derivation of order properties using only principles of intuitionistic logic.

Let us *define* $^*R$ to be the set of all pairs $r = \langle U_r, L_r \rangle$ of subsets of Q that satisfy 01, 06 and 07, where $r<c$ and $c<r$ mean that $c \in U_r$ and $c \in L_r$ respectively, and $c \leqslant r$ and $r \leqslant c$ have their positive meanings as in Exercise 19. It then follows from Exercises 18 and 19 that r satisfies 02, 02′, 03 and 04 and hence we could recover the negative characterisation of $\leqslant$. The advantage of the present approach is of course that we have fewer axioms to deal with. Notice also by Exercise 20 that $R_d \subseteq {}^*R$.

EXERCISE 21. If $r, s \in {}^*R$, show that

(a) $$\forall c(s<c \supset r<c)$$

together with 06 implies

(b) $\quad \forall c(c<r \supset c<s),$

and dually (b) and 07 together give (a). □

We now define $r \leqslant s$, for $r,s \in {}^*R$, to mean that either of the equivalent conditions of the last Exercise obtains.

EXERCISE 22. Prove that $r \leqslant s$ iff every rational upper bound of L_s is an upper bound of L_r, i.e.

$$\forall d(d<s \supset d \leqslant c) \supset \forall d(d<r \supset d \leqslant c).$$

Show that this is equivalent to the statement that every rational lower bound of U_r is a lower bound of U_s. □

If B is a subset of ${}^*R$, we put $B \leqslant s$ to mean that s is an upper bound of B, i.e. that

$$\forall t(t \in B \supset t \leqslant s).$$

Suppose that B is inhabited ($\exists s(s \in B)$) and has an upper bound. To define a least upper bound r_0 for B we have to give its upper and lower cuts. Writing $B<d$, for *rational* d, to mean that

$$\forall t(t \in B \supset t<d)$$

we put

$$r_0<c \quad \text{iff} \quad \exists d(B<d<c)$$
$$c<r_0 \quad \text{iff} \quad \exists d(c<d \wedge \sim(B<d)).$$

The first thing we have to prove about r_0 is that it is in ${}^*R$:–

Verification of 01: The upper cut of r_0 is inhabited: there exists an s with $B \leqslant s$, and by 01 applied to s there is some $d>s$. Then if $t \in B$ we get $t \leqslant s<d$, so $t<d$ by definition of $t \leqslant s$. This establishes $B<d$, so taking any $c>d$ puts $r_0<c$.

Dually, we use the fact that there is a $t \in B$. By 01 again there is a $d<t$. Then $B<d$ would imply $t<d$, in contradiction with 02. Hence $\sim(B<d)$, so any $c<d$ gives $c<r_0$. □

Verification of 06: Suppose $c<r_0$. Then for some $d, c<d$ and $\sim(B<d)$. We prove that $d \leqslant r_0$. For , if $r_0<e$, there is an e_0 with $B<e_0<e$. Now if

$e \leqslant d$, then any $t \in B$ would have $t < e_0 < d$, hence $t < d$ by 04. But that would imply $B < d$, contrary to $\sim(B < d)$. Thus if $r_0 < e$, we must have $d < e$, which means that $d \leqslant r_0$ as required.

Conversely, suppose $c < d \leqslant r_0$ for some d. Take any rational e with $c < e < d$. Then if $B < e$, we have $B < e < d$, implying $r_0 < d$, which in turn by $d \leqslant r_0$ gives the contradiction $d < d$. Thus it must be that $c < e$ and $\sim(B < e)$, giving $c < r_0$. □

Verification of 07: If $r_0 < c$ then $B < d < c$ for some d. To show that $r_0 \leqslant d$, take any $e < r_0$. Then for some e_0, $e < e_0$ and $\sim(B < e_0)$. But then $d \leqslant e$ would imply $B < d < e_0$, leading by 04 to the contradiction $B < e_0$. Hence we must have $e < d$ as required.

Conversely, if $r_0 \leqslant d < c$, take an e with $d < e < c$. If we can show $B < e$, this will yield our desideratum $r_0 < c$. So let $t \in B$. If $e_0 < t$ then by 03 $e_0 < e_1 < t$ for some e_1. But then $B < e_1$ would give the contradiction $t < e_1$. So we have $\sim(B < e_1)$, implying $e_0 < r_0$, which by $r_0 \leqslant d$ gives $e_0 < d$. This establishes $t \leqslant d$. But since $d < e$, 07 for t gives $t < e$ as required. □

The role of r_0 as least upper bound of B is given by the fact that for any $s \in {}^*R$,

$$B \leqslant s \quad \text{iff} \quad r_0 \leqslant s$$

PROOF. Suppose $B \leqslant s$. Then $s < c$ implies $s < d < c$ for some d (04). But then if $t \in B$ we get $t \leqslant s < d$, hence $t < d$. This shows that $B < d < c$, putting $r_0 < c$.

Conversely, assume $r_0 \leqslant s$, and let $t \in B$. Then if $s < c$ we have $r_0 < c$ and so for some $d, B < d < c$. Hence $t < d < c$, giving $t < c$ by 04. This proves $t \leqslant s$. □

EXERCISE 23. Show that 01 and the "close together" axiom 05 yield the property

05′ $$\forall n\, \exists c,d\left(c < r < d \wedge d - c < \frac{1}{n}\right)$$

where n is a symbol for positive integers (assume the classical theory of arithmetic for rationals).

Show that 05′, 03, and 04 together imply 05.

EXERCISE 24. Show that each of 06 and 07 implies that for each integer $n>0$, the set

$$\left\{\langle c, d\rangle : c<r<d \wedge d-c<\frac{1}{n}\right\}$$

is non-empty in the weak sense. That is,

$$05'': \qquad \forall n \sim\sim \exists c,d\left(c<r<d \wedge d-c<\frac{1}{n}\right).$$

EXERCISE 25. Construct examples of $r,s \in {}^*R$ satisfying

(i) 01, 04, 06, 05″, but not 07.

(ii) 01, 03, 07, 05″, but not 06. □

Let us return now to the result stated earlier that ${}^*R = R_d$ if De Morgan's law

$$\sim(\alpha \wedge \beta) \equiv (\sim\alpha \vee \sim\beta)$$

is valid. Since $R_d \subseteq {}^*R$, the various results given earlier imply that it suffices to show that any $r \in {}^*R$ satisfies 05. Given the present set up, the proof is quite brief. For any rational e we have (02)

$$\sim(e<r \wedge r<e)$$

and so De Morgan's law gives

$$\sim(e<r) \vee \sim(r<e)$$

which by the earlier analysis of the consequences of 03 and 04 is equivalent to

$$(r \leqslant e) \vee (e \leqslant r).$$

Now to derive 05, suppose $c<d$. Taking any e with $c<e<d$, we then have either $e \leqslant r$, hence $c<e \leqslant r$ and so $c<r$ by 06, or $r \leqslant e$ and so $r<d$ by 07. □

To date we have studiously avoided reference to the ordering $<$ for general members of ${}^*R$. In the classical case, the density of $\mathbb{Q}$ in $\mathbb{R}$ guarantees that $r<s$ just in case

$$\exists c(r<c<s),$$

and this last condition is used to define $<$ on R_d in general (cf. Exercise 28 below). It will not do however for ${}^*R$, and the procedure adopted there is to invoke the arithmetical structure of Q to put

$$r<s \quad \text{iff} \quad \exists d(d>0 \wedge r+d \leqslant s)$$

where $r+d$ is defined by specifying its upper and lower cuts by the (obvious?) clauses

$$c<r+d \quad \text{iff} \quad c-d<r$$

$$r+d<c \quad \text{iff} \quad r<c-d.$$

EXERCISE 26. Show that $r+d \in {}^*R$ if $r \in {}^*R$.

EXERCISE 27. Give examples of $r,s \in {}^*R$ with $r+d \leqslant s$ for some $d>0$ but for which $r<c<d$ fails for all rationals c.

EXERCISE 28. Use the above density condition to define $<$ on R_d in **CΩ-Set**, giving $\leqslant$ by either its positive or negative description, and $r \neq s$ by $(r<s) \vee (s<r)$. Show that in the topological case $\Omega = \Theta$ these lead to the Θ-valued relations with which we began this subsection.

EXERCISE 29. Let X be any of ω $\mathbb{Z}$, $\mathbb{Q}$, $\mathbb{R}$. Show that the standard rigid relations $\neq$, $<$ etc. on $\bar{\mathbf{X}}$ lift to the simple sheaf $\mathbf{X}^*$ to satisfy

$$[\![s \neq t]\!] = \bigsqcup \{[\![s \approx a]\!] \sqcap [\![t \approx b]\!] : a \neq b \in X\}$$

$$[\![s < t]\!] = \bigsqcup \{[\![s \approx a]\!] \sqcap [\![t \approx b]\!] : a < b \in X\}$$

etc. Investigate the properties of these Ω-valued order relations. □

Points

An important feature of the study of number systems in **CΩ-Set** is a generalisation of the notion of a *point* in a topological space. A given $i \in I$ determines the function $f_i : \Theta_I \to 2$ that has

$$f_i(V) = \begin{cases} 1 & \text{if} \quad i \in V \\ 0 & \text{if} \quad i \notin V, \end{cases}$$

f_i is an $\sqcap$-$\bigsqcup$ map, and so in general a *point of a* **CHA** Ω is defined to be an $\sqcap$-$\bigsqcup$ map of the form $\Omega \to 2$. The abstraction from Θ to Ω is a movement to view a generalised "space" as being made up of its *parts* (open sets) rather than its points (Lawvere [76]). In some classical topological spaces every point $\Theta_I \to 2$ is of the form f_i for some $i \in I$. Such spaces are called *sober* (all points are in focus). These include all Hausdorff spaces, so in particular $\mathbb{R}$ is sober.

There is a categorial duality between the category of sober spaces with continuous functions and the category of **CHA**'s with $\sqcap$-$\sqcup$ maps that gives a natural isomorphism between the former and **CHA**'s of the type Θ. For an arbitrary **CHA** Ω, let $\beta(\Omega)$ be the set of all points $\Omega \to 2$ of Ω. (A sober space is one having $I \cong \beta(\Theta_I)$). For $p \in \Omega$ let

$$\begin{aligned} V_p &= \{f \in \beta(\Omega) : f(p) = 1\} \\ &= \{f : \text{“}f \in p\text{”}\} \end{aligned}$$

be the set of points f that “belong to p”.

Exercise 30. Prove that

$$V_p \cap V_q = V_{p \sqcap q} \quad \text{all} \quad p, q \in \Omega$$

$$\bigcup_{p \in C} V_p = V_{\sqcup C} \quad \text{all} \quad C \subseteq \Omega$$ □

This result implies that the collection

$$\Theta_\Omega = \{V_p : p \in \Omega\}$$

is closed under finite intersections and arbitrary unions, i.e. is a topology on $\beta(\Omega)$.

Exercise 31. Given a point $g : \Theta_\Omega \to 2$ define $f_g : \Omega \to 2$ by

$$f_g(p) = g(V_p).$$

Show that $f_g \in \beta(\Omega)$ and

$$g(V_p) = 1 \quad \text{iff} \quad f_g \in V_p.$$ □

Thus we see that $\langle \beta(\Omega), \Theta_\Omega \rangle$ is a sober topological space. Moreover the previous exercise implies that the function $p \mapsto V_p$ is a surjective $\sqcap$-$\sqcup$ map (**CHA**-homomorphism). Ω will be said to *have enough points* if it satisfies

$$V_p = V_q \quad \text{only if} \quad p = q, \quad \text{all} \quad p, q \in \Omega.$$

This is an extensionality principle, asserting that if two parts have the same points in them (”$f \in p$ iff $f \in q$”) then they are equal. Obviously a topology Θ has enough points, and conversely the condition implies that the function $p \mapsto V_p$ is an isomorphism between Ω and Θ_Ω.

Thus the *spatial* **CHA**'s (the topologies) are precisely those that have enough points. At the other extreme there exist **CHA**'s that are quite pointless, and the associated sheaf categories of such structures can exhibit extremely pathological behaviour. For instance, Michael Fourman and Martin Hyland have constructed topoi along these lines that fail to satisfy such standard mathematical "facts" as "every complex number has a square root", "the equation $x^3+ax+b=0$ has a real solution for $a,b \in \mathbb{R}$", and "the unit interval [0, 1] is compact".

An account of the construction of number systems in topoi is given in Chapter 6 of Johnstone [77] and further details of the order and topological properties of R_d may be found in Stout [76] (cf. also Mulvey [74] for spatial sheaves). The major source of information in this area is the Proceedings of the Durham Conference on sheaf theory (Fourman, Mulvey, Scott [79]) which contains details of all the results that have been mentioned in this section without references.

CHAPTER 15

ADJOINTNESS AND QUANTIFIERS

"... *adjoints occur almost everywhere in many branches of Mathematics. ... a systematic use of all these adjunctions illuminates and clarifies these subjects.*"

Saunders Maclane

The isolation and explication of the notion of *adjointness* is perhaps the most profound contribution that category theory has made to the history of general mathematical ideas. In this final chapter we shall look at the nature of this concept, and demonstrate its ubiquity with a range of illustrations that encompass almost all concepts that we have discussed. We shall then see how it underlies the proof of the Fundamental Theorem of Topoi, and finally examine its role in a particular analysis of quantifiers in a topos.

15.1. Adjunctions

The basic data for an *adjoint situation*, or *adjunction*, comprise two categories, $\mathscr{C}$ and $\mathscr{D}$, and functors F and G between them

$$\mathscr{C} \underset{G}{\overset{F}{\rightleftarrows}} \mathscr{D}$$

in each direction, enabling an interchange of their objects and arrows. Given $\mathscr{C}$-object a and $\mathscr{D}$-object b we obtain

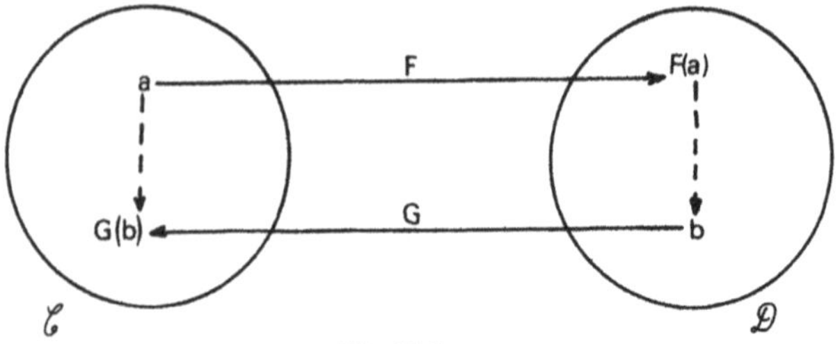

Fig. 15.1.

$G(b)$ in $\mathscr{C}$ and $F(a)$ in $\mathscr{D}$. Adjointness occurs when there is an exact correspondence of arrows between these objects in the directions indicated by the broken arrows in the picture, so that any passage from a to $G(b)$ in $\mathscr{C}$ is matched uniquely by a passage from $F(a)$ to b in $\mathscr{D}$. In other words we require for each a and b as shown, a bijection

(1) $$\theta_{ab}: \mathscr{D}(F(a), b) \cong \mathscr{C}(a, G(b))$$

between the set of $\mathscr{D}$-arrows of the form $F(a) \to b$ and the $\mathscr{C}$-arrows of the form $a \to G(b)$. Moreover the assignment of bijections θ_{ab} is to be "natural in a and b", which means that it preserves categorial structure as a and b vary. Specifically, the assignment to the pair $\langle a, b\rangle$ of the "hom-set" $\mathscr{D}(F(a), b)$ generates a functor from the product category $\mathscr{C}^{op} \times \mathscr{D}$ to **Set** (why $\mathscr{C}^{op}$ and not $\mathscr{C}$? Examine the details), while the assignment of $\mathscr{C}(a, G(b))$ establishes another such functor. We require that the θ_{ab}'s form the components of a natural transformation θ between these two functors.

When such a θ exists we call the triple $\langle F, G, \theta\rangle$ an *adjunction* from $\mathscr{C}$ to $\mathscr{D}$. F is then said to be *left adjoint* to G, denoted $F \dashv G$, while G is *right adjoint* to F, $G \vdash F$. The relationship between F and G given by θ as in (1) is presented schematically by

$$\frac{a \to G(b)}{F(a) \to b}$$

which displays the "left-right" distinction.

An adjoint situation is expressible in terms of the behaviour of special arrows associated with each object of $\mathscr{C}$ and $\mathscr{D}$:

Let a be a particular $\mathscr{C}$-object, and put $b = F(a)$ in (1). Applying θ (i.e. the appropriate component) to the identity arrow on $F(a)$ we obtain the $\mathscr{C}$-arrow $\eta_a = \theta(1_{F(a)})$, to be called the *unit* of a. Then for any b in $\mathscr{D}$, we know that any $g: a \to G(b)$ corresponds to a unique $f: F(a) \to b$ under θ_{ab}. Using the naturality of θ in a and b we find in fact that η_a enjoys a certain co-universal property, namely that to any such g there is exactly one such f such that

(2)

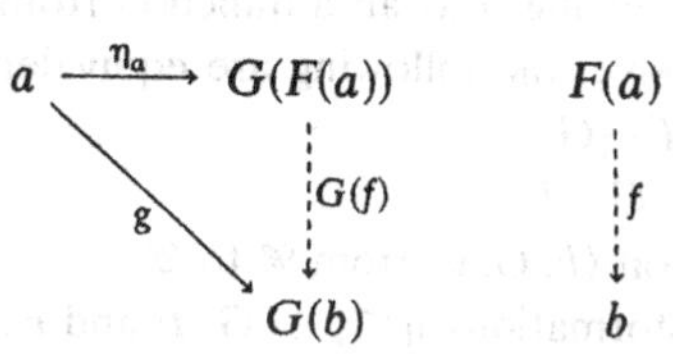

commutes. Indeed $g = \theta_{ab}(f)$, and so

(3) $$\theta_{ab}(f) = G(f) \circ \eta_a.$$

Naturality of θ implies also that

$$\begin{array}{ccc} a & \xrightarrow{\eta_a} & G(F(a)) \\ {\scriptstyle k}\downarrow & & \downarrow{\scriptstyle G(F(k))} \\ a' & \xrightarrow{\eta_{a'}} & G(F(a')) \end{array}$$

commutes for all such $\mathscr{C}$-arrows k, and so the η_a's form the components of a natural transformation $\eta : 1_{\mathscr{C}} \dot{\to} G \circ F$, called the *unit* of the adjunction.

Dually, let b be a particular $\mathscr{D}$ object and put $a = G(b)$ in (1). If τ is the inverse to the natural isomorphism θ ($\tau_{ab} = \theta_{ab}^{-1}$), apply τ to the identity arrow on $G(b)$ to get the *co-unit* $\varepsilon_b = \tau(1_{G(b)})$ of b. ε_b has the universal property that to any $\mathscr{D}$-arrow $f : F(a) \to b$ there is exactly one $\mathscr{C}$-arrow $g : a \to G(b)$ such that

(4)

$$\begin{array}{ccccc} F(G(b)) & \xrightarrow{\varepsilon_b} & b & \qquad & G(b) \\ {\scriptstyle F(g)}\uparrow & \nearrow{\scriptstyle f} & & & \uparrow{\scriptstyle g} \\ F(a) & & & & a \end{array}$$

commutes. Since $f = \tau_{ab}(g)$, we get

(5) $$\tau_{ab}(g) = \varepsilon_b \circ F(g),$$

while the ε_b's form the components of the natural transformation $\varepsilon : F \circ G \dot{\to} 1_{\mathscr{D}}$, the *co-unit* of the adjunction.

On the other hand, given natural transformations η and ε of this form, we could define natural transformations θ and τ by specifying their components by equations (3) and (5). If the universal properties of diagrams (2) and (4) hold, then θ_{ab} and τ_{ab} would be inverse to each other, hence each a bijection, giving θ as an adjunction from $\mathscr{C}$ to $\mathscr{D}$.

Thus, given F and G as above, the following are equivalent:

(a) F is left adjoint to G, $F \dashv G$

(b) G is right adjoint to F, $G \vdash F$

(c) there exists an adjunction $\langle F, G, \theta \rangle$ from $\mathscr{C}$ to $\mathscr{D}$

(d) there exist natural transformations $\eta : 1_{\mathscr{C}} \dot{\to} G \circ F$ and $\varepsilon : F \circ G \dot{\to} 1_{\mathscr{D}}$

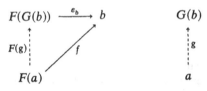

whose components have the universal properties of diagrams (2) and (4) above.

Diagrams (2) and (4) are instances of a more general phenomenon. Suppose that $G:\mathscr{D}\to\mathscr{C}$ is a functor and a an object of $\mathscr{C}$. Then a pair $\langle b,\eta\rangle$ consisting of a $\mathscr{D}$-object b and a $\mathscr{C}$-arrow $\eta:a\to G(b)$ is called *free over a with respect to G* iff for any $\mathscr{C}$-arrow of the form $g:a\to G(c)$ there is exactly one $\mathscr{D}$-arrow $f:b\to c$ such that

(6)

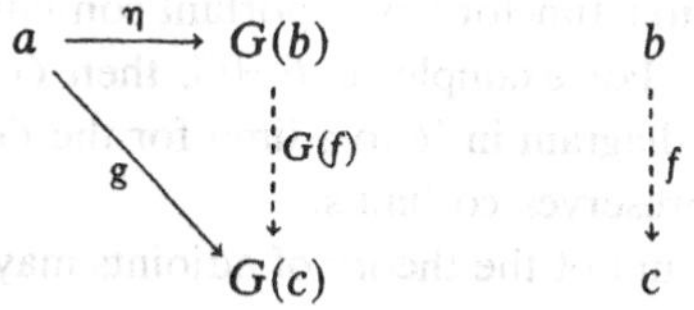

commutes.

Such a pair $\langle b,\eta\rangle$ is also known as a *universal arrow* from a to G.

Thus, whenever $F\dashv G$, the pair $\langle F(a),\eta_a\rangle$ is free over a with respect to G.

Dually, given a functor $F:\mathscr{C}\to\mathscr{D}$ and a $\mathscr{D}$-object b, a pair $\langle a,\varepsilon\rangle$, comprising a $\mathscr{C}$-object a and an arrow $\varepsilon:F(a)\to b$ is called *co-free over b with respect to F* if to each pair $\langle c,f\rangle$ comprising a $\mathscr{C}$-object c and an arrow $f:F(c)\to b$ there is a unique $g:c\to a$ in $\mathscr{C}$ such that

(7)

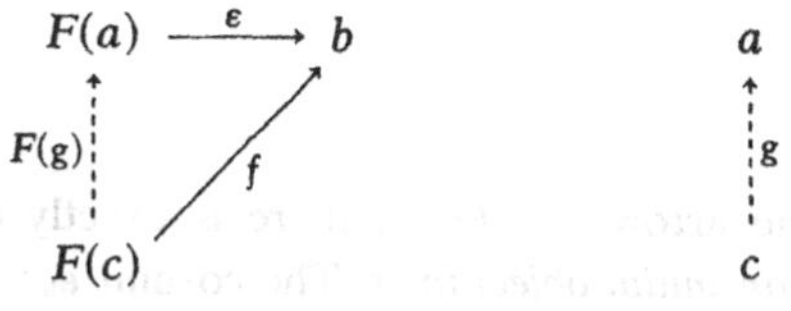

commutes. Such a pair is also called a universal arrow from F to b.

Exercise 1. Describe a right adjoint G to F in terms of pairs that are co-free over $\mathscr{D}$-objects with respect to F.

Exercise 2. Suppose that $\langle b,\eta\rangle$ is a universal arrow from a to $G:\mathscr{D}\to\mathscr{C}$. Show that the arrow $\eta:a\to G(b)$ is an initial object in the category $a\downarrow F$ whose objects are $\mathscr{C}$-arrows of the form $f:a\to G(c)$ and whose arrows are $\mathscr{D}$-arrows $g:c\to d$ such that

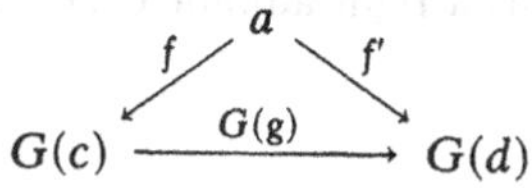

commutes.

EXERCISE 3. Dualise Exercise 2.

EXERCISE 4. Suppose that for every $\mathscr{C}$-object a, there is a universal arrow from a to $G:\mathscr{D}\to\mathscr{C}$. Construct a functor $F:\mathscr{C}\to\mathscr{D}$ such that $F\dashv G$.

EXERCISE 5. Dualise Exercise 4. □

The existence of an adjoint to a functor has important consequences for the properties of that functor. For example, if $F\dashv G$, then G preserves limits (i.e. maps the limit of a diagram in $\mathscr{D}$ to a limit for the G-image of that diagram in $\mathscr{C}$), while F preserves co-limits.

The details of this brief account of the theory of adjoints may be found in any standard text on category theory.

15.2. Some adjoint situations

Initial objects

Let $\mathscr{C}=\mathbf{1}$ be the category with one object, say 0, and G the unique functor $\mathscr{D}\to\mathbf{1}$. If $F:\mathbf{1}\to\mathscr{D}$ is left adjoint to G then for any b in $\mathscr{D}$,

$$\frac{0\to G(b)}{F(0)\to b}$$

since there is exactly one arrow $0\to G(b)$, there is exactly one arrow $F(0)\to b$. Hence $F(0)$ *is an initial object in* $\mathscr{D}$. The co-unit $\varepsilon_b:F(G(b))\to b$ is the unique arrow $F(0)\to b$.

EXERCISE 1. Show that $\mathscr{D}$ has a terminal object iff the functor $!:\mathscr{D}\to\mathbf{1}$ has a right adjoint. □

Products

Let $\Delta:\mathscr{C}\to\mathscr{C}\times\mathscr{C}$ be the diagonal functor taking a to $\langle a,a\rangle$ and $f:a\to b$ to $\langle f,f\rangle:\langle a,a\rangle\to\langle b,b\rangle$. Suppose Δ has a right adjoint $G:\mathscr{C}\times\mathscr{C}\to\mathscr{C}$.

Then we have

$$\frac{c\to G(x)}{\langle c,c\rangle\to x}$$

where c is in $\mathscr{C}$ and $x = \langle a, b\rangle$ is in $\mathscr{C}\times\mathscr{C}$. The co-unit $\varepsilon_x : \Delta(G(x)) \to \langle a, b\rangle$ is a pair of $\mathscr{C}$-arrows $p : G(x) \to a$ and $q : G(x) \to b$. Using the "co-freeness" property of ε_x, for any arrows $f : c \to a$, $g : c \to b$, there is a unique $h : c \to G(x)$ such that

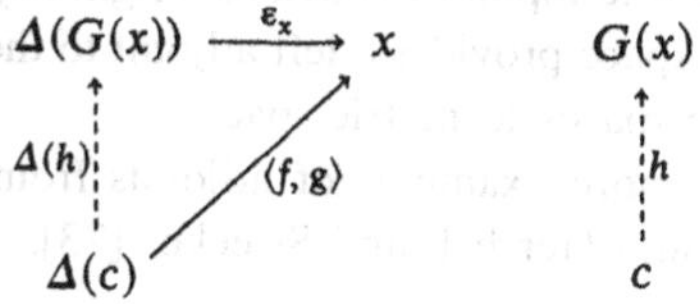

and hence

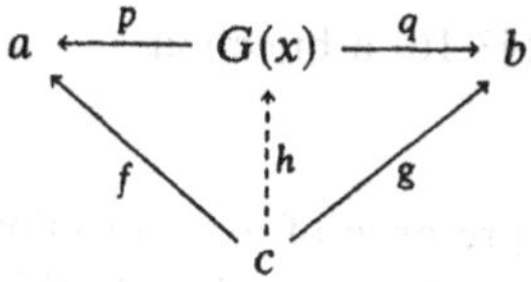

commutes. Thus $G(x)$ *is a product* $a\times b$ *of* a *and* b *with* ε_x *as the pair of associated projections.* We have the adjunction

$$\frac{c \to a\times b}{c \to a,\ c \to b}$$

The unit $\eta_c : c \to c\times c$ is the diagonal product arrow $\langle 1_c, 1_c\rangle$.

EXERCISE 2. Show that $\mathscr{C}$ has co-products iff $\Delta : \mathscr{C} \to \mathscr{C}\times\mathscr{C}$ has a left adjoint. □

It can be shown that the limit and co-limit of any type of diagram in a category $\mathscr{C}$ arise, when they exist, from right and left adjoints of a "diagonal" functor $\mathscr{C} \to \mathscr{C}^J$, where J is a canonical category having the "shape" of that diagram (for products, J is the discrete category $\{0, 1\}$). The unit for the left adjoint is the universal co-cone, the co-unit for the right adjoint is the universal cone.

Topology and algebra

There are many significant constructions that arise as adjoints to forgetful functors. The forgetful functor $U : \mathbf{Grp} \to \mathbf{Set}$ from groups to sets has as left adjoint the functor assigning to each set the free group generated by that set (here "free" has precisely the above meaning associated with units of an adjunction).

The construction of the field of quotients of an integral domain gives a functor left adjoint to the forgetful functor from the category of fields to the category of integral domains.

The specification of the discrete topology on a set gives a left adjoint to $U:\mathbf{Top} \to \mathbf{Set}$, while the indiscrete topology provides a right adjoint to U.

The completion of a metric space provides a left adjoint to the forgetful functor from complete metric spaces to metric spaces.

The reader will find many more examples of adjoints from topology and algebra in Maclane [71] and Herrlich and Strecker [73].

Exponentiation

If $\mathscr{C}$ has exponentials, then there is (§3.16) a bijection

$$\mathscr{C}(c \times a, b) \cong \mathscr{C}(c, b^a)$$

for all objects a, b, c, indicating the presence of an adjunction.

Let $F:\mathscr{C} \to \mathscr{C}$ be the right product functor $- \times a$ of §9.1 taking any c to $c \times a$. Then F has as right adjoint the functor $(\ \ )^a:\mathscr{C} \to \mathscr{C}$ taking any b to b^a and any arrow $f:c \to b$ to $f^a:c^a \to b^a$, which is the exponential *adjoint* to the composite $f \circ ev':c^a \times a \to c \to b$, i.e. the unique arrow for which

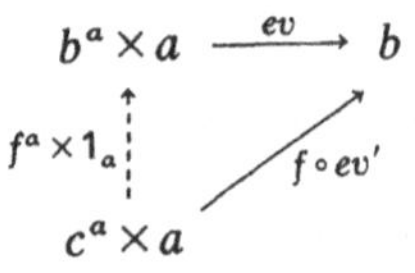

commutes.

The co-unit $\varepsilon_b:F(b^a) \to b$ is precisely the evaluation arrow $ev:b^a \times a \to b$, and its "co-freeness" property yields the axiom of exponentials given in §3.16.

The adjoint situation is

$$\frac{c \to b^a}{c \times a \to b}.$$

Thus *$\mathscr{C}$ has exponentials iff the functor $- \times a$ has a right adjoint for each $\mathscr{C}$-object a.*

Relative pseudo-complements

This is a special case of exponentials (cf. §8.3). In any r.p.c. lattice the condition

$$c \sqcap a \sqsubseteq b \quad \text{iff} \quad c \sqsubseteq a \Rightarrow b$$

yields the adjunction

$$\frac{c \to (a \Rightarrow b)}{c \sqcap a \to b}$$

A lattice is r.p.c. iff the functor $-\sqcap a$ *taking* c *to* $c \sqcap a$ *has a right adjoint for each* a.

Natural numbers objects (cf. Lawvere [69])

A $\mathscr{C}$ arrow f is *endo* (from "endomorphism") iff $\operatorname{dom} f = \operatorname{cod} f$, i.e. f has the form $f : a \to a$, or $a\circlearrowleft^{f}$. The category $\mathscr{C}\circlearrowleft$ has as objects the $\mathscr{C}$-endo's, with an arrow from $a\circlearrowleft^{f}$ to $b\circlearrowleft^{g}$ being a $\mathscr{C}$-arrow $h : a \to b$ such that

$$\begin{array}{ccc} a & \xrightarrow{h} & b \\ {\scriptstyle f}\downarrow & & \downarrow{\scriptstyle g} \\ a & \xrightarrow{h} & b \end{array}$$

i.e.

$$a\circlearrowleft^{f} \xrightarrow{h} b\circlearrowleft^{g}$$

commutes. Let $G : \mathscr{C}\circlearrowleft \to \mathscr{C}$ be the forgetful functor taking $f : a \to a$ to its domain a.

Suppose G has a left adjoint

$$\frac{a \to G(b)}{F(a) \to b},$$

and let the endo $F(1)$ be denoted $N\circlearrowleft^{\sigma}$ and the unit $\eta_1 : 1 \to G(F(1))$ denoted $O : 1 \to N$. The notation is of course intentional:

the freeness of $(F(1), \eta_1)$ over 1

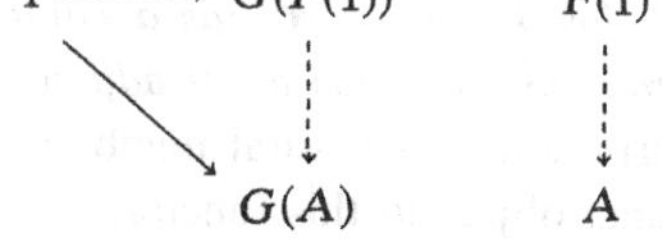

means that for any endo $A : a \xrightarrow{f} a$ and any $\mathscr{C}$-arrow $x : 1 \to a = G(A)$ there is a unique arrow $h : F(1) \to A$, i.e.

$$N\circlearrowleft^{\sigma} \xrightarrow{h} a\circlearrowleft^{f},$$

such that

$$\begin{array}{ccc} 1 & \xrightarrow{O} & N \\ & {}_{x}\searrow & \downarrow{h} \\ & & a \end{array}$$

and hence

$$\begin{array}{ccc} & & N \circlearrowleft^{\sigma} \\ & {}^{O}\nearrow & \\ 1 & & \downarrow{h} \\ & {}_{x}\searrow & \\ & & a \circlearrowleft^{f} \end{array}$$

commutes. Thus $(F(1), \eta_1)$ is a natural numbers object.

Conversely, if $\mathscr{C} \models \text{NNO}$, define $F: \mathscr{C} \to \mathscr{C}^{\circlearrowleft}$ to take a to the endo

$$a \times N \xrightarrow{1_a \times \sigma} a \times N$$

and $f: a \to b$ to $f \times 1_N$.

Then by the theorem 13.2.1 of Freyd, if $\mathscr{C}$ has exponentials, then for any endo $f: b \to b$ and any arrow $h_0: a \to b$ there is a unique h for which

$$\begin{array}{ccc} & & a \times N \circlearrowleft^{1_a \times \sigma} \\ & {}^{\langle 1_a, O_a \rangle}\nearrow & \\ a & & \downarrow{h} \\ & {}_{h_0}\searrow & \\ & & b \circlearrowleft^{f} \end{array}$$

commutes. We have the situation

$$\frac{a \xrightarrow{h_0} G(b \circlearrowleft^{f})}{F(a) \xrightarrow{h} b \circlearrowleft^{f}}$$

indicating that $F \dashv G$. The unit η_1 now becomes $\langle 1_1, O \rangle : 1 \to 1 \times N$ from which we recover $O: 1 \to N$ under the natural isomorphism $1 \times N \cong N$.

Altogether then, *a cartesian closed category $\mathscr{C}$ has a natural numbers object iff the forgetful functor from $\mathscr{C}^{\circlearrowleft}$ to $\mathscr{C}$ has a left adjoint.*

We also obtain the characterisation of a natural numbers object as a universal arrow from the terminal object to this functor.

Adjoints in posets

Let $(P, \sqsubseteq)$ and $(Q, \sqsubseteq)$ be posets. A functor from P to Q is a function $f: P \to Q$ that is monotonic, i.e. has

$$p \sqsubseteq q \quad \text{only if} \quad f(p) \sqsubseteq f(q).$$

Then $g:Q\to P$ will be right adjoint to f,

$$\frac{p\to g(r)}{f(p)\to r}$$

iff for all $p\in P$ and $r\in Q$,

$$p\sqsubseteq g(r) \quad \text{iff} \quad f(p)\sqsubseteq r.$$

On the other hand g will be left adjoint to f,

$$\frac{r\to f(p)}{g(r)\to p},$$

when

$$g(r)\sqsubseteq p \quad \text{iff} \quad r\sqsubseteq f(p).$$

For example, given a function $f:A\to B$, and subsets $X\subseteq A$, $Y\subseteq B$, we have

$$X\subseteq f^{-1}(Y) \quad \textit{iff} \quad f(X)\subseteq Y$$

and so the functor $f^{-1}:\mathscr{P}(B)\to\mathscr{P}(A)$ taking $Y\subseteq B$ to $f^{-1}(Y)$ is right adjoint to the functor $\mathscr{P}(f):\mathscr{P}(A)\to\mathscr{P}(B)$ of §9.1, that takes $X\subseteq A$ to $f(X)\subseteq B$.

As well as having a left adjoint, $\mathscr{P}(f)\dashv f^{-1}$, f^{-1} has a right adjoint

$$f^{+}:\mathscr{P}(A)\to\mathscr{P}(B)$$

given by $f^{+}(X)=\{y\in B: f^{-1}\{y\}\subseteq X\}$ where $f^{-1}\{y\}=\{x: f(x)=y\}$ is the inverse image of $\{y\}$. That $f^{-1}\dashv f^{+}$ follows from the fact that

$$f^{-1}(Y)\subseteq X \quad \text{iff} \quad Y\subseteq f^{+}(X).$$

Subobject classifier

The display (Lawvere [72])

$$\frac{d\to\Omega}{?\rightarrowtail d}$$

where $?\rightarrowtail d$ denotes an arbitrary subobject of d, indicates that the Ω-axiom expresses a property related to adjointness.

The functor $\text{Sub}:\mathscr{C}\to\textbf{Set}$ described in §9.1, Example 11, assigns to each object d the collection of subobjects of d, and to each arrow $f:c\to d$ the function $\text{Sub}(f):\text{Sub}(d)\to\text{Sub}(c)$ that takes each subobject of d to its pullback along f. As it stands, Sub is contravariant. However, by switching to the opposite category of $\mathscr{C}$ we can regard Sub as a covariant functor

$$\text{Sub}: \mathscr{C}^{op} \to \mathbf{Set}.$$

Now in the case $\mathscr{C} = \mathscr{E}$ (a topos) the arrow $true: 1 \to \Omega$ is a subobject of Ω and so corresponds to a function $\eta: 1 = \{0\} \to \text{Sub}(\Omega)$.

Now consider the diagram

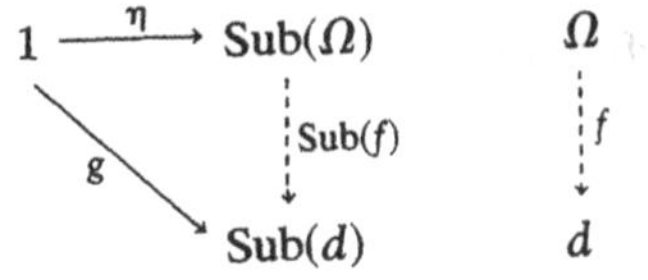

A function g as shown picks out a subobject $g_0: a \rightarrowtail d$ of d, for which we have a character χ_{g_0}, and pullback

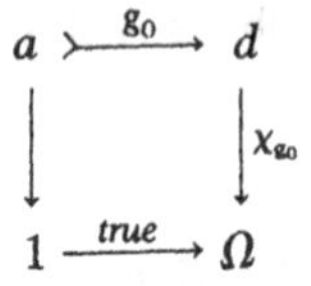

in $\mathscr{E}$. Thus $f = (\chi_{g_0})^{op}$ is an $\mathscr{E}^{op}$ arrow from Ω to d. Then $\text{Sub}(f)(= \text{Sub}(\chi_{g_0})$ originally) takes *true* to its pullback along χ_{g_0}, i.e. to the subobject g_0, and so the above triangle commutes. But by the uniqueness of the character of g_0, the only arrow along which *true* pulls back to give g_0 is χ_{g_0} and so the only $\mathscr{E}^{op}$ arrow for which the triangle commutes is $f = (\chi_{g_0})^{op}$.

Thus the pair $\langle \Omega, \eta \rangle$, i.e. $\langle \Omega, true: 1 \to \Omega \rangle$ is free over 1 with respect to Sub.

Conversely the freeness of $\langle \Omega, \eta \rangle$ implies that $\eta(0)$ classifies subobjects and so we can say that *any category* $\mathscr{C}$ *with pullbacks has a subobject classifier iff there exists a universal arrow from* 1 *to* $\text{Sub}: \mathscr{C}^{op} \to \mathbf{Set}$. (cf. Herrlich and Strecker [73], Theorem 30.14).

EXERCISE 1. Let $Rel(\text{-}, a): \mathscr{C} \to \mathbf{Set}$ take each $\mathscr{C}$-object b to the collection of all $\mathscr{C}$-arrows of the form $R \rightarrowtail b \times a$ ("relations" from b to a). For any $f: c \to b$, $Rel(f, a)$ maps $R \rightarrowtail b \times a$ to its pullback along $f \times 1_a$, so that $Rel(\text{-}, a)$ as defined is contravariant. Show that $\mathscr{C}$ (finitely complete) has power objects iff for each $\mathscr{C}$-object a, there is a universal arrow from 1 to

$$Rel(\text{-}, a): \mathscr{C}^{op} \to \mathbf{Set}.$$

EXERCISE 2. Can you characterise the partial arrow classifier $\eta_a: a \rightarrowtail \tilde{a}$ in terms of universal arrows? □

Notice that the Ω-axiom states that

$$\mathrm{Sub}(d) \cong \mathscr{C}(d, \Omega) \cong \mathscr{C}^{op}(\Omega, d)$$

and similarly we have

$$Rel(b, a) \cong \mathscr{C}(b, \Omega^a) \cong \mathscr{C}^{op}(\Omega^a, b),$$

and so the covariant $\mathscr{C}^{op} \to$ **Set** versions of Sub and $Rel(\text{-}, a)$ are naturally isomorphic to "hom-functors" of the form $\mathscr{C}(d, \text{-})$ (§9.1, Example (7)). In general a **Set**-valued functor isomorphic to a hom-functor is called *representable.* Representable functors are always characterised by their possession of objects free over 1 in **Set**.

15.3. The fundamental theorem

Let $\mathscr{C}$ be a category with pullbacks, and $f: a \to b$ a $\mathscr{C}$-arrow. Then f induces a "pulling-back" functor $f^*: \mathscr{C}\downarrow b \to \mathscr{C}\downarrow a$ which generalises the $f^{-1}: \mathscr{P}(B) \to \mathscr{P}(A)$ example of the last section. f^* acts as in the diagram

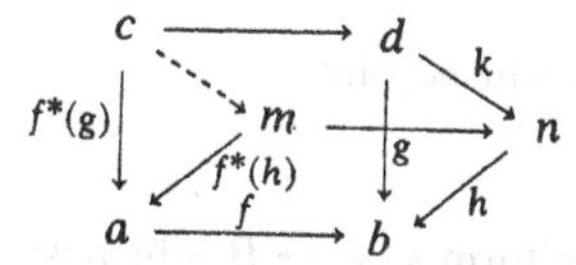

k is a $\mathscr{C}\downarrow b$ arrow from g to h, $f^*(g)$ and $f^*(h)$ are the pullbacks of g and h along f, yielding a unique arrow $c \to m$ as shown which we take as $f^*(k): f^*(g) \to f^*(h)$.

The "composing with f" functor

$$\Sigma_f: \mathscr{C}\downarrow a \to \mathscr{C}\downarrow b$$

takes object $g: c \to a$ to $f \circ g: c \to b$, and arrow

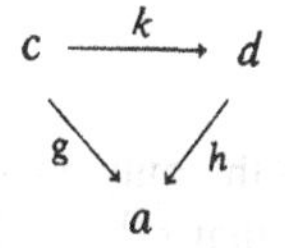

to

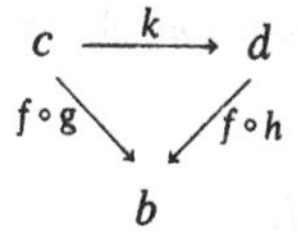

Now an arrow k

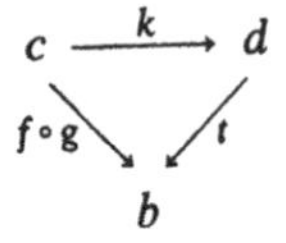

from $\Sigma_f(g)$ to $t: b \to d$ in $\mathscr{C} \downarrow b$ corresponds to a unique $\mathscr{C} \downarrow a$ arrow k'

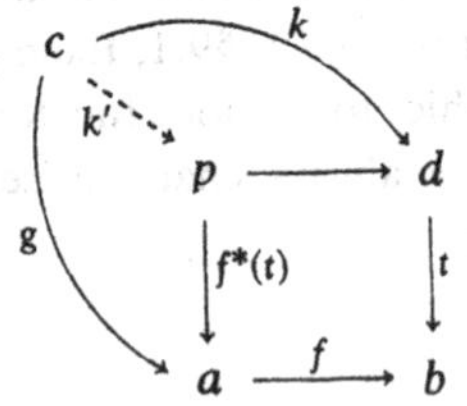

from g to $f^*(t)$, by the universal property of the pullback $f^*(t)$, and so we have the adjunction

$$\frac{g \to f^*(t)}{\Sigma_f(g) \to t}$$

showing $\Sigma_f \dashv f^*$.

For set functions, f^* also has a right adjoint

$$\Pi_f : \mathbf{Set} \downarrow A \to \mathbf{Set} \downarrow B.$$

Given $g: X \to A$, then $\Pi_f(g)$ has the form $k: Z \to B$, which we regard as a bundle over B. Thinking likewise of g, the stalk in Z over $b \in B$, i.e. $k^{-1}\{b\}$, is the set of all local sections of g defined on $f^{-1}\{b\} \subseteq A$.

Formally Z is the set of all pairs (b, h) such that h is a function with domain $f^{-1}\{b\}$, such that

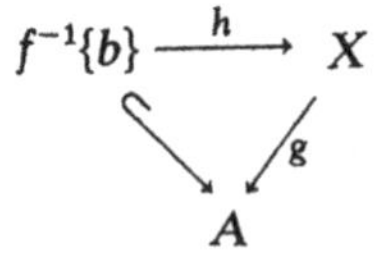

commutes. k is the projection to B.

Notice that if g is an inclusion $g: X \hookrightarrow A$ then the only possible section h as above is the inclusion $f^{-1}\{b\} \hookrightarrow X$, provided that $f^{-1}\{b\} \subseteq X$. Thus the stalk over b in Z is empty if not $f^{-1}\{b\} \subseteq X$, and has one element otherwise. Thus k can be identified with the inclusion of the set

$$\{b : f^{-1}\{b\} \subseteq X\} = f^+(X)$$

into B, and so the functor f^+ is a special case of Π_f.

Now given arrows $g: X \to A$ and $h: Y \to B$, consider

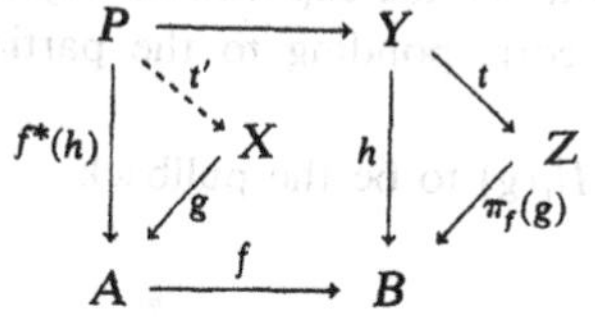

t is an arrow from h to $\Pi_f(g)$ in **Set** $\downarrow B$. $f^*(h)$, the pullback of h along f, is the projection to A of the set

$$P = \{\langle a, y\rangle : f(a) = h(y)\}.$$

Thus if $\langle a, y\rangle \in P$, y lies in the stalk over $f(a)$ in B, and so $t(y)$ is in the stalk over $f(a)$ of $\Pi_f(g)$. Thus $t(y)$ is a section s of g over $f^{-1}\{f(a)\}$, which includes a. Put $t'(\langle a, y\rangle) = s(a)$. Then t' is an arrow from $f^*(h)$ to g in **Set** $\downarrow A$.

In this way we establish a correspondence

$$\frac{h \xrightarrow{t} \Pi_f(g)}{f^*(h) \xrightarrow{t'} g}$$

which gives $f^* \dashv \Pi_f$.

EXERCISE. How do you go from $t': f^*(h) \to g$ to $t: h \to \Pi_f(g)$? □

The full statement of the *Fundamental Theorem of Topoi* (Freyd [72], Theorem 2.31) is this:

For any topos $\mathscr{E}$, and $\mathscr{E}$-object b, the comma category $\mathscr{E} \downarrow b$ is a topos, and for any arrow $f: a \to b$ the pulling-back functor $f^*: \mathscr{E} \downarrow b \to \mathscr{E} \downarrow a$ has both a left adjoint Σ_f and a right adjoint Π_f.

The existence of Σ_f requires only pullbacks. The construction of Π_f is special to topoi, in that it uses partial arrow classifiers (N.B. local sections are partial functions).

Given $f: a \to b$, let k be the unique arrow for which

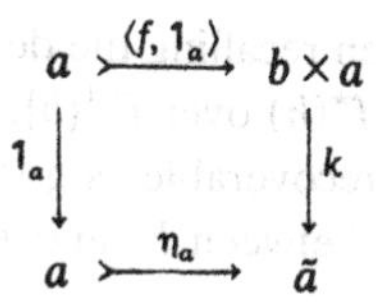

is a pullback, where now η_a denotes the partial arrow classifier of §11.8 (why is $\langle f, 1_a\rangle$ monic?). Let $h: b \to \tilde{a}^a$ be the exponential adjoint to k. (In **Set** h takes $b \in B$ to the arrow corresponding to the partial function $f^{-1}\{b\} \hookrightarrow A$ from A to A).

Then, for any $g: c \to a$, define $\Pi_f(g)$ to be the pullback

$$\begin{array}{ccc} \pi_f(c) & \longrightarrow & \tilde{c}^a \\ {\scriptstyle \pi_f(g)}\downarrow & & \downarrow{\scriptstyle \tilde{g}^a} \\ b & \xrightarrow{h} & \tilde{a}^a \end{array}$$

where $\tilde{g}$ is the unique arrow making the pullback

$$\begin{array}{ccc} c & \xrightarrow{\eta_c} & \tilde{c} \\ {\scriptstyle g}\downarrow & & \downarrow{\scriptstyle \tilde{g}} \\ a & \xrightarrow{\eta_a} & \tilde{a} \end{array}$$

and $\tilde{g}^a$ is the image of $\tilde{g}$ under the functor $(\ )^a: \mathscr{E} \to \mathscr{E}$.

It is left to the reader to show how this reflects the definition of Π_f in **Set**.

The Π_f functor is also used to verify that $\mathscr{E}\downarrow b$ has exponentials. Illustrating with **Set** once more, given objects $f: A \to B$ and $h: Y \to B$ in **Set** $\downarrow B$, their exponential is of the form $h^f: E \to B$. According to the description in Chapter 4, the stalk in E over b consists of all pairs $\langle b, t\rangle$ where $t: f^{-1}\{b\} \to Y$ makes

$$\begin{array}{ccc} f^{-1}\{b\} & \xrightarrow{t} & Y \\ & {\scriptstyle f}\searrow \quad \swarrow{\scriptstyle h} & \\ & B & \end{array}$$

commute. Now if we form the pullback $f^*(h)$

$$\begin{array}{ccccc} & & P & \xrightarrow{g} & Y \\ & {\scriptstyle t'}\nearrow & {\scriptstyle f^*(h)}\downarrow & & \downarrow{\scriptstyle h} \\ f^{-1}\{b\} & \hookrightarrow & A & \xrightarrow{f} & B \end{array}$$

and define t' as shown by $t'(a) = \langle a, t(a)\rangle$, then recalling the description of P given earlier, t' is seen to be a section of $f^*(h)$ over $f^{-1}\{b\}$, i.e. a germ at b of the bundle $\Pi_f(f^*(h))$. Moreover t is recoverable as $g \circ t'$, giving an exact correspondence, and an isomorphism, between h^f and $\Pi_f(f^*(h))$ in **Set**.

In $\mathscr{C}\downarrow b$ then, given $f: a \to b$ and $h: c \to b$ we find that $\Pi_f(f^*(h))$ serves as the exponential h^f. We can alternatively express this in the language of adjointness, since the product functor

$$-\times f : \mathscr{C}\downarrow b \to \mathscr{C}\downarrow b$$

is the composite functor of

$$\mathscr{C}\downarrow b \xrightarrow{f^*} \mathscr{C}\downarrow a \xrightarrow{\Sigma_f} \mathscr{C}\downarrow b.$$

This is because the product of h and f, $h\times f$, in $\mathscr{C}\downarrow b$ is their pullback

$$\begin{array}{ccc} p & \xrightarrow{f^*(h)} & a \\ \downarrow & & \downarrow f \\ c & \xrightarrow{h} & b \end{array}$$

$f\circ f^*(h) = \Sigma_f(f^*(h))$ in $\mathscr{C}$.

But each of f^* and Σ_f has a right adjoint, Π_f and f^* respectively, and their composite $\Pi_f \circ f^*$ provides a right adjoint to $-\times f$.

The details of the Fundamental Theorem may be found in Freyd [72] or Kock and Wraith [71].

15.4. Quantifiers

If $\mathfrak{A} = \langle A, \ldots\rangle$ is a first-order model, then a formula $\varphi(v_1, v_2)$ of index 2 determines the subset

$$X = \{\langle x, y\rangle : \mathfrak{A} \models \varphi[x, y]\}$$

of A^2. The formulae $\exists v_2\varphi$ and $\forall v_2\varphi$, being of index 1, determine in a corresponding fashion subsets of A. These can be defined in terms of X as

$$\exists_p(X) = \{x : \text{for some } y, \langle x, y\rangle \in X\}$$

$$\forall_p(X) = \{x : \text{for all } y, \langle x, y\rangle \in X\}.$$

The "p" refers to the first projection from A^2 to A, having $p(\langle x, y\rangle) = x$. $\exists_p(X)$ is in fact precisely the image $p(X)$ of X under p, and so we know that for any $X \subseteq A^2$ and $Y \subseteq A$,

$$X \subseteq p^{-1}(Y) \quad \textit{iff} \quad \exists_p(X) \subseteq Y,$$

i.e. $\exists_p : \mathcal{P}(A^2) \to \mathcal{P}(A)$ is left adjoint to the functor $p^{-1} : \mathcal{P}(A) \to \mathcal{P}(A^2)$ analysed in §15.2.

Since, for any $x \in A$, $p^{-1}\{x\} = \{\langle x, y\rangle : y \in A\}$ we see that

$$\forall_p(X) = \{x : p^{-1}\{x\} \subseteq X\} = p^+(X)$$

(cf. §15.2) and so we have

$$p^{-1}(Y) \subseteq X \quad \textit{iff} \quad Y \subseteq \forall_p(X)$$

and altogether $\exists_p \dashv p^{-1} \dashv \forall_p$.

In general then, for any $f : A \to B$, the left adjoint $\mathcal{P}(f)$ to $f^{-1} : \mathcal{P}(B) \to \mathcal{P}(A)$ will be renamed $\exists_f$, and the right adjoint f^+ will be denoted $\forall_f$. The link with the quantifiers is made explicit by the characterisations of $\exists_f(X) = f(X)$ and $\forall_f(X) = f^+(X)$ as

$$\exists_f(X) = \{y : \exists x(x \in X \text{ and } f(x) = y)\}$$

$$\forall_f(X) = \{y : \forall x(f(x) = y \text{ implies } x \in X)\}.$$

Moving now to a general topos $\mathscr{E}$, an arrow $f : a \to b$ induces a functor

$$f^{-1} : \mathrm{Sub}(b) \to \mathrm{Sub}(a)$$

that takes a subobject of b to its pullback along f (pullbacks preserve monics).

A left adjoint $\exists_f : \mathrm{Sub}(a) \to \mathrm{Sub}(b)$ to f^{-1} is obtained by defining $\exists_f(g)$, for $g : c \rightarrowtail a$ to be the image arrow $\mathrm{im}(f \circ g)$ of $f \circ g$, so we have

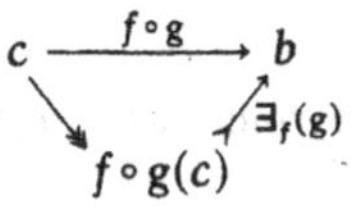

Using the fact that the image of an arrow is the smallest subobject through which it factors (Theorem 5.2.1) the reader may attempt the

EXERCISE 1. Show that $g \subseteq h$ implies $\exists_f(g) \subseteq \exists_f(h)$, i.e. $\exists_f$ is a functor.

EXERCISE 2. Analyse the adjoint situation

$$\frac{g \to f^{-1}(h)}{\exists_f(g) \to h}$$

for $g : c \rightarrowtail a$ and $h : d \rightarrowtail b$, that gives $\exists_f \dashv f^{-1}$. □

The right adjoint $\forall_f : \mathrm{Sub}(a) \to \mathrm{Sub}(b)$ to f^{-1} is obtained from the functor $\Pi_f : \mathscr{E} \downarrow a \to \mathscr{E} \downarrow b$ (recall that in **Set**, f^+ is a special case of Π_f).

$\forall_f$ assigns to the subobject $g: c \rightarrowtail a$ the subobject $\Pi_f(g)$. Strictly speaking, g, as a *subobject*, is an equivalence class of arrows. Any ambiguity however is taken care of by

EXERCISE 3. If $g \subseteq h$ then $\forall_f(g) \subseteq \forall_f(h)$, and so

EXERCISE 4. If $g \simeq h$ then $\forall_f(g) \simeq \forall_f(h)$. □

The adjunction

$$\frac{h \to \forall_f(g)}{f^{-1}(h) \to g}$$

showing $f^{-1} \dashv \forall_f$, derives from the fact that $f^* \dashv \Pi_f$.

By selecting a particular monic to represent each subobject, we obtain a functor $i_a: \mathrm{Sub}(a) \to \mathscr{C} \downarrow a$. In the opposite direction, $\sigma_a: \mathscr{C} \downarrow a \to \mathrm{Sub}(a)$ takes $g: c \to a$ to $\sigma_a(g) = \mathrm{im}\, g: g(c) \rightarrowtail a$, and an $\mathscr{C} \downarrow a$ arrow

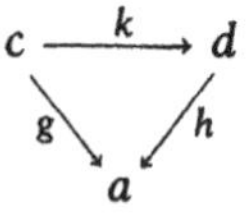

to the inclusion $\sigma_a(k)$,

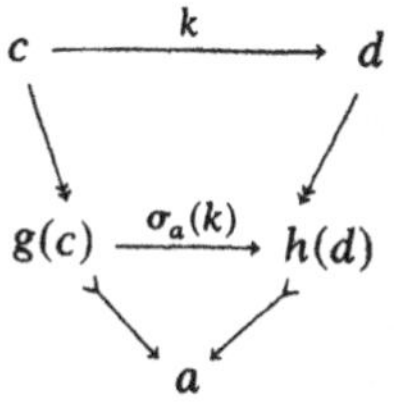

which exists because im g is the smallest subobject through which g factors. For the same reason, given $g: c \to a$ and $h: d \to a$ we have that

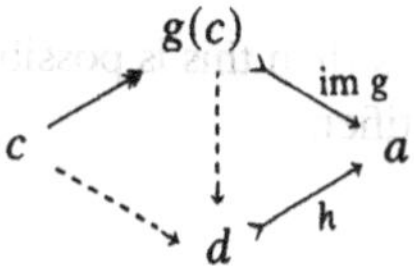

im g factors through h, i.e. $\sigma_a(g) \subseteq h$, precisely when g factors through h, i.e. precisely when there is an arrow

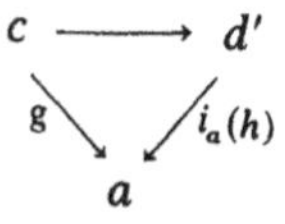

in $\mathscr{E}\downarrow a$. So we have the situation

$$\frac{g \to i_a(h)}{\sigma_a(g) \to h}$$

making σ_a left adjoint to i_a.

Putting the work of these last two sections together we have the "doctrinal diagram" of Kock and Wraith [71] for the arrow $f: a \to b$

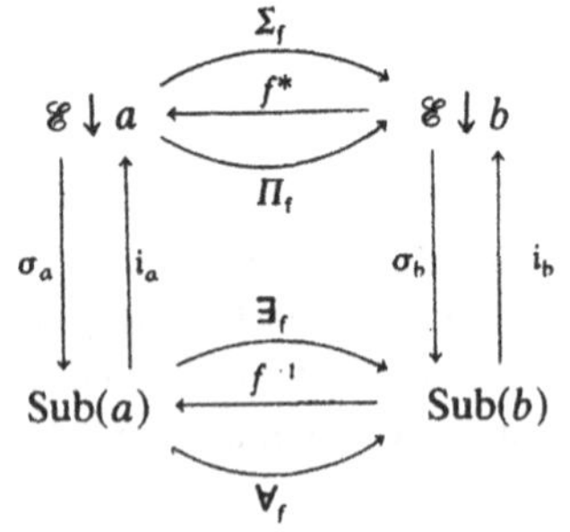

with

$$\exists_f \dashv f^{-1} \dashv \forall_f$$
$$\Sigma_f \dashv f^* \dashv \Pi_f$$
$$\sigma \dashv i$$

EXERCISE 5. Show that

$$\exists_f \circ \sigma_a = \sigma_b \circ \Sigma_f$$
$$i_b \circ \forall_f = \Pi_f \circ i_a$$
$$i_a \circ f^{-1} = f^* \circ i_b$$
$$f^{-1} \circ \sigma_b = \sigma_a \circ f^*$$

□

An even more general analysis of quantifiers than this is possible. Given a relation $R \subseteq A \times B$ in **Set** we define quantifiers

$$\exists_R : \mathscr{P}(A) \to \mathscr{P}(B)$$
$$\forall_R : \mathscr{P}(A) \to \mathscr{P}(B)$$

"along R" by

$$\exists_R(X) = \{y : \exists x(x \in X \text{ and } xRy)\}$$
$$\forall_R(X) = \{y : \forall x(xRy \text{ implies } x \in X)\}$$

Given an arrow $r: R \rightarrowtail a \times b$ in a topos there are actual arrows

$$\forall_r : \Omega^a \to \Omega^b$$

$$\exists_r : \Omega^a \to \Omega^b$$

which correspond *internally* to $\exists_R$ and $\forall_R$ in **Set**. Constructions for these are given by Street [74] and they are further analysed by Brockway [76]. In particular, for a given $f: a \to b$, applying these constructions to the relation

$$\langle 1_a, f \rangle : a \rightarrowtail a \times b$$

(the "graph" of f) yields arrows of the form $\Omega^a \to \Omega^b$ which are internal counterparts to the functors $\forall_f$ and $\exists_f$.

Specialising further by taking f to be the arrow $!: a \to 1$, we obtain arrows $\Omega^a \to \Omega^1$, which under the isomorphism $\Omega^1 \cong \Omega$ become the quantifier arrows

$$\forall_a : \Omega^a \to \Omega \qquad \exists_a : \Omega^a \to \Omega$$

used for the semantics in a topos of Chapter 11.

The functors $\forall_f$ and $\exists_f$, in the case that f is a projection, are used in the topos semantics developed by the Montréal school. More information about their basic properties is given by Reyes [74].

CHAPTER 16

LOGICAL GEOMETRY

> *"It is a very interesting fact that notions originally developed for the purposes of (abstract) algebraic geometry turn out to be intimately related to logic and model theory. Compared to other existing versions of algebraic logic, categorical logic has the distinction of being concerned with objects that appear in mathematical practice."*
>
> Michael Makkai and Gonzalo Reyes

The theory discussed in this book emerges from an interaction between sheaf theory and logic, and for the most part we have dwelt on the impact of the former on the conceptual framework of the latter. In this chapter we will consider ways in which the application has gone in the opposite direction. Specifically, we study the concept of a *geometric morphism*, a certain kind of functor between topoi that plays a central role in the work of the Grothendieck school (Artin et al. [SGA 4]). In their book *First Order Categorical Logic*, henceforth referred to as [MR], Makkai and Reyes have shown that this notion of morphism can be reformulated in logical terms, and that some important theorems of Pierre Deligne and Michael Barr about the existence of geometric morphisms can be derived by model-theoretic constructions. The essence of their approach is to associate a *theory* (set of axioms) with a given site, and identify functors defined on the site with models of this theory. Conversely, from a certain type of theory a site can be built by a method that adds a new dimension of mathematical significance to the well-known Lindenbaum-algebra construction (cf. §6.5).

These developments will be described below, with our main aim being to account for the fact that Deligne's theorem is actually equivalent to a version of the classical Gödel Completeness Theorem for **Set**-based semantics of first-order logic.

Model theory is both an independent science and an effective technique for studying mathematical structures and explaining their properties. The second of these aspects is perhaps most closely associated with the name of Abraham Robinson, who summarised it in the title of one of his papers – "Model theory as a framework for algebra" (Robinson [73]). Since Robinson was at Yale during the latter part of his career, this attitude has become known as "eastern" model theory, by contrast with the "western" approach, associated with Alfred Tarski at Berkeley, which focuses on the general properties of formal languages and their semantics. The work of Makkai and Reyes is in the eastern style, and constitutes "model theory as a framework for topos theory". One of the goals of this chapter is to exhibit their proof of Deligne's Theorem as a major exercise in applied mathematical logic.

The distinction between western and eastern model theory is given a syntactic expression by H. J. Keisler (cf. page 48 of Barwise [77]): the former is concerned with *all* formulae of first-order languages, while the latter emphasises *universal-existential* formulae – those of the form $\forall v_1 \ldots \forall v_n \exists w_1 \ldots \exists w_m \varphi$, with φ quantifier-free – since these suffice to axiomatise the main structures of classical algebra. We will see that the logic of geometric morphisms has an analogous syntactic emphasis, in that it is expressed by formulae, called "geometric" or "coherent", that have the form $\varphi \supset \psi$, where φ and ψ have no occurrence of the symbols $\sim, \supset, \forall$.

16.1. Preservation and reflection

In order to define geometric morphisms we need some general information about how the behaviour of a functor affects the existence of limits and colimits in its domain and codomain. So, let $F: \mathscr{C} \to \mathscr{D}$ be a functor between categories $\mathscr{C}$ and $\mathscr{D}$. F is said to *preserve monics* if, for any $\mathscr{C}$-arrow f, if f is monic in $\mathscr{C}$, then $F(f)$ is monic in $\mathscr{D}$. On the other hand F *reflects* monics if, for any $\mathscr{C}$-arrow f, if $F(f)$ is monic in $\mathscr{D}$ then f is monic in $\mathscr{C}$. Replacing "monic" by "epic" or "iso" here defines what it is for F to preserve or reflect these latter types of arrows.

Similarly, F is said to *preserve equalisers* if whenever e equalises f and g in $\mathscr{C}$, then $F(e)$ equalises $F(f)$ and $F(g)$ in $\mathscr{D}$. If the converse of this last implication always holds, then F is said to *reflect equalisers*. To describe reflection and preservation of categorial constructs in general, it is helpful to invoke the language of diagrams and limits of §3.11. Let D be a diagram in $\mathscr{C}$, comprising $\mathscr{C}$-objects $d_i, d_j, \ldots$ and $\mathscr{C}$-arrows $g: d_i \rightarrow d_j$. The action of F on D produces a diagram $F(D)$ in $\mathscr{D}$ comprising the $\mathscr{D}$-objects $F(d_i), F(d_j), \ldots$ and $\mathscr{D}$-arrows $F(g): F(d_i) \rightarrow F(d_j)$. F *preserves* D-*limits* if whenever $\{f_i : c \rightarrow d_i\}$ is a collection of arrows forming a limit (universal cone) for D in $\mathscr{C}$, then $\{F(f_i): F(c) \rightarrow F(d_i)\}$ is a limit for $F(D)$ in $\mathscr{D}$. On the other hand, if F always maps a colimit for D in $\mathscr{C}$ to a colimit for $F(D)$ in $\mathscr{D}$, then F *preserves colimits of* D. Reversing the implications in these last two definitions yields the notions of F *reflecting* limits and colimits, respectively, of D.

To be even more general we may simply say that if P is some categorial "property", then F *preserves* P if the image under F of an entity in $\mathscr{C}$ with property P has property P in $\mathscr{D}$, and F *reflects* P if whenever the F-image of an entity from $\mathscr{C}$ has P in $\mathscr{D}$, then that entity itself has P in $\mathscr{C}$.

EXERCISE 1. Show that any functor preserves identities, iso arrows, and commutative diagrams.

EXERCISE 2. If F preserves pullbacks, then F preserves monics. □

A functor $F: \mathscr{C} \rightarrow \mathscr{D}$ is *faithful* if it acts injectively on each "hom-set" $\mathscr{C}(a, b)$ (cf. Example 9.1.6). This means that for any pair $f, g: a \rightarrow b$ of $\mathscr{C}$-arrows with the same domain and codomain, if $F(f) = F(g)$ then $f = g$.

EXERCISE 3. Show that the forgetful functor **Grp** $\rightarrow$ **Set** is faithful but is not bijective on objects or on identity arrows.

EXERCISE 4. Show that a faithful functor reflects monics, epics (and hence iso's if its domain is a topos), and commutative diagrams.

EXERCISE 5. Suppose that $\mathscr{C}$ has an equaliser for any parallel pair of arrows. Show that a parallel pair are equal iff their equaliser is iso. Hence show that if F is a functor on $\mathscr{C}$ that preserves equalisers, then F reflects iso's only if F is faithful. □

It follows by these exercises that a functor, defined on a topos, which preserves equalisers is faithful if, and only if, it reflects iso arrows. There is another important variant of faithfulness, which is the notion of a functor that reflects inclusions of subobjects. To be precise, we need to assume that F preserves monics. Then if f and g are subobjects of a

$\mathscr{C}$-object d, $F(f)$ and $F(g)$ will serve as subobjects of $F(d)$ in $\mathscr{D}$. We say that F is *conservative* if whenever $F(f) \subseteq F(g)$ in Sub$(F(d))$, it follows that $f \subseteq g$ in Sub(d).

EXERCISE 6. Suppose that $\mathscr{C}$ has equalisers, and that these are preserved by F. Show that if F is conservative, then F is faithful.

EXERCISE 7. Suppose that $\mathscr{C}$ has pullbacks of all appropriate pairs of arrows, and that these pullbacks are preserved by F. Using the pullback characterisation of intersections (Theorem 7.1.2) show that F reflects iso's only if F is conservative. □

Thus it follows that for a functor which is defined on a topos and preserves equalisers and pullbacks, "faithful", "conservative", and "preserves iso's" are all equivalent.

We will be particularly concerned with functors that preserve all finite limits (i.e. limits of all finite diagrams). Such a functor is called *left exact*, while, dually, a *right exact* functor is one that preserves colimits of all finite diagrams. One that is both left and right exact is simply called *exact*. If a category $\mathscr{C}$ is finitely complete (i.e. has all finite limits, cf. §3.15), then it can be shown that for a functor F defined on $\mathscr{C}$ to be left exact it suffices either that F preserves terminal objects and pullbacks, or that F preserves terminal objects, equalisers, and products of pairs of $\mathscr{C}$-objects (Herrlich and Strecker [73], Theorem 24.2). The dual statement is left to the reader.

Since monics and epics are special cases of limits and colimits respectively (Exercise 3.13.9 and its dual), we see that exact functors preserve epi-monic factorisations. In view of Theorem 5.2.2, we then have the following important fact.

EXERCISE 8. If F is an exact functor between two topoi, then F preserves images of arrows, i.e. $F(\text{im}\, f)$ is $\text{im}(F(f))$. □

One context in which preservation of certain limits and colimits is guaranteed is that of an adjoint situation (§15.1).

THEOREM 1. *If $\langle F, G, \theta\rangle$ is an adjunction from $\mathscr{C}$ to $\mathscr{D}$, then the left adjoint F preserves all colimits of $\mathscr{C}$, while the right adjoint G preserves all $\mathscr{D}$-limits.*

PROOF. We outline the argument showing that F preserves colimits, giving enough of the construction to display the role of the adjunction, and leaving the fine detail as a worthy exercise for the reader.

Using the notation of §3.11, let D be a diagram in $\mathscr{C}$ that has a colimit $\{f_i : d_i \to c\}$. Since F preserves commutative diagrams, the collection $\{F(f_i): F(d_i) \to F(c)\}$ will be a cocone for the diagram $F(D)$ in $\mathscr{D}$. We wish to show that it is co-universal for $F(D)$. So, let $\{h_i : F(d_i) \to d\}$ be another cocone for $F(D)$ in $\mathscr{D}$, meaning that

$$\begin{array}{ccc} F(d_i) & \xrightarrow{F(g)} & F(d_j) \\ & {\scriptstyle h_i}\searrow \quad \swarrow{\scriptstyle h_j} & \\ & d & \end{array}$$

commutes for each arrow $g: d_i \to d_j$ in D. Applying the components $\theta_{d_i d}$ of θ we then obtain a family $\{\theta(h_i): d_i \to G(d)\}$ of $\mathscr{C}$-arrows which proves to be a cocone for D, since the naturalness of θ can be invoked to show that

$$\begin{array}{ccc} d_i & \xrightarrow{g} & d_j \\ & {\scriptstyle \theta(h_i)}\searrow \quad \swarrow{\scriptstyle \theta(h_j)} & \\ & G(d) & \end{array}$$

always commutes, where g is as above. But then as $\{f_i : d_i \to c\}$ is a colimit for D, there is a unique $\mathscr{C}$-arrow $f: c \to G(d)$ such that

$$\begin{array}{ccc} & d_i & \\ & {\scriptstyle f_i}\swarrow \quad \searrow{\scriptstyle \theta(h_i)} & \\ c & \overset{}{\underset{f}{\dashrightarrow}} & G(d) \end{array}$$

commutes for all d_i in D.

Applying the inverse of the component θ_{cd} to f, we obtain an arrow $k: F(c) \to d$ such that

$$\begin{array}{ccc} & F(d_i) & \\ & {\scriptstyle F(f_i)}\swarrow \quad \searrow{\scriptstyle h_i} & \\ F(c) & \underset{k}{\longrightarrow} & d \end{array}$$

always commutes. Indeed k is $\varepsilon_d \circ F(f)$, where $\varepsilon : F \circ G \to 1_{\mathscr{D}}$ is the counit of the adjunction. Moreover, the uniqueness of f and the injectivity of θ_{cd} lead us to conclude that k is the only arrow for which this last diagram always commutes (the couniversal property of the unit η of the adjunction expressed in (2) and (3) of §15.1 can be used to prove this). □

Thus we see that a left exact functor F which has a right adjoint must preserve all finite limits and all colimits (and hence be exact). Functors of this kind lie at the heart of the notion of geometric morphism, which we now proceed to define.

16.2. Geometric morphisms

Let X and Y be topological spaces, with Θ_X and Θ_Y their associated poset categories of open sets. A function $f:X\to Y$ is continuous precisely when each member of Θ_Y pulls back under f to a member of Θ_X, i.e. $V\in\Theta_Y$ only if $f^{-1}(V)\in\Theta_X$, where $f^{-1}(V)=\{x\in X: f(x)\in V\}$ (recall the discussion in Example 3.13.2 of the inverse image $f^{-1}(V)$ as a pullback). In this case, the map f^* taking V to $f^{-1}(V)$ becomes a functor $f^*:\Theta_Y\to\Theta_X$ which is an $\sqcap$-$\sqcup$ map of **CHA**'s (and which is a special case of the pulling-back functor $f^*:\mathscr{C}\downarrow b\to\mathscr{C}\downarrow a$ discussed in §15.3). As a functor, f^* has a right adjoint $f_*:\Theta_X\to\Theta_Y$ defined, for each $U\in\Theta_X$, by

$$f_*(U)=\bigsqcup\{V: f^{-1}(V)\subseteq U\}.$$

EXERCISE 1. Why is f^* left exact?

EXERCISE 2. Show that

$$f^*(V)\subseteq U \quad \text{iff} \quad V\subseteq f_*(U),$$

and hence

$$f^*\dashv f_* \qquad \square$$

A continuous function $f:X\to Y$ can be lifted to a pair (f^*,f_*) of adjoint functors between the topoi **Top**(Y) and **Top**(X) which generalises the above situation. First we define the functor $f^*:\mathbf{Top}(Y)\to\mathbf{Top}(X)$, as follows. If $g:A\to Y$ is a **Top**(Y)-object, i.e. a local homeomorphism into Y, we form the pullback h of g along f in **Set**, thus:

$$\begin{array}{ccc} X\times_Y A & \longrightarrow & A \\ {\scriptstyle h}\downarrow & & \downarrow{\scriptstyle g} \\ X & \xrightarrow[f]{} & Y \end{array}$$

The domain of h inherits the product topology of X and A, and h proves to be a local homeomorphism, hence a **Top**(X)-object. We put $f^*(g)=h$, and leave it to the reader to use the universal property of pullbacks to define f^* on **Top**(Y)-arrows (cf. §15.3) and to show f^* is left exact.

EXERCISE 3. Explain how Θ_Y can be regarded as a subcategory of **Top**(Y), and $f^*: \mathbf{Top}(Y) \to \mathbf{Top}(X)$ an extension of the $\sqcap$-$\bigsqcup$ map induced on Θ_Y by f. □

To define f_* we switch from sheaves of germs to sheaves of sections. We saw in §14.1 how **Top**(X) is equivalent to the topos **Sh**(X) whose objects are those contravariant functors $F: \Theta_X \to \mathbf{Set}$ which satisfy the axiom COM. But we have just seen that f gives rise to an $\sqcap$-$\bigsqcup$ map $\Theta_Y \to \Theta_X$, and so we can compose this with F to obtain $f_*(F): \Theta_Y \to \mathbf{Set}$. In other words, for $V \in \Theta_Y$, we put

$$f_*(F)(V) = F(f^{-1}(V)).$$

This definition of $f_*(F)$ turns out to produce a sheaf over Y, and gives rise to a functor $f_*: \mathbf{Sh}(X) \to \mathbf{Sh}(Y)$. Applying the equivalence of **Sh** and **Top** then leads to a functor from **Top**(X) to **Top**(Y) that proves to be right adjoint to f^*.

EXERCISE 4. Explain how this right adjoint can be construed as an extension of the function $f_*: \Theta_X \to \Theta_Y$ defined earlier.

EXERCISE 5. Let $f^*: \Omega \to \Omega'$ be an $\sqcap$-$\bigsqcup$ map between **CHA**'s. If **A** is an Ω-set, define an Ω'-set $f^*(\mathbf{A})$, based on the same **Set**-object as **A**, by putting

$$[\![x \approx y]\!]_{f^*(\mathbf{A})} = f^*([\![x \approx y]\!]_{\mathbf{A}}).$$

Using completions of Ω-sets (§14.7), show that this gives rise to a functor $f^*: \mathbf{Sh}(\Omega) \to \mathbf{Sh}(\Omega')$. Conversely, show that the process of "composing with $f^*: \Omega \to \Omega'$" gives rise to a functor $f_*: \mathbf{Sh}(\Omega') \to \mathbf{Sh}(\Omega)$ that has f^* as a left exact left adjoint (cf. Fourman and Scott [79], §6, for details of this construction). □

In view of the analysis thus far, we are led to the following definition: a *geometric morphism* $f: \mathscr{E}_1 \to \mathscr{E}_2$ of elementary topoi $\mathscr{E}_1$ and $\mathscr{E}_2$ is a pair (f^*, f_*) of functors of the form

$$\mathscr{E}_1 \underset{f_*}{\overset{f^*}{\rightleftarrows}} \mathscr{E}_2$$

such that f^* is left exact and left adjoint to f_*. f^* is called the *inverse image part*, and f_* the *direct image part*, of the geometric morphism.

As explained at the end of the last section, the conditions on the inverse image part f^* of a geometric morphism entail that it preserves finite limits and arbitrary colimits. This naturally generalises the notion of an $\sqcap$-$\sqcup$ map of **CHA**'s, and hence, ultimately, that of a continuous function between topological spaces.

In any adjoint situation, each functor determines the other up to natural isomorphism, in the sense that any two left adjoints of a given functor are naturally isomorphic to each other and dually (MacLane [71], Chap. IV, or Herrlich and Strecker [73], Cor. 27.4). In this sense each part of a geometric morphism uniquely determines the other.

Further examples of geometric morphisms

EXAMPLE 1. The inclusion functor $\mathbf{Sh}(I) \hookrightarrow \mathbf{St}(I)$ from the topos of sheaves of sections over a topological space I to the topos of presheaves over I (§14.1) is the direct image part of a geometric morphism whose inverse image part is the "sheafification" functor $F \longmapsto F_{p_F}$ (Exercise 14.1.9).

EXAMPLE 2. Example 1 extends to any elementary site $(\mathscr{E}, j)$. The inclusion $sh_j(\mathscr{E}) \hookrightarrow \mathscr{E}$ of the j-sheaves into $\mathscr{E}$ has as left adjoint the left exact sheafification functor $\mathscr{S}\!h_j : \mathscr{E} \to sh_j(\mathscr{E})$ mentioned in §14.4. In addition to the references given there, details may also be found in Tierney [73], Johnstone [77] §3.3, and Veit [81]. The latter gives the construction of $\mathscr{S}\!h_j$ and a proof of its left exactness by means of the internal logic of the site.

EXAMPLE 3. The fundamental Theorem of Topoi (§15.3) states that if $f: a \to b$ is any arrow in an elementary topos $\mathscr{E}$, then the pulling-back functor $f^*: \mathscr{E}\downarrow b \to \mathscr{E}\downarrow a$ has a right adjoint Π_f. The pair (f^*, Π_f) form a geometric morphism from $\mathscr{E}\downarrow a$ to $\mathscr{E}\downarrow b$.

EXAMPLE 4. If $\mathscr{E}_1$ and $\mathscr{E}_2$ are topoi, the projection functor $\mathscr{E}_1 \times \mathscr{E}_2 \to \mathscr{E}_1$ is left exact and left adjoint to the functor taking the $\mathscr{E}_1$-object a to $(a, 1)$.

EXAMPLE 5. **Kan Extensions.** Let $\mathscr{C}$ and $\mathscr{D}$ be two categories, whose nature will be qualified below. A given functor $F: \mathscr{C} \to \mathscr{D}$ induces a functor $F : \mathbf{St}(\mathscr{D}) \to \mathbf{St}(\mathscr{C})$ between pre-sheaf categories which takes the $\mathbf{St}(\mathscr{D})$-object $G: \mathscr{D} \to \mathbf{Set}$ to $G \circ F: \mathscr{C} \to \mathbf{Set}$, and the arrow $\tau: G \dot{\to} G'$ to $\sigma: G \circ F \dot{\to} G \circ F'$ where the component σ_c is $\tau_{F(c)}$. There is a general theory, due to Daniel Kan, that produces a left adjoint $F^{\cdot}:\mathbf{St}(\mathscr{C})\to\mathbf{St}(\mathscr{D})$ to $F_{\cdot}$. Full details are given in [MR], p. 38 (cf. also MacLane [71], Ch. X, and Verdier [SGA4], Exp. I, §5). We will describe the construction of $F^{\cdot}(G)$ for a $\mathbf{St}(\mathscr{C})$-object $G:\mathscr{C}\to\mathbf{Set}$. $F^{\cdot}(G)$ is called the *left Kan extension* of G along F.

If d is a $\mathscr{D}$-object, $F^{\cdot}(G)(d)$ will be an object in **Set**, realised as a colimit of a diagram. First we define a category $d\downarrow F$ whose objects are the pairs (c,f) such that c is a $\mathscr{C}$-object and f a $\mathscr{D}$-arrow of the form $d\to F(c)$. An arrow from (c,f) to (c',f') in $d\downarrow F$ is a $\mathscr{C}$-arrow $g:c\to c'$ such that the diagram

$$\begin{array}{ccc} & d & \\ {}^{f}\swarrow & & \searrow^{f'} \\ F(c) & \xrightarrow[F(g)]{} & F(c') \end{array}$$

commutes. There is a "forgetful" functor $U:d\downarrow F^{\mathrm{op}}\to\mathscr{C}^{\mathrm{op}}$ given by $U(c,f)=c$, $U(g)=g$. The image of $G\circ U$ is then a diagram in **Set**. $F^{\cdot}(G)(d)$ is defined as the colimit of this diagram.

Of course this definition depends on the existence of the colimit in question, and to guarantee this we have to limit the "size" of $\mathscr{C}$ and $\mathscr{D}$. The category **Set** is *bicomplete*, in the sense that it has limits and colimits of all small diagrams (cf. MacLane [71], Ch. V, or Herrlich and Strecker [73], §23). The adjective "small" is applied to a collection which is a set, i.e. a **Set**-object, rather than a proper class (§1.1). Thus a diagram is small if its collection of objects and arrows forms a set, and the same definition of smallness applies to a category. Of course many of the categories we deal with are not small (e.g. **Set**, **Top**(X), **Sh**(X), **St**($\mathscr{C}$), Ω-**Set**, etc.). But they often satisfy the weaker condition of *local* smallness, which means that for any two objects a and b, the collection of all arrows from a to b in the category is small.

Now if $\mathscr{C}$ is a small category, and $\mathscr{D}$ is locally small, then the category $d\downarrow F$ above will be small, and hence the image of $G\circ U$ will be a small diagram in **Set**. Under these conditions then, the functor $F^{\cdot}$ is well-defined, and proves to be left adjoint to $F_{\cdot}$, and left exact if $\mathscr{C}$ has finite limits that are preserved by F ([MR], p. 39).

To sum up: if $\mathscr{C}$ is a finitely complete small category, $\mathscr{D}$ is locally small, and $F:\mathscr{C}\to\mathscr{D}$ is left exact, then the pair $(F^{\cdot},F_{\cdot})$ form a geometric morphism from $\mathbf{St}(\mathscr{D})$ to $\mathbf{St}(\mathscr{C})$.

We will take up this construction again below in relation to Grothendieck topoi. □

A geometric morphism $f:\mathscr{E}_1\to\mathscr{E}_2$ is called *surjective* if its inverse image part $f^*:\mathscr{E}_2\to\mathscr{E}_1$ is a faithful functor. By the work of the previous section, this is equivalent to requiring that f^* be conservative, or that it reflect iso's. The justification for the terminology is contained in the following exercises.

EXERCISE 6. Let $f:X\to Y$ be a continuous function that is surjective, i.e. $\operatorname{Im} f = Y$. If

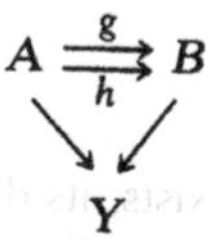

are two parallel $\mathbf{Top}(Y)$-arrows such that $f^*(g)=f^*(h)$ in $\mathbf{Top}(X)$, show that $g\circ k = h\circ k$, where $k:X\times_Y A\to A$ is the pullback of f along $A\to Y$. Noting that k is onto, conclude that $f^*:\mathbf{Top}(Y)\to\mathbf{Top}(X)$ is faithful.

EXERCISE 7. Show, with the help of 5.3.1, that the construction of Exercise 6 works for any arrow $f:a\to b$ in any elementary topos $\mathscr{E}$, in the sense that if f is $\mathscr{E}$-epic then the geometric morphism $\mathscr{E}\downarrow a\to\mathscr{E}\downarrow b$ given in Example 3 above is surjective.

EXERCISE 8. If $f:a\to b$ is an $\mathscr{E}$-arrow, show that in $\mathscr{E}\downarrow a$, $f^*(\operatorname{im} f)$ is an iso arrow. Hence show, conversely to the last exercise, that if $f^*:\mathscr{E}\downarrow b\to \mathscr{E}\downarrow a$ reflects iso's, then f is an epic arrow in $\mathscr{E}$. □

If $\mathscr{E}$ is a topos, then an *$\mathscr{E}$-topos* is a pair $(\mathscr{E}_1, f_1)$ comprising a topos $\mathscr{E}_1$ and a geometric morphism $f_1:\mathscr{E}_1\to\mathscr{E}$. A morphism $f:\mathscr{E}_1\to\mathscr{E}_2$ of $\mathscr{E}$-topoi is a geometric morphism which makes the diagram

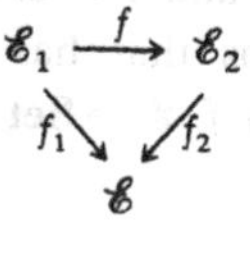

commute up to natural isomorphism, i.e. the functors $f_{2*}\circ f_*$ and f_{1*} are naturally isomorphic, as are $f^*\circ f_2^*$ and f_1^*.

An $\mathscr{E}$-topos is said to be *defined over* $\mathscr{E}$, and the arrow f in the above diagram is called a geometric morphism *over* $\mathscr{E}$. A topos defined over **Set** will be called an **S**-topos. The extent to which **Set** determines the structure of an **S**-topos can be seen by examining the reasons behind the fact that for any topos $\mathscr{E}$ there is, up to natural isomorphism, *at most* one geometric morphism $f:\mathscr{E}\to\mathbf{Set}$. This is because the adjunction of f^* and f_* provides, for each $\mathscr{E}$-object b, a bijection

$$\theta_b:\mathscr{E}(f^*(1),b)\cong\mathbf{Set}(1,f_*(b))$$

which is natural in b. But in **Set**, arrows of the form $1\to f_*(b)$ correspond bijectively to elements of the set $f_*(b)$. Also f^*, being left exact, preserves terminal objects, so that $f^*(1)$ is terminal in $\mathscr{E}$. In this way we obtain a bijection

$$\mathscr{E}(1,b)\cong f_*(b)$$

natural in b. Hence if such a geometric morphism exists, its direct image part f_* is determined up to natural isomorphism as the functor $\mathscr{E}(1,-)$ (Example 9.1.7). Since f_* is thus determined, its left adjoint f^* is too.

By pursuing this analysis of f, we can find sufficient conditions for $\mathscr{E}$ to be an **S**-topos. First, for any two $\mathscr{E}$-objects a and b, $\mathscr{E}$-arrows of the form $a\to b$ correspond bijectively with those of the form $1\times a\to b$, via the isomorphism $1\times a\cong a$ (Exercise 3.8.4), and hence bijectively with those of the form $1\to b^a$, by exponentiation (cf. the discussion of the "name" of an arrow in §4.1). Therefore there is a bijection between $\mathscr{E}(a,b)$ and $\mathscr{E}(1,b^a)$ and so, as above, one between $\mathscr{E}(a,b)$ and the **Set**-object $f_*(b^a)$. It follows that $\mathscr{E}(a,b)$ is a set, and that $\mathscr{E}$ is a locally small category, in the sense defined previously in our discussion of Kan extensions.

Secondly, the preservation properties of the inverse image part f^* allow us to conclude that $\mathscr{E}$ has arbitrary *set-indexed copowers* of 1. This means that any collection $\{1_s:s\in S\}$ of terminal $\mathscr{E}$-objects, indexed by a *set* S, has a coproduct in $\mathscr{E}$. For, in **Set** S is $\varinjlim_{s\in S}\{s\}$, and so as f^* preserves colimits, $f^*(S)$ is $\varinjlim_{s\in S}f^*(\{s\})$. But $\{s\}$ is terminal in **Set**, and f^* left exact, so $f^*(\{s\})\cong 1_s$, implying that $f^*(S)$ is a coproduct of $\{1_s\colon s\in S\}$ as desired.

Thus we see that an **S**-topos is locally small and has arbitrary set-indexed copowers of 1. But if $\mathscr{E}$ is any topos that has these two properties, we can *define* a geometric morphism $f:\mathscr{E}\to\mathbf{Set}$ by putting $f_*(b)=\mathscr{E}(1,b)$ and $f^*(S)=\varinjlim_{s\in S}1_s$.

EXERCISE 9. Show that for any topos $\mathscr{E}$ there is at most one geometric morphism $\mathscr{E}\to\mathbf{Finset}$, and that it exists iff $\mathscr{E}(a,b)$ is finite for all $\mathscr{E}$-objects a and b. □

There is a particularly direct way of showing that **Top**(X) is always an **S**-topos. If $\{*\}$ is a one-point space with the *discrete* topology in which all subsets are open (this is the only possible topology on $\{*\}$), then the unique function $X \to \{*\}$ is continuous, and so induces a geometric morphism $\mathbf{Top}(X) \to \mathbf{Top}(\{*\})$. But a $\mathbf{Top}(\{*\})$-object is a topological space Y for which $Y \to \{*\}$ is a local homeomorphism. This, however, is only possible when Y itself has the discrete topology, and the latter is determined as soon as we are given the underlying set of Y. Hence $\mathbf{Top}(\{*\})$ is an isomorphic copy of **Set**.

EXERCISE 10. For any **CHA** Ω, show that there is an ⊓-⊔ map $2 \to \Omega$. Hence show that $\mathbf{Sh}(\Omega)$ is an **S**-topos. □

It is notable that the existence of set-indexed copowers of 1 in a topos $\mathscr{E}$ implies that the **HA** $\mathrm{Sub}_{\mathscr{E}}(1)$ (or, isomorphically, $\mathscr{E}(1, \Omega)$) is complete (this was mentioned at the end of §14.7). The proof is as follows.

EXERCISE 11. Let $\{a_s \rightarrowtail 1 : s \in S\}$ be a set of subojbects of 1 in $\mathscr{E}$, with characteristic arrow $\chi_s : 1 \to \Omega$ for each $s \in S$. Show that the support of the subobject whose characteristic arrow is the coproduct of the χ_s's is a join of the a_s's in Sub(1). □

Geometric morphisms of Grothendieck topoi

To discuss these, we are going to modify our earlier notation and terminology a little. Let $\mathbf{C} = (\mathscr{C}, Cov)$ be a site (§14.3), consisting of a pretopology Cov on a category $\mathscr{C}$. The full subcategory of the pre-sheaf category $\mathbf{St}(\mathscr{C})$ generated by the sheaves over $\mathbf{C}$ will now be denoted $\mathbf{Sh}(\mathbf{C})$ instead of $\mathbf{Sh}(Cov)$. $\mathbf{C}$ will be called a *small* site if $\mathscr{C}$ is a small category. The name "Grothendieck topos" will be reserved for categories equivalent to those of the form $\mathbf{Sh}(\mathbf{C})$ for *small* sites $\mathbf{C}$. Moreover we will assume throughout that *all sites are finitely complete*, i.e. have all finite limits.

For small sites $\mathbf{C}$, $\mathbf{Sh}(\mathbf{C})$ satisfies the two conditions given above that suffice to make it an **S**-topos. The existence of set-indexed copowers of 1 is just a special case of the fact that $\mathbf{Sh}(\mathbf{C})$ is *bicomplete* in the sense that every *small* diagram has a limit and a colimit. This fact derives ultimately from the bicompleteness of **Set** itself, which allows all set-indexed limits and colimits to be constructed "component-wise" in the pre-sheaf category $\mathbf{St}(\mathscr{C})$ (cf. §9.3, or MacLane [71], V.3). Then if D is a small diagram in $\mathbf{Sh}(\mathbf{C})$, the limit of D in $\mathbf{St}(\mathscr{C})$ proves to be a sheaf, and hence a D-limit in $\mathbf{Sh}(\mathbf{C})$. On the other hand the colimit for D in $\mathbf{St}(\mathscr{C})$ is transferred by the colimit preserving sheafification functor $\mathbf{St}(\mathscr{C}) \to \mathbf{Sh}(\mathbf{C})$ to a colimit for D in $\mathbf{Sh}(\mathbf{C})$.

For local smallness of $\mathbf{Sh}(\mathbf{C})$ we note first that the axioms of ZF set theory allow us to form the product $\varprojlim_{i\in I} A_i$ of a collection of sets A_i, indexed by a set I, as the **Set**-object

$$\{f: f \text{ is a function } \& \operatorname{dom} f = I \ \& \ f(i) \in A_i \text{ for all } i \in I\}.$$

Now an arrow $\tau: F \rightarrowtail G$ in $\mathbf{Sh}(\mathbf{C})$ is a natural transformation, and hence is a function assigning to each $\mathscr{C}$-object c a set-function $\tau_c : F(c) \to G(c)$, i.e. a member of the set $\mathbf{Set}(F(c), G(c))$. But if $\mathscr{C}$ is small, then the collection $|\mathscr{C}|$ of $\mathscr{C}$-objects is small, so the collection $\mathbf{Sh}(\mathbf{C})(F, G)$ of $\mathbf{Sh}(\mathbf{C})$-arrows from F to G is included in the set

$$\varprojlim_{c\in|\mathscr{C}|} \mathbf{Set}(F(c), G(c))$$

and thus is itself small.

Assuming only that $\mathscr{C}$ is locally small, a functor $E_{\mathbf{C}}: \mathscr{C} \to \mathbf{Sh}(\mathbf{C})$, known as the *canonical* functor ([SGA4], II 4.4), can be defined as the composite of two other functors $\mathscr{Y}: \mathscr{C} \to \mathbf{St}(\mathscr{C})$ and $Sh: \mathbf{St}(\mathscr{C}) \to \mathbf{Sh}(\mathbf{C})$. The second of these is the sheafification or "associated-sheaf" functor that forms the inverse image part of the geometric morphism whose direct image part is the inclusion $\mathbf{Sh}(\mathbf{C}) \hookrightarrow \mathbf{St}(\mathscr{C})$. For a detailed account of Sh the reader is referred to the work of Verdier [SGA4] II.2, [MR]1.2, or Schubert [72], §20.3.

The functor $\mathscr{Y}$ is the dual form of the fundamental *Yoneda functor*. It takes the $\mathscr{C}$-object c to the contravariant hom-functor $\mathscr{C}(-, c): \mathscr{C} \to \mathbf{Set}$ of Example 9.1.10, and the $\mathscr{C}$-arrow $f: c \to d$ to the natural transformation $\mathscr{C}(-, f): \mathscr{C}(-, c) \rightarrowtail \mathscr{C}(-, d)$ where, for any $\mathscr{C}$-object a, the component assigned to a by $\mathscr{C}(-, f)$ is the "composing with f" function $\mathscr{C}(a, f): \mathscr{C}(a, c) \to \mathscr{C}(a, d)$. Note that the local smallness of $\mathscr{C}$ is essential here in order for the functor $\mathscr{Y}(c)$, i.e. $\mathscr{C}(-, c)$, to have its values in **Set**.

Underlying the definition of $\mathscr{Y}$ is a very important piece of category theory known as the *Yoneda Lemma* (MacLane [71] III §2, Herrlich and Strecker [73] §30). In its dual form it states that for any $\mathscr{C}$-object c and presheaf $F: \mathscr{C}^{\mathrm{op}} \to \mathbf{Set}$, there is a bijection

$$\mathbf{St}(\mathscr{Y}(c), F) \cong F(c)$$

between $\mathbf{St}(\mathscr{C})$-arrows (i.e. natural transformations) from $\mathscr{C}(-, c)$ to F and elements of the set $F(c)$.

EXERCISE 12. If $x \in F(c)$ and d is any $\mathscr{C}$-object, show that the equation

$$x_d(f) = F(f)(x)$$

defines a function $x_d : \mathscr{C}(d, c) \to F(d)$. Show that the x_d's form the components of a natural transformation $\mathscr{Y}(c) \dot{\to} F$, and that this construction gives the bijection asserted above.

Formulate precisely, and prove, the condition that this bijection be "natural" in c and F. □

In particular, the Yoneda Lemma implies, for any $\mathscr{C}$-objects c and d, that

$$\mathbf{St}(\mathscr{Y}(c), \mathscr{Y}(d)) \cong \mathscr{C}(c, d),$$

so that $\mathscr{Y}$ acts bijectively on hom-sets. It is also injective on objects, and so embeds $\mathscr{C}$ isomorphically into $\mathbf{St}(\mathscr{C})$, making it possible to identify c and $\mathscr{Y}(c)$, and regard $\mathscr{C}$ as a full subcategory of $\mathbf{St}(\mathscr{C})$.

Now in a cocomplete topos, the existence of set-indexed coproducts allows us to form the *union* of any set $\{G_x \rightarrowtail F : x \in X\}$ of subobjects of an object F, by defining $\bigcup_X G_x \rightarrowtail F$ to be the image arrow of the coproduct arrow $(\varinjlim G_x) \to F$ (thereby extending the formation of unions given by Theorem 3 of §7.1). This construction enables us to make the topos itself into a site! A set $\{F_x \xrightarrow{f_x} F : x \in X\}$ of arrows is defined to be a cover of F if, in Sub(F), $\bigcup_X \operatorname{im} f_x$ is 1_F (and so $\bigcup_X f_x(F_x) \cong F$). Equivalently, the definition requires that the coproduct arrow $[\operatorname{im} f_x]$ of the arrows $\operatorname{im} f_x : f_x(F_x) \rightarrowtail F$ be epic.

This notion of cover defines the *canonical* pre-topology, which in the case of a Grothendieck topos $\mathbf{Sh}(\mathbf{C})$ proves to have the property that all the hom-functors $\mathscr{Y}(c)$ are sheaves, so that the Yoneda functor maps $\mathscr{C}$ into $\mathbf{Sh}(\mathbf{C})$. There is another way of defining canonical covers in $\mathbf{Sh}(\mathbf{C})$ which is formally simpler to express and avoids reference to colimits. We say that $C = \{F_x \xrightarrow{f_x} F : x \in X\}$ is an *epimorphic family* if, for any pair $f, g : F \to G$ of parallel arrows with domain F, if $f \circ f_x = g \circ f_x$ for all $x \in X$, then $f = g$.

EXERCISE 13. Show that C as above is an epimorphic family iff the coproduct $[f_x] : \varinjlim F_x \to F$ of the f_x's is epic.

EXERCISE 14. Show that the epic parts $F_x \twoheadrightarrow f_x(F_x)$ of the arrows f_x give rise to an epic arrow $\varinjlim F_x \twoheadrightarrow \varinjlim f_x(F_x)$ which factors $[f_x]$ through $[\operatorname{im} f_x] : \varinjlim f_x(F_x) \to F$. Hence show that $[\operatorname{im} f_x]$ is epic iff $[f_x]$ is epic, and so the canonical covers are precisely the epimorphic families. □

To place the canonical pretopology in broader perspective, we need to examine the general conditions under which $\mathcal{Y}(c)$ is a sheaf over **C**. To do this, we reformulate the sheaf axiom COM of §14.3 in the terms given by the Yoneda Lemma. Let $F:\mathscr{C}\to\mathbf{Set}$ be a presheaf, and $\{a_x \xrightarrow{f_x} a : x\in X\}$ a cover of the site **C**. Instead of dealing with elements $s_x\in F(a_x)$ we deal, via Exercise 12, with arrows $s_x:\mathcal{Y}(a_x)\to F$. Compatibility of a selection of such "elements" s_x for each $x\in X$ requires that for all $x, y\in X$ we have that $s_x\circ\mathcal{Y}(f)=s_y\circ\mathcal{Y}(g)$, where f and g are the pullback in $\mathscr{C}$ of f_x and f_y:

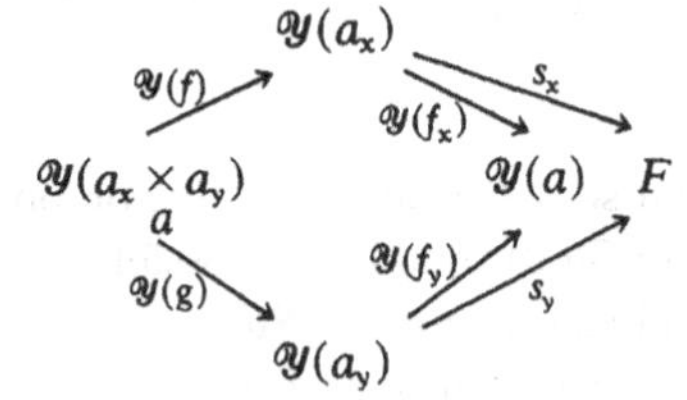

Fulfillment of COM for this situation requires a unique arrow $\mathcal{Y}(a)\to F$ that for all $x, y\in X$ makes this diagram commute.

Now if F is of the form $\mathcal{Y}(c)$, the fact that $\mathcal{Y}$ is injective on objects and bijective on hom-sets allows us to pull the above diagram back into $\mathscr{C}$ itself. This leads to the following notion.

A collection $C=\{a_x \xrightarrow{f_x} a : x\in X\}$ of $\mathscr{C}$-arrows is called an *effectively epimorphic family* if for any $\mathscr{C}$-object c, and for any collection $D=\{a_x \xrightarrow{g_x} c : x\in X\}$ of $\mathscr{C}$-arrows such that for all $x, y\in X$ we have

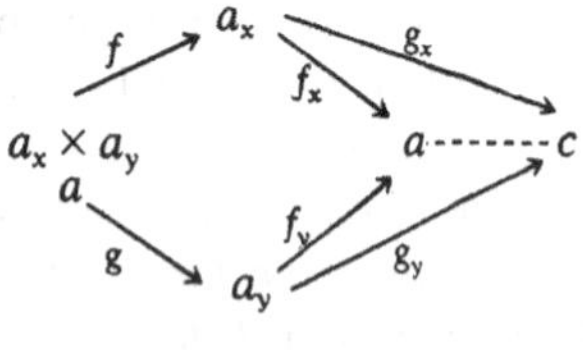

$g_x\circ f=g_y\circ g$, where f and g are the pullback of f_x and f_y, there is a *unique* $\mathscr{C}$-arrow $g:a\to c$ such that $g\circ f_x=g_x$ for all $x\in X$.

A collection D satisfying the hypothesis of this definition will be called *compatible* with C. Thus the definition requires that any collection compatible with C is factored through C by a unique arrow.

Exercise 15. Show that an effectively epimorphic family is epimorphic.

Exercise 16. If C is the empty set of arrows with condomain a, show that C is effectively epimorphic iff a is an initial object. □

It is apparent from our discussion that for a site **C** in which every cover is effectively epimorphic, the hom-functors are all sheaves, and so $\mathcal{Y}$ embeds **C** in **Sh**(**C**). Such a pretopology is called *precanonical.* In the case of a general finitely complete category $\mathscr{C}$, an effectively epimorphic family is called *stable* (or universal) if its pullback along any arrow is also effectively epimorphic. The stable effectively epimorphic families form a precanonical pretopology on $\mathscr{C}$ that includes any other precanonical one ([MR], Proposition 1.1.9). Hence it is known as the *canonical* pretopology on $\mathscr{C}$.

In a Grothendieck topos **Sh**(**C**), the stable effectively epimorphic families prove to be precisely the epimorphic families as defined prior to Exercise 13 ([MR], Proposition 3.4.11). Whenever we refer to **Sh**(**C**) as a site, we will thus be referring to epimorphic families as covers. The canonical functor $E:\mathbf{Sh}(\mathbf{C}) \to \mathbf{Sh}(\mathbf{Sh}(\mathbf{C}))$ from **Sh**(**C**) to the category of sheaves on the site **Sh**(**C**) will then just be the Yoneda embedding. It turns out that E is an equivalence, so that **Sh**(**C**) and **Sh**(**Sh**(**C**)) are equivalent categories in the sense of §9.2, allowing us to think of any Grothendieck topos as being the topos of sheaves on a canonical site. The proof of this fact is part of a number of fundamental characterisations of Grothendieck topoi that may be found in [SGA4], IV.1, or [MR], Theorem 1.4.5. The fact itself is needed to show that geometric morphisms between Grothendieck topoi are determined by certain "continuous morphisms" between sites, as we shall now see.

If $\mathbf{C} = (\mathscr{C}, Cov)$ and $\mathbf{D} = (\mathscr{D}, Cov')$ are sites, a *continuous morphism* $F:\mathbf{C} \to \mathbf{D}$ is a functor $F:\mathscr{C} \to \mathscr{D}$ that is left exact (remember sites are presumed to be finitely complete) and preserves covers, i.e. has $\{f_x : x \in X\} \in Cov(c)$ only if $\{F(f_x): x \in X\} \in Cov'(F(c))$. For example, if $f: V \to W$ is a continuous function of topological spaces, then $f^*: \Theta_W \to \Theta_v$ preserves open covers in the usual topological sense. Similarly, an $\sqcap$-$\sqcup$ map $f^*: \Omega \to \Omega'$ between **CHA**'s is continuous with respect to the definition of Cov_Ω introduced just prior to Exercise 14.7.11 – indeed left exactness amounts to preservation of $\sqcap$, and preservation of members of Cov_Ω means preservation of $\sqcup$.

The examples indicate that the concept of continuous morphism of sites generalises that of continuous function of topological spaces, and hence is linked to the notion of geometric morphism. Indeed, if $f^*: \mathbf{Sh}(\mathbf{C}) \to \mathscr{E}$ is the inverse image part of a geometric morphism of Grothendieck topoi, then f^* is continuous with respect to the associated canonical sites. This is because in that context the notion of epimorphic family is characterised by colimits (viz. coproducts and epic arrows), and colimits are preserved by f^*. Moreover, the canonical functor $E_{\mathbf{C}}: \mathscr{C} \to \mathbf{Sh}(\mathbf{C})$ proves to be continuous. In fact, $E_{\mathbf{C}}$ both preserves and *reflects* covers in the sense that $\{f_x : x \in X\} \in Cov(c)$ in $\mathbf{C}$ if and only if $\{E_{\mathbf{C}}(f_x) : x \in X\}$ is an epimorphic family in $\mathbf{Sh}(\mathbf{C})$ ([SGA4], II.4.4, and [MR], Proposition 1.3.3). Thus we can compose $E_{\mathbf{C}}$ and f^* to get a continuous morphism $\mathbf{C} \to \mathscr{E}$. Conversely, and more importantly, every geometric morphism $\mathscr{E} \to \mathbf{Sh}(\mathbf{C})$ can be obtained uniquely as an extension of a continuous morphism of this type. To show this we need the following result.

THEOREM 1. *Let $F: \mathbf{C} \to \mathbf{D}$ be a continuous morphism of sites, with $\mathbf{C}$ small and $\mathbf{D}$ locally small. Then there is a geometric morphism $f: \mathbf{Sh}(\mathbf{D}) \to \mathbf{Sh}(\mathbf{C})$ such that the diagram*

$$\begin{array}{ccc} \mathbf{C} & \xrightarrow{E_{\mathbf{C}}} & \mathbf{Sh}(\mathbf{C}) \\ {\scriptstyle F}\downarrow & & \downarrow{\scriptstyle f^*} \\ \mathbf{D} & \xrightarrow[E_{\mathbf{D}}]{} & \mathbf{Sh}(\mathbf{D}) \end{array}$$

commutes. Moreover there is, up to natural isomorphism, at most one continuous $\mathbf{Sh}(\mathbf{C}) \to \mathbf{Sh}(\mathbf{D})$ that makes this diagram commute, so that f is unique up to natural isomorphism.

This theorem is proven in Proposition 1.2 of Expose III of [SGA4]. In [MR], the reference is Theorem 1.3.10, with the uniqueness clause coming from 1.3.12. We will do no more here than outline the definition of f.

Recall, from the discussion of Kan extensions in Example 5 of our list of geometric morphisms, that F induces a functor $F_{.}: \mathbf{St}(\mathscr{D}) \to \mathbf{St}(\mathscr{C})$ that has a left exact left adjoint $F^{\cdot}$. Now consider the diagram

$$\begin{array}{ccccc} \mathscr{C} & \xrightarrow{\mathscr{Y}_{\mathscr{C}}} & \mathbf{St}(\mathscr{C}) & \underset{\mathscr{I}_{\mathbf{C}}}{\overset{Sh_{\mathbf{C}}}{\rightleftarrows}} & \mathbf{Sh}(\mathbf{C}) \\ {\scriptstyle F}\downarrow & & {\scriptstyle F^{\cdot}}\downarrow\uparrow{\scriptstyle F_{.}} & & \\ \mathscr{D} & \xrightarrow{\mathscr{Y}_{\mathscr{D}}} & \mathbf{St}(\mathscr{D}) & \underset{\mathscr{I}_{\mathbf{D}}}{\overset{Sh_{\mathbf{D}}}{\rightleftarrows}} & \mathbf{Sh}(\mathbf{D}) \end{array}$$

Here, $\mathscr{Y}$ denotes a Yoneda functor, Sh a sheafification functor, and $\mathscr{I}$ an inclusion. f^* is defined to be $Sh_{\mathbf{D}} \circ F^{\cdot} \circ \mathscr{I}_{\mathbf{C}}$, and f_* is $Sh_{\mathbf{C}} \circ F_{.} \circ \mathscr{I}_{\mathbf{D}}$. (Since, in any adjoint situation, each adjoint determines the other up to natural isomorphism, the uniqueness of f^* implies that of f_*, and hence of f.)

If we now apply Theorem 1 in the case that $\mathbf{D}$ is itself a Grothendieck topos $\mathscr{E}$, with the canonical pretopology, then $E_{\mathbf{D}}$ is an equivalence whose "inverse" $\mathbf{Sh}(\mathscr{E}) \to \mathscr{E}$ may be composed with f^* to yield a continuous morphism $\mathbf{Sh}(\mathbf{C}) \to \mathscr{E}$. This leads to the following central result.

THEOREM 2. (Reduction Theorem). *If* $\mathbf{C}$ *is a small size, and* $\mathscr{E}$ *a Grothendieck topos, then for any continuous morphism* $F:\mathbf{C} \to \mathscr{E}$ *there exists a continuous* $f^*:\mathbf{Sh}(\mathbf{C}) \to \mathscr{E}$*, unique up to natural isomorphism, such that*

$$\begin{array}{ccc} \mathbf{C} & \xrightarrow{E_{\mathbf{C}}} & \mathbf{Sh}(\mathbf{C}) \\ & {}_{F}\searrow & \downarrow{\scriptstyle f^*} \\ & & \mathscr{E} \end{array}$$

commutes. Moreover f^* *is the inverse image part of a (thereby unique up to natural isomorphism) geometric morphism* $f:\mathscr{E} \to \mathbf{Sh}(\mathbf{C})$. □

Thus we see that any geometric morphism $f:\mathscr{E} \to \mathbf{Sh}(\mathbf{C})$ is determined uniquely up to natural isomorphism by the continuous functor $f^* \circ E_{\mathbf{C}}:\mathbf{C} \to \mathscr{E}$, and by this result that the construction of geometric morphisms between Grothendieck topoi reduces to the construction of continuous morphisms defined on small sites. In the next section, the later notion will be reformulated in terms of models of logical theories.

As a final topic on this theme we consider the question as to when the functor f^* in Theorem 2 is faithful, so that the associated geometric morphism is surjective. To discuss this we need to know the fact that the $E_{\mathbf{C}}$-image of $\mathbf{C}$ in $\mathbf{Sh}(\mathbf{C})$ forms a set of *generators* for $\mathbf{Sh}(\mathbf{C})$. This means that for any $\mathbf{Sh}(\mathbf{C})$-object H, the family of arrows from objects of the form $E_{\mathbf{C}}(c)$ to H is epimorphic. In other words, if $\sigma, \tau: H \rightarrow G$ are *distinct* arrows in $\mathbf{Sh}(\mathbf{C})$, then there is a $\mathscr{C}$-object c and an arrow $\rho: E_{\mathbf{C}}(c) \rightarrow H$ such that $\sigma \circ \rho \neq \tau \circ \rho$.

To prove this, observe that if $\sigma \neq \tau$, then for some c, and some $x \in H(c)$, $\sigma_c(x) \neq \tau_c(x)$. But by the Yoneda Lemma (Exercise 12), x determines an arrow $\rho': \mathscr{Y}(c) \rightarrow H$ such that $\rho'_c(\mathbf{1}_c) = x$, and so $\sigma \circ \rho' \neq \tau \circ \rho'$. Then by the co-universal property associated with the left adjoint sheafification functor $Sh: \mathbf{St}(\mathscr{C}) \to \mathbf{Sh}(\mathbf{C})$ (cf. (2) of §15.1), ρ' factors uniquely

$$\begin{array}{ccc} \mathscr{Y}(c) & \longrightarrow & Sh(\mathscr{Y}(c)) \\ & {}_{\rho'}\searrow & \downarrow{\scriptstyle \rho} \\ & & H \end{array}$$

through an arrow $\rho: E_{\mathbf{C}}(c) \to H$ (using the fact that the right adjoint of Sh is the inclusion) which must then have $\sigma \circ \rho \neq \tau \circ \rho$.

We see then that in $\mathbf{Sh}(\mathbf{C})$, every object is "covered" by a family of objects of the form $E_{\mathbf{C}}(c)$. This generating role of these objects gives rise to the following result, whose proof may be found in [MR], Lemma 1.3.8.

LEMMA. *If* $e: K \rightarrowtail E_{\mathbf{C}}(c)$ *is monic in* $Sh(\mathbf{C})$, *then there is an epimorphic family* $\{E_{\mathbf{C}}(c_x) \xrightarrow{h_x} K: x \in X\}$ *such that each composite* $e \circ h_x$ is $E_{\mathbf{C}}(g_x)$ *for some* $\mathscr{C}$*-arrow* $g_x: c_x \to c$. □

THEOREM 3. *Let* $F: \mathbf{C} \to \mathscr{E}$ *be a continuous morphism as in Theorem 2. Then the extension* $f^*: \mathbf{Sh}(\mathbf{C}) \to \mathscr{E}$ *of* F *along* $E_{\mathbf{C}}$ *is faithful if, for any set* $\{g_x: x \in X\}$ *of* $\mathscr{C}$*-arrows with a common codomain,* $\{F(g_x): x \in X\}$ *is epimorphic in* $\mathscr{E}$ *only if* $\{E_{\mathbf{C}}(g_x): x \in X\}$ *is epimorphic in* $\mathbf{Sh}(\mathbf{C})$.

PROOF. Let $\sigma, \tau: H \to G$ be a pair of $\mathbf{Sh}(\mathbf{C})$-arrows such that $f^*(\sigma) = f^*(\tau)$. If $\sigma \neq \tau$, then by what we have just seen, there is a $\mathscr{C}$-object c and an arrow $\rho: E_{\mathbf{C}}(c) \to H$ such that $\sigma \circ \rho \neq \tau \circ \rho$. Let $e: K \rightarrowtail E_{\mathbf{C}}(c)$ be the equaliser in $\mathbf{Sh}(\mathbf{C})$ of $\sigma \circ \rho$ and $\tau \circ \rho$. By the Lemma there is an epimorphic family of arrows $h_x: E_{\mathbf{C}}(c_x) \to K$, for all x in some set X, such that each $e \circ h_x$ is $E_{\mathbf{C}}(g_x)$ for some $g_x: c_x \to c$. Since f^* is continuous, $\{f^*(h_x): x \in X\}$ is epimorphic in $\mathscr{E}$. But since f^* is left exact, $f^*(e)$ equalises $f^*(\sigma \circ \rho)$ and $f^*(\tau \circ \rho)$ in $\mathscr{E}$, and these last two arrows are equal, since $f^*(\sigma) = f^*(\tau)$ and f^* preserves composites. Therefore $f^*(e)$ is iso, from which it follows readily that $\{f^*(e) \circ f^*(h_x): x \in X\}$ is an epimorphic family. But $f^*(e) \circ f^*(h_x) = f^*(e \circ h_x) = f^*(E_{\mathbf{C}}(g_x)) = F(g_x)$, so the hypothesis of the Theorem implies that $\{E_{\mathbf{C}}(g_x): x \in X\}$ is epimorphic. However $(\sigma \circ \rho) \circ E_{\mathbf{C}}(g_x) = (\sigma \circ \rho) \circ e \circ h_x = (\tau \circ \rho) \circ e \circ h_x = (\tau \circ \rho) \circ E_{\mathbf{C}}(g_x)$ (by definition of e), so this entails that $\sigma \circ \rho = \tau \circ \rho$ – contrary to hypothesis. Thus our assumption that $\sigma \neq \tau$ must be false. □

COROLLARY 4. *If* F *reflects covers, then* f^* *is faithful.*

PROOF. This follows immediately from the fact that $E_{\mathbf{C}}$ preserves covers, i.e. if $\{g_x: x \in X\}$ is a cover in $\mathbf{C}$ then $\{E_{\mathbf{C}}(g_x): x \in X\}$ is a cover in $\mathbf{Sh}(\mathbf{C})$. □

Points

If Y is a topological space, then a point $y \in Y$ determines a continuous function $\{*\} \to Y$, where $\{*\}$ is the one-point space. Since $\mathbf{Top}(\{*\})$ is isomorphic to $\mathbf{Set}$, this in turn gives rise to a geometric morphism $p_y: \mathbf{Set} \to \mathbf{Top}(Y)$.

EXERCISE 17. Show that the inverse image functor p_y^* takes each **Top**(Y)-object to its stalk over y, and each arrow to its restriction to this stalk.

EXERCISE 18. Show that for any **CHA** Ω, an $\sqcap$-$\sqcup$ map $\Omega \to 2$ (i.e. a point of Ω in the sense of §14.8) gives rise to a geometric morphism from **Set** to **Sh**(Ω). □

In view of these examples we define a *point* of an **S**-topos $\mathscr{E}$ to be a geometric morphism p: **Set** $\to \mathscr{E}$. By left exactness, a subobject $a \rightarrowtail 1$ of 1 in $\mathscr{E}$ will be mapped by p^* to a subobject of 1 in **Set**, so $p^*(a) \in \{0, 1\}$. As p^* also preserves colimits, we obtain thereby an $\sqcap$-$\sqcup$ map $\Omega_{\mathscr{E}} \to 2$, where, in the notation of §14.7, $\Omega_{\mathscr{E}}$ is the **CHA** $\text{Sub}_{\mathscr{E}}(1)$ of subobjects of 1 in $\mathscr{E}$. Thus a point of $\mathscr{E}$ gives rise to a point of $\Omega_{\mathscr{E}}$ (recall from Exercise 11 that constraining $\mathscr{E}$ to be an **S**-topos ensures that $\Omega_{\mathscr{E}}$ is a *complete* **HA**).

In the topological case, subobjects of 1 in **Top**(Y) correspond to open subsets of Y, and Sub(1) can be identified with Θ_Y (cf. §4.5). If Y is sober, in the sense (defined in §14.8) that every **CHA**-point $f : \Theta_Y \to 2$ is of the form

$$f(V) = \begin{cases} 1 & \text{if } y \in V, \\ 0 & \text{if } y \notin V \end{cases}$$

for some $y \in Y$, then the geometric points **Set** $\to$ **Top**(Y) are precisely those that arise from elements of Y in the above manner.

More generally, we can define a topology on the class of points of an **S**-topos $\mathscr{E}$ by taking as opens the collections

$$V_a = \{p : p^*(a) = 1\}$$

for each $a \rightarrowtail 1$ in $\text{Sub}_{\mathscr{E}}(1)$. In the case of **Top**(Y), this produces a space topologically isomorphic to the sober space $\beta(\Theta_Y)$ of all points of Θ_Y (called the "soberification" of Y – cf. Wraith [75], §4, and Johnstone [77], §7.2).

Now if P is a class of points of $\mathscr{E}$, we call P *sufficient* if any $\mathscr{E}$-arrow f with the property that $p^*(f)$ is iso in **Set** for all $p \in P$ must itself be iso in $\mathscr{E}$. In other words, whenever f is not iso in $\mathscr{E}$, then there is at least one $p \in P$ such that $p^*(f)$ is not iso in **Set**. By the work of §16.1, the reader should recognise that this concept is linked to those of conservative and faithful functors.

EXERCISE 19. P is sufficient iff for any parallel pair $f, g : a \to b$ of $\mathscr{E}$-arrows, if $p^*(f) = p^*(g)$ for all $p \in P$, then $f = g$.

EXERCISE 20. P is sufficient iff for any two subobjects f, g of any $\mathscr{E}$-object, if $p^*(f) \subseteq p^*(g)$ for all $p \in P$, then $f \subseteq g$.

EXERCISE 21. There exists a sufficient class of $\mathscr{E}$-points if and only if the class of all $\mathscr{E}$-points is sufficient. □

We say that $\mathscr{E}$ *has enough points* if the class of all points of $\mathscr{E}$ is sufficient. In the case of **Top**(Y), a pair f, g of parallel arrows

$$\begin{array}{ccc} A & \overset{f}{\underset{g}{\rightrightarrows}} & B \\ & \searrow \quad \swarrow & \\ & Y & \end{array}$$

are equal if and only if they agree on the stalk of A over each point $y \in Y$. By Exercises 17 and 19 then, it is clear that the topos **Top**(Y) has enough points, and indeed that the set $\{p_y : y \in Y\}$ of points is sufficient.

The question as to when a topos $\mathscr{E}$ has enough points has some interesting answers in the case that $\mathscr{E}$ is the Grothendieck topos **Sh**(**C**) of sheaves over a small site **C**. First there is the fact that if **Sh**(**C**) does have enough points, then it has a sufficient *set* of points. The proof of this ([SGA4], IV 6.5(b), Johnstone [77], 7.17) is too involved to give here, but an inkling of why such a size reduction is plausible comes from the knowledge that, with the aid of the Yoneda Lemma, it can be shown that any functor from **C** to **Set** is constructible as the colimit of a diagram in $\mathbf{Set}^{\mathscr{C}}$ whose objects are hom-functors on $\mathscr{C}$. Since $\mathscr{C}$ is small, the class of all such hom-functors is small. But any geometric morphism **Set** $\rightarrow$ **Sh**(**C**) is determined by a continuous functor from **C** to **Set** (**Set** is of course a Grothendieck topos, being equivalent to **Sh**($\{*\}$)).

Now a set P of points of **Sh**(**C**) can be combined into a single geometric morphism $\pi : \mathbf{Set}^P \rightarrow \mathbf{Sh}(\mathbf{C})$. Here $\mathbf{Set}^P$ is the Boolean topos of set-valued functions $f : P \rightarrow \mathbf{Set}$ on the discrete category P, and is equivalent to **Bn**(P) (§9.3). Alternatively, by §14.1.II, viewing P as a discrete poset makes $\mathbf{Set}^P$ equivalent to **Sh**(P), where P becomes a space under the discrete topology $\Theta_P = \mathscr{P}(P)$. Yet another way of looking at this category is to identify it as the Grothendieck topos **Sh**(Ω), defined in §14.7, where we take the **CHA** Ω to be the Boolean power-set algebra $\mathscr{P}(P)$.

To define π, it suffices by the Reduction Theorem (Theorem 2) to specify its inverse image part $\pi^* : \mathbf{Sh}(\mathbf{C}) \rightarrow \mathbf{Set}^P$ as a continuous morphism, and indeed it would be enough to specify the continuous morphism

$\pi^* \circ E_{\mathbf{C}} : \mathbf{C} \to \mathbf{Set}^P$. In a similar vein, we can regard each geometric morphism $p \in P$ as being a continuous morphism $p : \mathbf{C} \to \mathbf{Set}$ that extends, uniquely up to isomorphism, to a continuous $p^* : \mathbf{Sh}(\mathbf{C}) \to \mathbf{Set}$ making

$$\begin{array}{ccc} \mathbf{C} & \xrightarrow{E_{\mathbf{C}}} & \mathbf{Sh}(\mathbf{C}) \\ & \searrow^{p} & \downarrow{p^*} \\ & & \mathbf{Set} \end{array}$$

commute.

$\mathbf{Set}^P$ is the *P-indexed power* of **Set**, i.e. the "P-fold product of **Set** with itself", having projection (evaluation) functors $ev_p : \mathbf{Set}^P \to \mathbf{Set}$, for each $p \in P$, where $ev_p(f) = f(p)$, and $ev_p(\sigma) = \sigma_p$ for each $\mathbf{Set}^P$-arrow $\sigma : f \dot{\to} g$. π^* is then the product arrow of $\{p^* : p \in P\}$, i.e. the unique functor making

$$\begin{array}{ccc} \mathbf{Sh}(\mathbf{C}) & \xrightarrow{\pi^*} & \mathbf{Set}^P \\ & \searrow^{p^*} & \downarrow{ev_p} \\ & & \mathbf{Set} \end{array}$$

commute for all $p \in P$. Thus $\pi^*(F) : P \to \mathbf{Set}$ is the function that takes p to $p^*(F)$, while $\pi^*(\tau) : \pi^*(F) \to \pi^*(G)$ is the natural transformation with pth component $p^*(\tau) : p^*(F) \to p^*(G)$.

Our earlier remark about the link between sufficiency and faithfulness can now be made precise:

EXERCISE 22. P is sufficient iff π^* is faithful. □

In order for π^* to determine a geometric morphism, it must be continuous, and in particular preserve canonical covers, i.e. epimorphic families.

LEMMA. *A set $A = \{f^x \xrightarrow{\sigma^x} f : x \in X\}$ of $\mathbf{Set}^P$-arrows is epimorphic in $\mathbf{Set}^P$ iff for each $p \in P$ the set $ev_p(A) = \{ev_p(\sigma^x) : x \in X\}$ is epimorphic in* **Set**.

PROOF. We prove necessity, the converse being more straightforward. Note that to define an arrow $\sigma : f \dot{\to} g$ in $\mathbf{Set}^P$ requires us just to specify a function $\sigma_p : f(p) \to g(p)$ for each $p \in P$. As P is a discrete category (i.e. has only identity arrows), σ is then automatically natural in p, so *any* P-indexed collection of functions $f(p) \to g(p)$ defines an arrow.

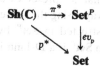

Suppose that A is epimorphic, and take $p \in P$. Let $k, l : f(p) \to B$ be arrows in **Set** such that $k \circ ev_p(\sigma^x) = l \circ ev_p(\sigma^x)$ for all $x \in X$. We need to show that $k = l$.

Define a $\mathbf{Set}^P$-object $g : P \to \mathbf{Set}$ by putting

$$g(r) = \begin{cases} B & \text{if } r = p, \\ f(r) & \text{if } r \neq p, \end{cases}$$

and define arrows $\tau, \rho : f \to g$ by putting $\tau_p = k$, $\rho_p = l$, and $\tau_r = \rho_r = \mathrm{id}_{f(r)}$ for $r \neq p$. Then $\tau \circ \sigma^x = \rho \circ \sigma^x$ for all $x \in X$. Since A is epimorphic, it follows that $\tau = \rho$, and so $\tau_p = \rho_p$ as desired. □

Now if C is a cover in **Sh(C)**, then for each p, continuity of p^* implies that $p^*(C)$, i.e. $ev_p(\pi^*(C))$ is epimorphic in **Set**. Hence, by the Lemma, $\pi^*(C)$ is epimorphic in $\mathbf{Set}^P$. This shows that π^* preserves covers. Left exactness of π^* is established in a similar way, using the left-exactness of each p^*, and the fact that limits are constructed in $\mathbf{Set}^P$ by pointwise evaluation, i.e. a cone U for a diagram D in $\mathbf{Set}^P$ is a D-limit if $ev_p\ (U)$ is an $ev_p(D)$-limit in **Set** for all $p \in P$.

EXERCISE 23. **Sh(C)** has enough points iff there exists a set P and a surjective geometric morphism $\mathbf{Set}^P \to \mathbf{Sh(C)}$. □

The question of faithfulness of π^* can also be approached in terms of the criterion given in Corollary 4. If $\pi : \mathbf{C} \to \mathbf{Set}^P$ is the continuous morphism $\pi^* \circ E_{\mathbf{C}}$, then the criterion is that π reflects covers, i.e. if C is a set of $\mathscr{C}$-arrows with a common codomain, and $\pi(C)$ is an epimorphic family in $\mathbf{Set}^P$, then C is a cover in **C**. But $\pi(C)$ will be epimorphic iff $ev_p(\pi(C))$ is epimorphic in **Set** for all $p \in P$. Since we have

$$\begin{array}{ccc} \mathbf{C} & \xrightarrow{E_{\mathbf{C}}} & \mathbf{Sh(C)} \\ {\scriptstyle \pi}\downarrow & \swarrow{\scriptstyle \pi^*} & \downarrow{\scriptstyle p^*} \\ \mathbf{Set}^P & \xrightarrow[ev_p]{} & \mathbf{Set} \end{array}$$

$ev_p \circ \pi = E_{\mathbf{C}} \circ p^* = p$, this leads to the following result.

THEOREM 5. ([SGA4], IV.6.5(a)). *A set P of points of* **Sh(C)** *is sufficient if, and only if, for any set C of $\mathscr{C}$-arrows that is not a cover in* **C** *there exists some $p \in P$ such that $p(C)$ is not epimorphic in* **Set**. □

This brings the theory of geometric morphisms to a point from which logical methods can be applied to give a proof of a theorem, due to Pierre Deligne ([SGA4], VI.9) about sufficiency of points for topoi that are called *coherent*. The definition of these categories can be motivated in part by the fundamental topological concept of *compactness*.

In a topological space I, a subset $A \subseteq I$ is *compact* if every open cover of A, i.e. every $C \subseteq \Theta$ such that $A \subseteq \bigcup C$, has a finite subcover, i.e. there is a finite subset C_0 of C such that $A \subseteq \bigcup C_0$. If a member V of Θ is compact, then the topological site (Θ, Cov_Θ) (Exercise 14.3.1) can be modified by changing $Cov_\Theta(V)$ to the set of *finite* open covers $C_0 \subseteq \Theta_V$, without altering the associated class of sheaves. This is seen as follows.

EXERCISE 24. Let F be a presheaf on I that fulfills the sheaf condition COM with respect to all finite open covers of an open set V. Show that if V is compact, then F fulfills COM with respect to all open covers of V. □

A site $(\mathscr{C}, Cov)$ will be called *finitary* if $\mathscr{C}$ is a small finitely complete category and every member of $Cov(c)$ is finite, for all $\mathscr{C}$-objects c. A *coherent* topos is a category that is equivalent to $\mathbf{Sh}(\mathbf{C})$ for some finitary site $\mathbf{C}$. The significance of this class of categories cannot really be conveyed here, except to say that it includes many of the sheaf categories of algebraic geometry to which the theory of Grothendieck topoi is addressed.

DELIGNE'S THEOREM. *Every coherent topos has enough points.* □

This theorem does not hold for all Grothendieck topoi. Several examples have been given of such categories that do not have enough points. One due to Deligne, constructed out of measure spaces, appears in [SGA4], IV.7.4. Wraith [75], Corollary 7.6, shows that for a "Hausdorff" topological space I in which no singletons are open (e.g. the real line $\mathbb{R}$ is such a space), the Boolean topos $sh_{\neg\neg}(\mathbf{Top}(I))$ of double negation sheaves on I has no points at all! (cf. also Johnstone [77], 7.12(iii)). A particularly apposite example is given by Barr [74], using atomless Boolean algebras, which we will now study.

Now an *atom* in a poset with a zero (minimum) element 0 is an element $a \neq 0$ such that there is no non-zero element strictly less than a (i.e. if $y \sqsubseteq a$, then $y = 0$ or $y = a$). A poset is *atomic* if for every non-zero element x there is an atom a such that $a \sqsubseteq x$. For any set P, the complete **BA** $\mathscr{P}(P)$ of all subsets of P is atomic, the atoms being the singletons $\{p\}$

corresponding to the points $p \in P$. Conversely any atomic complete **BA B** is isomorphic to $\mathscr{P}(P_{\mathbf{B}})$, where $P_{\mathbf{B}}$ is the set of all atoms in **B**. The isomorphism assigns to each **B**-element b the set $\{p \in P_{\mathbf{B}} : p \sqsubseteq b\}$.

EXERCISE 25. Show that in any **BA**, an element $a \neq 0$ is an atom iff for any y, $a \sqsubseteq y$ or $a \sqsubseteq y'$.

EXERCISE 26. Let $\mathscr{E}$ be an **S**-topos in which $\mathrm{Sub}_{\mathscr{E}}(1)$ is a Boolean algebra. If $p : \mathbf{Set} \to \mathscr{E}$ is a geometric morphism, show that the $\sqcap$-$\bigsqcup$ map $p^* : \mathrm{Sub}(1) \to 2$ induced by the inverse image part of p preserves Boolean complements, and thus preserves meets $\bigsqcap$. Hence show that $\bigsqcap\{f : p^*(f) = 1\}$ is an atom in Sub(1). □

Now let **B** be a complete Boolean algebra that has no atoms at all (e.g. the algebra of "regular" open subsets of the real line – Mendelson [70], 5.48). As Barr suggests, **B** may be thought of as a "set without points". But in the Grothendieck topos **Sh**(**B**), or equivalently **CB-Set**, Sub(1) is in fact isomorphic to **B** itself. This can be seen from the fact that in **CB-Set**, elements of **B** correspond to global elements of the subobject classifier, and the latter correspond to subobjects of 1 (cf. Exercise 14.7.46). (Alternatively, note that in **CB-Set**, the terminal object is **B** itself, and associate each subobject of **B** with its join in **B**.) Thus it follows by Exercise 26 that the topos **Sh**(**B**) does not have any points.

Returning to Deligne's Theorem, it follows from all that we have said that if $\mathscr{E}$ is a coherent topos, then there is a set P and a surjective geometric morphism $\pi : \mathbf{Sh}(\mathscr{P}(P)) \to \mathscr{E}$ (since $\mathbf{Sh}(\mathscr{P}(P))$ is a Boolean topos, π is sometimes called a "Boolean-valued point"). In this form the theorem has an appropriate generalisation to Grothendieck topoi (first conjectured by Lawvere, and proven in Barr [74]), obtained by abandoning the atomicity requirement on complete **BA**'s.

BARR'S THEOREM. *If $\mathscr{E}$ is a Grothendieck topos, then there is a complete Boolean algebra* **B** *and a surjective geometric morphism* $\mathbf{Sh}(\mathbf{B}) \to \mathscr{E}$. □

This section has been a descriptive sketch of what is an extensive mathematical theory, and has only attempted to reproduce enough of it to allow a statement of the theorems of Deligne and Barr and an explanation of their model-theoretic content (to follow). A deeper understanding of this theory may be gained from Chapter 1 of [MR]. Its ultimate source is, of course, the monumental treatise [SGA4].

16.3. Internal logic

In this section we introduce the ideas of *many-sorted* languages and structures and show how to use them to express the internal structure of a category.

A model $\mathfrak{A} = \langle A, \ldots \rangle$ for an elementary language, as described in §11.2, consists of a single set A that carries certain operations $g : A^n \to A$, relations $R \subseteq A^n$, and distinguished elements $c \in A$. The corresponding first-order language has a single set $\{v_1, v_2, \ldots\}$ of individual variables that "range over A". But it is common in mathematics to deal with operations whose various arguments are of different *sorts*, i.e. come from different specified sets. A classic (two-sorted) example is the notion of a *vector space*, which involves a set V of "vectors", a set S of "scalars", and an operation of the form $S \times V \to V$ of "scalar multiplication of vectors". We formalise this sort of situation as follows.

Let $\mathcal{S}$ be a class, whose members will be called *sorts*. The basic alphabet for elementary languages of §11.2 is now adapted to an alphabet for $\mathcal{S}$-sorted languages by retaining the symbols $\wedge$, $\vee$, $\sim$, $\supset$, $\forall$, $\exists$, $\approx$,), (, and replacing the single list of individual variables by a denumerable set V_a of such variables for each $a \in \mathcal{S}$, with V_a disjoint from V_b whenever $a \neq b$. We often write v: a, and say "v is of sort a", when $v \in V_a$.

An $\mathcal{S}$-sorted language $\mathcal{L}$ is a collection of operation and relation symbols, and individual constants, such that:

(1) each relation symbol $\mathbf{R}$ has assigned to it a natural number n, called its *number of places*, and a sequence $\langle a_1, \ldots, a_n \rangle$ of sorts. We write $\mathbf{R}: \langle a_1, \ldots, a_n \rangle$ to indicate this;

(2) each operation symbol $\mathbf{g}$ has an assigned *number of places n*, and a sequence $\langle a_1, \ldots, a_{n+1} \rangle$ of sorts. We indicate this by $\mathbf{g}: \langle a_1, \ldots, a_n \rangle \to a_{n+1}$;

(3) each individual constant $\mathbf{c}$ is assigned a sort $a \in \mathcal{S}$, indicated by $\mathbf{c}: a$ (this could be seen as a special case of (2) – an individual constant is a 0-placed operation symbol).

Terms and formulae of $\mathcal{L}$ are defined inductively as usual, with additional qualifications relating to the sort of each term. Thus variables and constants of sort a are terms of sort a, and if $\mathbf{g}: \langle a_1, \ldots, a_n \rangle \to a_{n+1}$, and $t_1, \ldots, t_n$ are terms of respective sorts $a_1, \ldots, a_n$, then $\mathbf{g}(t_1, \ldots, t_n)$ is a term of sort a_{n+1}. Atomic formulae are those of the form $(t \approx u)$, where t and u are terms of the same sort, and of the form $\mathbf{R}(t_1, \ldots, t_n)$, where if $\mathbf{R}: \langle a_1, \ldots, a_n \rangle$ then $t_1: a_1, \ldots, t_n: a_n$. Other $\mathcal{L}$-formulae are built up from the atomic ones in the standard manner. We also include two *atomic sentences*, denoted $\top$ and $\bot$, in any language $\mathcal{L}$.

If $\mathscr{E}$ is an elementary topos, then an $\mathscr{E}$-model for an $\mathscr{S}$-sorted language $\mathscr{L}$ is a function $\mathfrak{A}$ with domain $\mathscr{S} \cup \mathscr{L}$ such that

(1) for each sort $a \in \mathscr{S}$, $\mathfrak{A}(a)$ is an $\mathscr{E}$-object;

(2) for each operation symbol $\mathbf{g}:\langle a_1, \ldots, a_n\rangle \to a_{n+1}$ in $\mathscr{L}$, $\mathfrak{A}(\mathbf{g})$ is an $\mathscr{E}$-arrow from $\mathfrak{A}(a_1)\times\cdots\times\mathfrak{A}(a_n)$ to $\mathfrak{A}(a_{n+1})$;

(3) for each relation symbol $\mathbf{R}:\langle a_1, \ldots, a_n\rangle$ in $\mathscr{L}$, $\mathfrak{A}(\mathbf{R})$ is a subobject of $\mathfrak{A}(a_1)\times\cdots\times\mathfrak{A}(a_n)$;

(4) for each individual constant $\mathbf{c}$: a, $\mathfrak{A}(\mathbf{c})$ is an arrow $1 \to \mathfrak{A}(a)$, i.e. a "global element" of $\mathfrak{A}(a)$.

We will use the notation $\mathfrak{A}:\mathscr{L} \to \mathscr{E}$ to indicate that $\mathfrak{A}$ is an $\mathscr{E}$-model for $\mathscr{L}$.

It is important to realise that this definition of model departs from that of §11.4 in that we now allow $\mathfrak{A}(a)$ to be any $\mathscr{E}$-object, including the initial object 0, or any other $\mathscr{E}$-object d that may have no global elements $1 \to d$ at all. This takes us into the domain of "free" logic (§11.8), but instead of using objects of partial elements, and existence predicates, we are following the approach of the Montreal school ([MR], Chapter 2), in which the notion of "model" directly abstracts the classical Tarskian one, while the standard rules of inference undergo restriction.

If $\mathbf{v} = \langle v_1, \ldots, v_m\rangle$ is a sequence of distinct variables, with $v_i: a_i$, we let $\mathfrak{A}(\mathbf{v})$ be $\mathfrak{A}(a_1)\times\cdots\times\mathfrak{A}(a_m)$. We also adopt the convention of declaring that if $\mathbf{v}$ is the *empty* sequence of variables then $\mathfrak{A}(\mathbf{v}) = 1$ (n.b., 1 is the product of the empty diagram). This is relevant to the interpretation of sentences (see below).

If t is a term of sort a, and $\mathbf{v} = \langle v_1, \ldots, v_m\rangle$ is *appropriate to* t in the sense that all of the variables of t occur in the list $\mathbf{v}$, then an $\mathscr{E}$-arrow $\mathfrak{A}^{\mathbf{v}}(t):\mathfrak{A}(\mathbf{v}) \to \mathfrak{A}(a)$ is defined inductively as follows.

(1) If t is the variable v_i, $\mathfrak{A}^{\mathbf{v}}(t)$ is the projection arrow $\mathfrak{A}(\mathbf{v}) \to \mathfrak{A}(a_i)$.

(2) If t is the constant $\mathbf{c}$, $\mathfrak{A}^{\mathbf{v}}(t)$ is the composite of

$$\mathfrak{A}(\mathbf{v}) \to 1 \xrightarrow{\mathfrak{A}(\mathbf{c})} \mathfrak{A}(a).$$

(3) If t is $\mathbf{g}(t_{i_1}, \ldots, t_{i_n})$, where $\mathbf{g}:\langle a_{i_1}, \ldots, a_{i_n}\rangle \to a$, then we inductively define $\mathfrak{A}^{\mathbf{v}}(t)$ to be the composite of

$$\mathfrak{A}(\mathbf{v}) \xrightarrow{f} \mathfrak{A}(a_{i_1})\times\cdots\times\mathfrak{A}(a_{i_n}) \xrightarrow{\mathfrak{A}(\mathbf{g})} \mathfrak{A}(a),$$

where f is the product arrow $\langle\mathfrak{A}^{\mathbf{v}}(t_{i_1}), \ldots, \mathfrak{A}^{\mathbf{v}}(t_{i_n})\rangle$.

If φ is an $\mathscr{L}$-formula, and the list $\mathbf{v}$ is *appropriate to* φ in that all *free* variables of φ appear in $\mathbf{v}$, then φ is interpreted by the model $\mathfrak{A}$ as a subobject $\mathfrak{A}^{\mathbf{v}}(\varphi)$ of $\mathfrak{A}(\mathbf{v})$. We often present this subobject as $\mathfrak{A}^{\mathbf{v}}(\varphi) \rightarrowtail \mathfrak{A}(\mathbf{v})$, so that the symbol "$\mathfrak{A}^{\mathbf{v}}(\varphi)$" tends to be associated with an object of $\mathscr{E}$, even though strictly speaking it denotes a subobject, whose domain is only determined up to isomorphism. The inductive definition of $\mathfrak{A}^{\mathbf{v}}(\varphi)$ is as follows.

(1) $\mathfrak{A}^{\mathbf{v}}(\top)$ is the maximum subobject $1:\mathfrak{A}(\mathbf{v}) \to \mathfrak{A}(\mathbf{v})$, i.e. the subobject whose character is $true!:\mathfrak{A}(\mathbf{v}) \to \Omega$.

$\mathfrak{A}^{\mathbf{v}}(\bot)$ is the minimum subobject $0 \to \mathfrak{A}(\mathbf{v})$, with character $false!:\mathfrak{A}(\mathbf{v}) \to \Omega$.

(2) If t and u are terms of sort a, $\mathfrak{A}^{\mathbf{v}}(t \approx u)$ is the equaliser of

$$\mathfrak{A}(\mathbf{v}) \underset{\mathfrak{A}^{\mathbf{v}}(u)}{\overset{\mathfrak{A}^{\mathbf{v}}(t)}{\rightrightarrows}} \mathfrak{A}(a).$$

(3) If φ is $\mathbf{R}(t_{i_1}, \ldots, t_{i_n})$, then $\mathfrak{A}^{\mathbf{v}}(\varphi)$ is the pullback

$$\begin{array}{ccc} \mathfrak{A}(\varphi) & \rightarrowtail & \mathfrak{A}(\mathbf{v}) \\ \downarrow & & \downarrow f \\ \mathfrak{A}(\mathbf{R}) & \rightarrowtail & \varprojlim_{j} \mathfrak{A}(a_{i_j}) \end{array}$$

where $\mathbf{R}:\langle a_{i_1}, \ldots, a_{i_n}\rangle$ and f is $\langle \mathfrak{A}^{\mathbf{v}}(t_{i_1}), \ldots, \mathfrak{A}^{\mathbf{v}}(t_{i_n})\rangle$.

(4) The connectives $\wedge, \vee, \sim, \supset$ are interpreted as the operations $\cap, \cup, -, \Rightarrow$ in the Heyting algebra $\mathrm{Sub}_{\mathscr{E}}(\mathfrak{A}(\mathbf{v}))$ (cf. §§7.1, 7.5).

(5) The quantifiers $\forall, \exists$, are interpreted by the functors $\forall_f$, $\exists_f:\mathrm{Sub}(\mathrm{dom}\, f) \to \mathrm{Sub}(\mathrm{cod}\, f)$ associated with an $\mathscr{E}$-arrow f, as defined in §15.4. If φ is $\exists w\psi$, or $\forall w\psi$, then all free variables of ψ appear in the list $\mathbf{v}, w = \langle v_1, \ldots, v_m, w\rangle$. Then if $pr:\mathfrak{A}(\mathbf{v}, w) \to \mathfrak{A}(\mathbf{v})$ is the evident projection, we put

$$\mathfrak{A}^{\mathbf{v}}(\exists w\psi) = \exists_{pr}(\mathfrak{A}^{\mathbf{v},w}(\psi)),$$
$$\mathfrak{A}^{\mathbf{v}}(\forall w\psi) = \forall_{pr}(\mathfrak{A}^{\mathbf{v},w}(\psi))$$

(cf. the beginning of §15.4 for motivation).

Note that if w is the only free variable of ψ, we need to allow that $\mathbf{v}$ be the empty sequence here. But in that case, pr can be identified with the unique arrow $\mathfrak{A}(w) \to 1$, so that the sentences $\exists w\psi$ and $\forall w\psi$ are interpreted as subobjects of 1.

Now if φ is any $\mathscr{L}$-formula, and $\mathbf{v}$ is the (possibly empty) sequence consisting of all and only the free variables of φ, we say that φ is ***true in*** $\mathfrak{A}$, or that $\mathfrak{A}$ is an *$\mathscr{E}$-model of* φ, denoted $\mathfrak{A} \models \varphi$, if $\mathfrak{A}^{\mathbf{v}}(\varphi)$ is the maximum subobject of $\mathfrak{A}(\mathbf{v})$ (i.e. if $\mathfrak{A}^{\mathbf{v}}(\varphi)$ is $\mathfrak{A}^{\mathbf{v}}(\top)$). If $\mathbb{T}$ is a class of formulae, then $\mathfrak{A}$ is a $\mathbb{T}$-model, $\mathfrak{A} \models \mathbb{T}$, if every member of $\mathbb{T}$ is true in $\mathfrak{A}$.

We may tend to drop the symbol "$\mathbf{v}$" from "$\mathfrak{A}^{\mathbf{v}}(\varphi)$" if the intention is clear, and especially if $\mathbf{v}$ is the list of all free variables of φ.

Exercise 1. Develop the notion of a many-sorted model in **Set** along the Tarskian set-theoretic lines of §11.2, allowing for the presence of empty sorts, and defining a satisfaction relation

$$\mathfrak{A} \models \varphi[x_1, \ldots, x_m].$$

Show that in these terms the categorial notion $\mathfrak{A}^{\mathbf{v}}(\varphi)$ corresponds to the set

$$\{\langle x_1, \ldots, x_m\rangle : \mathfrak{A} \models \varphi[x_1, \ldots, x_m]\}.$$

Exercise 2 (Substitution). Let $\mathbf{v} = \langle v_1, \ldots, v_m\rangle$ be appropriate to a term t of sort a. Let u be a term of the same sort as v_i, and let $\mathbf{u}$ be a sequence appropriate to the term $t(v_i/u)$. Define $\mathfrak{A}\,|v_i/u| : \mathfrak{A}(\mathbf{u}) \to \mathfrak{A}(\mathbf{v})$ to be the product arrow

$$\langle \mathfrak{A}^{\mathbf{u}}(v_1), \ldots, \mathfrak{A}^{\mathbf{u}}(v_{i-1}), \mathfrak{A}^{\mathbf{u}}(u), \mathfrak{A}^{\mathbf{u}}(v_{i+1}), \ldots, \mathfrak{A}^{\mathbf{u}}(v_m)\rangle.$$

(i) Show that

$$\begin{array}{ccc} & \mathfrak{A}(\mathbf{u}) & \\ {\scriptstyle \mathfrak{A}|v_i/u|}\swarrow & & \searrow{\scriptstyle \mathfrak{A}^{\mathbf{u}}(t(v_i/u))} \\ \mathfrak{A}(\mathbf{v}) & \xrightarrow[\mathfrak{A}^{\mathbf{v}}(t)]{} & \mathfrak{A}(a) \end{array}$$

commutes.

(ii) If $\mathbf{v}$ is appropriate to φ, v_i free for u in φ, and $\mathbf{u}$ is appropriate to $\varphi(v_i/u)$, show that $\mathfrak{A}^{\mathbf{u}}(\varphi(v_i/u))$ is the pullback

$$\begin{array}{ccc} \mathfrak{A}^{\mathbf{u}}(\varphi(v_i/u)) & \rightarrowtail & \mathfrak{A}(\mathbf{u}) \\ \downarrow & & \downarrow{\scriptstyle \mathfrak{A}|v_i/u|} \\ \mathfrak{A}^{\mathbf{v}}(\varphi) & \rightarrowtail & \mathfrak{A}(\mathbf{v}) \end{array}$$

of $\mathfrak{A}^{\mathbf{v}}(\varphi)$ along $\mathfrak{A}\,|v_i/u|$.

Exercise 3. $\mathfrak{A} \models \varphi \wedge \psi$ iff $\mathfrak{A} \models \varphi$ and $\mathfrak{A} \models \psi$.

Exercise 4. $\mathfrak{A} \models \varphi \supset \psi$ iff $\mathfrak{A}(\varphi) \subseteq \mathfrak{A}(\psi)$, and hence $\mathfrak{A} \models \varphi \equiv \psi$ iff $\mathfrak{A}(\varphi) = \mathfrak{A}(\psi)$.

EXERCISE 5. $\mathfrak{A}(\varphi)=\mathfrak{A}(\top\supset\varphi)$.

EXERCISE 6. $\mathfrak{A}(\sim\varphi)=\mathfrak{A}(\varphi\supset\bot)$. □

The general existence in $\mathscr{E}$ of the interpretation $\mathfrak{A}(\varphi)$ of the formula φ depends on the possibility of performing certain categorial constructions in $\mathfrak{A}$. For instance, the interpretation of universal quantifiers requires the functors $\forall_f$, whose definition in §15.3 used properties that are very special to topoi. On the other hand, the definition of "$\mathscr{L}$-model" itself refers only to products and their subobjects. Indeed if $\mathscr{L}$ has only *one-placed* operation symbols, and no relation symbols or constants, we can construct $\mathscr{L}$-models in any category $\mathscr{C}$. $\mathscr{C}$ would have to have finite products for $\mathfrak{A}(\mathbf{v})$ to exist for all sequences $\mathbf{v}$ of variables, including a terminal object (empty product) for the case that $\mathbf{v}$ is the empty sequence. If $\mathscr{C}$ also had equalisers, then all *equations*, i.e. atomic identities $(t\approx u)$, would have interpretations in a $\mathscr{C}$-model $\mathfrak{A}$. Since Sub(d) is always a poset with a maximum element 1_d (§4.1), we could then talk about the truth in $\mathfrak{A}$ of such equations. But if a category has a terminal object, equalisers, and a product for any pair of objects, then it has all finite limits (cf. §3.15). In sum then, provided that we assume that $\mathscr{C}$ is finitely complete, we can at least construct $\mathscr{C}$-models of equational logic.

The general question as to what categorial structure needs to be present for various types of $\mathscr{L}$-formulae to be interpretable is discussed in Reyes [74], [MR], and Kock and Reyes [77], and leads to notions of "Heyting" and "logical" categories. Similar work is carried out by Volger [75].

The language of a category

Let $\mathscr{C}$ be a finitely complete category. We associate with $\mathscr{C}$ a many-sorted language $\mathscr{L}_{\mathscr{C}}$ and a *canonical* $\mathscr{C}$-model $\mathfrak{A}_{\mathscr{C}}:\mathscr{L}_{\mathscr{C}}\to\mathscr{C}$ of $\mathscr{L}_{\mathscr{C}}$:

(1) the collection of sorts of $\mathscr{L}_{\mathscr{C}}$ is the class $|\mathscr{C}|$ of $\mathscr{C}$-objects, i.e. each $\mathscr{C}$-object *is* a sort;

(2) each $\mathscr{C}$-arrow $f:a\to b$ is declared to be a one-placed operation symbol, with associated sequence $\langle a,b\rangle$ of sorts. These are the only operation symbols of $\mathscr{L}_{\mathscr{C}}$, and there are no constants or relation symbols;

(3) the model $\mathfrak{A}_{\mathscr{C}}$ is simply the identity function on $|\mathscr{C}|\cup\mathscr{L}_{\mathscr{C}}$. Thus if a is a sort, $\mathfrak{A}_{\mathscr{C}}(a)$ is a as a $\mathscr{C}$-object, and if f is an operation symbol, $\mathfrak{A}_{\mathscr{C}}(f)$ is f as a $\mathscr{C}$-arrow $\mathfrak{A}_{\mathscr{C}}(a)\to\mathfrak{A}_{\mathscr{C}}(b)$.

Now if $\mathscr{C}$ is a finitely complete category, the truth of certain equations $(t\approx u)$ in $\mathfrak{A}_{\mathscr{C}}$ can be used to characterise the structure of $\mathscr{C}$ as a category.

To see this, consider the question as to whether a triangle

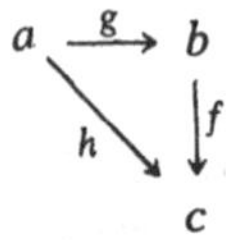

of $\mathscr{C}$-arrows commutes. If v is a variable of sort a. Then $\mathfrak{A}^v_{\mathscr{C}}(g(v))$ is just g, and correspondingly for the term $h(v)$, while $\mathfrak{A}^v_{\mathscr{C}}(f(g(v)))$ is $f \circ g$. Thus the equation $(f(g(v)) \approx h(v))$ is interpreted by $\mathfrak{A}^v_{\mathscr{C}}$ as the equaliser of $f \circ g$ and h. Since parallel arrows are equal if, and only if, their equaliser is iso, we get

$$\mathfrak{A}_{\mathscr{C}} \models (f(g(v)) \approx h(v)) \quad \text{iff} \quad f \circ g = h,$$

and, in particular,

$$\mathfrak{A}_{\mathscr{C}} \models (f(g(v)) \approx f \circ g(v)).$$

Exercise 7. Let $f: a \to a$ be an endo $\mathscr{C}$-arrow. Show that if v: a, then

$$\mathfrak{A}_{\mathscr{C}} \models (f(v) \approx v) \quad \text{iff} \quad f = \mathbf{1}_a,$$

and so

$$\mathfrak{A}_{\mathscr{C}} \models (\mathbf{1}_a(v) \approx v).$$ □

Now if $\mathfrak{A}: \mathscr{L}_{\mathscr{C}} \to \mathscr{D}$ is an $\mathscr{L}_{\mathscr{C}}$-model in a category $\mathscr{D}$, then $\mathfrak{A}$ assigns a $\mathscr{D}$-object $\mathfrak{A}(a)$ to each $\mathscr{C}$-object (i.e. $\mathscr{L}_{\mathscr{C}}$-sort) a, and a $\mathscr{D}$-arrow $\mathfrak{A}(f): \mathfrak{A}(a) \to \mathfrak{A}(b)$ to each $\mathscr{C}$-arrow ($\mathscr{L}_{\mathscr{C}}$ operation symbol) $f: a \to b$. Thus $\mathfrak{A}$ is *exactly* the same type of function as is a functor $\mathfrak{A}: \mathscr{C} \to \mathscr{D}$. To actually qualify as a functor, $\mathfrak{A}$ is required to preserve identity arrows and commutative triangles. Since these two notions have been expressed as $\mathscr{L}_{\mathscr{C}}$-equations, we can repeat the above arguments and exercises in $\mathscr{D}$ to show that the truth in $\mathfrak{A}$ of these equations exactly captures the required preservation property. Given a triangle f, g, h of $\mathscr{C}$-arrows, let

$$\mathbf{id}(f) \quad \text{be} \quad (f(v) \approx v),$$

and

$$\mathbf{com}(f, g, h) \quad \text{be} \quad (f(g(v)) \approx h(v)).$$

(In each case v is a variable of the required sort to make the formula well-formed, so to be precise **id**(f) is a formula schema, representing a different formula for each choice of v. We will in future gloss over this point).

THEOREM 1. *If $\mathscr{D}$ is finitely complete, then a $\mathscr{D}$-model $\mathfrak{A}:\mathscr{L}_{\mathscr{C}} \to \mathscr{D}$ for $\mathscr{L}_{\mathscr{C}}$ is a functor $\mathfrak{A}:\mathscr{C} \to \mathscr{D}$ if, and only if, for all $\mathscr{C}$-objects a, and all composable pairs f, g of $\mathscr{C}$-arrows, the equations* $\mathbf{id}(1_a)$ *and* $\mathbf{com}(f, g, f \circ g)$ *are true in* $\mathfrak{A}$. □

This result displays the essential idea of the logical characterisation of categorial properties (the reader familiar with model theory will recognise it as a variant of the "method of diagrams"). Note that the result does not depend on the existence of any limits in $\mathscr{C}$.

Continuing in this vein, we develop logical axioms for products, equalisers etc. This will involve us in the use of existential quantifiers, and hence the subobject functors $\exists_f$. So, from now on we will assume that the category in which our model exists is a topos (although this is stronger than is needed for $\exists_f$ to exist).

Recall that $\exists_f:\mathrm{Sub}(a) \to \mathrm{Sub}(b)$ takes $g: c \rightarrowtail a$ to the image arrow of $f \circ g$, so

$$\begin{array}{ccc} c & \overset{g}{\rightarrowtail} & a \\ \downarrow & & \downarrow f \\ f\circ g(c) & \underset{\exists_f(g)}{\rightarrowtail} & b \end{array}$$

$\exists_f(g)$ is the smallest subobject of b through which $f \circ g$ factors (Theorem 5.2.1). The interplay between $\exists$ and im is very much to the fore in the next series of exercises, for which we assume that $\mathfrak{A}$ is the canonical $\mathscr{L}_{\mathscr{E}}$-model $\mathscr{L}_{\mathscr{E}} \to \mathscr{E}$ in a topos $\mathscr{E}$.

EXERCISE 8. Let a be an $\mathscr{E}$-object, and v, w variables of sort a.

(1) Show that $\mathfrak{A}(\exists v(v \approx v))$ is the support $\sup(a) \rightarrowtail 1$ of a (§12.1), and hence that $\mathfrak{A} \vDash \exists v(v \approx v)$ iff the unique arrow $a \to 1$ is epic.

(2) Show that the two projection arrows $a \times a \to a$ are equal iff $a \to 1$ is monic. Hence show that this last arrow is monic iff $\mathfrak{A} \vDash (v \approx w)$.

(3) Let $\mathbf{term}(a)$ be the conjunction of the formulae $\exists v(v \approx v)$ and $(v \approx w)$. Show that

$$\mathfrak{A} \vDash \mathbf{term}(a) \quad \text{iff} \quad a \text{ is a terminal object}$$ □

The formula $\mathbf{term}(a)$ may be regarded as expressing "there exist a unique v of sort a".

EXERCISE 9. If $f: a \to b$ is an arrow, let $\mathbf{mon}(f)$ be the formula $(f(v) \approx f(w) \supset v \approx w)$. Show that

$$\mathfrak{A} \vDash \mathbf{mon}(f) \quad \text{iff} \quad f \text{ is monic.}$$

EXERCISE 10. Given $f: a \to b$, $v: a$, and $w: b$, show that the "graph" $\langle 1_a, f\rangle: a \rightarrowtail a \times b$ of f is $\mathfrak{A}(f(v) \approx w)$. Hence show that $\operatorname{im} f: f(a) \rightarrowtail b$ is $\mathfrak{A}(\mathbf{ep}(f))$, where $\mathbf{ep}(f)$ is the formula $\exists v(f(v) \approx w)$. Thus

$$\mathfrak{A} \models \mathbf{ep}(f) \quad \text{iff} \quad f \text{ is epic.} \qquad \square$$

Thus the condition "f is iso" is characterised by the truth of $\mathbf{mon}(f) \wedge \mathbf{ep}(f)$, a formula that expresses "there is a unique v such that $f(v) = w$".

Next we consider equalisers. If $i: e \rightarrowtail a$ equalises $f, g: a \to b$, then i, as a subobject, is precisely $\mathfrak{A}(f(v) \approx g(v))$. On the other hand, since i is monic it can be identified with $\operatorname{im} i$, and hence (Exercise 10) with $\mathfrak{A}(\mathbf{ep}(i))$, so that

$$\mathfrak{A} \models (f(v) \approx g(v)) \equiv \exists w(i(w) \approx v).$$

Now if the arrow h in

$$c \xrightarrow{h} a \overset{f}{\underset{g}{\rightrightarrows}} b$$

is monic and has $f \circ h = g \circ h$, then h, or equivalently $\operatorname{im} h$, is a subobject of the equaliser of f and g, which means that $\mathfrak{A}(\mathbf{ep}(h)) \subseteq \mathfrak{A}(f(v) \approx g(v))$. Therefore, if the converse of this last inclusion holds, h itself is an equaliser of f and g. These observations lead to the following result.

EXERCISE 11. Given $f, g: a \to b$, and $h: c \to a$, let $\mathbf{equ}(h, f, g)$ be the conjunction of the three formulae

$$\mathbf{mon}(h),$$
$$f(h(w)) \approx g(h(w)),$$
$$f(v) \approx g(v) \supset \exists w(h(w) \approx v).$$

Then

$$\mathfrak{A} \models \mathbf{equ}(h, f, g) \quad \text{iff} \quad h \text{ is an equaliser of } f \text{ and } g. \qquad \square$$

For the case of products, given $f: c \to a$ and $g: c \to b$, then c will be a product of a and b, with f and g as projection arrows, precisely when the product arrow $\langle f, g\rangle$

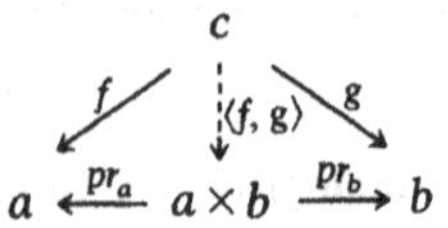

is iso (cf. §3.8), i.e. monic and epic. Ostensibly then we could express this by the formulae **mon** and **ep** applied to the arrow $\langle f, g\rangle$. But it is desirable that we have a formula that explicitly refers only to f and g. After all, the notation "$\langle f, g\rangle$" does not refer to a uniquely determined arrow (unlike "$f \circ g$"), but is only unique up to isomorphism and depends on the choice of the product $a \times b$ and projections pr_a and pr_b. Thus we reduce the two desired properties of $\langle f, g\rangle$ to properties of f and g.

EXERCISE 12. Let f, g, pr_a, pr_b be as above.

(1) Show that

$$\mathfrak{A}(\langle f, g\rangle(v) \approx \langle f, g\rangle(w)) = \mathfrak{A}(f(v) \approx f(w) \wedge g(v) \approx g(w)).$$

(2) Show that the graph $\langle \mathbf{1}_c, \langle f, g\rangle\rangle : c \to c \times (a \times b)$ of $\langle f, g\rangle$ is $\mathfrak{A}(f(v) \approx w \wedge g(v) \approx z)$, where $v : c$, $w : a$, $z : b$.

(3) Let **prod**(f, g) be the conjunction of the formulae

$$(f(v) \approx f(w) \wedge g(v) \approx g(w)) \supset v \approx w$$
$$\exists v (f(v) \approx w \wedge g(v) \approx z).$$

Show that

$\mathfrak{A} \models$ **prod**(f, g) iff c is a product of a and b with projections f and g

(cf. Exercises 9 and 10 above). □

By adapting these exercises to a model of the form $\mathfrak{A} : \mathscr{L}_{\mathscr{C}} \to \mathscr{E}$, we can extend Theorem 1 above to characterise left exactness of $\mathfrak{A}$, as a functor $\mathscr{C} \to \mathscr{E}$, in terms of the truth in $\mathfrak{A}$ of the formulae of the type **term**, **equ**, and **prod** determined by the terminal objects, equalisers, and products of pairs of objects in $\mathscr{C}$ (left exactness being equivalent to preservation of these particular limits). Our use of this logical characterisation will be in the context of continuous morphisms from a site on $\mathscr{C}$ to a topos $\mathscr{E}$ with its canonical pretopology. In the latter case, a set $C = \{f_x : x \in X\}$ of $\mathscr{E}$-arrows with the same codomain c is a cover of c iff $\bigcup_X \operatorname{im} f_X$ is the maximum element $\mathbf{1}_c$ of $\mathrm{Sub}_{\mathscr{E}}(c)$. But in the canonical model $\mathfrak{A} : \mathscr{L}_{\mathscr{E}} \to \mathscr{E}$, $\operatorname{im} f_x$ is $\mathfrak{A}(\mathbf{ep}(f_x))$, where $\mathbf{ep}(f_x)$ is the formula $\exists v_x (f_x(v_x) \approx v)$ for v_x: dom f_x and v: c. Moreover, *if X is finite*, we can form the disjunction

$$\bigvee_{x \in X} \mathbf{ep}(f_x)$$

as an $\mathscr{L}_{\mathscr{E}}$-formula. Since $\mathfrak{A}$ interprets disjunction as union in Sub(c), we have that $\mathfrak{A}(\bigvee \mathbf{ep}(f_x)) = \bigcup \operatorname{im} f_x$. Thus it follows in this case that C is a cover for the canonical site on $\mathscr{E}$ if, and only if, $\mathfrak{A} \models$ **cov**(C), where **cov**(C)

is the formula

$$\bigvee_{x\in X} \exists v_x (f_x(v_x) \approx v).$$

The restriction to finite X is of course because formulae in the first-order languages we are currently using are finite sequences of symbols, and we are disbarred from disjoining infinitely many formulae at once (the possibility of allowing this will be taken up later). So our present theory is appropriate to the case of *finitary* sites, in which all covers are finite.

If $\mathbf{C}=(\mathscr{C}, Cov)$ is a finitary site, we define a collection $\mathbb{T}_{\mathbf{C}}$ of formulae of the canonical language $\mathscr{L}_{\mathscr{C}}$ of the small category $\mathscr{C}$. $\mathbb{T}_{\mathbf{C}}$ is called the *theory* of the site $\mathbf{C}$, and consists of

(1) $\mathbf{id}(1_a)$, for each $\mathscr{C}$-object a;
(2) $\mathbf{com}(f, g, f\circ g)$ for each composable pair f, g of $\mathscr{C}$-arrows;
(3) $\mathbf{term}(a)$, for each terminal object a in $\mathscr{C}$;
(4) $\mathbf{equ}(h, f, g)$, for each equaliser h of a parallel pair f, g of $\mathscr{C}$-arrows;
(5) $\mathbf{prod}(f, g)$, for each pair of $\mathscr{C}$-arrows with $\text{dom}\, f = \text{dom}\, g$ that forms a product diagram in $\mathscr{C}$;
(6) $\mathbf{cov}(C)$, for each cover C in $\mathbf{C}$.

Notice that since $\mathbf{C}$ is small, so too is $\mathbb{T}_{\mathbf{C}}$.

THEOREM 2. *If $\mathscr{E}$ is a Grothendieck topos, and $\mathbf{C}$ a finitary site, then an $\mathscr{E}$-model $\mathfrak{A}:\mathscr{L}_{\mathscr{C}}\to\mathscr{E}$ for the canonical language of $\mathscr{C}$ is a continuous morphism $\mathfrak{A}:\mathbf{C}\to\mathscr{E}$ if, and only if, $\mathfrak{A}\models\mathbb{T}_{\mathbf{C}}$.* □

In view of the Reduction Theorem 16.2.2, we now see from Theorem 2 that the existence of geometric morphisms $\mathscr{E}\to\mathbf{Sh}(\mathbf{C})$ reduces to the existence of $\mathscr{E}$-models of $\mathbb{T}_{\mathbf{C}}$. In particular, points of the form $\mathbf{Set}\to\mathbf{Sh}(\mathbf{C})$ correspond to classical **Set**-based models of $\mathbb{T}_{\mathbf{C}}$. Since Deligne's Theorem is about the existence of sufficiently many points, while the classical Completeness Theorem is about the existence of sufficiently many **Set**-models (a falsifying one for each non-theorem), we begin to see why, and how, these two basic results are related.

The exacting reader will be dissatisfied with the gap between Theorems 1 and 2 of this section and the given arguments and exercises for **id**, **com**, **term**, **prod**, **equ**, and **cov** that lie behind them. The latter were stated in terms of canonical models $\mathscr{L}_{\mathscr{C}}\to\mathscr{C}$, whereas the Theorems refer to models $\mathscr{L}_{\mathscr{C}}\to\mathscr{E}$ of $\mathscr{L}_{\mathscr{C}}$ in other categories than $\mathscr{C}$. The only comment made about the connection was that the arguments and exercises given for $\mathscr{C}$ could be "repeated" in $\mathscr{E}$. This can be made precise by observing that a model $\mathfrak{A}: \mathscr{L}_{\mathscr{C}} \to \mathscr{E}$ can be regarded as a function $\mathfrak{A}: \mathscr{L}_{\mathscr{C}} \to \mathscr{L}_{\mathscr{E}}$ between canonical languages, takings sorts and operation symbols of $\mathscr{L}_{\mathscr{C}}$ to the corresponding entities in $\mathscr{L}_{\mathscr{E}}$. This induces a translation $(-)^{\mathfrak{A}}$ of $\mathscr{L}_{\mathscr{C}}$-formulae φ to $\mathscr{L}_{\mathscr{E}}$-formulae $\varphi^{\mathfrak{A}}$, obtained by replacing each operation symbol $\mathbf{g}$ in φ by $\mathfrak{A}(\mathbf{g})$, and regarding variables of sort a in φ as being of sort $\mathfrak{A}(a)$ in $\varphi^{\mathfrak{A}}$. It is then readily seen that each "axiom" associated with a diagram D in $\mathscr{C}$ translates under $(-)^{\mathfrak{A}}$ to the axiom associated with the image diagram $\mathfrak{A}(D)$ in $\mathscr{E}$. In other words, $(\mathbf{term}(a))^{\mathfrak{A}} = \mathbf{term}(\mathfrak{A}(a))$, $(\mathbf{cov}(C))^{\mathfrak{A}} = \mathbf{cov}(\mathfrak{A}(C))$, and so on. It is also straightforward to show that the interpretation of any $\mathscr{L}_{\mathscr{C}}$-formula φ in $\mathfrak{A}$ is the same as the interpretation of its translate $\varphi^{\mathfrak{A}}$ in the canonical $\mathscr{E}$-model $\mathfrak{A}_{\mathscr{E}}: \mathscr{L}_{\mathscr{E}} \to \mathscr{E}$. That is, we have ([MR] Proposition 3.5.1)

$$\mathfrak{A}^{\mathrm{v}}(\varphi) = \mathfrak{A}^{\mathrm{v}}_{\mathscr{E}}(\varphi^{\mathfrak{A}}),$$

and so

$$\mathfrak{A} \models \varphi \quad \text{iff} \quad \mathfrak{A}_{\mathscr{E}} \models \varphi^{\mathfrak{A}}.$$

Now suppose that D is one of the types of diagram in $\mathscr{C}$ that we have been considering (finite limit, cover etc.), with its categorial property P characterised by the $\mathfrak{A}$-truth of some $\mathscr{L}_{\mathscr{C}}$-formula φ_P (where φ_P has one of the forms **term**, **prod**, **cov** etc.). Then it follows that the same property for $\mathfrak{A}(D)$ in $\mathscr{E}$ is characterised by the truth in $\mathfrak{A}_{\mathscr{E}}$ of $(\varphi_P)^{\mathfrak{A}}$. In view of the last equation, this establishes the principle ([MR], Metatheorem 3.5.2) that

$$\mathfrak{A} \text{ preserves the property } P \text{ of } D \quad \text{iff} \quad \mathfrak{A} \models \varphi_P.$$

16.4. Geometric logic

A formula will be called *positive-existential* if, in addition to atomic formulae, it contains no logical symbols other than $\top$, $\bot$, $\wedge$, $\vee$, $\exists$. The class of all positive-existential $\mathscr{L}$-formulae will be denoted $\mathscr{L}^{\mathrm{g}}$. A *geometric*, or *coherent* L-formula is one of the form $\varphi \supset \psi$, where φ and ψ are in $\mathscr{L}^{\mathrm{g}}$. Since any φ can be identified with $(\top \supset \varphi)$, in the sense that $\mathfrak{A}(\varphi) = \mathfrak{A}(\top \supset \varphi)$ (Ex. 16.3.5), each positive existential formula can be regarded as being geometric. Also in this sense, the negation of an $\mathscr{L}^{\mathrm{g}}$-formula is geometric, as in general $\sim\varphi$ is equivalent to $(\varphi \supset \bot)$. A *set* $\mathbb{T}$ of geometric formulae will be called a *geometric theory*.

The concept of a geometric theory is central to our present context, as all members of the theory $\mathbb{T}_C$ of a finitary site are geometric. Moreover, all formulae of this type have their "truth-value" preserved by the inverse image parts of geometric morphisms. To see this, let $f:\mathscr{F}\to\mathscr{E}$ be a geometric morphism of topoi, and $\mathfrak{A}:\mathscr{L}\to\mathscr{E}$ an $\mathscr{E}$-model for some language $\mathscr{L}$. We define an $\mathscr{F}$-model $f^*\mathfrak{A}$ of $\mathscr{L}$, which, as a function on the collection of sorts, operation symbols, and relation symbols of $\mathscr{L}$, is just the composite of f^* and $\mathfrak{A}$. Thus for each $\mathscr{L}$-sort a, the $\mathscr{E}$-object $f^*\mathfrak{A}(a)$ is $f^*(\mathfrak{A}(a))$, the result of applying the functor f^* to the $\mathscr{E}$-object $\mathfrak{A}(a)$, and so on. The fact that f^* preserves products and subobjects and is functorial ensures that the definition of "model" is thereby satisfied.

THEOREM 1. (1) *For any positive-existential formula* φ,

$$(f^*\mathfrak{A})^{\mathbf{v}}(\varphi)=f^*(\mathfrak{A}^{\mathbf{v}}(\varphi)).$$

(2) *If* θ *is geometric, then*

$$\mathfrak{A}\models\theta \quad \textit{implies} \quad f^*\mathfrak{A}\models\theta.$$

(3) *If* f^* *is faithful, then for geometric* θ,

$$\mathfrak{A}\models\theta \quad \textit{iff} \quad f^*\mathfrak{A}\models\theta.$$

PROOF. (1) (Outline). This is proven by induction over the formation of φ. The essential point is that, being an exact functor, f^* preserves all finite limits and colimits, and hence preserves all the categorial structure involved in interpreting a positive-existential formula.

First of all, preservation of monics ensures that $f^*\mathfrak{A}^{\mathbf{v}}(\varphi)$ is a subobject of $f^*\mathfrak{A}(\mathbf{v})$. Functoriality of f^* and preservation of products makes $f^*\mathfrak{A}(\mathbf{v})=f^*(\mathfrak{A}(\mathbf{v}))$ for sequences of variables, and $f^*\mathfrak{A}^{\mathbf{v}}(t)=f^*(\mathfrak{A}^{\mathbf{v}}(t))$ for terms t to which $\mathbf{v}$ is appropriate.

Preservation of terminal and initial objects ensures that the desired result holds when φ is $\top$ or $\bot$, while the cases of the other atomic formulae use equalisers, products, and pullbacks. Pullbacks are used in the inductive case for $\wedge$, and coproducts and images (Ex. 16.1.8) are needed for $\vee$. Finally, preservation of images (and projection arrows) is needed for the inductive case of $\exists$.

(2) If θ is $\varphi\supset\psi$, where φ and ψ are in $\mathscr{L}^{g}$, and $\mathfrak{A}\models\theta$, then $\mathfrak{A}(\varphi)\subseteq\mathfrak{A}(\psi)$, so that there is an arrow h factoring $\mathfrak{A}(\varphi)$ through $\mathfrak{A}(\psi)$ in $\mathscr{E}$, i.e. $\mathfrak{A}(\varphi)=\mathfrak{A}(\psi)\circ h$. But then as f^* is a functor, $f^*(h)$ factors $f^*(\mathfrak{A}(\varphi))$ through $f^*(\mathfrak{A}(\psi))$. Hence by (1), we have $f^*\mathfrak{A}(\varphi)\subseteq f^*\mathfrak{A}(\psi)$, so that $f^*\mathfrak{A}\models\varphi\supset\psi$.

(3) If f^* is faithful, or equivalently conservative (§16.1), then f^* reflects

subobjects, so that, by (1), $f^*\mathfrak{A}(\varphi) \subseteq f^*\mathfrak{A}(\psi)$ only if $\mathfrak{A}(\varphi) \subseteq \mathfrak{A}(\psi)$. In combination with (2), the result then follows. □

Now if $\mathscr{E}$ is a *coherent* topos, and $\mathfrak{A}$ an $\mathscr{E}$-model of a geometric theory $\mathbb{T}$, then for any point $p : \mathbf{Set} \to \mathscr{E}$ it follows that $p^*\mathfrak{A}$ is a **Set**-model of $\mathbb{T}$. On the other hand, if $\mathfrak{A} \nvDash \mathbb{T}$ then there is some (geometric) formula $(\varphi \supset \psi)$ in $\mathbb{T}$ such that $\mathfrak{A}(\varphi) \nsubseteq \mathfrak{A}(\psi)$. But then, by Deligne's Theorem there exists a point p of $\mathscr{E}$ such that $p^*(\mathfrak{A}(\varphi)) \nsubseteq p^*(\mathfrak{A}(\psi))$, so that $p^*\mathfrak{A} \nvDash (\varphi \supset \psi)$. In this way, truth of geometric formulae in $\mathscr{E}$ reduces to the question of their truth in standard set-theoretic models. We have

THEOREM 2. *If $\mathfrak{A}$ is an $\mathscr{L}$-model in a coherent topos $\mathscr{E}$, and $\mathbb{T}$ a geometric theory, then*

$$\mathfrak{A} \vDash \mathbb{T} \text{ in } \mathscr{E} \quad \text{iff} \quad \text{for all points } p \text{ of } \mathscr{E},\ p^*\mathfrak{A} \vDash \mathbb{T} \text{ in } \mathbf{Set}.$$

EXERCISE 1. By appropriate choice of $\mathscr{L}$, $\mathfrak{A}$, and $\mathbb{T}$, deduce Deligne's Theorem from Theorem 2. □

The Theorem of Barr on the existence of Boolean-valued points for Grothendieck topoi also leads to an important metatheorem about models of geometric theories. Let us write $\mathbb{T} \vdash_c \theta$ to mean that θ is derivable from $\mathbb{T}$ by classical logic. This notion is defined by admitting proof sequences for θ that may contain as "axioms" members of $\mathbb{T}$ and classically valid axioms like $\varphi \vee {\sim}\varphi$. There is a standard completeness Theorem to the effect that $\mathbb{T} \vdash_c \varphi$ iff φ is true in every **Set**-model of $\mathbb{T}$ (Henkin [49]). But then from Barr's Theorem, we get

THEOREM 3. *If $\mathbb{T}$ and θ are geometric, and $\mathbb{T} \vdash_c \theta$, then θ is true in every $\mathbb{T}$-model in every Grothendieck topos.*

PROOF. Let $\mathfrak{A} : \mathscr{L} \to \mathscr{E}$ be a model of $\mathbb{T}$, with $\mathscr{E}$ a Grothendieck topos. By Barr's Theorem there exists a surjective geometric morphism of the form $f : \mathbf{Sh}(\mathbf{B}) \to \mathscr{E}$ for some complete **BA B**. Then as $\mathbb{T}$ is geometric, Theorem 1(2) implies that $f^*\mathfrak{A} \vDash \mathbb{T}$. But the laws of classical logic hold in the *Boolean* topos $\mathbf{Sh}(\mathbf{B})$, and so as $\mathbb{T} \vdash_c \theta$, $f^*\mathfrak{A} \vDash \theta$. Since f^* is faithful, Theorem 1(3) then gives $\mathfrak{A} \vDash \theta$. □

EXERCISE 2. Show that the restriction of Theorem 3 to coherent topoi follows from Deligne's Theorem. □

One interpretation of Theorem 3 is that for geometric formulae, anything inferrable by classical logic is inferrable by the weaker intuitionistic logic, and so we gain no new geometric theorems by using principles that are not intuitionistically valid (note of course that $\varphi \vee \sim \varphi$ is not geometric). But the importance of the result resides in its use in lifting mathematical constructions from **Set** to non-Boolean topoi. For example, suppose $\mathbb{T}$ consists of the axioms for the notion of a group. Then to show that $\mathbb{T}$-models in Grothendieck topoi have a certain property, then provided that the property can be expressed by a geometric formula θ, it suffices to show that all standard groups, i.e. all $\mathbb{T}$-models in **Set**, satisfy θ. But for the latter we have at our disposal the power of classical logic, and the techniques of standard group theory.

An application of this method to the "Galois theory of local rings" is given by Wraith [79].

Proof theory

By a *sequent* we mean an expression $\Gamma \supset \psi$, where Γ is a finite set of formulae, and ψ a single formula. A sequent is *geometric* if all members of $\Gamma \cup \{\psi\}$ are positive-existential.

A sequent is not a formula, since Γ is not, but if $\Gamma = \{\varphi_1, \ldots, \varphi_n\}$, then $\Gamma \supset \psi$ is "virtually the same thing as" the formula $(\varphi_1 \wedge .. \wedge \varphi_n) \supset \psi$. Thus if this last formula is true in a model $\mathfrak{A}$, we will say that the sequent $\Gamma \supset \psi$ is *true in* $\mathfrak{A}$. A set $\mathbb{T}$ of sequents will be called a *theory*, just as for a set of formulae.

It is clear that the notions of geometric sequent and geometric formula can be interchanged, and we will tend to do this at times. The point of introducing sequents at all is to provide a convenient notation for expressing axioms and rules of inference that enable us to derive geometric formulae. Given a theory $\mathbb{T}$, and sequent θ, we are going to define the relation "θ is provable from $\mathbb{T}$", denoted $\mathbb{T} \vdash \theta$. The aim of this proof-theoretic approach will be to obtain θ from $\mathbb{T}$ by operations on sequents that depend only on their syntactic form (i.e. on the nature of the symbols that occur in them), and not on any semantic notions of "truth", "implication", etc.

In the rules to follow, the union $\Gamma \cup \Delta$ of sets Γ and Δ will be written Γ, Δ, or Γ, φ if $\Delta = \{\varphi\}$. If $\Gamma = \{\varphi_1, \ldots, \varphi_n\}$, then $\bigwedge \Gamma$ denotes the conjunction $\varphi_1 \wedge .. \wedge \varphi_n$, while $\bigvee \Gamma$ is the disjunction $\varphi_1 \vee .. \vee \varphi_n$. If Γ is the empty set, $\bigvee \Gamma$ is $\bot$, while $\bigwedge \Gamma$ is $\top$, so that $\Gamma \supset \psi$ is identified with $\top \supset \psi$, or simply ψ in conformity with our conventions stated earlier.

We write $\Gamma(v_1, \ldots, v_n)$ to indicate that any free variable occurring in any member of Γ is amongst $v_1, \ldots, v_n$. $\Gamma(t_1, \ldots, t_n)$ denotes the set of formulae obtained by uniformly substituting the term t_i for v_i throughout Γ.

Given a set $\mathbb{T}$ of geometric sequents, $\mathbb{T}^{=}$ will denote the union of $\mathbb{T}$ and all the following

Axioms of Identity

$$v \approx v,$$

$$v \approx w \supset w \approx v,$$

$$v \approx w, \varphi \supset \varphi(v/w),$$

where v and w are variables of the same sort, and φ is atomic.

We can now set out the axiom system for geometric sequents developed by Makkai and Reyes ([MR], §5.2), which we will call GL.

Axiom

$$\Gamma \supset \psi, \quad \text{if} \quad \psi \in \Gamma.$$

Rules of inference

The rules all have the form

$$\frac{\{\theta_i : i \in I\}}{\theta},$$

the intended meaning being that the sequent θ is derivable if all of the sequents θ_i have been derived, i.e. the *conclusion* θ is a consequence of the *premisses* θ_i.

$(R\bigwedge_1)$ $$\frac{\Delta, \bigwedge\Gamma, \varphi \supset \psi}{\Delta, \bigwedge\Gamma \supset \psi}, \quad \text{if } \varphi \in \Gamma,$$

$(R\bigwedge_2)$ $$\frac{\Delta, \bigwedge\Gamma \supset \psi}{\Delta \supset \psi}, \quad \text{if } \Gamma \subseteq \Delta,$$

$(R\bigvee_1)$ $$\frac{\Delta, \varphi, \bigvee\Gamma \supset \psi}{\Delta, \varphi \supset \psi}, \quad \text{if } \varphi \in \Gamma,$$

and all free variables occurring in Γ also occur free in the conclusion.

$$(R\bigvee_2) \qquad \frac{\{\Delta, \bigvee\Gamma, \varphi \supset \psi : \varphi \in \Gamma\}}{\Delta, \bigvee\Gamma \supset \psi},$$

$$(R\exists_1) \qquad \frac{\Delta, \varphi(v/t), \exists w\varphi(v/w) \supset \psi}{\Delta, \varphi(v/t) \supset \psi},$$

$$(R\exists_2) \qquad \frac{\Delta, \exists w\varphi(v/w), \varphi \supset \psi}{\Delta, \exists w\varphi(v/w) \supset \psi},$$

if v does not occur free in the conclusion.

$$(R\mathbb{T}) \qquad \frac{\Delta, \Gamma(t_1, \ldots, t_n), \varphi(t_1, \ldots, t_n) \supset \psi}{\Delta, \Gamma(t_1, \ldots, t_n) \supset \psi},$$

provided that all free variables in the premiss occur free in the conclusion, and for some $v_1, \ldots, v_n$, $\Gamma(v_1, \ldots, v_n) \supset \varphi(v_1, \ldots, v_n)$ belongs to $\mathbb{T}^-$.

Note that the last rule $R\mathbb{T}$ depends on the particular theory $\mathbb{T}$. The restriction on free variables in $R\bigvee_1$, and $R\mathbb{T}$ are necessary for these rules to be truth-preserving, as our models now may involve "empty" objects (cf. the discussion of Detachment in §11.9).

We say that geometric sequent θ is *derivable from* $\mathbb{T}$ in the system GL, which we denote simply by $\mathbb{T} \vdash \theta$, if there is a proof sequence for θ from $\mathbb{T}$, i.e. a finite list of sequents ending in θ and such that each member of the list is either an axiom or a consequence of earlier members of the list by one of the above rules. It can be shown (cf. [MR] Theorem 3.2.8), that all of the axioms, including the Axioms of Identity, are true in any model in any topos, and that the rules of GL preserve this property. Hence the

Soundness Theorem. *If $\mathbb{T} \vdash \theta$, and if $\mathfrak{A}$ is a model of $\mathbb{T}$ in a topos $\mathscr{E}$, then $\mathfrak{A} \models \theta$.* □

The converse of Soundness asserts that if $\mathbb{T} \nvdash \theta$ (i.e. if θ is not GL-derivable from $\mathbb{T}$), then there is a $\mathbb{T}$-model in some topos that falsifies θ. If such a model can be found in a Grothendieck topos, then, in view of the discussion of the logical significance of Barr's Theorem (Theorem 3 above), one must exist in **Set**. Indeed we have the

Classical Completeness Theorem (cf. [MR], 5.2.3(b)). *If $\mathbb{T} \nvdash \theta$, then there is a* **Set**-*model $\mathfrak{A}$ such that $\mathfrak{A} \models \mathbb{T}$ and $\mathfrak{A} \not\models \theta$.*

There is a systematic technique for proving theorems of this kind. It is known as the *Henkin method*, after the work of Leon Henkin [49], who introduced it as a way of proving completeness for systems of the type described in §11.3. The basic idea is as follows.

If θ is $\Gamma \supset \varphi$, then since $\mathbb{T} \nvdash \theta$, the GL Axiom implies that $\varphi \notin \Gamma$. We then attempt to expand Γ to a set Σ of formulae that still does not include φ and which will be the full theory of the desired $\mathbb{T}$-model $\mathfrak{A}$, i.e. will have

$$\psi \in \Sigma \quad \text{iff} \quad \mathfrak{A} \models \psi.$$

Given Σ, the model is defined through specification of an equivalence relation on the terms of a given sort by

$$t \sim u \quad \text{iff} \quad (t \approx u) \in \Sigma.$$

The resulting equivalence classes $\hat{t}$ then become the individuals of the given sort, and relation and operation symbols are interpreted by putting

$$\mathfrak{A}(\mathbf{R})(\hat{t}_1, \ldots, \hat{t}_n) \quad \text{iff} \quad \mathbf{R}(t_1, \ldots, t_n) \in \Sigma,$$
$$\mathfrak{A}(\mathbf{g})(\hat{t}_1, \ldots, \hat{t}_n) = \hat{t} \quad \text{iff} \quad (\mathbf{g}(t_1, \ldots, t_n) \approx t) \in \Sigma.$$

Note that since $\mathbb{T}$ is a set, by ignoring any symbols extraneous to $\mathbb{T} \cup \{\theta\}$ we can assume we are dealing with a *small* language. This guarantees that the $\mathfrak{A}$-individuals of a given sort form a set, so that $\mathfrak{A}$ is indeed **Set**-based.

Now if Σ is to correspond to a model in this way, then it must satisfy certain closure properties, e.g. $\bigwedge \Delta \in \Sigma$ iff $\Delta \subseteq \Sigma$; if $(\Delta \supset \psi) \in \mathbb{T}^-$ and $\Delta \subseteq \Sigma$ then $\psi \in \Sigma$; if $\Sigma \vdash \theta$ then $\theta \in \Sigma$, and so on. Reflection on the desired properties of $\mathfrak{A}$ tells us exactly what properties Σ must have. The procedure then is to work through an enumeration of the formulae of the language, deciding of each formula in turn whether or not to add it into Σ, in such a way that the end result is as desired. In trying to do this we discover what rules of inference our axiom system needs to admit. If these rules are in turn truth-preserving, so that the Soundness Theorem is fulfilled, then the whole procedure becomes viable, and actually gives a systematic technique for constructing an axiomatisation of the class of "true" or "valid" formulae determined by a given notion of "model".

The reader will find a construction and proof of this Henkin type in almost any standard text on mathematical logic. A significant point for us to note about the method here is that it is entirely independent of category theory.

It follows from Classical Completeness that in order for $\mathbb{T}$ to have a set-theoretic model at all, it is sufficient that $\mathbb{T}$ be *consistent*, which means that $\mathbb{T} \nvdash \bot$. But since our proof theory is finitary, i.e. proof sequences are of finite length, it is the case in general that if $\mathbb{T} \vdash \theta$, then $\mathbb{T}_0 \vdash \theta$ for some finite subset $\mathbb{T}_0$ of $\mathbb{T}$. Consequently, in order for $\mathbb{T}$ to be consistent, and therefore have a **Set**-model, it suffices that each finite subset of $\mathbb{T}$ be consistent. Since Soundness implies that any theory having a model must be consistent, we have the following fundamental feature of finitary logic.

COMPACTNESS THEOREM. *If every finite subset of* $\mathbb{T}$ *has a* **Set**-*model, so too does* $\mathbb{T}$ □

Proof of Deligne's Theorem

We now apply Classical Completeness to the theory of a finitary site **C** to show that the coherent topos **Sh(C)** has enough points. For this we use the criterion stated as Theorem 16.2.5.

Let $C = \{f_x : x \in X\}$ be a set of **C**-arrows with the same codomain c, such that $C \notin Cov(c)$. We need to construct a continuous morphism $p : \mathbf{C} \to \mathbf{Set}$, i.e. by Theorem 16.3.2 a **Set**-model of the theory $\mathbb{T}_{\mathbf{C}}$, such that $p(C)$ is not epimorphic in **Set**.

Now let $\mathscr{L} = \mathscr{L}_{\mathbf{C}} \cup \{\mathbf{c}\}$, where $\mathscr{L}_{\mathbf{C}}$ is the language of the category underlying **C**, and **c** is a new individual constant of sort c. Let v be a variable of sort c, and for each $x \in X$ let v_x be a variable of sort dom f_x. Let $\varphi_x(v)$ be the $\mathscr{L}$-formula $\exists v_x(f_x(v_x) \approx v)$, and let $\varphi_x(\mathbf{c})$ be $\exists v_x(f_x(v_x) \approx \mathbf{c})$. Put

$$\mathbb{T} = \mathbb{T}_{\mathbf{C}} \cup \{\sim\varphi_x(\mathbf{c}) : x \in X\}.$$

Then $\mathbb{T}$ is geometric, and is a set in view of the smallness of the site **C**. We will show below that $\mathbb{T}$ is consistent, and therefore by Classical Completeness that there is a set-theoretic model $\mathfrak{A}_C : \mathscr{L} \to \mathbf{Set}$ such that $\mathfrak{A}_C \models \mathbb{T}$. But then $\mathfrak{A}_C$ is a model of $\mathbb{T}_{\mathbf{C}}$, and so determines a continuous morphism $\mathbf{C} \to \mathbf{Set}$, and hence a point of **Sh(C)**. If A is the set $\mathfrak{A}_C(c)$, then $\mathfrak{A}_C$ interprets the constant **c** as an element $a \in A$, and interprets each $\mathscr{L}_{\mathbf{C}}$-operation-symbol f_x as a function $g_x : \mathfrak{A}_C(\mathrm{dom}\, f_x) \to A$. But for each $x \in X$,

$$\mathfrak{A}_C \models \sim\exists v_x(f_x(v_x) \approx \mathbf{c}),$$

which means that $a \notin \mathrm{Im}\, g_x$, Thus $A \neq \bigcup_X \mathrm{Im}\, g_x$, showing that the family $\mathfrak{A}_C(C) = \{g_x : x \in X\}$ is not epimorphic in **Set**, as desired.

To prove that $\mathbb{T}$ is consistent, it is enough to show that each finite subset of $\mathbb{T}$ is consistent. Hence it is enough to show $\mathbb{T}_0 = \mathbb{T}_{\mathbf{C}} \cup \{\sim\varphi_x(\mathbf{c}) : x \in X_0\}$ is consistent for any finite $X_0 \subseteq X$. To this end, let

$E_{\mathbf{C}}$ be the canonical functor $\mathbf{C} \to \mathbf{Sh}(\mathbf{C})$. Then $E_{\mathbf{C}}$ is a continuous morphism, and so (16.3.2) serves as an $\mathscr{L}_{\mathbf{C}}$-model in $\mathbf{Sh}(\mathbf{C})$ of $\mathbb{T}_{\mathbf{C}}$. Moreover $E_{\mathbf{C}}$ reflects covers, so that $E_{\mathbf{C}}(C)$ is not epimorphic in $\mathbf{Sh}(\mathbf{C})$. Thus $\bigcup\{\text{im}\, E_{\mathbf{C}}(f_x)\colon x \in X\}$ is not the maximum subobject of $E_{\mathbf{C}}(c)$, and hence $\bigcup\{\text{im}\, E_{\mathbf{C}}(f_x)\colon x \in X_0\}$ is not the maximum either. But $E_{\mathbf{C}}$, as a model, interprets f_x, as a symbol, as the arrow $E_{\mathbf{C}}(f_x)$, and so

$$E_{\mathbf{C}} \nvDash \bigvee\{\varphi_x(v)\colon x \in X_0\}.$$

As $E_{\mathbf{C}} \vDash \mathbb{T}_{\mathbf{C}}$, it follows by Soundness that $\mathbb{T}_{\mathbf{C}} \nvdash \bigvee\{\varphi_x(v)\colon x \in X_0\}$. Hence by Classical Completeness it follows that there is a model $\mathfrak{A}\colon \mathscr{L}_{\mathbf{C}} \to \mathbf{Set}$ such that $\mathfrak{A} \vDash \mathbb{T}_{\mathbf{C}}$ and

$$\mathfrak{A} \nvDash \bigvee\{\varphi_x(v)\colon x \in X_0\}.$$

But this means that $\bigcup\{\text{Im}\,\mathfrak{A}(f_x)\colon x \in X_0\} \neq \mathfrak{A}(c)$, so that there is an element $a \in \mathfrak{A}(c)$ such that $a \notin \text{Im}\,\mathfrak{A}(f_x)$, for all $x \in X_0$. Then defining $\mathfrak{A}(\mathbf{c}) = a$ allows us to extend $\mathfrak{A}$ to become an $\mathscr{L}$-model in which all members of $\mathbb{T}_0$ are true. But if $\mathbb{T}_0$ has a model it must, by Soundness, be consistent as desired.

This finishes the proof that $\mathfrak{A}_C$ exists and has the required property that $\mathfrak{A}_C(C)$ is not epimorphic in **Set**. If we define P to be the set of such models $\mathfrak{A}_C$ for all sets C of $\mathbf{C}$-arrows that are not covers, then by Theorem 16.2.5, P is a sufficient set of points for $\mathbf{Sh}(\mathbf{C})$, and Deligne's Theorem is proved. □

Infinitary generalisation

In defining the theory $\mathbb{T}_{\mathbf{C}}$ we noted that the finiteness of first-order formulae restricted us to finitary sites, and that if we wanted to treat the general case we would have to be able to form disjunctions of infinite sets of formulae. There is no technical obstacle to doing this. We add to the inductive rules for generating formulae the condition that for any set Γ of formulae that has altogether finitely many free variables occurring in its members, there is a formula $\bigvee \Gamma$ with the set-theoretic semantics (§§11.3, 11.4).

$$\mathfrak{A} \vDash \bigvee \Gamma[x_1, \ldots, x_m] \quad \text{iff} \quad \text{for some } \varphi \in \Gamma,\ \mathfrak{A} \vDash \varphi[x_1, \ldots, x_m].$$

In any topos in which Sub(d) is always a complete lattice (which includes any Grothendieck topos), we can interpret infinitary disjunction by

$$\mathfrak{A}^{\mathbf{v}}(\bigvee \Gamma) = \bigcup\{\mathfrak{A}^{\mathbf{v}}(\varphi)\colon \varphi \in \Gamma\}.$$

We denote by $\mathscr{L}^{\infty}$ the class (in fact proper class) of infinitary formulae generated from $\mathscr{L}$ by allowing formation of $\bigvee\Gamma$ for sets Γ as above (cf. Barwise [75], §III.1 for a careful presentation of the syntax of infinitary formulae). $\mathscr{L}^{g\infty}$ denotes the positive-existential members of $\mathscr{L}^{\infty}$, i.e. those with no occurrence of the symbols $\forall$, $\supset$, $\sim$. The definition of derivability for $\mathscr{L}^{\infty}$ can no longer be given in terms of proof sequences of finite length, since even though a sequent $\Gamma \supset \varphi$ will continue to have Γ as a finite set of formulae, members of Γ can themselves have infinitely many subformulae, and so an instance of the rule $R\bigvee_2$ may well involve infinitely many premisses. Thus we will now stipulate that the relation $\mathbb{T} \vdash \theta$ holds when θ belongs to the smallest collection of formulae that contains $\mathbb{T}$ as well as all axioms and is closed under the rules of inference of the system GL. In other words, $\{\theta : \mathbb{T} \vdash \theta\}$ is the intersection of all collections that have these properties. (For finitary logic, this definition is equivalent to the one given in terms of finite proof sequences). For any $\mathbb{T}$-model $\mathfrak{A}$ in a topos, $\{\theta\colon \mathfrak{A} \models \theta\}$ is such a collection, and so contains all θ such that $\mathbb{T} \vdash \theta$. Hence we retain the Soundness Theorem.

If **C** is any small site, then by extending the definition of $\mathbb{T}_{\mathbf{C}}$ to include $\bigvee\{\exists v_x(f_x(v_x) \approx v)\colon x \in X\}$ for any cover $\{f_x\colon x \in X\}$ in **C**, we obtain the theory of **C** as a set of $\mathscr{L}^{\infty}_{\mathbf{C}}$-formulae. However we cannot now use $\mathbb{T}_{\mathbf{C}}$ in the way we did for Deligne's Theorem. Infinitary formulae do not enjoy the properties in **Set** that finitary ones do. To see this, let φ be the formula $\bigvee\{v \approx \mathbf{c}_n\colon n \in \omega\}$, where $\mathbf{c}_0, \mathbf{c}_1, \ldots$ is an infinite list of distinct individual constants. If $\mathbf{d}$ is a constant distinct from all the $\mathbf{c}_n$'s, then $\{\varphi\} \cup \{\sim(\mathbf{d} \approx \mathbf{c}_n) : n \in \omega\}$ cannot have a **Set**-model, even though each of its finite subsets does. Thus the Compactness Theorem fails for infinitary logic. Moreover, the Completeness Theorem no longer holds. It can be shown ([MR], p. 162) that if we admit disjunctions of *countable* sets Γ only, then a countable set of geometric sequents has a **Set**-model if it is consistent with respect to GL. On the other hand there exist uncountable sets of infinitary formulae that are consistent but have no **Set**-model at all (cf. Scott [65] for an example).

It was shown by Mansfield [72] that an Infinitary Completeness Theorem can be obtained if we replace standard set-theoretic models by **B**-valued models, for complete Boolean algebras **B**. Such a model interprets a formula $\varphi(v_1, \ldots, v_n)$ as a function of the form $A^n \to \mathbf{B}$, where A is the set of individuals of the model. This is very similar to the notion of model in the topos Ω-**Set** (for $\Omega = \mathbf{B}$) outlined at the end of §11.9.

Makkai and Reyes have adapted Mansfield's approach to their axioms for many-sorted geometric logic without existence assumptions. They show that for any set $\mathbb{T}$ of geometric sequents there is a complete **BA** $\mathbf{B}_{\mathbb{T}}$ and a model $\mathfrak{A}_{\mathbb{T}}$ of $\mathbb{T}$ in the Grothedieck topos $\mathbf{Sh}(\mathbf{B}_{\mathbb{T}})$ of sheaves over $\mathbf{B}_{\mathbb{T}}$ such that

$$\mathbb{T} \vdash \theta \quad \text{iff} \quad \mathfrak{A}_{\mathbb{T}} \models \theta$$

for "suitable" θ (see below).

It could be held that the **B**-valued approach recovers Completeness by generalising the notion of "model". But from the point of view of categorial logic one could say that the notion of model is invariant, in that the definition of $\mathscr{E}$-model is the same for all topoi $\mathscr{E}$, including $\mathscr{E} = \mathbf{Set}$, but that in order to obtain Completeness we have to allow the category in which the model lives to change as we change the theory $\mathbb{T}$.

Let us now sketch the definition of $\mathfrak{A}_{\mathbb{T}}$, and see how it can be used to prove Barr's Theorem. Given a geometric theory $\mathbb{T}$ in $\mathscr{L}^{\infty}$, i.e. a set $\mathbb{T} \subseteq \mathscr{L}^{g\infty}$, let L be any sub*set* of $\mathscr{L}^{g\infty}$ that contains $\mathbb{T}$ and is closed under (i) subformulae, and (ii) substitution for free variables of terms whose variables all occur in L. A subclass of $\mathscr{L}^{g\infty}$ satisfying (i) and (ii) is called a *fragment*. Since $\mathbb{T}$ is small, there do in fact exist small fragments containing $\mathbb{T}$. Let P be the collection of all sequents $\Gamma \supset \varphi$ of formulae from L such that $\mathbb{T} \nvdash \Gamma \supset \varphi$. If $p = (\Gamma \supset \varphi)$ is in P, we write Γ_p for Γ and φ_p for φ. Then a partial ordering on P is given by

$$p \sqsubseteq q \quad \text{iff} \quad \Gamma_p \subseteq \Gamma_q \quad \text{and} \quad \varphi_p = \varphi_q.$$

The Boolean algebra we want is obtained by applying double negation to the **CHA** $\mathbf{P}^+$ of hereditary subsets of $\mathbf{P} = (P, \sqsubseteq)$ (cf. §8.4). For each $S \in \mathbf{P}^+$, let S^* be $\neg\neg S$. Then $\mathbf{B}_{\mathbb{T}}$ is $\{S^*: S \in \mathbf{P}^+\}$, the lattice of *regular* elements of $\mathbf{P}^+$, and in general is a complete **BA** in which $\sqcap$ is set-theoretic intersection, $\bigsqcup X$ is $(\bigcup X)^*$, and $\neg$ is the Boolean complement (cf. Rasiowa and Sikorski [63], §IV.6). Since L is small, $\mathbf{B}_{\mathbb{T}}$ is too, and $\mathbf{Sh}(\mathbf{B}_{\mathbb{T}})$ is a Grothendieck topos.

For each formula $\varphi \in L$, put

$$P(\varphi) = \{p \in P: \varphi \in \Gamma_p\},$$

and for each term t occurring in L, put

$$P(t) = \{p \in P: \text{every variable of } t \text{ occurs free in } P\}.$$

Then $P(\varphi)$ and $P(t)$ are hereditary in P. Let $[\![t]\!] = P(t)^*$.

If $\mathfrak{A}_{\mathbb{T}}(a)$ denotes the set of L-terms of sort a, then a $\mathbf{B}_{\mathbb{T}}$-valued equality relation on $\mathfrak{A}_{\mathbb{T}}(a)$ is given by

$$[\![t \approx u]\!] = [\![t]\!] \cap [\![u]\!] \cap P(t \approx u)^*.$$

This makes $\mathfrak{A}_{\mathbb{T}}(a)$ into an object in the topos $\mathbf{B}_{\mathbb{T}}$-**Set** of $\mathbf{B}_{\mathbb{T}}$-valued sets (§11.9).

Similarly, for an n-placed relation symbol $\mathbf{R}\colon \langle a_1, \ldots, a_n\rangle$ of L, $\mathfrak{A}_{\mathbb{T}}(\mathbf{R})$ is defined as that function $\mathfrak{A}_{\mathbb{T}}(a_1)\times\cdots\times_{\mathbb{T}}(a_n)\to\mathbf{B}_{\mathbb{T}}$ given by

$$\mathfrak{A}_{\mathbb{T}}(\mathbf{R})\langle t_1, \ldots, t_n\rangle = [\![t_1]\!]\cap\ldots\cap[\![t_n]\!]\cap P(\mathbf{R}(t_1, \ldots, t_n))^*.$$

If $\mathbf{g}\colon \langle a_1, \ldots, a_n\rangle \to a_{n+1}$ is an n-placed operation symbol, then according to the definition of arrows and products in $\mathbf{B}_{\mathbb{T}}$-**Set** of §11.9, $\mathfrak{A}_{\mathbb{T}}(\mathbf{g})$ is to be a function from $\mathfrak{A}_{\mathbb{T}}(\mathrm{a}_1)\times\cdots\times\mathfrak{A}_{\mathbb{T}}(a_{n+1})$ to $\mathbf{B}_{\mathbb{T}}$. The definition is

$$\mathfrak{A}_{\mathbb{T}}(\mathbf{g})\langle t_1, \ldots, t_n, u\rangle = [\![t_1]\!]\cap\ldots\cap[\![t_n]\!]\cap[\![\mathbf{g}(t_1, \ldots, t_n)\approx u]\!].$$

This construction defines a $\mathbb{T}$-model in $\mathbf{B}_{\mathbb{T}}$-**Set** such that for any sequent θ of formulae in L,

$$\mathfrak{A}_{\mathbb{T}} \models \theta \quad \text{iff} \quad \mathbb{T} \vdash \theta$$

([MR], §§4.1, 4.2, 5.1, 5.2). But from the work of Denis Higgs referred to in §14.7 we know that there is an equivalence between **B-Set** and **Sh(B)**, and this allows us to realise $\mathfrak{A}_{\mathbb{T}}$ as a model in $\mathbf{Sh}(\mathbf{B}_{\mathbb{T}})$ as desired.

With regard to the Theorem of Barr, the construction is applied to the case that $\mathbb{T}$ is the theory $\mathbb{T}_{\mathbf{C}}$ of a small site $\mathbf{C}$ to give a model $\mathfrak{A}_{\mathbb{T}_{\mathbf{C}}}\colon \mathscr{L}_{\mathbf{C}} \to \mathbf{Sh}(\mathbf{B}_{\mathbb{T}_{\mathbf{C}}})$ such that $\mathfrak{A}_{\mathbb{T}_{\mathbf{C}}} \models \mathbb{T}_{\mathbf{C}}$. Then $\mathfrak{A}_{\mathbb{T}_{\mathbf{C}}}$ is a continuous morphism from $\mathbf{C}$ to $\mathbf{Sh}(\mathbf{B}_{\mathbb{T}_{\mathbf{C}}})$. We may choose our small fragment L of $\mathscr{L}_{\mathbf{C}}^{g\infty}$ to include the formula $\mathbf{cov}(C)$, i.e. $\bigvee\{\exists v_x(f_x(v_x)\approx v)\colon x\in X\}$ for every set $C=\{f_x\colon x\in X\}$ of $\mathbf{C}$-arrows with a common codomain. Then if $\{\mathfrak{A}_{\mathbb{T}_{\mathbf{C}}}(f_x)\colon x\in X\}$ is epimorphic in $\mathbf{Sh}(\mathbf{B}_{\mathbb{T}_{\mathbf{C}}})$, we have $\mathfrak{A}_{\mathbb{T}_{\mathbf{C}}} \models \mathbf{cov}(C)$, and so $\mathbb{T}_{\mathbf{C}} \vdash \mathbf{cov}(C)$ by the above construction. Since $E_{\mathbf{C}}\colon \mathbf{C}\to\mathbf{Sh}(\mathbf{C})$ is a model of $\mathbb{T}_{\mathbf{C}}$ in $\mathbf{Sh}(\mathbf{C})$, Soundness then implies that $E_{\mathbf{C}} \models \mathbf{cov}(C)$, which means that $\{E_{\mathbf{C}}(f_x)\colon x\in X\}$ is epimorphic in $\mathbf{Sh}(\mathbf{C})$. By Theorem 16.2.3, it follows that the geometric morphism $\mathbf{Sh}(\mathbf{B}_{\mathbb{T}_{\mathbf{C}}})\to\mathbf{Sh}(\mathbf{C})$ determined by $\mathfrak{A}_{\mathbb{T}_{\mathbf{C}}}$ is surjective.

16.5. Theories as sites

To derive Deligne's Theorem from Classical Completeness, a theory $\mathbb{T}_{\mathbf{C}}$ was associated with each finitary site $\mathbf{C}$. To make the converse derivation, the association will be reversed. Given a geometric theory $\mathbb{T}$ of finitary formulae, a site $\mathbf{C}_{\mathbb{T}}$ will be constructed such that models of $\mathbb{T}$ in a Grothendieck topos $\mathscr{E}$ correspond to continuous morphisms $\mathbf{C}_{\mathbb{T}}\to\mathscr{E}$. In particular, the canonical functor $E_{\mathbf{C}_{\mathbb{T}}}\colon \mathbf{C}_{\mathbb{T}}\to\mathbf{Sh}(\mathbf{C}_{\mathbb{T}})$ becomes a $\mathbb{T}$-model in $\mathbf{Sh}(\mathbf{C}_{\mathbb{T}})$ satisfying

$$E_{\mathbf{C}_{\mathbb{T}}} \models \varphi \quad \text{iff} \quad \mathbb{T} \vdash \varphi.$$

Application of Deligne's Theorem to $\mathbf{Sh}(\mathbf{C}_{\mathbb{T}})$ then yields Classical Completeness for $\mathbb{T}$.

The construction of $\mathbf{C}_{\mathbb{T}}$ is an elegant development of the "Lindenbaum algebra" notion outlined for propositional logic in §§6.5 and 8.3. To present the construction we will work from now within the class $\mathscr{L}^g$ of positive-existential finitary $\mathscr{L}$-formulae, where $\mathscr{L}$ is the language of the geometric theory $\mathbb{T}$.

Two $\mathscr{L}^g$-formulae φ and ψ will be called *provably equivalent* relative to $\mathbb{T}$ in GL if $\mathbb{T} \vdash \varphi \supset \psi$ and $\mathbb{T} \vdash \psi \supset \varphi$. This defines an equivalence relation on $\mathscr{L}^g$, for which the equivalence class of φ will be denoted $[\varphi]$. In the Lindenbaum algebra, equivalence classes are partially ordered by putting

$$[\varphi] \sqsubseteq [\psi] \quad \text{iff} \quad \mathbb{T} \vdash \varphi \supset \psi,$$

but in the case of $\mathbf{C}_{\mathbb{T}}$, these equivalence classes are going to be *arrows*, rather than objects. The objects, on the other hand, are to be classes of formulae under the equivalence relation determined by "changes of variables". To define this relation, consider two variable-sequences $\mathbf{v} = \langle v_1, \ldots, v_m \rangle$ and $\mathbf{v}' = \langle v'_1, \ldots, v'_m \rangle$ of the same length, with each v_i of the same sort as the corresponding v'_i. Let $\mathbf{v} \to \mathbf{v}'$ denote the function which associates v'_i with v_i. Acting on a formula $\varphi(\mathbf{v})$, $\mathbf{v} \to \mathbf{v}'$ produces the formula, denoted $\varphi(\mathbf{v}/\mathbf{v}')$ or simply $\varphi(\mathbf{v}')$, which is obtained by replacing every free occurrence of v_i in φ by v'_i. The *change of variables* $\mathbf{v} \to \mathbf{v}'$ is *acceptable* for $\varphi(\mathbf{v})$ if each v_i is free for v'_i in φ (cf. §11.3). An equivalence relation is then given by putting $\varphi \sim \psi$ iff ψ is the result of applying some acceptable change of variables to φ. The class $\{\psi : \varphi \sim \psi\}$ will be denoted $\{\varphi\}$. These classes are the objects of a category $\mathscr{C}_{\mathbb{T}}$. In dealing with these objects, it is useful to know that $\{\varphi\} \subseteq [\varphi]$, i.e. that if $\varphi \sim \psi$ then $\mathbb{T} \vdash \varphi \supset \psi$ and $\mathbb{T} \vdash \psi \supset \varphi$. Hence, by Soundness, if $\mathfrak{A} \models \mathbb{T}$ and $\varphi \sim \psi$, then $\mathfrak{A}(\varphi) = \mathfrak{A}(\psi)$.

The $\mathscr{C}_{\mathbb{T}}$-arrows from $\{\varphi(\mathbf{v})\}$ to $\{\psi(\mathbf{w})\}$ are the provable equivalence classes $[\alpha(\mathbf{v}', \mathbf{w}')]$ of formulae $\alpha(\mathbf{v}', \mathbf{w}')$, where $\mathbf{v}'$ and $\mathbf{w}'$ are disjoint sequences of variables having $\mathbf{v} \to \mathbf{v}'$ acceptable for $\varphi(\mathbf{v})$ and $\mathbf{w} \to \mathbf{w}'$ acceptable for $\psi(\mathbf{w})$, such that the following three geometric formulae are derivable from $\mathbb{T}$:

(α1) $\quad \alpha(\mathbf{v}', \mathbf{w}') \supset \varphi(\mathbf{v}') \vee \psi(\mathbf{w}')$,

(α2) $\quad \varphi(\mathbf{v}') \supset \exists \mathbf{w}' \alpha(\mathbf{v}', \mathbf{w}')$,

(α3) $\quad \alpha(\mathbf{v}', \mathbf{w}') \wedge \alpha(\mathbf{v}', \mathbf{w}'') \supset \mathbf{w}' \approx \mathbf{w}''$.

The notation being used here has $\exists \mathbf{v}\varphi$ abbreviating $\exists v_1 \ldots \exists v_m \varphi$, and $(\mathbf{v} \approx \mathbf{v}')$ abbreviating $(v_1 \approx v_1') \wedge \ldots \wedge (v_m \approx v_m')$, and so on. Note that since there are infinitely many variables of each sort, any two objects $\{\varphi(\mathbf{v})\}$ and $\{\psi(\mathbf{w})\}$ can be represented, by suitable changes of variables, in such a way that $\mathbf{v}$ and $\mathbf{w}$ are disjoint sequences, so that arrows can be taken in the form $[\alpha(\mathbf{v}, \mathbf{w})]$. This kind of relettering can be extended to all the objects and arrows of a finite diagram, and we will sometimes assume this has already been done in what follows.

To understand the definition of $\mathscr{C}_{\mathbb{T}}$-arrow, observe that in a **Set**-model $\mathfrak{A}$, $\mathfrak{A}(\alpha)$ will be the graph of a function from $\mathfrak{A}(\varphi)$ to $\mathfrak{A}(\psi)$. This interpretation motivates much of the structure of $\mathscr{C}_{\mathbb{T}}$. A formalisation of the interpretation is given in the next exercise, which we will make use of later on.

EXERCISE 1. Let $\mathfrak{A}: \mathscr{L} \to \mathscr{E}$ be a model in a topos, and let $\varphi(\mathbf{v})$, $\psi(\mathbf{w})$, $\alpha(\mathbf{v}, \mathbf{w})$ be formulae such that the formulae

(α1) $\qquad \alpha(\mathbf{v}, \mathbf{w}) \supset \varphi(\mathbf{v}) \wedge \psi(\mathbf{w})$,

(α2) $\qquad \varphi(\mathbf{v}) \supset \exists \mathbf{w} \alpha(\mathbf{v}, \mathbf{w})$,

(α3) $\qquad \alpha(\mathbf{v}, \mathbf{w}) \wedge \alpha(\mathbf{v}, \mathbf{w}') \supset \mathbf{w} \approx \mathbf{w}'$

are true in $\mathfrak{A}$. Assume that $\mathbf{v}$ and $\mathbf{w}$ are disjoint, so that if $\mathbf{z}$ is the sequence $\mathbf{v}, \mathbf{w}$, then $\mathfrak{A}(\mathbf{z})$ can be identified with $\mathfrak{A}(\mathbf{v}) \times \mathfrak{A}(\mathbf{w})$.

(1) Show that the product arrow $h: \mathfrak{A}^{\mathbf{v}}(\varphi) \times \mathfrak{A}^{\mathbf{w}}(\psi) \to \mathfrak{A}(\mathbf{z})$ is monic, and so determines a subobject of $\mathfrak{A}(\mathbf{z})$.

(2) Use the $\mathfrak{A}$-truth of (α1) to show that there is a monic $k: \mathfrak{A}^{\mathbf{z}}(\alpha) \rightarrowtail \mathfrak{A}^{\mathbf{v}}(\varphi) \times \mathfrak{A}^{\mathbf{w}}(\psi)$ factoring $\mathfrak{A}^{\mathbf{z}}(\alpha) \rightarrowtail \mathfrak{A}(\mathbf{z})$ through h.

(3) Let g be $pr \circ k$, where

$$\begin{array}{ccc} \mathfrak{A}^{\mathbf{z}}(\alpha) & \rightarrowtail & \mathfrak{A}^{\mathbf{v}}(\varphi) \times \mathfrak{A}^{\mathbf{w}}(\psi) \\ & {\scriptstyle g} \searrow & \downarrow {\scriptstyle pr} \\ & & \mathfrak{A}^{\mathbf{v}}(\varphi) \end{array}$$

pr is the projection. Use the truth of (α2) and (α3) to deduce that g is iso in $\mathscr{E}$.

(4) Using g^{-1}, construct an arrow $f_\alpha: \mathfrak{A}^{\mathbf{v}}(\varphi) \to \mathfrak{A}^{\mathbf{w}}(\psi)$ such that $\langle 1, f_\alpha \rangle: \mathfrak{A}^{\mathbf{v}}(\varphi) \rightarrowtail \mathfrak{A}^{\mathbf{v}}(\varphi) \times \mathfrak{A}^{\mathbf{w}}(\psi)$ and $\mathfrak{A}^{\mathbf{z}}(\alpha) \rightarrowtail \mathfrak{A}(\mathbf{z})$ are equal as subobjects of $\mathfrak{A}(\mathbf{z})$.

(5) Hence show that there is a *unique* arrow $\mathfrak{A}^{\mathbf{v}}(\varphi) \to \mathfrak{A}^{\mathbf{w}}(\psi)$ whose "graph" is $\mathfrak{A}^{\mathbf{z}}(\alpha) \rightarrowtail \mathfrak{A}(\mathbf{z})$. □

To specify the structure of $\mathscr{C}_{\mathbb{T}}$ as a category, the identity arrow on $\{\varphi(\mathbf{v})\}$ is defined to be $[\varphi(\mathbf{v})\wedge(\mathbf{v}\approx\mathbf{v}')]:\{\varphi(\mathbf{v})\}\to\{\varphi(\mathbf{v}')\}$. The composite of $[\alpha(\mathbf{v},\mathbf{w})]:\{\varphi(\mathbf{v})\}\to\{\psi(\mathbf{w})\}$, and $[\beta(\mathbf{w},\mathbf{z})]:\{\psi(\mathbf{w})\}\to\{\chi(\mathbf{z})\}$ is given by $[\exists\mathbf{w}(\alpha(\mathbf{v},\mathbf{w})\wedge\beta(\mathbf{w},\mathbf{z}))]$ (assuming that $\mathbf{v}$, $\mathbf{w}$, and $\mathbf{z}$ have been chosen to be mutually disjoint).

EXERCISE 2. Verify the category axioms for $\mathscr{C}_{\mathbb{T}}$.

EXERCISE 3. Show that for any formula φ, $[\varphi]$ is the one and only arrow from $\{\varphi\}$ to $\{\top\}$.

EXERCISE 4. Show that $\{\varphi(\mathbf{v})\}$ and $\{\psi(\mathbf{w})\}$ have product object $\{\varphi(\mathbf{v})\wedge\psi(\mathbf{w})\}$ with the projection to $\{\varphi(\mathbf{v}')\}$ being $[\varphi(\mathbf{v})\wedge\psi(\mathbf{w})\wedge(\mathbf{v}\approx\mathbf{v}')]$, and similarly for the projection to $\{\psi(\mathbf{w}')\}$.

EXERCISE 5. Show that arrows $[\alpha_i(\mathbf{v}_i,\mathbf{z})]:\{\varphi_i(\mathbf{v}_i)\}\to\{\chi(\mathbf{z})\}$, for $i=1,2$, have a pullback whose domain is $\{\exists\mathbf{z}(\alpha_1(\mathbf{v}_1,\mathbf{z})\wedge\alpha_2(\mathbf{v}_2,\mathbf{z}))\}$.

EXERCISE 6. Show that $[\alpha(\mathbf{v},\mathbf{w})]$ is monic iff $\mathbb{T}\vdash\alpha(\mathbf{v},\mathbf{w})\wedge\alpha(\mathbf{v}',\mathbf{w})\supset\mathbf{v}\approx\mathbf{v}'$. □

From Exercises 3 and 4 it follows that $\mathscr{C}_{\mathbb{T}}$ has a terminal object and pullbacks, and therefore has all finite limits. Note, by Exercise 4, that if $\mathbf{v}=\langle v_1,\ldots,v_m\rangle$, then $\{\mathbf{v}\approx\mathbf{v}\}$, i.e. $\{(v_1\approx v_1)\wedge\ldots\wedge(v_m\approx v_m)\}$, is the product of the objects $\{v_i\approx v_i\}$. If $\mathbf{v}$ is a sequence appropriate to φ, we write $\varphi^{\mathbf{v}}$ for the formula $\varphi\wedge(\mathbf{v}\approx\mathbf{v})$. We then have a subobject $\{\varphi^{\mathbf{v}}\}\rightarrowtail\{\mathbf{v}\approx\mathbf{v}\}$ of $\{\mathbf{v}\approx\mathbf{v}\}$ given by $[\varphi\wedge(\mathbf{v}\approx\mathbf{v}')]$. This subobject may be denoted simply as $\{\varphi^{\mathbf{v}}\}$, and whenever it is presented without naming the arrow, it will be the arrow just mentioned that is intended.

EXERCISE 7. Show that if $\mathbf{v}$ is appropriate to $(t_1\approx t_2)$, and w is a variable of the same sort as t_1 and t_2, then $\{(t_1\approx t_2)^{\mathbf{v}}\}\rightarrowtail\{\mathbf{v}\approx\mathbf{v}\}$ equalises the arrows $[(t_i\approx w)\wedge(\mathbf{v}\approx\mathbf{v})]:\{\mathbf{v}\approx\mathbf{v}\}\to\{w\approx w\}$, for $i=1,2$.

EXERCISE 8. Show that $\{(\varphi_1\wedge\varphi_2)^{\mathbf{v}}\}$ is $\{\varphi_1^{\mathbf{v}}\}\cap\{\varphi_2^{\mathbf{v}}\}$. □

To make $\mathscr{C}_{\mathbb{T}}$ into a site, a *finite* set $C=\{[\alpha_x(\mathbf{v}_x,\mathbf{w})]: x\in X\}$ of arrows, where $[\alpha_x]$ is of the form $\{\varphi_x(\mathbf{v}_x)\}\to\{\psi(\mathbf{w})\}$, is defined to be *provably epimorphic* if

$$\mathbb{T}\vdash\psi(\mathbf{w})\to\bigvee\{\exists\mathbf{v}_x\alpha_x(\mathbf{v}_x,\mathbf{w}): x\in X\}.$$

In particular, if C is the empty set, this means that $\mathbb{T} \vdash (\psi(\mathbf{w}) \supset \bot)$, i.e. $\mathbb{T} \vdash {\sim}\psi(\mathbf{w})$.

The reader may care to verify that the finite provably epimorphic families form a pretopology on $\mathscr{C}_{\mathbb{T}}$, and this gives us a *finitary* site $\mathbf{C}_{\mathbb{T}}$, and hence a *coherent* topos $\mathbf{Sh}(\mathbf{C}_{\mathbb{T}})$.

EXERCISE 9. Suppose that $\mathbf{v}$ is appropriate to $\varphi_1 \vee \varphi_2$. For $i = 1, 2$, let $f_i : \{\varphi_i^{\mathbf{v}}\} \to \{(\varphi_1 \vee \varphi_2)^{\mathbf{v}}\}$ be $[\varphi_i \wedge (\mathbf{v} \approx \mathbf{v}')]$. Show that $\{f_1, f_2\}$ is provably epimorphic. Hence show that if $F : \mathbf{C}_{\mathbb{T}} \to \mathscr{E}$ is a continuous morphism, where $\mathscr{E}$ is a Gorthendieck topos with its canonical pretopology, then $F(\{(\varphi_1 \vee \varphi_2)^{\mathbf{v}}\})$ is $F(\{\varphi_1^{\mathbf{v}}\}) \cup F(\{\varphi_2^{\mathbf{v}}\})$.

EXERCISE 10. Let $\mathbf{v}$ be appropriate to the formula $\exists w\varphi$, with w a variable not occurring in $\mathbf{v}$, and let $\mathbf{z}$ be the sequence $\mathbf{v}, w$. Let g_φ be the arrow $[\varphi^{\mathbf{z}} \wedge (\mathbf{v} \approx \mathbf{v}')] : \{\varphi^{\mathbf{z}}\} \to \{(\exists w\varphi)^{\mathbf{v}}\}$. Show that $\{g_\varphi\}$ is provably epimorphic, and hence that if F is as in Exercise 9, then $F(g_\varphi)$ is an epic arrow in $\mathscr{E}$.

EXERCISE 11. Let $\mathbf{v}$, w, φ, $\mathbf{z}$, and g_φ be as in the last Exercise. Show that the diagram

$$\begin{array}{ccc} \{\varphi^{\mathbf{z}}\} & \rightarrowtail & \{\mathbf{z} \approx \mathbf{z}\} \\ {\scriptstyle g_\varphi}\downarrow & & \downarrow{\scriptstyle pr} \\ \{(\exists w\varphi)^{\mathbf{v}}\} & \rightarrowtail & \{\mathbf{v} \approx \mathbf{v}\} \end{array}$$

commutes, where pr is the evident projection. Hence show that if $F : \mathbf{C}_{\mathbb{T}} \to \mathscr{E}$ is continuous, as in Exercise 9, then F takes this diagram to an epi-monic factorisation of the F-image of $\{\varphi^{\mathbf{z}}\} \rightarrowtail \{\mathbf{z} \approx \mathbf{z}\} \xrightarrow{pr} \{\mathbf{v} \approx \mathbf{v}\}$. Using the left exactness of F, deduce from this that

$$F(\{(\exists w\varphi)^{\mathbf{v}}\}) = \exists_{F(pr)}(F\{\varphi^{\mathbf{z}}\}). \qquad \square$$

These last exercises indicate that if $F : \mathbf{C}_{\mathbb{T}} \to \mathscr{E}$ is a continuous morphism from $\mathbf{C}_{\mathbb{T}}$ to a Grothendieck topos then F preserves some of the structure relevant to the interpretation of formulae. Indeed we can use F to define a $\mathbb{T}$-model $\mathfrak{A}_F : \mathscr{L} \to \mathscr{E}$, where $\mathscr{L}$ is the language of $\mathbb{T}$, as follows.

If a is an $\mathscr{L}$-sort, we choose a variable $v : a$, and put $\mathfrak{A}_F(a) = F(\{v \approx v\})$. Since $\{v \approx v\} = \{v' \approx v'\}$ whenever v' is any other variable of sort a, the definition is unambiguous. If $\mathbf{v} = \langle v_1, \ldots, v_m \rangle$, with $v_i : a_i$, then $\mathfrak{A}_F(\mathbf{v})$ is $F(\{v_1 \approx v_1\}) \times \cdots \times F(\{v_m \approx v_m\})$. But F preserves products, so then $\mathfrak{A}_F(\mathbf{v})$ is $F(\{\mathbf{v} \approx \mathbf{v}\})$. Hence if $\mathbf{g} : \langle a_{i_1}, \ldots, a_{i_n} \rangle \to a$ is an n-placed operation symbol, we can put $\mathfrak{A}_F(\mathbf{g}) = F(\bar{g})$, where $\bar{g}$ is $[\mathbf{g}(v_{i_1}, \ldots, v_{i_n}) \approx v]: \{\mathbf{v}' \approx \mathbf{v}'\} \to \{v \approx v\}$, where v: a, and $\mathbf{v}' = \langle v_{i_1}, \ldots, v_{i_n} \rangle$. If $\mathbf{c}$ is an individual constant of sort a, we take $\mathfrak{A}_F(\mathbf{c})$ to be $F([\mathbf{c} \approx v]: \{\top\} \to \{v \approx v\})$. Finally, if $\mathbf{R}$: $\langle a_{i_1}, \ldots, a_{i_n} \rangle$ is an n-placed relation symbol, we put $\mathfrak{A}_F(\mathbf{R}) = F(\{\mathbf{R}(v_{i_1}, \ldots, v_{i_n})\} \rightarrowtail \{\mathbf{v}' \approx \mathbf{v}'\})$, noting that F preserves monics.

EXERCISE 12. Suppose that $\mathbf{v}$ is appropriate to the term t: a, and let w: a. Show that $\mathfrak{A}_F^{\mathbf{v}}(t)$ is the F-image of

$$[(t \approx w) \wedge (\mathbf{v} \approx \mathbf{v})]: \{\mathbf{v} \approx \mathbf{v}\} \to \{w \approx w\}. \qquad \square$$

THEOREM 1. *If φ is in $\mathcal{L}^g$, then for any sequence $\mathbf{v} = \langle v_1, \ldots, v_m \rangle$ appropriate to φ,*

$$\mathfrak{A}_F^{\mathbf{v}}(\varphi) = F(\{\varphi^{\mathbf{v}}\})$$

PROOF. By induction on the formation of φ.

(1) If φ is $\top$, $\mathfrak{A}_F^{\mathbf{v}}(\varphi)$ is the maximum subobject of $\mathfrak{A}_F(\mathbf{v})$, i.e. of $F(\{\mathbf{v} \approx \mathbf{v}\})$. But $\{\top^{\mathbf{v}}\} \rightarrowtail \{\mathbf{v} \approx \mathbf{v}\}$ is iso, and F preserves iso's, hence the result holds in this case.

(2) If φ is $\perp$, $\mathfrak{A}_F^{\mathbf{v}}(\varphi)$ is the minimum subobject $0 \rightarrowtail F(\{\mathbf{v} \approx \mathbf{v}\})$. But since $\top \vdash \perp^{\mathbf{v}} \supset \perp$, the empty set is provably epimorphic and covers $\{\perp^{\mathbf{v}}\}$ in $\mathbf{C}_{\mathbb{T}}$. By continuity of F then, the empty set covers $F(\{\perp^{\mathbf{v}}\})$ in $\mathscr{E}$. But canonical covers in $\mathscr{E}$ are effectively epimorphic, and so by Exercise 16.2.16, $F(\{\perp^{\mathbf{v}}\})$ is initial in $\mathscr{E}$, as desired.

(3) If φ is $(t_1 \approx t_2)$, then $\mathfrak{A}_F^{\mathbf{v}}(\varphi)$ equalises $\mathfrak{A}_F^{\mathbf{v}}(t_1)$ and $\mathfrak{A}_F^{\mathbf{v}}(t_2)$. Since F preserves equalisers, the result follows by Exercises 7 and 12.

(4) If φ is $\mathbf{R}(v_{i_1}, \ldots, v_{i_n})$, then with $\mathbf{v}' = \langle v_{i_1}, \ldots, v_{i_n} \rangle$, there is a pullback in $\mathscr{C}_{\mathbb{T}}$ of the form

$$\begin{array}{ccc} \{\varphi^{\mathbf{v}}\} & \rightarrowtail & \{\mathbf{v} \approx \mathbf{v}\} \\ \downarrow & & \downarrow{\scriptstyle pr} \\ \{\varphi\} & \rightarrowtail & \{\mathbf{v}' \approx \mathbf{v}'\} \end{array}$$

But F preserves pullbacks, and the F-image of the bottom arrow is, by definition, $\mathfrak{A}_F(\mathbf{R})$. Hence the F-image of the top arrow is $\mathfrak{A}_F^{\mathbf{v}}(\varphi)$, by definition of the latter.

(5) If φ is $(\varphi_1 \wedge \varphi_2)$, and the result holds for φ_1 and φ_2, then $\mathfrak{A}_F^{\mathbf{v}}(\varphi)$ is

$F(\{\varphi_1^{\mathbf{v}}\}) \cap F(\{\varphi_2^{\mathbf{v}}\})$. But by Exercise 8 there is a pullback

$$\begin{array}{ccc} \{(\varphi_1 \wedge \varphi_2)^{\mathbf{v}}\} & \rightarrowtail & \{\varphi_2^{\mathbf{v}}\} \\ \downarrow & & \downarrow \\ \{\varphi_1^{\mathbf{v}}\} & \rightarrowtail & \{\mathbf{v} \approx \mathbf{v}\} \end{array}$$

in $\mathscr{C}_{\mathbb{T}}$, and F preserves pullbacks, and so $\mathfrak{A}_F^{\mathbf{v}}(\varphi)$ is $F(\{(\varphi_1 \wedge \varphi_2)^{\mathbf{v}}\})$.

(6) If φ is $\varphi_1 \vee \varphi_2$, use Exercise 9 in a similar manner to the previous case.

(7) If φ is $\exists w\psi$, then $\mathfrak{A}_F^{\mathbf{v}}(\varphi)$ is $\exists_p(\mathfrak{A}_F^{\mathbf{z}}(\psi))$, where $\mathbf{z}$ is $\mathbf{v}, w$ and $p : \mathfrak{A}_F(\mathbf{z}) \to \mathfrak{A}_F(\mathbf{v})$ is the projection. Hence if the Theorem holds for ψ, $\mathfrak{A}_F^{\mathbf{v}}(\varphi)$ is $\exists_p(\{\varphi^{\mathbf{z}}\})$. But by left exactness, p is $F(pr : \{\mathbf{z} \approx \mathbf{z}\} \to \{\mathbf{v} \approx \mathbf{v}\})$, and so the desired conclusion follows by Exercise 11. □

COROLLARY 2. *For any geometric $\mathscr{L}$-formula θ,*

$$\mathbb{T} \vdash \theta \quad \textit{implies} \quad \mathfrak{A}_F \models \theta.$$

In particular, $\mathfrak{A}_F \models \mathbb{T}$.

PROOF. Let θ be $\varphi \supset \psi$, and let $\mathbf{v}$ be the sequence of all variables that have a free occurrence in θ. Then if $\mathbb{T} \vdash \theta$, we have $\mathbb{T} \vdash \varphi \supset \varphi \wedge \psi$ (by the Axiom and rules $R\bigwedge_2$ and $R\mathbb{T}$ of GL). From this it follows readily that

$$\begin{array}{ccc} \{\varphi^{\mathbf{v}}\} & \searrow & \\ {\scriptstyle [\varphi \wedge \mathbf{v} \approx \mathbf{v}']} \downarrow & & \{\mathbf{v} \approx \mathbf{v}\} \\ \{\psi^{\mathbf{v}}\} & \nearrow & \end{array}$$

commutes in $\mathscr{C}_{\mathbb{T}}$, i.e. $\{\varphi^{\mathbf{v}}\} \subseteq \{\psi^{\mathbf{v}}\}$. Since F preserves monics and commutative triangles, with the aid of the Theorem we then have $\mathfrak{A}_F^{\mathbf{v}}(\varphi) \subseteq \mathfrak{A}_F^{\mathbf{v}}(\psi)$ in $\mathscr{E}$, so that $\mathfrak{A}_F \models \varphi \supset \psi$, as desired.

Finally, since $\mathbb{T} \vdash \theta$ whenever $\theta \in \mathbb{T}$ (by the GL Axiom and $R\mathbb{T}$), this makes every member of $\mathbb{T}$ true in $\mathfrak{A}_F$. □

Now the definition of $\mathfrak{A}_F$ can be applied in the case that F is the canonical continuous morphism $E_{\mathbf{C}_{\mathbb{T}}} : \mathbf{C}_{\mathbb{T}} \to \mathbf{Sh}(\mathbf{C}_{\mathbb{T}})$ to yield a model of $\mathbb{T}$ in $\mathbf{Sh}(\mathbf{C}_{\mathbb{T}})$, which we will denote $\mathfrak{A}_{\mathbb{T}}$ (so that the subscripting will not become ridiculous). Moreover, the pretopology defining $\mathbf{C}_{\mathbb{T}}$ is *precanonical* in the sense of §16.2. For if $C = \{[\alpha_x(\mathbf{v}_x, \mathbf{w})]\ x \in X\}$: is provably epimorphic, and $D = \{[\beta_x(\mathbf{v}_x, \mathbf{z})]: x \in X\}$ is a family that is compatible with

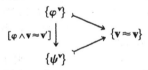

C, in the sense defined prior to Exercise 16.2.15, then

$$g = [\bigvee \{\exists \mathbf{v}_x(\alpha_x(\mathbf{v}_x, \mathbf{w}) \wedge \beta_x(\mathbf{v}_x, \mathbf{z})) \colon x \in X\}]$$

proves to be the one and only arrow

$$\begin{array}{ccc} \{\varphi_x(\mathbf{v}_x)\} & & \\ {\scriptstyle[\alpha_x]}\downarrow & \searrow^{[\beta_x]} & \\ & & \{\chi(\mathbf{z})\} \\ \{\psi(\mathbf{w})\} & \dashrightarrow_{g} & \end{array}$$

that factors each $[\beta_x]$ through the corresponding $[\alpha_x]$ (cf. Johnstone [77], p. 245). Thus each cover in $\mathbf{C}_{\mathbb{T}}$ is an effectively epimorphic family, so that the canonical functor $E_{\mathbf{C}_{\mathbb{T}}}$ is actually the Yoneda embedding, which thereby makes $\mathbf{C}_{\mathbb{T}}$ (isomorphic to) a full subcategory of $\mathbf{Sh}(\mathbf{C}_{\mathbb{T}})$. This allows us to sharpen the last Corollary in the case of $\mathfrak{A}_{\mathbb{T}}$.

THEOREM 3. *For geometric* θ,

$$\mathbb{T} \vdash \theta \quad \textit{iff} \quad \mathfrak{A}_{\mathbb{T}} \models \theta.$$

PROOF. If $\mathfrak{A}_{\mathbb{T}} \models \varphi \supset \psi$, where $\varphi, \psi \in \mathscr{L}^g$, then $\mathfrak{A}^{\mathbf{v}}_{\mathbb{T}}(\varphi) \subseteq \mathfrak{A}^{\mathbf{v}}_{\mathbb{T}}(\psi)$, so that, by Theorem 1, we have a factoring

$$\begin{array}{ccc} E_{\mathbf{C}_{\mathbb{T}}}(\{\varphi^{\mathbf{v}}\}) & & \\ \downarrow & \searrow & \\ & & E_{\mathbf{C}_{\mathbb{T}}}(\{\mathbf{v} \approx \mathbf{v}\}) \\ & \nearrow & \\ E_{\mathbf{C}_{\mathbb{T}}}(\{\psi^{\mathbf{v}}\}) & & \end{array}$$

in $\mathbf{Sh}(\mathbf{C}_{\mathbb{T}})$. But the Yoneda embedding is injective on objects, and bijective on hom-sets, and so this last diagram pulls back to a factoring

$$\begin{array}{ccc} \{\varphi^{\mathbf{v}}\} & & \\ {\scriptstyle[\alpha]}\downarrow & \searrow & \\ & & \{\mathbf{v} \approx \mathbf{v}\} \\ & \nearrow & \\ \{\psi^{\mathbf{v}}\} & & \end{array}$$

of $\{\varphi^{\mathbf{v}}\}$ through $\{\psi^{\mathbf{v}}\}$ in $\mathscr{C}_{\mathbb{T}}$. Applying acceptable reletterings $\mathbf{v} \to \mathbf{v}'$ and $\mathbf{v} \to \mathbf{v}''$ to $\varphi^{\mathbf{v}}$ and $\psi^{\mathbf{v}}$ respectively, the definition of composition in $\mathscr{C}_{\mathbb{T}}$ then implies that

$$\mathbb{T} \vdash \varphi(\mathbf{v}') \wedge (\mathbf{v}' \approx \mathbf{v}) \supset \exists \mathbf{v}''(\alpha(\mathbf{v}', \mathbf{v}'') \wedge \psi(\mathbf{v}'') \wedge (\mathbf{v}'' \approx \mathbf{v})),$$

from which we can obtain

$$\mathbb{T} \vdash \varphi(\mathbf{v}) \supset \psi(\mathbf{v}).$$ □

At last we are in a position to show that Deligne's Theorem implies the Classical Completeness Theorem for finitary geometric theories $\mathbb{T}$. For, if $\mathbb{T} \nvdash \varphi \supset \psi$, where $\varphi, \psi \in \mathcal{L}^g$, the last Theorem implies that $\mathfrak{A}_{\mathbb{T}}(\varphi) \nsubseteq \mathfrak{A}_{\mathbb{T}}(\psi)$ in the model $\mathfrak{A}_{\mathbb{T}}$ in the coherent topos $\mathbf{Sh}(\mathbf{C}_{\mathbb{T}})$. By Deligne's Theorem (and Exercise 16.2.20), there is therefore a point $p: \mathbf{Set} \to \mathbf{Sh}(\mathbf{C}_{\mathbb{T}})$ such that $p^*(\mathfrak{A}_{\mathbb{T}}(\varphi)) \nsubseteq p^*(\mathfrak{A}_{\mathbb{T}}(\psi))$. But then, as in §16.4, p gives rise to a model $p^*\mathfrak{A}_{\mathbb{T}}: \mathcal{L} \to \mathbf{Set}$ which, by Theorem 16.4.1(1), has $p^*\mathfrak{A}_{\mathbb{T}}(\varphi) \nsubseteq p^*\mathfrak{A}_{\mathbb{T}}(\psi)$. Moreover, as $\mathbb{T}$ consists of geometric formulae, 16.4.1(2) implies that $p^*\mathfrak{A}_{\mathbb{T}} \models \mathbb{T}$. Thus there exists a $\mathbb{T}$-model in $\mathbf{Set}$ in which $(\varphi \supset \psi)$ is not true. □

As a final, cautionary, note on this topic we observe that the above derivation is founded entirely on the structure of $\mathbf{C}_{\mathbb{T}}$, and hence ultimately on the properties of the relation of $\mathbb{T}$-derivability. In many cases one can most quickly confirm that $\mathbb{T} \vdash \varphi$ by observing that θ is true in all $\mathbf{Set}$ models of $\mathbb{T}$ and then appealing to Classical Completeness. But of course if we want to use this approach to prove Completeness, it has to be shown directly in each case that there is a proof sequence for θ within the axiom system in question.

Classifying topoi

What is the relationship between a finitary geometric theory $\mathbb{T}$ and the theory $\mathbb{T}_{\mathbf{C}_{\mathbb{T}}}$ of its associated finitary site $\mathbf{C}_{\mathbb{T}}$? Introducing the notation "$\mathfrak{A}: \mathbb{T} \to \mathscr{E}$" to mean "$\mathfrak{A}$ is a model of $\mathbb{T}$ in $\mathscr{E}$", we can say from our earlier work (Theorem 16.3.2) that models $\mathbb{T}_{\mathbf{C}_{\mathbb{T}}} \to \mathscr{E}$ in a Grothendieck topos $\mathscr{E}$ correspond precisely to continuous morphisms $\mathbf{C}_{\mathbb{T}} \to \mathscr{E}$. We have also seen in this Section (Corollary 2) that such morphisms determine models $\mathbb{T} \to \mathscr{E}$ of $\mathbb{T}$ in $\mathscr{E}$. We will now show that the converse is true, i.e. that every $\mathbb{T} \to \mathscr{E}$ arises in this way from a unique continuous $\mathbf{C}_{\mathbb{T}} \to \mathscr{E}$. In this sense, the theories $\mathbb{T}$ and $\mathbb{T}_{\mathbf{C}_{\mathbb{T}}}$ have exactly the same models in Grothendieck topoi.

Given a model $\mathfrak{A}: \mathcal{L} \to \mathscr{E}$ such that $\mathfrak{A} \models \mathbb{T}$, we define a continuous morphism $F_{\mathfrak{A}}: \mathbf{C}_{\mathbb{T}} \to \mathscr{E}$. For each $\mathcal{L}^g$-formula φ, $\mathfrak{A}(\varphi)$ is a subobject of $\mathfrak{A}(\mathbf{v})$, where $\mathbf{v}$ is the list of free variables of φ. Identifying $\mathfrak{A}(\varphi)$ with its domain, so that we can regard it as an $\mathscr{E}$-object, we put $F_{\mathfrak{A}}(\{\varphi\}) = \mathfrak{A}(\varphi)$ (strictly speaking this determines $F_{\mathfrak{A}}$ "up to isomorphism" only). Note that if $\{\varphi\} = \{\psi\}$, then $[\varphi] = [\psi]$ and so, as $\mathfrak{A} \models \mathbb{T}$, we have $\mathfrak{A}(\varphi) = \mathfrak{A}(\psi)$ by Soundness. Hence $F_{\mathfrak{A}}$ is unambiguously defined on objects.

If $[\alpha(\mathbf{v}, \mathbf{w})]: \{\varphi(\mathbf{v})\} \to \{\psi(\mathbf{w})\}$ is a $\mathscr{C}_{\mathbb{T}}$-arrow, then the geometric formulae $(\alpha 1)$, $(\alpha 2)$, and $(\alpha 3)$, whose $\mathbb{T}$-derivability is implied by the definition of "$\mathscr{C}_{\mathbb{T}}$-arrow", must be true in $\mathfrak{A}$. Exercise 1 then yields a unique $\mathscr{E}$-arrow $f_\alpha : \mathfrak{A}(\varphi) \to \mathfrak{A}(\psi)$ whose graph is $\mathfrak{A}(\alpha)$. We put $F_{\mathfrak{A}}([\alpha]) = f_\alpha$. Again a Soundness argument confirms that if $[\alpha] = [\beta]$ then $f_\alpha = f_\beta$. It is left to the reader to verify that $F_{\mathfrak{A}}$ preserves identities and commutative triangles, and so is a functor from $\mathscr{C}_{\mathbb{T}}$ to $\mathscr{E}$.

EXERCISE 13. Given $[\alpha(\mathbf{v}, \mathbf{w})]$ and f_α as above, show that $\operatorname{im} f_\alpha : f_\alpha(\mathfrak{A}(\varphi)) \rightarrowtail \mathfrak{A}(\psi)$ is equal, as a subobject of $\mathfrak{A}(\psi)$, to $\mathfrak{A}(\exists \mathbf{v}\alpha(\mathbf{v}, \mathbf{w})) \rightarrowtail \mathfrak{A}(\psi)$, where the latter monic derives, via (α1) and $R\exists_2$, from the fact that $\mathfrak{A} \models \exists \mathbf{v}\alpha(\mathbf{v}, \mathbf{w}) \supset \psi$. □

Now if $C = \{[\alpha_x(\mathbf{v}_x, \mathbf{w})] \colon x \in X\}$ is a provably epimorphic family in $\mathbf{C}_{\mathbb{T}}$, with $[\alpha_x] \colon \{\varphi_x(\mathbf{v}_x)\} \to \{\psi(\mathbf{w})\}$, then in the $\mathbb{T}$-model $\mathfrak{A}$ we have

$$\mathfrak{A} \models \psi(\mathbf{w}) \supset \bigvee \{\exists \mathbf{v}_x \alpha_x(\mathbf{v}_x, \mathbf{w}) \colon x \in X\},$$

and so

$$\mathfrak{A}(\psi) \subseteq \bigcup \{\mathfrak{A}(\exists \mathbf{v}_x \alpha_x(\mathbf{v}_x, \mathbf{w})) \colon x \in X\}.$$

From the last Exercise it then follows that

$$\mathfrak{A}(\psi) \subseteq \bigcup \{f_{\alpha_x}(\mathfrak{A}(\varphi_x)) : x \in X\},$$

so that $\{f_{\alpha_x} \colon x \in X\}$ is a canonical cover in $\mathscr{E}$. Thus $F_{\mathfrak{A}}$ preserves covers. Since $\mathfrak{A}(\top)$ is 1 (by definition), and $\{\top\}$ is terminal in $\mathbf{C}_{\mathbb{T}}$ (Exercise 3), $F_{\mathfrak{A}}$ preserves terminals. Finally, we leave it to the reader once more to confirm that $F_{\mathfrak{A}}$ preserves pullbacks, and hence complete the proof that $F_{\mathfrak{A}}$ is a continuous morphism.

If we now use $F_{\mathfrak{A}}$ to construct a model $\mathfrak{A}_{F_{\mathfrak{A}}} : \mathscr{L} \to \mathscr{E}$ as above, then Theorem 1 implies that $\mathfrak{A}_{F_{\mathfrak{A}}}(\varphi) = F_{\mathfrak{A}}(\{\varphi\}) = \mathfrak{A}(\varphi)$. Indeed for any $\mathscr{L}$-sort a, if $v \colon a$ then $\mathfrak{A}_{F_{\mathfrak{A}}}(a)$ is $F_{\mathfrak{A}}(\{v \approx v\})$, i.e. $\mathfrak{A}(v \approx v)$, which we identify with the domain of the identity arrow on $\mathfrak{A}(a)$. If $\mathbf{g}$ is an operation symbol, and $[\alpha(\mathbf{v}, \mathbf{w})]$ is $[\mathbf{g}(\mathbf{v}) \approx w]$, then $\mathfrak{A}(\alpha)$ is the graph of $\mathfrak{A}(\mathbf{g})$, so that $\mathfrak{A}(\mathbf{g}) = F_{\mathfrak{A}}([\alpha]) = \mathfrak{A}_F(\mathbf{g})$. Similarly, $\mathfrak{A}_{F_{\mathfrak{A}}}(\mathbf{R})$ is the same subobject of $\mathfrak{A}(\mathbf{v})$ as is $\mathfrak{A}(\mathbf{R})$, and so $\mathfrak{A}_{F_{\mathfrak{A}}}$ and $\mathfrak{A}$ prove to be the same model.

On the other hand, starting from F we find that $F(\{\varphi\}) = \mathfrak{A}_F(\varphi) = F_{\mathfrak{A}_F}(\{\varphi\})$. Since $\mathfrak{A}_F(\varphi)$, as a object, is only defined up to isomorphism, we find that the functors F and $F_{\mathfrak{A}_F}$ are naturally isomorphic. In this sense we obtain an exact correspondence between $\mathscr{E}$-models of $\mathbb{T}$ and continuous morphisms $\mathbf{C}_{\mathbb{T}} \to \mathscr{E}$.

Let us now return to the co-universal property of the canonical morphism $E_{\mathbf{C}}:\mathbf{C}\to\mathbf{Sh}(\mathbf{C})$ of a small site $\mathbf{C}$, as expressed by the diagram

$$\begin{array}{ccc} \mathbf{C} & \xrightarrow{E_{\mathbf{C}}} & \mathbf{Sh}(\mathbf{C}) \\ & {\scriptstyle F}\searrow & \downarrow{\scriptstyle F^*} \\ & & \mathscr{E} \end{array}$$

(cf. Theorem 16.2.2). The diagram tells us that every continuous morphism defined on $\mathbf{C}$ extends along $E_{\mathbf{C}}$ to a continuous morphism on $\mathbf{Sh}(\mathbf{C})$ that is unique up to a natural isomorphism. This property gives rise to the following diagram of $\mathbb{T}$-models

$$\begin{array}{ccc} \mathbb{T} & \xrightarrow{\mathfrak{A}_{\mathbb{T}}} & \mathbf{Sh}(\mathbf{C}_{\mathbb{T}}) \\ & {\scriptstyle \mathfrak{A}}\searrow & \downarrow{\scriptstyle F^*_{\mathfrak{A}}} \\ & & \mathscr{E} \end{array}$$

This diagram conveys that for any $\mathbb{T}$-model $\mathfrak{A}$ there is a unique (up to natural isomorphism) continuous $F^*_{\mathfrak{A}}:\mathbf{Sh}(\mathbf{C}_{\mathbb{T}})\to\mathscr{E}$, given by

$$\begin{array}{ccc} \mathbf{C}_{\mathbb{T}} & \xrightarrow{E_{\mathbf{C}_{\mathbb{T}}}} & \mathbf{Sh}(\mathbf{C}_{\mathbb{T}}) \\ & {\scriptstyle F_{\mathfrak{A}}}\searrow & \downarrow{\scriptstyle F^*_{\mathfrak{A}}} \\ & & \mathscr{E} \end{array}$$

such that the $\mathscr{E}$-model $F^*_{\mathfrak{A}}\mathfrak{A}_{\mathbb{T}}$, defined as for Theorem 16.4.1, is $\mathfrak{A}$ itself.

This characteristic property of $\mathfrak{A}_{\mathbb{T}}$ is what is meant by the notion of a *classifying topos* for a (possibly infinitary) geometric theory $\mathbb{T}$. To define this concept in general we fix a "base" Grothendieck topos $\mathscr{E}$, and consider pairs $(\mathscr{F},\mathfrak{A})$ consisting of a Grothendieck $\mathscr{E}$-topos $\mathscr{F}$ and an $\mathscr{F}$-model $\mathfrak{A}$ of $\mathbb{T}$ (we call $\mathfrak{A}$ a $\mathbb{T}$-model *over* $\mathscr{E}$). Then we say that $(\mathscr{E}[\mathbb{T}], \mathfrak{A}_{\mathbb{T}})$ is a *classifying* $\mathscr{E}$*-topos for* $\mathbb{T}$ with *generic* model $\mathfrak{A}_{\mathbb{T}}$, if it is universal among such pairs, i.e. if for any pair $(\mathscr{F},\mathfrak{A})$ as above, there is a geometric morphism $f:\mathscr{F}\to\mathscr{E}[\mathbb{T}]$ over $\mathscr{E}$ unique up to natural isomorphism such that $\mathfrak{A}=f^*\mathfrak{A}_{\mathbb{T}}$. Thus a generic $\mathbb{T}$-model over $\mathscr{E}$ has the property that every other $\mathbb{T}$-model over $\mathscr{E}$ arises by pulling the generic model back along some (unique) geometric morphism. Since inverse image functors preserve geometric formulae, it follows that the geometric formulae true in $\mathfrak{A}_{\mathbb{T}}$ are precisely those that are true in all $\mathbb{T}$-models over $\mathscr{E}$. The notation $\mathscr{E}[\mathbb{T}]$ is intended to convey the idea that the classifying $\mathscr{E}$-topos for $\mathbb{T}$ is generated by "adjoining a generic $\mathbb{T}$-model to $\mathscr{E}$".

If $\mathbf{C}$ is a small site, then models of the (infinitary) theory $\mathbb{T}_{\mathbf{C}}$ in

Grothendieck topoi correspond to geometric morphisms into **Sh**(**C**), and so by the first of the above three diagrams, $E_{\mathbf{C}}:\mathbf{C}\to\mathbf{Sh}(\mathbf{C})$ is a generic $\mathbb{T}_{\mathbf{C}}$-model, making **Sh**(**C**) a classifying topos for $\mathbb{T}_{\mathbf{C}}$. Since Grothendieck topoi are all defined over **Set**, we can thus express **Sh**(**C**) as $\mathbf{Set}[\mathbb{T}_{\mathbf{C}}]$. Hence each Grothendieck topos arises by adjoining to **Set** a generic model of a geometric theory.

In the converse direction, if $\mathbb{T}$ is a finitary geometric theory, then the construction of $\mathfrak{A}_{\mathbb{T}}$ from $E_{\mathbf{C}_{\mathbb{T}}}:\mathbf{C}_{\mathbb{T}}\to\mathbf{Sh}(\mathbf{C}_{\mathbb{T}})$ provides a generic $\mathbb{T}$-model, making $\mathbf{Sh}(\mathbf{C}_{\mathbb{T}})$ the classifying **S**-topos $\mathbf{Set}[\mathbb{T}]$ for $\mathbb{T}$. Hence the coherent topoi are *precisely* the classifying **S**-topoi for finitary geometric theories. This analysis can be extended to any infinitary geometric theory $\mathbb{T}$, to show that $\mathbf{Set}[\mathbb{T}]$ exists, but this requires a great deal more work. Amongst other things, the category $\mathscr{C}_{\mathbb{T}}$ has to be "enlarged" to include coproducts and "quotients of equivalence relations". However that is a story that we shall have to leave for the reader to pursue in Chapters 8 and 9 of [MR].

The conclusion of this work is that the concepts of "Grothendieck topos" and "classifying topos of a geometric theory" are coextensive. This has particular relevance in Algebraic Geometry, where some of the most important categories ("Zariski" topos, "Etale" topos) which form the focus of the work of the Grothendieck school turn out to be the classifying topoi for certain naturally occurring algebraic theories (cf. [MR], Chapter 9, Wraith [79]).

Forcing topologies

Let $\mathscr{L}$ be a language that has altogether finitely many sorts, relation and operation symbols, and constants, and let $\mathbb{T}$ be a finite geometric $\mathscr{L}$-theory. Then it can be shown that for any elementary topos $\mathscr{E}$ with a natural numbers object there exists a classifying topos $\mathscr{E}[\mathbb{T}]$ for models of $\mathbb{T}$ in $\mathscr{E}$-topoi. The proof of this is given by Tierney [76] (cf. also Johnstone [77], 6.56). We will not attempt to reproduce it here, but will briefly discuss an aspect of the construction which uses topologies $j:\Omega\to\Omega$ in an interesting way to produce $\mathbb{T}$-models.

If $I\rightarrowtail\Omega$ is a subobject of Ω in an elementary topos $\mathscr{E}$ then some work of Diaconescu [75] (cf. Johnstone [77], 3.58) shows that there is a smallest subobject $J\rightarrowtail\Omega$ containing I such that the characteristic arrow $\Omega\to\Omega$ of J is a topology. This characteristic arrow is called the topology *generated by I*. If $m:a\rightarrowtail b$ is any $\mathscr{E}$-monic, and $I_m\rightarrowtail\Omega$ the image of $\chi_m:b\to\Omega$, we denote the topology generated by I_m by j_m. Then the inclusion functor $sh_{j_m}(\mathscr{E}) \hookrightarrow \mathscr{E}$ has the property that for any geometric morphism $f:\mathscr{F} \to \mathscr{E}$ there exists a factorisation of the form

$$\begin{array}{ccc} \mathscr{F} & & \\ \downarrow & \searrow^{f} & \\ sh_{j_m}(\mathscr{E}) & \hookrightarrow & \mathscr{E} \end{array}$$

if, and only if, $f^*(m)$ is iso in $\mathscr{F}$ (Tierney [76], p. 212; Johnstone [77], 4.19). In particular the sheafification functor $\mathscr{S}\!\mathscr{h}_{j_m}:\mathscr{E} \to sh_{j_m}(\mathscr{E})$ makes $\mathscr{S}\!\mathscr{h}_{j_m}(m)$ iso. In view of this universal property, Tierney calls j_m the topology that *forces* m to be iso.

This notion can be extended to finitely many monics $m_1, \ldots, m_n$: the topology generated by $I_{m_1} \cup .. \cup I_{m_n} \rightarrowtail \Omega$ forces all of $m_1, \ldots, m_n$ to be iso.

Now let $\mathfrak{A}:\mathscr{L} \to \mathscr{E}$ be an $\mathscr{L}$-model in $\mathscr{E}$. Then a geometric $\mathscr{L}$-formula $(\varphi \supset \psi)(\mathbf{v})$ is true in $\mathfrak{A}$ iff the monic m in the following pullback is iso

$$\begin{array}{ccc} \mathfrak{A}(\varphi) \cap \mathfrak{A}(\psi) & \rightarrowtail & \mathfrak{A}(\psi) \\ {\scriptstyle m}\downarrow & & \downarrow \\ \mathfrak{A}(\varphi) & \rightarrowtail & \mathfrak{A}(\mathbf{v}) \end{array}$$

Thus if $m_1, \ldots, m_n$ are all these monics corresponding to the members of $\mathbb{T}$, and $j_{\mathbb{T}}$ is the topology that forces $m_1, \ldots, m_n$ to be iso, then $j_{\mathbb{T}}$ forces $\mathfrak{A}$ to become a $\mathbb{T}$-model in $sh_{j_{\mathbb{T}}}(\mathscr{E})$. For any geometric morphism $f:\mathscr{F} \to \mathscr{E}$, f factors through $sh_{j_{\mathbb{T}}}(\mathscr{E}) \hookrightarrow \mathscr{E}$ iff $f^*\mathfrak{A}$ is a model of $\mathbb{T}$ in $\mathscr{F}$.

This forcing construction is not special to models of first-order languages. Tierney observes that "given any diagram D in a topos $\mathscr{E}$, we can force any appropriate finite configuration (or even not necessarily finite if properly indexed over a base topos) in D to become a limit or colimit".

Rings and fields

We end this chapter by pointing the reader in the direction of literature that applies logical aspects of geometric morphisms to some familiar algebraic notions.

In classical algebra, a *commutative ring with unity* (henceforth simply called a *ring*) can be defined as a structure $(R, +, 0, \times, 1)$, consisting of a set R carrying two commutative binary operations $+$ and $\times$, and two distinguished elements 0 and 1, such that

(i) $(R, +, 0)$ is a group;

(ii) $(R, \times, 1)$ is a monoid; and

(iii) $x \times (y+z) = (x \times y) + (x \times z)$, for all $x, y, z \in R$.

Standard examples of rings are of course the number systems $\mathbb{Z}$, $\mathbb{Q}$, $\mathbb{R}$, $\mathbb{C}$, with $+$, $\times$, 0, 1 having their usual arithmetical meanings. The notion of a ring can be expressed in the first-order language having symbols for $+$, $\times$, 0, 1 by a finite set of *equations*.

A ring is a *field* if $R-\{0\}$ is a group under $\times$ with identity 1. This requires that $0 \neq 1$ (a geometric condition), and that any $x \neq 0$ has an inverse under $\times$. Writing $U(x)$ to mean that $\exists y(x \times y = 1)$, then in **Set** this last condition can be expressed by any of the following three classically equivalent assertions.

(1) $(x=0) \vee U(x)$;

(2) $\sim(x=0) \supset U(x)$;

(3) $\sim U(x) \supset x=0$.

These conditions are not generally equivalent for rings (i.e. models of the ring axioms) in non-Boolean topoi. Since (1) is expressed by a geometric formula, rings satisfying it are called *geometric fields*. (2) and (3) define, respectively, the notions of "field of fractions" and "residue field" (Johnstone [77], 6.64).

Another possible field axiom, considered by Kock [76], is

(4) $$\sim\left(\bigwedge_{i=1}^{n} (x_i = 0)\right) \supset \bigvee_{i=1}^{n} U(x_i).$$

This in turn implies the geometric condition

(5) $$U(x+y) \supset U(x) \vee U(y),$$

which defines the notion of a *local ring*. In general, (5) is weaker than (4), but Kock proves the significant fact that the generic local ring is a field in the sense of (4). If $\mathbb{T}_l$ is the geometric theory consisting of the ring axioms together with (5), then by the generic local ring is meant the generic model $\mathfrak{A}_{\mathbb{T}_l}$ in the classifying **S**-topos for $\mathbb{T}_l$. Now the geometric formulae true in all local rings in **S**-topoi are precisely those true in $\mathfrak{A}_{\mathbb{T}_l}$, i.e. those deducible from $\mathbb{T}_l$. Since, as Kock proves, $\mathfrak{A}_{\mathbb{T}_l}$ satisfies (4), the Soundness Theorem then yields the following metalogical principle:

If θ is geometric and $\mathbb{T}_l$, (4) $\vdash \varphi$, then $\mathbb{T}_l \vdash \varphi$.

Thus in deriving a geometric consequence of $\mathbb{T}_l$ we can invoke the assistance of the stronger, non-geometric, condition (4). Indeed (4) can be replaced in this argument by any axiom which is satisfied by the generic local ring. Further results along these lines, and a detailed analysis of a dozen or so possible axioms for fields and local rings, are given by Johnstone [77i] and [77], §§6.5, 6.6.

There is now in existence a vast literature about the *representation* of rings, and other algebraic structures, by global sections of sheaves (cf. Pierce [67], Dauns and Hofmann [68], Hofmann [72], Hofmann and Liukkonen [74]). Suppose that $p : A \to I$ is a sheaf (local homeomorphism) over a topological space I such that each stalk $p^{-1}(\{i\})$ is a ring in its own right under operations $+_i$, $\times_i$, and identities 0_i and 1_i. If $f, g : I \to A$ are global sections of the sheaf, then we can define sections $f+g$, $f\times g$ by putting

$$f+g(i) = f(i) +_i g(i),$$
$$f\times g(i) = f(i) \times_i g(i), \quad \text{all } i \in I.$$

If $f+g$, $f\times g$, and the sections 0 and 1 having $0(i) = 0_i$, $1(i) = 1_i$, are continuous whenever f and g are continuous, then p is called a *sheaf of rings* over the space I. In this situation, and with these definitions, the set of continuous global sections of p forms a ring. The aim of representation theory is to show that a given ring is isomorphic to the ring of continuous global sections of some sheaf of rings. An important result in this direction concerns *regular* rings, which are those satisfying

$$\forall x \exists y (x \times y \times x = x).$$

Regular rings include fields (let y be the $\times$-inverse of x). But every regular ring can be represented as the ring of continuous sections of a sheaf in which each stalk is a field! (Pierce [67], §10). This phenomenon gives rise to "transfer principles", in which properties of fields are shown to hold for regular rings by showing that they are preserved by the representation. An early paper on this theme is MacIntyre [73], concerned with transferring a property called "model completeness".

More generally we can study sheaves whose stalks are all **Set**-models of some theory $\mathbb{T}$, and seek to show that a Set-model $\mathfrak{A}$ of some other theory $\mathbb{T}_0$ is isomorphic to the structure of continuous sections of a sheaf of $\mathbb{T}$-models over some space I. In this situation $\mathfrak{A}$ may also be regarded as a model in the topos **Top**(I) of sheaves over I. Its behaviour as a **Top**(I)-model may differ from that which it exhibits as a **Set**-model. In particular, any geometric formula true in each stalk will be true in the

Top(I)-model $\mathfrak{A}$ (Fourman and Scott [79], 6.9). Thus if a regular ring R is represented by a sheaf of fields over I, then R becomes a **Top**(I)-model of the geometric field axioms, i.e. the regular ring R "is" a field from the point of view of the mathematical universe **Top**(I) (cf. Fourman and Scott [79], p. 367).

This theme is taken up in the thesis of Louillis [79], who adapted some of the work of classical model theory to categories of sheaves. The papers of Coste [79], Bunge and Reyes [81] and Bunge [81] present major advances in the use of geometric morphisms to transfer model-theoretic properties from the theory of the stalks to the theory of the global continuous sections of sheaves. A survey of more classical applications of model theory in sheaves is given by Burris and Werner [79].

REFERENCES

The following list is confined to articles and books that have been referred to in the preceding text. A fuller bibliography of topos theory is to be found in Johnstone [77].

M. A. Arbib and E. G. Manes,

[75] *Arrows, Structures, and Functors,* Academic Press, 1975.

M. Artin, A. Grothendieck and J. L. Verdier,

[SGA4] *Théorie des topos et cohomologie étale des schémas,* Vols., 1, 2, and 3, Lecture Notes in Mathematics, Vol. 269 (1972), Vol. 270 (1972) and Vol. 305 (1973), Springer-Verlag.

Michael Barr,

[74] Toposes without points, *J. of Pure and Applied Algebra,* 5, 1974, 265–280.

Jon Barwise,

[75] *Admissible Sets and Structures,* Springer-Verlag, 1975.

[77] *Handbook of Mathematical Logic* (ed.), North-Holland, 1977.

J. L. Bell and A. B. Slomson,

[69] *Models and Ultraproducts,* North-Holland, 1969.

E. W. Beth,

[56] Semantic construction of Intuitionistic logic, Mededelingen der Koninklijke Nederlandse Akademie van Wetenschappen, 1956, 357–388.

[59] *The Foundations of Mathematics,* North-Holland, 1959.

D. Bergamini,

[65] *Mathematics,* Time-Life International, 1965.

E. Bishop,

[67] *Foundations of Constructive Analysis,* McGraw-Hill, 1967.

André Boileau,

[75] *Types Versus Topos,* Thèse de Philosophiae Doctor, Université de Montréal, 1975.

M. J. Brockway,

[76] *Topoi and Logic,* M.Sc. Thesis, Victoria University of Wellington, 1976.

T. G. Brook,
[74] *Finiteness: Another Aspect of Topoi*, Ph.D. Thesis, Australian National University, 1974.

L. E. J. Brouwer,
[75] *Collected Works, Vol. 1*, (A. Heyting, ed.), North-Holland, 1975.

Marta C. Bunge,
[74] Topos Theory and Souslin's Hypothesis, *J. of Pure and Applied Algebra*, 4, 1974, 159–187.
[81] Sheaves and prime model extensions, *J. of Algebra*, 68, 1981, 79–96.

Marta Bunge and Gonzalo E. Reyes,
[81] Boolean spectra and model completions, *Fundamenta Mathematicae*, CXIII, 1981, 165–173

Charles W. Burden,
[78] The Hahn–Banach theorem in a category of sheaves, *J. of Pure and Applied Algebra*, 17, 1980, 25–34.

Stanley Burris and Heinrich Werner,
[79] Sheaf constructions and their elementary properties, *Transactions of the American Mathematical Society*, 248, 1979, 269–309.

Fritjof Capra,
[76] *The Tao of Physics*, Fontana, 1976.

R. Carnap,
[47] *Meaning and Necessity*, University of Chicago Press, 1947.

C. C. Chang and H. J. Keisler,
[73] *Model Theory*, North-Holland, 1973.

Paul Cohen,
[66] *Set Theory and the Continuum Hypothesis*, Benjamin, 1966.

J. C. Cole,
[73] Categories of Sets and Models of Set Theory, in *The Proceedings of the Bertrand Russell Memorial Logic Conference*, Denmark 1971; School of Mathematics, Leeds, 1973, 351–399.

Michel Coste,
[79] Localisation, spectra and sheaf representation, in: Fourman, Mulvey and Scott [79], 212–238.

Johns Dauns and Karl Heinrich Hofmann,
[68] Representation of rings by sections, *Memoirs of the American Mathematical Society*, 83, 1968.

R. Dedekind,
[88] Was sind und was sollen die Zahlen, (1888). English translation by W. W. Beman in *Essays on the Theory of Numbers*, Chicago, Open Court, 1901.

R. Diaconescu,
[75] Axiom of choice and complementation, *Proc. of the American Mathematical Society*, 51 (1975), 176–178.
[75] Change of base for toposes with generators, *J. of Pure and Applied Algebra*, 6, 1975, 191–218.

Michael Dummett,
[59] A propositional calculus with denumerable matrix, *J. of Symbolic Logic*, 24, 1959, 97–106.
[77] *Elements of Intuitionism*, Oxford University Press, 1977.

S. Eilenberg and S. Maclane,
[45] General theory of natural equivalences, *Transactions of the American Mathematical Society*, 58, 1945, 231–294.

J. Fang,
[70] *Bourbaki*, Paideia Press, 1970.

M. C. Fitting,
[69] *Intuitionistic Logic Model Theory and Forcing*, North-Holland, 1969.

Michael P. Fourman,
[74] Connections between category theory and logic, D. Phil. thesis, Oxford University, 1974.
[77] The Logic of Topoi, in: Barwise [77], 1053–1090

M. P. Fourman, C. J. Mulvey and D. S. Scott,
[79] *Applications of Sheaf Theory to Algebra, Analysis and Topology* (eds.), Lecture Notes in Mathematics Springer-Verlag, 1979.

M. P. Fourman and D. S. Scott,
[79] Sheaves and logic, in: Fourman, Mulvey and Scott [79], 302–401.

Peter Freyd,
[72] Aspects of topoi, *Bulletin of the Australian Mathematical Society*, 7, 1972, 1–76.

K. Gödel,
[30] Die Vollständigkeit der Axiome des logischen Funktionenkalküls, *Monatshefte Math. Phys.*, 37, 349–360.
[40] *The Consistency of the Axiom of Choice and of the Generalised Continuum Hypothesis with the Axioms of Set Theory*, Annals of Mathematics Studies 3 (Princeton Univ. Press).

Robert Goldblatt,

[77] Grothendieck Topology As Geometric Modality, *Zeitschr. f. math. Logik und Grundlagen d. Math.*, Bd. 27, 1981, 495–529.

William S. Hatcher,

[68] *Foundations of Mathematics*, W. B. Saunders Co., 1968.

L. A. Henkin,

[49] The completeness of the first-order functional calculus, *J. of Symbolic Logic*, 14, 1949, 159–166.

A. Heyting,

[66] *Intuitionism*, 2nd, revised edition, North-Holland, 1966.

Horst Herrlich and George E. Strecker,

[73] *Category Theory*, Allyn and Bacon, 1973.

D. Higgs,

[73] A category approach to Boolean-valued set theory (unpublished).

Karl Heinrich Hofmann,

[72] Representations of algebras by continuous sections, *Bulletin of the American Mathematical Society*, 78, 1972, 291–373.

Karl Heinrich Hofmann and John R. Liukkonen,

[74] Recent advances in the representation theory of rings and C^*-algebras by continuous sections, *Memoirs. of the American Mathematical Society*, 148, 1974.

P. T. Johnstone,

[77] *Topos Theory*, Academic Press, 1977.

[77(i)] Rings, fields, and spectra, *J. of Algebra*, 49, 1977, 238–260.

Anders Kock,

[76] Universal projective geometry via topos theory, *J. of Pure and Applied Algebra*, 9, 1976, 1–24.

A. Kock, P. Lecouturier, and C. J. Mikkelson,

[75] Some Topos Theoretic Concepts of Finiteness, in Lawvere [75(i)], 209–283.

A. Kock and G. E. Reyes,

[77] Doctrines in categorical logic, in: Barwise [77], 283–313.

A. Kock and G. C. Wraith,

[71] *Elementary Toposes*, Aarhus Lecture Note Series, No. 30, 1971.

S. A. Kripke,

[65] Semantical Analysis of Intuitionistic Logic I, in *Formal Systems and Recursive Functions*, ed. by J. N. Crossley and M. A. E. Dummett, North-Holland, 1965, 92–130.

F. W. Lawvere,

[64] An elementary theory of the category of sets, *Proc. of the National Academy of Sciences of the U.S.A.*, 51, 1964, 1506–1510.

[66] The Category of Categories as a Foundation for Mathematics, *Proc. of the Conference on Categorical Algebra*, (La Jolla, 1965), Springer-Verlag 1966, 1–20.

[69] Adjointness in foundations, *Dialectica*, 23, 1969, 281–296.

[70] Qunatifiers and Sheaves, *Actes des Congrés International des Mathématiques*, 1970, tome 1, 329–334.

[72] Introduction and ed. for *Toposes, Algebraic Geometry and Logic*, Lecture Notes in Mathematics, Vol. 274, Springer-Verlag, 1972.

[75] Continuously variable sets; Algebraic Geometry = Geometric Logic, in *Logic Colloquium '73*, ed. by H. E. Rose and J. C. Shepherdson, North-Holland, 1975, 135–156.

[75(i)] Introduction and ed. (with C. Maurer and G. C. Wraith) for *Model Theory and Topoi*, Lecture Notes in Mathematics, Vol. 445, Springer–Verlag, 1975.

[76] Variable Quantities and Variable Structures in Topoi, in *Algebra, Topology and Category Theory*, A Collection of papers in Honor of Samuel Eilenberg, Academic Press, 1976.

E. J. Lemmon,

[65] *Beginning Logic*, Nelson, 1965.

George Loullis,

[79] Sheaves and Boolean valued model theory, *J. of Symbolic Logic*, 44, 1979, 153–183.

Angus MacIntyre,

[73] Model-completeness for sheaves of structures, *Fundamenta Mathematicae*, LXXXI, 1973, 73–89.

Saunders Maclane,

[71] *Categories for the Working Mathematician*, Graduate Texts in Mathematics 5, Springer-Verlag, 1971.

[75] Sets, Topoi, and Internal Logic in Categories, in *Logic Colloquium '73*, ed. by H. E. Rose and J. C. Shepherdson, North-Holland, 1975, 119–134.

Michael Makkai and Gonzalo E. Reyes,

[MR] *First Order Categorical Logic*, Lecture Notes in Mathematics 611, Springer-Verlag, 1977.

E. G. Manes,
[75] *Category Theory Applied to Computation and Control,* Lecture Notes in Computer Science, Vol. 25, Springer-Verlag, 1975 (ed.).

R. Mansfield,
[72] The completeness theorem for infinitary logic, *J. of Symbolic Logic,* 37, 1972, 31–34.

J. C. C. McKinsey and A. Tarski,
[44] The Algebra of Topology, *Annals of Mathematics,* 45, 1944, 141–191.
[46] On Closed Elements in Closure Algebras, *Annals of Mathematics,* 47, 1946, 122–162.
[48] Some Theorems About the Sentential Calculi of Lewis and Langford, *J. of Symbolic Logic,* 13, 1948, 1–15.

Elliot Mendelson,
[70] *Boolean Algebra, Schaum's Outline Series,* McGraw-Hill, 1970.

William Mitchell,
[72] Boolean Topoi and the Theory of Sets, *J. of Pure and Applied Algebra,* 9, 1972, 261–274.

A. Mostowski,
[49] An undecidable arithmetical statement, *Fundamenta Mathematicae,* 36, 1949, 143–164.
[51] On the rules of proof in the pure functional calculus of the First Order, *J. of Symbolic Logic,* 16, 1951, 107–111.

Christopher Mulvey,
[74] Intuitionistic algebra and representations of rings, *Memoirs of the American Mathematical Society,* Vol. 148, 1974, 3–57.

Gerhard Osius,
[74] Categorical set theory: a characterisation of the category of Sets, *J. of Pure and Applied Algebra,* 4, 1974, 79–119.
[75] Logical and Set Theoretical Tools in Elementary Topoi, in: Lawvere [75(i)], 297–346.
[75(i)] A Note on Kripke–Joyal Semantics for the Internal Language of Topoi, in: Lawvere [75(i)], 349–354.

Robert Paré
[74] Co-limits in topoi, *Bulletin of the American Mathematical Society,* 80, 1974, 556–561.

Anna Michaelides Penk,
[75] Two forms of the Axiom of Choice for an elementary topos, *J. of Symbolic Logic,* 40, 1975, 197–212.

R. S. Pierce,

[67] Modules over commutative regular rings, *Memoirs of the American Mathematical Society*, 70, 1967.

H. Rasiowa,

[74] *An Algebraic Approach to Non-Classical Logics*, North-Holland, 1974.

H. Rasiowa and R. Sikorski,

[63] *The Mathematics of Metamathematics*, Polish Scientific Publishers, 1963.

Gonzalo E. Reyes,

[74] From Sheaves to Logic, in *Studies in Algebraic Logic*, ed. by Aubert Diagneault, MAA Studies in Mathematics, Volume 9, 1974.

[76] *Theorie Des Modeles et Faisceaux*, Rapport no. 63, Juin 1976, Institut de Mathématique Pure et Appliquée, Université Catholique de Louvain.

Abraham Robinson,

[69] Germs, in *Applications of Model Theory to Algebra, Analysis and Probability*, ed. by W. A. J. Luxemburg; Holt, Rinehart, and Winston, 1969, 138–149.

[73] Model theory as a framework for algebra, in: Studies in Model Theory, M. D. Morley (ed.), The Mathematical Association of America, 1973, 134–157.

[75] Concerning Progress in the Philosophy of Mathematics, in *Logic Colloquium '73*, ed. by H. E. Rose and J. C. Shepherdson, North-Holland, 1975, 41–52.

Monique Robitaille-Giguère

[75] Modèles d'un Catégorie Logique dans des Topos de Pré faisceaux et d'Ensembles de Heyting, Memoire de la Maitrise ès Sciences, Université de montréal, 1975.

Dana Scott,

[65] Logic with denumerably long formulas and finite strings of quantifiers, in: *The Theory of Models*, J. W. Addison et al. (eds.), North-Holland (2nd reprint 1972), 329–341.

[67] Existence and Description in Formal Logic, in *Bertrand Russell: Philosopher of the Century*, ed. by Ralph Schoenman, George Allen and Unwin, 1967.

[68] Extending the topological interpretation to intuitionistic analysis, *Compositio Mathematica*, 20, 1968, 194–210.

[70] Extending the Topological Interpretation to Intuitionistic

Analysis, II, in *Proof Theory and Intuitionism*, A. Kino, J. Myhill and R. E. Vesley (eds.), North-Holland, 1970, 235–255.

[70(i)] Advice on Modal Logic, in *Philosophical Problems in Logic*, ed. by Karel Lambert, Reidel, 1970, 143–173.

Krister Segerberg,

[68] Propositional logics related to Heyting's and Johansson's, *Theoria*, 34, 1968, 26–61.

Dana J. Schlomiuk,

[74] Topos di Grothendieck e topos di Lawvere e Tierney, *Rendiconti de Matematica*, 7, 1974, 513–553.

Horst Schubert,

[72] *Categories*, Springer-Verlag, 1972.

J. Seebach, L. Seebach, and L. Steen,

[70] What is a sheaf?, *American Mathematical Monthly*, 77, 1970, 681–703.

John Staples

[71] On constructive fields, *Proc. of London Mathematical Society*, Series 3, vol. 23 (1971), 753–768.

Lawrence N. Stout

[76] Topological properties of the real numbers object in a topos, *Cahiers de Topologie et Geometrie Differentielle*, vol. XVII-3, 1976, 295–326.

M. H. Stone,

[37] Topological representation of distributive lattices and Brouwerian logics, *Casopis pro Pestovani Matematiky a Fysiky*, 67, 1937, 1–25.

Ross Street,

[74] Lectures in Topos Theory, unpublished notes, Monash University, January 1974.

Stanislaw Surma,

[73] *Studies in the History of Mathematical Logic*, Polish Academy of Sciences, 1973.

A. Tarski,

[36] Der Wahrheitsbegriff in den formalisierten Sprachen, *Studia Philosophica*, 1, 1936, 261–405. (English translation in *Logic, Semantics and Metamathematics*, Oxford, 1956, 152–278).

[38] Der Aussagenkalkül und die Topologie, *Fundamenta Mathematicae*, 31, 1938, 103–134. (English translation: ibid, 421–454).

René Thom,

[71] "Modern" mathematics: an educational and philosophic error? *American Scientist*, Nov–Dec 1971, 695–699.

Richmond H. Thomason,

[68] On the strong semantical completeness of the intuitionistic predicate calculus, *J. of Symbolic Logic*, 33, 1968, 1–7.

Myles Tierney,

[72] Sheaf Theory and the Continuum Hypothesis, in: Lawvere [72], 13–42.

[73] Axiomatic sheaf theory: some constructions and applications, in: *Proceedings of the C.I.M.E. Conference on Categories and Commutative Algebra*, Varenna 1971, Edizioni Cremonese, 1973, 249–325.

[76] Forcing topologies and classifying topoi, in *Algebra, Topology and Category Theory: A Collection of Papers in Honor of Samuel Eilenberg*, A. Heller and M. Tierney (eds.), Academic Press, 1976, 211–219.

John Unterecker,

[73] Foreword to *Einstein and Beckett* by Edwin Schlossberg, Links Books, 1973.

D. Van Dalen,

[78] An interpretation of intuitionistic analysis, *Anals of Mathematical Logic*, vol. 13, 1978, 1–43.

Barbara Veit,

[81] A proof of the associated sheaf theorem by means of categorical logic, *J. of Symbolic Logic*, 46, 1981, 45–55.

Hugo Volger,

[75] Completeness theorem for logical categories, and Logical categories, semantical categories and topoi, in: F. W. Lawvere [75(i)], 51–100.

G. C. Wraith,

[75] Lectures on Elementary Topoi, in, Lawvere [75(i)], 114–206.

[79] Generic Galois theory of local rings, in: Fourman, Mulvey and Scott [79], 739–767.

CATALOGUE OF NOTATION

Relations

Objects (elements, sets, structured sets, algebras, spaces)

F_f	stack of sections	360
p_F	sheaf of germs of sections	362
$\mathbf{C}_X$	Heyting-valued set of continuous partial functions	388
$\mathbf{C}_k$	Heyting-valued set of continuous local sections	389
$F_{\mathbf{A}}$	sheaf corresponding to complete Ω-set	391
$\mathbf{A}_F$	Ω-set corresponding to sheaf	392
F_X	sheaf of continuous partial functions	392
$\dot{\Omega}$	sheaf of truth-values	400
$\bar{\mathbf{X}}$	rigid Ω-set	406
$\mathbf{X}^*$	simple sheaf	406
$\mathbf{A}_a$	sheaf of partial elements	410
N	natural numbers object	302
N^+	positive integers object	414
Z	integers object	414
Q	rationals object	414
R_c	Cauchy reals	414
R_d	Dedekind reals	414ff.
C	object of complex numbers	423
$^*R$	order-complete reals	423, 431
$\mathbb{Z}, \mathbb{Q}, \mathbb{R}, \mathbb{C}$	classical number systems	413
$\beta(\Omega)$	soberification	436

Arrows (functions, functors)

$f:A \to B$	function	17
$f:a \to b$	arrow	24
$f:A \rightsquigarrow B$	partial function	268
$f:a \rightsquigarrow b$	partial arrow	268
$1 \to a$	element	78
$1 \rightsquigarrow a$	partial element	268
$A \hookrightarrow B$	inclusion function	19
$h:r \hookrightarrow s$	inclusion of relations	322
id_A	identity function	19
1_b	identity arrow	25
$g \circ f$	composition	20, 24
$\rightarrowtail$	monic	38
$\twoheadrightarrow$	epic	39
f^{-1}	inverse arrow	40
f^{op}	dual arrow	45
$f:a \cong b$	iso arrow	40
0_a	unique arrow $0 \to a$	44

$!_a$	unique arrow $1 \to a$	44
p_A	projection to set	46
pr_a	projection to object	47
pr_j^m	j-th projection $a^m \to a$	53
$\langle f, g\rangle$	product arrow	47
$f \times g$	product of arrows	50
$\dashrightarrow$	uniquely existing arrow	52
i_a	injection	54
$[f, g]$	coproduct arrow	54
$f+g$	coproduct of arrows	55
f_R	natural map	62
ev	evaluation arrow	70–1
ev_a	evaluation on Ω^a	106
$\in_a$	membership relation	104
$\hat{g}$	exponential adjoint	71–2
$\ulcorner f \urcorner$	name of f	78
χ_A	characteristic function	79
χ_f	characteristic arrow (character)	81
true	subobject classifier	80–1
$\top$	*true*	81
$\ulcorner f \urcorner$	unit of HA (largest truth-value)	277
$true_a$	$= \top_a = true \circ !_a$	83
$true_j$	classifier for j-sheaves	369, 380
$1 \to \Omega$	truth-value	94
false	character of $0 \to 1$	117
$\perp$	*false*	117
	least truth-value	277
$\neg$	negation truth-arrow	127, 139
$\neg \circ \neg$	double negation	184, 379
$\cap$	conjunction	127, 139
$\cup$	disjunction	128, 139
$\Rightarrow$	implication	128, 139
$\circledcirc$	ordering on Ω	139
$\mathrm{im}\, f$	image of f	112
$f[g]$	image under f	320
Ω^f	internal images	320
$\{x : \varphi\}$	subobject determined by "property"	107
f^*	epic part of f	112
$F : \mathscr{C} \to \mathscr{D}$	functor	195
$\mathscr{P}, \bar{\mathscr{P}}$	powerset functors	195, 197
$\mathscr{P}(f)$	f-images functor	195
$1_{\mathscr{C}}$	identity functor	195
$-\times a, a\times -$	product functors	196, 200

Systems, languages, axioms, rules

PO-$P3$, $P3A$, $P3B$	Peano Postulates	348–351
$F1, F2$	Freyd Postulates	352
COM	'patching' of compatible elements (sections)	362, 376, 391
$\delta(\mathbf{r})$	Dedekind cut axioms	415, 423
$0(r)$	order-axioms for cuts	427, 430

Logical symbols

$\vee, \wedge, \sim$	'and', 'or', 'not'	126
$\supset$	'implies'	128
$\equiv$	'if and only if'	274
∇	'It is locally the case that'	382
$\approx$	identity/equality	232, 274
$\approxeq$	equivalence	274
$\forall$	'for all'	231
$\exists$	'there exists'	231
$\exists!$	'there exists exactly one'	233
$\mathbf{E}(t)$	't exists'	267
$\mathsf{I}v$	'the unique v such that'	287, 404
$\varphi'(v)$	function-value term	288
ε	membership	305
$v\varepsilon\{u: \varphi\}$	class abstract	309
$\varphi(v_{i_1},\ldots,v_{i_n})$	open formula	235
$\varphi(v/t)$	substitution	237

INDEX OF DEFINITIONS

A CATALOG OF SELECTED

DOVER BOOKS

IN SCIENCE AND MATHEMATICS

Astronomy

CHARIOTS FOR APOLLO: The NASA History of Manned Lunar Spacecraft to 1969, Courtney G. Brooks, James M. Grimwood, and Loyd S. Swenson, Jr. This illustrated history by a trio of experts is the definitive reference on the Apollo spacecraft and lunar modules. It traces the vehicles' design, development, and operation in space. More than 100 photographs and illustrations. 576pp. 6 3/4 x 9 1/4. 0-486-46756-2

EXPLORING THE MOON THROUGH BINOCULARS AND SMALL TELESCOPES, Ernest H. Cherrington, Jr. Informative, profusely illustrated guide to locating and identifying craters, rills, seas, mountains, other lunar features. Newly revised and updated with special section of new photos. Over 100 photos and diagrams. 240pp. 8 1/4 x 11. 0-486-24491-1

WHERE NO MAN HAS GONE BEFORE: A History of NASA's Apollo Lunar Expeditions, William David Compton. Introduction by Paul Dickson. This official NASA history traces behind-the-scenes conflicts and cooperation between scientists and engineers. The first half concerns preparations for the Moon landings, and the second half documents the flights that followed Apollo 11. 1989 edition. 432pp. 7 x 10. 0-486-47888-2

APOLLO EXPEDITIONS TO THE MOON: The NASA History, Edited by Edgar M. Cortright. Official NASA publication marks the 40th anniversary of the first lunar landing and features essays by project participants recalling engineering and administrative challenges. Accessible, jargon-free accounts, highlighted by numerous illustrations. 336pp. 8 3/8 x 10 7/8. 0-486-47175-6

ON MARS: Exploration of the Red Planet, 1958-1978--The NASA History, Edward Clinton Ezell and Linda Neuman Ezell. NASA's official history chronicles the start of our explorations of our planetary neighbor. It recounts cooperation among government, industry, and academia, and it features dozens of photos from Viking cameras. 560pp. 6 3/4 x 9 1/4. 0-486-46757-0

ARISTARCHUS OF SAMOS: The Ancient Copernicus, Sir Thomas Heath. Heath's history of astronomy ranges from Homer and Hesiod to Aristarchus and includes quotes from numerous thinkers, compilers, and scholasticists from Thales and Anaximander through Pythagoras, Plato, Aristotle, and Heraclides. 34 figures. 448pp. 5 3/8 x 8 1/2. 0-486-43886-4

AN INTRODUCTION TO CELESTIAL MECHANICS, Forest Ray Moulton. Classic text still unsurpassed in presentation of fundamental principles. Covers rectilinear motion, central forces, problems of two and three bodies, much more. Includes over 200 problems, some with answers. 437pp. 5 3/8 x 8 1/2. 0-486-64687-4

BEYOND THE ATMOSPHERE: Early Years of Space Science, Homer E. Newell. This exciting survey is the work of a top NASA administrator who chronicles technological advances, the relationship of space science to general science, and the space program's social, political, and economic contexts. 528pp. 6 3/4 x 9 1/4. 0-486-47464-X

STAR LORE: Myths, Legends, and Facts, William Tyler Olcott. Captivating retellings of the origins and histories of ancient star groups include Pegasus, Ursa Major, Pleiades, signs of the zodiac, and other constellations. "Classic." – *Sky & Telescope.* 58 illustrations. 544pp. 5 3/8 x 8 1/2. 0-486-43581-4

A COMPLETE MANUAL OF AMATEUR ASTRONOMY: Tools and Techniques for Astronomical Observations, P. Clay Sherrod with Thomas L. Koed. Concise, highly readable book discusses the selection, set-up, and maintenance of a telescope; amateur studies of the sun; lunar topography and occultations; and more. 124 figures. 26 halftones. 37 tables. 335pp. 6 1/2 x 9 1/4. 0-486-42820-6

Chemistry

MOLECULAR COLLISION THEORY, M. S. Child. This high-level monograph offers an analytical treatment of classical scattering by a central force, quantum scattering by a central force, elastic scattering phase shifts, and semi-classical elastic scattering. 1974 edition. 310pp. 5 3/8 x 8 1/2. 0-486-69437-2

HANDBOOK OF COMPUTATIONAL QUANTUM CHEMISTRY, David B. Cook. This comprehensive text provides upper-level undergraduates and graduate students with an accessible introduction to the implementation of quantum ideas in molecular modeling, exploring practical applications alongside theoretical explanations. 1998 edition. 832pp. 5 3/8 x 8 1/2. 0-486-44307-8

RADIOACTIVE SUBSTANCES, Marie Curie. The celebrated scientist's thesis, which directly preceded her 1903 Nobel Prize, discusses establishing atomic character of radioactivity; extraction from pitchblende of polonium and radium; isolation of pure radium chloride; more. 96pp. 5 3/8 x 8 1/2. 0-486-42550-9

CHEMICAL MAGIC, Leonard A. Ford. Classic guide provides intriguing entertainment while elucidating sound scientific principles, with more than 100 unusual stunts: cold fire, dust explosions, a nylon rope trick, a disappearing beaker, much more. 128pp. 5 3/8 x 8 1/2. 0-486-67628-5

ALCHEMY, E. J. Holmyard. Classic study by noted authority covers 2,000 years of alchemical history: religious, mystical overtones; apparatus; signs, symbols, and secret terms; advent of scientific method, much more. Illustrated. 320pp. 5 3/8 x 8 1/2. 0-486-26298-7

CHEMICAL KINETICS AND REACTION DYNAMICS, Paul L. Houston. This text teaches the principles underlying modern chemical kinetics in a clear, direct fashion, using several examples to enhance basic understanding. Solutions to selected problems. 2001 edition. 352pp. 8 3/8 x 11. 0-486-45334-0

PROBLEMS AND SOLUTIONS IN QUANTUM CHEMISTRY AND PHYSICS, Charles S. Johnson and Lee G. Pedersen. Unusually varied problems, with detailed solutions, cover of quantum mechanics, wave mechanics, angular momentum, molecular spectroscopy, scattering theory, more. 280 problems, plus 139 supplementary exercises. 430pp. 6 1/2 x 9 1/4. 0-486-65236-X

ELEMENTS OF CHEMISTRY, Antoine Lavoisier. Monumental classic by the founder of modern chemistry features first explicit statement of law of conservation of matter in chemical change, and more. Facsimile reprint of original (1790) Kerr translation. 539pp. 5 3/8 x 8 1/2. 0-486-64624-6

MAGNETISM AND TRANSITION METAL COMPLEXES, F. E. Mabbs and D. J. Machin. A detailed view of the calculation methods involved in the magnetic properties of transition metal complexes, this volume offers sufficient background for original work in the field. 1973 edition. 240pp. 5 3/8 x 8 1/2. 0-486-46284-6

GENERAL CHEMISTRY, Linus Pauling. Revised third edition of classic first-year text by Nobel laureate. Atomic and molecular structure, quantum mechanics, statistical mechanics, thermodynamics correlated with descriptive chemistry. Problems. 992pp. 5 3/8 x 8 1/2. 0-486-65622-5

ELECTROLYTE SOLUTIONS: Second Revised Edition, R. A. Robinson and R. H. Stokes. Classic text deals primarily with measurement, interpretation of conductance, chemical potential, and diffusion in electrolyte solutions. Detailed theoretical interpretations, plus extensive tables of thermodynamic and transport properties. 1970 edition. 590pp. 5 3/8 x 8 1/2. 0-486-42225-9

Engineering

FUNDAMENTALS OF ASTRODYNAMICS, Roger R. Bate, Donald D. Mueller, and Jerry E. White. Teaching text developed by U.S. Air Force Academy develops the basic two-body and n-body equations of motion; orbit determination; classical orbital elements, coordinate transformations; differential correction; more. 1971 edition. 455pp. 5 3/8 x 8 1/2. 0-486-60061-0

INTRODUCTION TO CONTINUUM MECHANICS FOR ENGINEERS: Revised Edition, Ray M. Bowen. This self-contained text introduces classical continuum models within a modern framework. Its numerous exercises illustrate the governing principles, linearizations, and other approximations that constitute classical continuum models. 2007 edition. 320pp. 6 1/8 x 9 1/4. 0-486-47460-7

ENGINEERING MECHANICS FOR STRUCTURES, Louis L. Bucciarelli. This text explores the mechanics of solids and statics as well as the strength of materials and elasticity theory. Its many design exercises encourage creative initiative and systems thinking. 2009 edition. 320pp. 6 1/8 x 9 1/4. 0-486-46855-0

FEEDBACK CONTROL THEORY, John C. Doyle, Bruce A. Francis and Allen R. Tannenbaum. This excellent introduction to feedback control system design offers a theoretical approach that captures the essential issues and can be applied to a wide range of practical problems. 1992 edition. 224pp. 6 1/2 x 9 1/4. 0-486-46933-6

THE FORCES OF MATTER, Michael Faraday. These lectures by a famous inventor offer an easy-to-understand introduction to the interactions of the universe's physical forces. Six essays explore gravitation, cohesion, chemical affinity, heat, magnetism, and electricity. 1993 edition. 96pp. 5 3/8 x 8 1/2. 0-486-47482-8

DYNAMICS, Lawrence E. Goodman and William H. Warner. Beginning engineering text introduces calculus of vectors, particle motion, dynamics of particle systems and plane rigid bodies, technical applications in plane motions, and more. Exercises and answers in every chapter. 619pp. 5 3/8 x 8 1/2. 0-486-42006-X

ADAPTIVE FILTERING PREDICTION AND CONTROL, Graham C. Goodwin and Kwai Sang Sin. This unified survey focuses on linear discrete-time systems and explores natural extensions to nonlinear systems. It emphasizes discrete-time systems, summarizing theoretical and practical aspects of a large class of adaptive algorithms. 1984 edition. 560pp. 6 1/2 x 9 1/4. 0-486-46932-8

INDUCTANCE CALCULATIONS, Frederick W. Grover. This authoritative reference enables the design of virtually every type of inductor. It features a single simple formula for each type of inductor, together with tables containing essential numerical factors. 1946 edition. 304pp. 5 3/8 x 8 1/2. 0-486-47440-2

THERMODYNAMICS: Foundations and Applications, Elias P. Gyftopoulos and Gian Paolo Beretta. Designed by two MIT professors, this authoritative text discusses basic concepts and applications in detail, emphasizing generality, definitions, and logical consistency. More than 300 solved problems cover realistic energy systems and processes. 800pp. 6 1/8 x 9 1/4. 0-486-43932-1

THE FINITE ELEMENT METHOD: Linear Static and Dynamic Finite Element Analysis, Thomas J. R. Hughes. Text for students without in-depth mathematical training, this text includes a comprehensive presentation and analysis of algorithms of time-dependent phenomena plus beam, plate, and shell theories. Solution guide available upon request. 672pp. 6 1/2 x 9 1/4. 0-486-41181-8

HELICOPTER THEORY, Wayne Johnson. Monumental engineering text covers vertical flight, forward flight, performance, mathematics of rotating systems, rotary wing dynamics and aerodynamics, aeroelasticity, stability and control, stall, noise, and more. 189 illustrations. 1980 edition. 1089pp. 5 5/8 x 8 1/4. 0-486-68230-7

MATHEMATICAL HANDBOOK FOR SCIENTISTS AND ENGINEERS: Definitions, Theorems, and Formulas for Reference and Review, Granino A. Korn and Theresa M. Korn. Convenient access to information from every area of mathematics: Fourier transforms, Z transforms, linear and nonlinear programming, calculus of variations, random-process theory, special functions, combinatorial analysis, game theory, much more. 1152pp. 5 3/8 x 8 1/2. 0-486-41147-8

A HEAT TRANSFER TEXTBOOK: Fourth Edition, John H. Lienhard V and John H. Lienhard IV. This introduction to heat and mass transfer for engineering students features worked examples and end-of-chapter exercises. Worked examples and end-of-chapter exercises appear throughout the book, along with well-drawn, illuminating figures. 768pp. 7 x 9 1/4. 0-486-47931-5

BASIC ELECTRICITY, U.S. Bureau of Naval Personnel. Originally a training course; best nontechnical coverage. Topics include batteries, circuits, conductors, AC and DC, inductance and capacitance, generators, motors, transformers, amplifiers, etc. Many questions with answers. 349 illustrations. 1969 edition. 448pp. 6 1/2 x 9 1/4. 0-486-20973-3

BASIC ELECTRONICS, U.S. Bureau of Naval Personnel. Clear, well-illustrated introduction to electronic equipment covers numerous essential topics: electron tubes, semiconductors, electronic power supplies, tuned circuits, amplifiers, receivers, ranging and navigation systems, computers, antennas, more. 560 illustrations. 567pp. 6 1/2 x 9 1/4. 0-486-21076-6

BASIC WING AND AIRFOIL THEORY, Alan Pope. This self-contained treatment by a pioneer in the study of wind effects covers flow functions, airfoil construction and pressure distribution, finite and monoplane wings, and many other subjects. 1951 edition. 320pp. 5 3/8 x 8 1/2. 0-486-47188-8

SYNTHETIC FUELS, Ronald F. Probstein and R. Edwin Hicks. This unified presentation examines the methods and processes for converting coal, oil, shale, tar sands, and various forms of biomass into liquid, gaseous, and clean solid fuels. 1982 edition. 512pp. 6 1/8 x 9 1/4. 0-486-44977-7

THEORY OF ELASTIC STABILITY, Stephen P. Timoshenko and James M. Gere. Written by world-renowned authorities on mechanics, this classic ranges from theoretical explanations of 2- and 3-D stress and strain to practical applications such as torsion, bending, and thermal stress. 1961 edition. 560pp. 5 3/8 x 8 1/2. 0-486-47207-8

PRINCIPLES OF DIGITAL COMMUNICATION AND CODING, Andrew J. Viterbi and Jim K. Omura. This classic by two digital communications experts is geared toward students of communications theory and to designers of channels, links, terminals, modems, or networks used to transmit and receive digital messages. 1979 edition. 576pp. 6 1/8 x 9 1/4. 0-486-46901-8

LINEAR SYSTEM THEORY: The State Space Approach, Lotfi A. Zadeh and Charles A. Desoer. Written by two pioneers in the field, this exploration of the state space approach focuses on problems of stability and control, plus connections between this approach and classical techniques. 1963 edition. 656pp. 6 1/8 x 9 1/4. 0-486-46663-9

Mathematics-Bestsellers

HANDBOOK OF MATHEMATICAL FUNCTIONS: with Formulas, Graphs, and Mathematical Tables, Edited by Milton Abramowitz and Irene A. Stegun. A classic resource for working with special functions, standard trig, and exponential logarithmic definitions and extensions, it features 29 sets of tables, some to as high as 20 places. 1046pp. 8 x 10 1/2. 0-486-61272-4

ABSTRACT AND CONCRETE CATEGORIES: The Joy of Cats, Jiri Adamek, Horst Herrlich, and George E. Strecker. This up-to-date introductory treatment employs category theory to explore the theory of structures. Its unique approach stresses concrete categories and presents a systematic view of factorization structures. Numerous examples. 1990 edition, updated 2004. 528pp. 6 1/8 x 9 1/4. 0-486-46934-4

MATHEMATICS: Its Content, Methods and Meaning, A. D. Aleksandrov, A. N. Kolmogorov, and M. A. Lavrent'ev. Major survey offers comprehensive, coherent discussions of analytic geometry, algebra, differential equations, calculus of variations, functions of a complex variable, prime numbers, linear and non-Euclidean geometry, topology, functional analysis, more. 1963 edition. 1120pp. 5 3/8 x 8 1/2. 0-486-40916-3

INTRODUCTION TO VECTORS AND TENSORS: Second Edition--Two Volumes Bound as One, Ray M. Bowen and C.-C. Wang. Convenient single-volume compilation of two texts offers both introduction and in-depth survey. Geared toward engineering and science students rather than mathematicians, it focuses on physics and engineering applications. 1976 edition. 560pp. 6 1/2 x 9 1/4. 0-486-46914-X

AN INTRODUCTION TO ORTHOGONAL POLYNOMIALS, Theodore S. Chihara. Concise introduction covers general elementary theory, including the representation theorem and distribution functions, continued fractions and chain sequences, the recurrence formula, special functions, and some specific systems. 1978 edition. 272pp. 5 3/8 x 8 1/2. 0-486-47929-3

ADVANCED MATHEMATICS FOR ENGINEERS AND SCIENTISTS, Paul DuChateau. This primary text and supplemental reference focuses on linear algebra, calculus, and ordinary differential equations. Additional topics include partial differential equations and approximation methods. Includes solved problems. 1992 edition. 400pp. 7 1/2 x 9 1/4. 0-486-47930-7

PARTIAL DIFFERENTIAL EQUATIONS FOR SCIENTISTS AND ENGINEERS, Stanley J. Farlow. Practical text shows how to formulate and solve partial differential equations. Coverage of diffusion-type problems, hyperbolic-type problems, elliptic-type problems, numerical and approximate methods. Solution guide available upon request. 1982 edition. 414pp. 6 1/8 x 9 1/4. 0-486-67620-X

VARIATIONAL PRINCIPLES AND FREE-BOUNDARY PROBLEMS, Avner Friedman. Advanced graduate-level text examines variational methods in partial differential equations and illustrates their applications to free-boundary problems. Features detailed statements of standard theory of elliptic and parabolic operators. 1982 edition. 720pp. 6 1/8 x 9 1/4. 0-486-47853-X

LINEAR ANALYSIS AND REPRESENTATION THEORY, Steven A. Gaal. Unified treatment covers topics from the theory of operators and operator algebras on Hilbert spaces; integration and representation theory for topological groups; and the theory of Lie algebras, Lie groups, and transform groups. 1973 edition. 704pp. 6 1/8 x 9 1/4. 0-486-47851-3

Mathematics–Logic and Problem Solving

Mathematics–Algebra and Calculus

VECTOR CALCULUS, Peter Baxandall and Hans Liebeck. This introductory text offers a rigorous, comprehensive treatment. Classical theorems of vector calculus are amply illustrated with figures, worked examples, physical applications, and exercises with hints and answers. 1986 edition. 560pp. 5 3/8 x 8 1/2. 0-486-46620-5

ADVANCED CALCULUS: An Introduction to Classical Analysis, Louis Brand. A course in analysis that focuses on the functions of a real variable, this text introduces the basic concepts in their simplest setting and illustrates its teachings with numerous examples, theorems, and proofs. 1955 edition. 592pp. 5 3/8 x 8 1/2. 0-486-44548-8

ADVANCED CALCULUS, Avner Friedman. Intended for students who have already completed a one-year course in elementary calculus, this two-part treatment advances from functions of one variable to those of several variables. Solutions. 1971 edition. 432pp. 5 3/8 x 8 1/2. 0-486-45795-8

METHODS OF MATHEMATICS APPLIED TO CALCULUS, PROBABILITY, AND STATISTICS, Richard W. Hamming. This 4-part treatment begins with algebra and analytic geometry and proceeds to an exploration of the calculus of algebraic functions and transcendental functions and applications. 1985 edition. Includes 310 figures and 18 tables. 880pp. 6 1/2 x 9 1/4. 0-486-43945-3

BASIC ALGEBRA I: Second Edition, Nathan Jacobson. A classic text and standard reference for a generation, this volume covers all undergraduate algebra topics, including groups, rings, modules, Galois theory, polynomials, linear algebra, and associative algebra. 1985 edition. 528pp. 6 1/8 x 9 1/4. 0-486-47189-6

BASIC ALGEBRA II: Second Edition, Nathan Jacobson. This classic text and standard reference comprises all subjects of a first-year graduate-level course, including in-depth coverage of groups and polynomials and extensive use of categories and functors. 1989 edition. 704pp. 6 1/8 x 9 1/4. 0-486-47187-X

CALCULUS: An Intuitive and Physical Approach (Second Edition), Morris Kline. Application-oriented introduction relates the subject as closely as possible to science with explorations of the derivative; differentiation and integration of the powers of x; theorems on differentiation, antidifferentiation; the chain rule; trigonometric functions; more. Examples. 1967 edition. 960pp. 6 1/2 x 9 1/4. 0-486-40453-6

ABSTRACT ALGEBRA AND SOLUTION BY RADICALS, John E. Maxfield and Margaret W. Maxfield. Accessible advanced undergraduate-level text starts with groups, rings, fields, and polynomials and advances to Galois theory, radicals and roots of unity, and solution by radicals. Numerous examples, illustrations, exercises, appendixes. 1971 edition. 224pp. 6 1/8 x 9 1/4. 0-486-47723-1

AN INTRODUCTION TO THE THEORY OF LINEAR SPACES, Georgi E. Shilov. Translated by Richard A. Silverman. Introductory treatment offers a clear exposition of algebra, geometry, and analysis as parts of an integrated whole rather than separate subjects. Numerous examples illustrate many different fields, and problems include hints or answers. 1961 edition. 320pp. 5 3/8 x 8 1/2. 0-486-63070-6

LINEAR ALGEBRA, Georgi E. Shilov. Covers determinants, linear spaces, systems of linear equations, linear functions of a vector argument, coordinate transformations, the canonical form of the matrix of a linear operator, bilinear and quadratic forms, and more. 387pp. 5 3/8 x 8 1/2. 0-486-63518-X

Mathematics–Probability and Statistics

BASIC PROBABILITY THEORY, Robert B. Ash. This text emphasizes the probabilistic way of thinking, rather than measure-theoretic concepts. Geared toward advanced undergraduates and graduate students, it features solutions to some of the problems. 1970 edition. 352pp. 5 3/8 x 8 1/2. 0-486-46628-0

PRINCIPLES OF STATISTICS, M. G. Bulmer. Concise description of classical statistics, from basic dice probabilities to modern regression analysis. Equal stress on theory and applications. Moderate difficulty; only basic calculus required. Includes problems with answers. 252pp. 5 5/8 x 8 1/4. 0-486-63760-3

OUTLINE OF BASIC STATISTICS: Dictionary and Formulas, John E. Freund and Frank J. Williams. Handy guide includes a 70-page outline of essential statistical formulas covering grouped and ungrouped data, finite populations, probability, and more, plus over 1,000 clear, concise definitions of statistical terms. 1966 edition. 208pp. 5 3/8 x 8 1/2. 0-486-47769-X

GOOD THINKING: The Foundations of Probability and Its Applications, Irving J. Good. This in-depth treatment of probability theory by a famous British statistician explores Keynesian principles and surveys such topics as Bayesian rationality, corroboration, hypothesis testing, and mathematical tools for induction and simplicity. 1983 edition. 352pp. 5 3/8 x 8 1/2. 0-486-47438-0

INTRODUCTION TO PROBABILITY THEORY WITH CONTEMPORARY APPLICATIONS, Lester L. Helms. Extensive discussions and clear examples, written in plain language, expose students to the rules and methods of probability. Exercises foster problem-solving skills, and all problems feature step-by-step solutions. 1997 edition. 368pp. 6 1/2 x 9 1/4. 0-486-47418-6

CHANCE, LUCK, AND STATISTICS, Horace C. Levinson. In simple, non-technical language, this volume explores the fundamentals governing chance and applies them to sports, government, and business. "Clear and lively ... remarkably accurate." – *Scientific Monthly.* 384pp. 5 3/8 x 8 1/2. 0-486-41997-5

FIFTY CHALLENGING PROBLEMS IN PROBABILITY WITH SOLUTIONS, Frederick Mosteller. Remarkable puzzlers, graded in difficulty, illustrate elementary and advanced aspects of probability. These problems were selected for originality, general interest, or because they demonstrate valuable techniques. Also includes detailed solutions. 88pp. 5 3/8 x 8 1/2. 0-486-65355-2

EXPERIMENTAL STATISTICS, Mary Gibbons Natrella. A handbook for those seeking engineering information and quantitative data for designing, developing, constructing, and testing equipment. Covers the planning of experiments, the analyzing of extreme-value data; and more. 1966 edition. Index. Includes 52 figures and 76 tables. 560pp. 8 3/8 x 11. 0-486-43937-2

STOCHASTIC MODELING: Analysis and Simulation, Barry L. Nelson. Coherent introduction to techniques also offers a guide to the mathematical, numerical, and simulation tools of systems analysis. Includes formulation of models, analysis, and interpretation of results. 1995 edition. 336pp. 6 1/8 x 9 1/4. 0-486-47770-3

INTRODUCTION TO BIOSTATISTICS: Second Edition, Robert R. Sokal and F. James Rohlf. Suitable for undergraduates with a minimal background in mathematics, this introduction ranges from descriptive statistics to fundamental distributions and the testing of hypotheses. Includes numerous worked-out problems and examples. 1987 edition. 384pp. 6 1/8 x 9 1/4. 0-486-46961-1

Mathematics–Geometry and Topology

PROBLEMS AND SOLUTIONS IN EUCLIDEAN GEOMETRY, M. N. Aref and William Wernick. Based on classical principles, this book is intended for a second course in Euclidean geometry and can be used as a refresher. More than 200 problems include hints and solutions. 1968 edition. 272pp. 5 3/8 x 8 1/2. 0-486-47720-7

TOPOLOGY OF 3-MANIFOLDS AND RELATED TOPICS, Edited by M. K. Fort, Jr. With a New Introduction by Daniel Silver. Summaries and full reports from a 1961 conference discuss decompositions and subsets of 3-space; n-manifolds; knot theory; the Poincaré conjecture; and periodic maps and isotopies. Familiarity with algebraic topology required. 1962 edition. 272pp. 6 1/8 x 9 1/4. 0-486-47753-3

POINT SET TOPOLOGY, Steven A. Gaal. Suitable for a complete course in topology, this text also functions as a self-contained treatment for independent study. Additional enrichment materials make it equally valuable as a reference. 1964 edition. 336pp. 5 3/8 x 8 1/2. 0-486-47222-1

INVITATION TO GEOMETRY, Z. A. Melzak. Intended for students of many different backgrounds with only a modest knowledge of mathematics, this text features self-contained chapters that can be adapted to several types of geometry courses. 1983 edition. 240pp. 5 3/8 x 8 1/2. 0-486-46626-4

TOPOLOGY AND GEOMETRY FOR PHYSICISTS, Charles Nash and Siddhartha Sen. Written by physicists for physics students, this text assumes no detailed background in topology or geometry. Topics include differential forms, homotopy, homology, cohomology, fiber bundles, connection and covariant derivatives, and Morse theory. 1983 edition. 320pp. 5 3/8 x 8 1/2. 0-486-47852-1

BEYOND GEOMETRY: Classic Papers from Riemann to Einstein, Edited with an Introduction and Notes by Peter Pesic. This is the only English-language collection of these 8 accessible essays. They trace seminal ideas about the foundations of geometry that led to Einstein's general theory of relativity. 224pp. 6 1/8 x 9 1/4. 0-486-45350-2

GEOMETRY FROM EUCLID TO KNOTS, Saul Stahl. This text provides a historical perspective on plane geometry and covers non-neutral Euclidean geometry, circles and regular polygons, projective geometry, symmetries, inversions, informal topology, and more. Includes 1,000 practice problems. Solutions available. 2003 edition. 480pp. 6 1/8 x 9 1/4. 0-486-47459-3

TOPOLOGICAL VECTOR SPACES, DISTRIBUTIONS AND KERNELS, François Trèves. Extending beyond the boundaries of Hilbert and Banach space theory, this text focuses on key aspects of functional analysis, particularly in regard to solving partial differential equations. 1967 edition. 592pp. 5 3/8 x 8 1/2. 0-486-45352-9

INTRODUCTION TO PROJECTIVE GEOMETRY, C. R. Wylie, Jr. This introductory volume offers strong reinforcement for its teachings, with detailed examples and numerous theorems, proofs, and exercises, plus complete answers to all odd-numbered end-of-chapter problems. 1970 edition. 576pp. 6 1/8 x 9 1/4. 0-486-46895-X

FOUNDATIONS OF GEOMETRY, C. R. Wylie, Jr. Geared toward students preparing to teach high school mathematics, this text explores the principles of Euclidean and non-Euclidean geometry and covers both generalities and specifics of the axiomatic method. 1964 edition. 352pp. 6 x 9. 0-486-47214-0

Mathematics-History

THE WORKS OF ARCHIMEDES, Archimedes. Translated by Sir Thomas Heath. Complete works of ancient geometer feature such topics as the famous problems of the ratio of the areas of a cylinder and an inscribed sphere; the properties of conoids, spheroids, and spirals; more. 326pp. 5 3/8 x 8 1/2. 0-486-42084-1

THE HISTORICAL ROOTS OF ELEMENTARY MATHEMATICS, Lucas N. H. Bunt, Phillip S. Jones, and Jack D. Bedient. Exciting, hands-on approach to understanding fundamental underpinnings of modern arithmetic, algebra, geometry and number systems examines their origins in early Egyptian, Babylonian, and Greek sources. 336pp. 5 3/8 x 8 1/2. 0-486-25563-8

THE THIRTEEN BOOKS OF EUCLID'S ELEMENTS, Euclid. Contains complete English text of all 13 books of the Elements plus critical apparatus analyzing each definition, postulate, and proposition in great detail. Covers textual and linguistic matters; mathematical analyses of Euclid's ideas; classical, medieval, Renaissance and modern commentators; refutations, supports, extrapolations, reinterpretations and historical notes. 995 figures. Total of 1,425pp. All books 5 3/8 x 8 1/2.
Vol. I: 443pp. 0-486-60088-2
Vol. II: 464pp. 0-486-60089-0
Vol. III: 546pp. 0-486-60090-4

A HISTORY OF GREEK MATHEMATICS, Sir Thomas Heath. This authoritative two-volume set that covers the essentials of mathematics and features every landmark innovation and every important figure, including Euclid, Apollonius, and others. 5 3/8 x 8 1/2. Vol. I: 461pp. 0-486-24073-8
Vol. II: 597pp. 0-486-24074-6

A MANUAL OF GREEK MATHEMATICS, Sir Thomas L. Heath. This concise but thorough history encompasses the enduring contributions of the ancient Greek mathematicians whose works form the basis of most modern mathematics. Discusses Pythagorean arithmetic, Plato, Euclid, more. 1931 edition. 576pp. 5 3/8 x 8 1/2.
0-486-43231-9

CHINESE MATHEMATICS IN THE THIRTEENTH CENTURY, Ulrich Libbrecht. An exploration of the 13th-century mathematician Ch'in, this fascinating book combines what is known of the mathematician's life with a history of his only extant work, the Shu-shu chiu-chang. 1973 edition. 592pp. 5 3/8 x 8 1/2.
0-486-44619-0

PHILOSOPHY OF MATHEMATICS AND DEDUCTIVE STRUCTURE IN EUCLID'S ELEMENTS, Ian Mueller. This text provides an understanding of the classical Greek conception of mathematics as expressed in Euclid's Elements. It focuses on philosophical, foundational, and logical questions and features helpful appendixes. 400pp. 6 1/2 x 9 1/4. 0-486-45300-6

BEYOND GEOMETRY: Classic Papers from Riemann to Einstein, Edited with an Introduction and Notes by Peter Pesic. This is the only English-language collection of these 8 accessible essays. They trace seminal ideas about the foundations of geometry that led to Einstein's general theory of relativity. 224pp. 6 1/8 x 9 1/4. 0-486-45350-2

HISTORY OF MATHEMATICS, David E. Smith. Two-volume history – from Egyptian papyri and medieval maps to modern graphs and diagrams. Non-technical chronological survey with thousands of biographical notes, critical evaluations, and contemporary opinions on over 1,100 mathematicians. 5 3/8 x 8 1/2.
Vol. I: 618pp. 0-486-20429-4
Vol. II: 736pp. 0-486-20430-8

Physics

THEORETICAL NUCLEAR PHYSICS, John M. Blatt and Victor F. Weisskopf. An uncommonly clear and cogent investigation and correlation of key aspects of theoretical nuclear physics by leading experts: the nucleus, nuclear forces, nuclear spectroscopy, two-, three- and four-body problems, nuclear reactions, beta-decay and nuclear shell structure. 896pp. 5 3/8 x 8 1/2. 0-486-66827-4

QUANTUM THEORY, David Bohm. This advanced undergraduate-level text presents the quantum theory in terms of qualitative and imaginative concepts, followed by specific applications worked out in mathematical detail. 655pp. 5 3/8 x 8 1/2. 0-486-65969-0

ATOMIC PHYSICS AND HUMAN KNOWLEDGE, Niels Bohr. Articles and speeches by the Nobel Prize–winning physicist, dating from 1934 to 1958, offer philosophical explorations of the relevance of atomic physics to many areas of human endeavor. 1961 edition. 112pp. 5 3/8 x 8 1/2. 0-486-47928-5

COSMOLOGY, Hermann Bondi. A co-developer of the steady-state theory explores his conception of the expanding universe. This historic book was among the first to present cosmology as a separate branch of physics. 1961 edition. 192pp. 5 3/8 x 8 1/2. 0-486-47483-6

LECTURES ON QUANTUM MECHANICS, Paul A. M. Dirac. Four concise, brilliant lectures on mathematical methods in quantum mechanics from Nobel Prize-winning quantum pioneer build on idea of visualizing quantum theory through the use of classical mechanics. 96pp. 5 3/8 x 8 1/2. 0-486-41713-1

THE PRINCIPLE OF RELATIVITY, Albert Einstein and Frances A. Davis. Eleven papers that forged the general and special theories of relativity include seven papers by Einstein, two by Lorentz, and one each by Minkowski and Weyl. 1923 edition. 240pp. 5 3/8 x 8 1/2. 0-486-60081-5

PHYSICS OF WAVES, William C. Elmore and Mark A. Heald. Ideal as a classroom text or for individual study, this unique one-volume overview of classical wave theory covers wave phenomena of acoustics, optics, electromagnetic radiations, and more. 477pp. 5 3/8 x 8 1/2. 0-486-64926-1

THERMODYNAMICS, Enrico Fermi. In this classic of modern science, the Nobel Laureate presents a clear treatment of systems, the First and Second Laws of Thermodynamics, entropy, thermodynamic potentials, and much more. Calculus required. 160pp. 5 3/8 x 8 1/2. 0-486-60361-X

QUANTUM THEORY OF MANY-PARTICLE SYSTEMS, Alexander L. Fetter and John Dirk Walecka. Self-contained treatment of nonrelativistic many-particle systems discusses both formalism and applications in terms of ground-state (zero-temperature) formalism, finite-temperature formalism, canonical transformations, and applications to physical systems. 1971 edition. 640pp. 5 3/8 x 8 1/2. 0-486-42827-3

QUANTUM MECHANICS AND PATH INTEGRALS: Emended Edition, Richard P. Feynman and Albert R. Hibbs. Emended by Daniel F. Styer. The Nobel Prize–winning physicist presents unique insights into his theory and its applications. Feynman starts with fundamentals and advances to the perturbation method, quantum electrodynamics, and statistical mechanics. 1965 edition, emended in 2005. 384pp. 6 1/8 x 9 1/4. 0-486-47722-3

Physics

INTRODUCTION TO MODERN OPTICS, Grant R. Fowles. A complete basic undergraduate course in modern optics for students in physics, technology, and engineering. The first half deals with classical physical optics; the second, quantum nature of light. Solutions. 336pp. 5 3/8 x 8 1/2. 0-486-65957-7

THE QUANTUM THEORY OF RADIATION: Third Edition, W. Heitler. The first comprehensive treatment of quantum physics in any language, this classic introduction to basic theory remains highly recommended and widely used, both as a text and as a reference. 1954 edition. 464pp. 5 3/8 x 8 1/2. 0-486-64558-4

QUANTUM FIELD THEORY, Claude Itzykson and Jean-Bernard Zuber. This comprehensive text begins with the standard quantization of electrodynamics and perturbative renormalization, advancing to functional methods, relativistic bound states, broken symmetries, nonabelian gauge fields, and asymptotic behavior. 1980 edition. 752pp. 6 1/2 x 9 1/4. 0-486-44568-2

FOUNDATIONS OF POTENTIAL THERY, Oliver D. Kellogg. Introduction to fundamentals of potential functions covers the force of gravity, fields of force, potentials, harmonic functions, electric images and Green's function, sequences of harmonic functions, fundamental existence theorems, and much more. 400pp. 5 3/8 x 8 1/2. 0-486-60144-7

FUNDAMENTALS OF MATHEMATICAL PHYSICS, Edgar A. Kraut. Indispensable for students of modern physics, this text provides the necessary background in mathematics to study the concepts of electromagnetic theory and quantum mechanics. 1967 edition. 480pp. 6 1/2 x 9 1/4. 0-486-45809-1

GEOMETRY AND LIGHT: The Science of Invisibility, Ulf Leonhardt and Thomas Philbin. Suitable for advanced undergraduate and graduate students of engineering, physics, and mathematics and scientific researchers of all types, this is the first authoritative text on invisibility and the science behind it. More than 100 full-color illustrations, plus exercises with solutions. 2010 edition. 288pp. 7 x 9 1/4. 0-486-47693-6

QUANTUM MECHANICS: New Approaches to Selected Topics, Harry J. Lipkin. Acclaimed as "excellent" (*Nature*) and "very original and refreshing" (*Physics Today*), these studies examine the Mössbauer effect, many-body quantum mechanics, scattering theory, Feynman diagrams, and relativistic quantum mechanics. 1973 edition. 480pp. 5 3/8 x 8 1/2. 0-486-45893-8

THEORY OF HEAT, James Clerk Maxwell. This classic sets forth the fundamentals of thermodynamics and kinetic theory simply enough to be understood by beginners, yet with enough subtlety to appeal to more advanced readers, too. 352pp. 5 3/8 x 8 1/2. 0-486-41735-2

QUANTUM MECHANICS, Albert Messiah. Subjects include formalism and its interpretation, analysis of simple systems, symmetries and invariance, methods of approximation, elements of relativistic quantum mechanics, much more. "Strongly recommended." – *American Journal of Physics.* 1152pp. 5 3/8 x 8 1/2. 0-486-40924-4

RELATIVISTIC QUANTUM FIELDS, Charles Nash. This graduate-level text contains techniques for performing calculations in quantum field theory. It focuses chiefly on the dimensional method and the renormalization group methods. Additional topics include functional integration and differentiation. 1978 edition. 240pp. 5 3/8 x 8 1/2. 0-486-47752-5

Physics

MATHEMATICAL TOOLS FOR PHYSICS, James Nearing. Encouraging students' development of intuition, this original work begins with a review of basic mathematics and advances to infinite series, complex algebra, differential equations, Fourier series, and more. 2010 edition. 496pp. 6 1/8 x 9 1/4. 0-486-48212-X

TREATISE ON THERMODYNAMICS, Max Planck. Great classic, still one of the best introductions to thermodynamics. Fundamentals, first and second principles of thermodynamics, applications to special states of equilibrium, more. Numerous worked examples. 1917 edition. 297pp. 5 3/8 x 8. 0-486-66371-X

AN INTRODUCTION TO RELATIVISTIC QUANTUM FIELD THEORY, Silvan S. Schweber. Complete, systematic, and self-contained, this text introduces modern quantum field theory. "Combines thorough knowledge with a high degree of didactic ability and a delightful style." – *Mathematical Reviews*. 1961 edition. 928pp. 5 3/8 x 8 1/2. 0-486-44228-4

THE ELECTROMAGNETIC FIELD, Albert Shadowitz. Comprehensive undergraduate text covers basics of electric and magnetic fields, building up to electromagnetic theory. Related topics include relativity theory. Over 900 problems, some with solutions. 1975 edition. 768pp. 5 5/8 x 8 1/4. 0-486-65660-8

THE PRINCIPLES OF STATISTICAL MECHANICS, Richard C. Tolman. Definitive treatise offers a concise exposition of classical statistical mechanics and a thorough elucidation of quantum statistical mechanics, plus applications of statistical mechanics to thermodynamic behavior. 1930 edition. 704pp. 5 5/8 x 8 1/4. 0-486-63896-0

INTRODUCTION TO THE PHYSICS OF FLUIDS AND SOLIDS, James S. Trefil. This interesting, informative survey by a well-known science author ranges from classical physics and geophysical topics, from the rings of Saturn and the rotation of the galaxy to underground nuclear tests. 1975 edition. 320pp. 5 3/8 x 8 1/2. 0-486-47437-2

STATISTICAL PHYSICS, Gregory H. Wannier. Classic text combines thermodynamics, statistical mechanics, and kinetic theory in one unified presentation. Topics include equilibrium statistics of special systems, kinetic theory, transport coefficients, and fluctuations. Problems with solutions. 1966 edition. 532pp. 5 3/8 x 8 1/2. 0-486-65401-X

SPACE, TIME, MATTER, Hermann Weyl. Excellent introduction probes deeply into Euclidean space, Riemann's space, Einstein's general relativity, gravitational waves and energy, and laws of conservation. "A classic of physics." – *British Journal for Philosophy and Science*. 330pp. 5 3/8 x 8 1/2. 0-486-60267-2

RANDOM VIBRATIONS: Theory and Practice, Paul H. Wirsching, Thomas L. Paez and Keith Ortiz. Comprehensive text and reference covers topics in probability, statistics, and random processes, plus methods for analyzing and controlling random vibrations. Suitable for graduate students and mechanical, structural, and aerospace engineers. 1995 edition. 464pp. 5 3/8 x 8 1/2. 0-486-45015-5

PHYSICS OF SHOCK WAVES AND HIGH-TEMPERATURE HYDRO DYNAMIC PHENOMENA, Ya B. Zel'dovich and Yu P. Raizer. Physical, chemical processes in gases at high temperatures are focus of outstanding text, which combines material from gas dynamics, shock-wave theory, thermodynamics and statistical physics, other fields. 284 illustrations. 1966–1967 edition. 944pp. 6 1/8 x 9 1/4. 0-486-42002-7